Physics of
New Laser Sources

NATO ASI Series

Advanced Science Institutes Series

A series presenting the results of activities sponsored by the NATO Science Committee, which aims at the dissemination of advanced scientific and technological knowledge, with a view to strengthening links between scientific communities.

The series is published by an international board of publishers in conjunction with the NATO Scientific Affairs Division

A	**Life Sciences**	Plenum Publishing Corporation
B	**Physics**	New York and London
C	**Mathematical and Physical Sciences**	D. Reidel Publishing Company Dordrecht, Boston, and Lancaster
D	**Behavioral and Social Sciences**	Martinus Nijhoff Publishers
E	**Engineering and Materials Sciences**	The Hague, Boston, and Lancaster
F	**Computer and Systems Sciences**	Springer-Verlag
G	**Ecological Sciences**	Berlin, Heidelberg, New York, and Tokyo

Recent Volumes in this Series

Series B: Physics

Physics of
New Laser Sources

Edited by

Neal B. Abraham
Bryn Mawr College
Bryn Mawr, Pennsylvania

F. T. Arecchi
University of Florence and
National Institute of Optics
Florence, Italy

Aram Mooradian
Lincoln Laboratory
Massachusetts Institute of Technology
Lexington, Massachusetts

and

Alberto Sona
C.I.S.E. (Center of Information, Studies and Experiments)
Segrate, Milan, Italy

SPRINGER SCIENCE+BUSINESS MEDIA, LLC

Proceedings of a NATO Advanced Study Institute on
Physics of New Laser Sources,
held July 11–21, 1984,
in San Miniato, Tuscany, Italy

Library of Congress Cataloging in Publication Data

NATO Advanced Study Institute on Physics of New Laser Sources (1984: San
    Miniato, Italy)
    Physics of new laser sources.

    (NATO ASI series. Series B, Physics; v. 132)
    "Proceedings of a NATO Advanced Study Institute on Physics of New Laser
Sources, held July 11–21, 1984, in San Miniato, Tuscany, Italy"—T.p. verso.
    "Published in cooperation with NATO Scientific Affairs Division."
    Bibliography: p.
    Includes index.
    1. Lasers—Congresses. 2. Quantum electronics—Congresses. I. Abraham,
Neal B. II. North Atlantic Treaty Organization. Scientific Affairs Division. III.
Title. IV. Series.
QC685.N39   1984   535.5'8   85-24386
ISBN 978-1-4757-6189-4   ISBN 978-1-4757-6187-0 (eBook)
DOI 10.1007/978-1-4757-6187-0

© 1985 Springer Science+Business Media New York 1985
Originally published by Plenum Press, New York in 1985

PREFACE

This volume contains the lectures and seminars
presented at the NATO Advanced Study Institute on
"Physics of New Laser Sources", the twelfth course of
the Europhysics School of Quantum Electronics, held
under the supervision of the Quantum Electronics
Division of the European Physical Society. The
Institute was held at Centro "I Cappuccini" San
Miniato, Tuscany, July 11-21, 1984.
The Europhysics School of Quantum Electronics was
started in 1970 with the aim of providing instruction
for young researchers and advanced students already
engaged in the area of quantum electronics or for those
wishing to switch into this area after working
previously in other areas. From the outset, the School
has been under the direction of Prof. F. T. Arecchi,
then at the University of Pavia, now at the University
of Florence, and Dr. D. Roess of Heraeus, Hanau. In
1981, Prof. H. Walther, University of Munich and
Max-Planck Institut fur Quantenoptik joined as
co-director. Each year the Directors choose a subject
of particular interest, alternating fundamental topics
with technological ones, and ask colleagues
specifically competent in the chosen areas to take the
scientific responsibility for that course.

The past courses were devoted to the following themes:
1) 1971: "Physical and Technical Measurements with
           Lasers"
2) 1972: "Nonlinear Optics and Short Pulses"
3) 1973: "Laser Frontiers: Short Wavelength and
           High Powers"
4) 1974: "Cooperative Phenomena in Multicomponent
           Systems"
5) 1975: "Molecular Spectroscopy and
           Photochemistry with Lasers"
6) 1976: "Coherent Optical Engineering"

7) 1977: "Coherence in Spectroscopy and Modern
           Physics"
8) 1979: "Lasers in Biology and Medicine"
9) 1980: "Physical Processes in Laser Material
           Interactions"
10) 1981: "Advances in Laser Spectroscopy"
11) 1982: "Laser Applications to Chemistry"
12) 1984: "Physics of New Laser Sources"

The first five courses were held in Erice, Sicily, at the Centre for Scientific Culture "Ettore Majorana", the next four were held at Villa Le Pianore, Camaiore, Tuscany. Beginning in 1981, thanks to generosity of Cassa di Risparmio di San Miniato, we have held the school in a conference center resulting from the restoration of an ancient convent in the countryside, very close to the small city of San Miniato, that is, just in the heart of Tuscany (equidistant from Florence, Pisa and Siena).

Prof. F. T. Arecchi, Prof. N. B. Abraham of Bryn Mawr College, Prof. A. Sona of CISE and Prof. A. Mooradian of the Massachussetts Institute of Technology undertook the scientific direction of the present courses, selecting the specific topics and lectures.

This Institute was classified by the NATO Scientific Affairs Division as part of its "Double Jump" Program because its topics seemed likely to be of interest to scientists from national and industrial laboratories as well as serving the needs of the academic community. In fact, twenty-four years after the operation of the first laser, we are witnessing widespread implementation of laser devices and laser technologies in industrial processes and commercial systems. The needs of the laser user-community have stimulated the search for new classes of laser sources with some exotic characteristics· in comparison with the now "conventional" characteristics of the first generation of lasers.

New laser sources are often rapidly incorporated into certain research and technological applications because they are designed for specific uses. However, the broader scientific community is often much slower to adopt new systems because they are of uncertain reliability and, more importantly, because the scientists are unfamiliar with the capabilities of these new sources. Through this ASI we intended to provide an understanding of the advantages and potential applications of maturing new sources and to provide the basis for participants to evaluate and explore uses of even more novel sources currently under design or development.

The "double jump" character perceived by the NATO sponsors also attracted many contributions from industries which provided either lecturers from among their experts or fellowships for other participants. Hence, in addition to the support of the NATO Scientific Affairs Division, we wish to acknowledge the contributions and cosponsorship of the Institute by the following organizations:

Allied Corporation
Apollo Lasers
AT&T Bell Laboratories
Cassa di Risparmio di San Miniato
Coherent GmbH
Coherent, Inc
Consiglio Nazionale delle Ricerche
CSELT
ENEA
FIAR
Fondazione Ugo Bordoni
Hughes Aircraft Co.
IBM - Italy
Istituto Nazionale di Ottica
Lambda-Physik
Laser Focus
Laser-Optronic GmbH
Nippon Electronic Co.
Quantel
Selenia
Spectra Physics GmbH
USAF European Office of Aerospace Research and
    Development
U.S. Army European Research Office
W.C. Heraus GmbH

In addition, the Deutscher Akademischer Austauschdienst contributed three fellowships for the participation of German graduate students. Finally, we offer our special thanks to Mrs. M. Petrone of the Istituto Nazionale di Ottica for her invaluable assistance in the organization of the Institute, and to her, Miss A. Camnasio of Servizio Documentazione, CISE, and Miss A. C. Arecchi for their assistance during the course itself. In addition, thanks are due to the staff of Cassa di Risparmio di San Miniato for the operation of the conference center and to Mrs. I. Arecchi for the organization of the social events. Finally we thank J. R. Tredicce for his able assistance in the final editing of the manuscript.

We note particularly that this volume includes not only specially written tutorial lectures as is the customary procedure, but in certain appropriate instances reprints of selected articles are used to present the highlights of fields or applications. This approach, which amounts to a guided tour of the most recent research literature, may ultimately become the standard route to efficient surveys of this type without the labor of preparing redundant camera-ready manuscripts.

N. B. Abraham
F. T. Arecchi
A. Mooradian
A. Sona

CONTENTS

TUNABLE SHORT-WAVELENGTH LASER SOURCES

B. P. Stoicheff

Department of Physics
University of Toronto
Toronto, Ontario, Canada, M5S 1A7

INTRODUCTION

The availability of tunable laser radiation in the visible and infrared wavelength regions has made possible many important advances in physics, chemistry, and biology. At the present time, the ultraviolet (UV) region of the spectrum, and in particular the vacuum ultraviolet (VUV, from 200 to 100 nm) and extreme ultraviolet (XUV, from 100 to ∿20 nm) regions lack tunable lasers. In fact, only a few lasers have been made to operate at these short wavelengths, in spite of considerable efforts being made in the past decade. The excimer lasers ArF (193 nm), Xe_2 (∿170 nm), and Ar_2 (∿120 nm), and the H_2 laser (∿110 nm) have been available for some time now, but these emit at discrete wavelengths or are tunable only over their relatively narrow bandwidths. More recent efforts have resulted in stimulated VUV emission by the anti-Stokes Raman process in I and Br[1] and by 2-photon excitation of H_2[2], and in the XUV region by 4-photon excitation in Kr (∿93 nm)[3]. Other techniques being explored include recombination processes[4] and excitation of ions[5] such as Li^+; and as we have learned at this school, in principle, the free-electron laser could operate at these short wavelengths.

The difficulty in producing stimulated emission in the VUV and XUV regions is well-known, and arises from the basic relation that, the probability for spontaneous emission, A, varies as $\nu^3 B$, where B is the probability for stimulated emission. Thus, losses in excited state population due to spontaneous emission increase rapidly at short wavelengths, and put severe demands on pumping sources in order to achieve inverted population. Nevertheless much effort in this direction is being made in many laboratories, and we hope that new VUV and XUV lasers will soon be reported.

In the meantime, there has been substantial progress in producing VUV and XUV radiation by harmonic generation and by frequency mixing of laser radiation in the rare gases and in metal vapors. The resulting radiation is coherent, monochromatic, directional, and tunable over broad regions, and thus has all the characteristics of laser radiation except for high intensity. Nevertheless, the intensities achieved to date are sufficient for many applications.

It is my purpose to review these accomplishments, to sketch the basic theoretical concepts, describe the experimental methods used in generating tunable, coherent VUV and XUV radiation, and to summarize the wavelength ranges over which such radiation is now available. Finally, I will briefly discuss some recent spectroscopic applications of this VUV and XUV radiation.

THEORY OF NONLINEAR PROCESSES: HARMONIC GENERATION AND FREQUENCY MIXING

Laser-driven VUV sources are based on third harmonic generation (THG) or 4-wave sum and difference mixing (4-WSM or 4-WDM). The process itself is dependent on the nonlinear susceptibilities of atomic or molecular systems when irradiated by intense laser radiation (Armstrong et al[6]). It is well-known that the polarization of a medium in the presence of a monochromatic field $\bar{E}(r,t) = \Sigma_i E(\omega_i)$ can be written as

$$\bar{P}(\omega_i) = \chi^{(1)}(\omega_i) \cdot \bar{E}(\omega_i) + \sum_{j,k} \chi^{(2)}(\omega_i = \omega_j + \omega_k) \cdot \bar{E}(\omega_j) \cdot \bar{E}(\omega_k)$$

$$+ \sum_{jkl} \chi^{(3)}(\omega_i = \omega_j + \omega_k + \omega_l) \cdot \bar{E}(\omega_j) \cdot \bar{E}(\omega_k) \cdot \bar{E}(\omega_l) + \ldots.$$

where $\chi^{(n)}$ are the susceptibility tensors of nth order. The lowest order term producing nonlinear effects is $\chi^{(2)}$. However, this tensor has nonzero components only in noncentro-symmetric systems: isotropic media such as cubic crystals, liquids, and gases do not exhibit quadratic nonlinearities. For third order processes such as THG and 4-WSM/4-WDM we need be concerned only with $\chi^{(3)}$, whose principal term may be written

$$\chi^{(3)}(\omega_0 = \omega_1 + \omega_2 + \omega_3) = \frac{3e^4}{4\hbar^3} \frac{<g|\mu|a><a|\mu|b><b|\mu|c><c|\mu|g>}{(\Omega_{cg} - \omega_1 - \omega_2 - \omega_3)(\Omega_{bg} - \omega_1 - \omega_2)(\Omega_{ag} - \omega_1)} \quad (1)$$

Here, $<g|\mu|a>$ is the electric dipole matrix element between the ground state $|g>$ and an excited state $|a>$, having a lifetime Γ_a, and $\Omega_{ag} = \omega_{ag} - i\Gamma_a/2$ is the energy difference (Fig. 1) between states $|a>$ and $|g>$, e is the electronic charge and $\hbar = h/2\pi$, with h being Planck's constant.

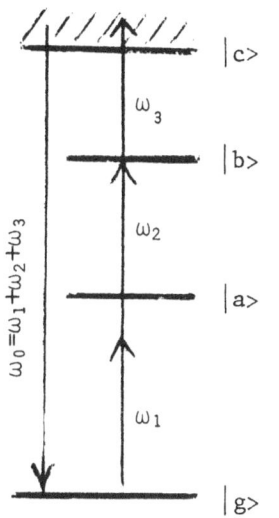

$\omega_0 = \omega_1 + \omega_2 + \omega_3$

Fig.1 4-WSM process $\omega_0 = \omega_1 + \omega_2 + \omega_3$ with a 2-photon resonance transition $\Omega_{bg} - \omega_1 - \omega_2 = 0$.

Equation (1) shows that $\chi^{(3)}$ will be resonantly enhanced whenever the applied frequencies, ω_1, ω_2, ω_3 are such that the real part of the resonance denominator vanishes, namely when $(\Omega_{ag} - \omega_1) = 0$, or $(\Omega_{bg} - \omega_1 - \omega_2) = 0$, or $(\Omega_{cg} - \omega_1 - \omega_2 - \omega_3) = 0$, corresponding to one, two, or three photon resonance, respectively. If any of ω_1, ω_2, ω_3 is set equal to a resonance frequency (Ω_{ag} etc.), $\chi^{(3)}$ will be enhanced but the incident radiation will be strongly absorbed. Similarly if $\omega_0 = \omega_1 + \omega_2 + \omega_3$ equals a resonance frequency, the generated radiation will be absorbed. If however, $\omega_1 + \omega_2$ is equal to a 2-photon resonance (Ω_{bg}), the incident radiation at $\omega_1 + \omega_2$ is expected to be only weakly absorbed by the 2-photon transition, while the resonance enhancement of $\chi^{(3)}$ could be just as strong as for the 1-photon resonances.

For third harmonic generation (THG), $\chi^{(3)}$ simplifies to

$$\chi^{(3)}(\omega_0 = 3\omega) = \frac{3e^4}{4\hbar^3} \frac{<g|\mu|a> \text{ etc.}}{(\Omega_{cg} - 3\omega)(\Omega_{bg} - 2\omega)(\Omega_{ag} - \omega)} \tag{2}$$

When 2ω approaches resonance, $\chi^{(3)}$ undergoes strong ($> 10^4$) enhancement. For efficient THG, collinear phase-matching is necessary, that is, the refractive index $n(3\omega) = n(\omega)$ in order to yield a maximum effective interaction length. With focused incident radiation, THG can be observed only in negatively dispersive media (Bjorklund[7]). Tunability is achieved by varying the incident frequency ω.

For generating tunable radiation by 4-WSM or 4-WDM, the process $\omega_0 = 2\omega_1 \pm \omega_2$ is of interest. $\chi^{(3)}$ then becomes

$$\chi^{(3)}(\omega_0 = 2\omega_1 \pm \omega_2) = \frac{3e^4}{4\hbar^3} \frac{<g|\mu|a> \text{ etc.}}{(\Omega_{cg} - 2\omega_1 \mp \omega_2)(\Omega_{bg} - 2\omega_1)(\Omega_{ag} - \omega_1)} \tag{3}$$

Strong enhancement is again achieved by tuning $2\omega_1$ to a parity-allowed 2-photon resonance, Ω_{bg}. Tunability (and further enhancement) is then obtained by selecting ω_2 so that $2\omega_1 + \omega_2$ corresponds to the ionization continuum or to broad autoionizing levels above the ionization limit (Hodgson et al.[8]). More detailed treatments of the relevant theory including phase-matching, saturation effects, and conversion efficiencies have been given by Vidal[9] and Jamroz and Stoicheff[10].

EXPERIMENTAL TECHNIQUES

Obviously, any medium used for generating radiation at wavelengths below 200 nm must be transparent to such radiation. This condition is generally met by the rare gases and some metal vapors. New and Ward[11] were the first to demonstrate THG in gases. Subsequently, Harris and Miles[12] showed that relatively high conversion efficiency of THG and 4-WSM could be obtained by using phase-matched metal vapors as nonlinear media, and that efficiency could be improved further by resonance enhancement.

Frequency conversion into the VUV and XUV regions has been achieved by a variety of laser systems[8,10]. Powerful pulsed lasers such as ruby, Nd:YAG, Nd in glass, flashlamp pumped dye (FPD), rare gas excimer, and rare gas halide exciplex lasers provide the primary coherent radiation. In some systems, tunable radiation from dye lasers (> 320 nm) is used directly, and in others laser radiation (> 400 nm) is doubled once or twice in nonlinear crystals to produce coherent radiation in the UV to about 200 nm. Subsequently, the coherent UV radiation is converted to coherent VUV and XUV by THG or frequency mixing in rare gases or metal vapors. These methods all use a cell or heat pipe to contain the nonlinear gas. Windows of LiF are used to the limit of transmittance at $\sim$104 nm; for generation to shorter wavelengths, pinholes with differential pumping are used. Recently, pulsed supersonic jets have been introduced for harmonic generation with rare gases[13,14]. Specific atomic (and molecular) systems are selected because of their large third order nonlinear susceptibility, suitability of energy levels for resonance enhancement, and low absorption at the desired VUV or XUV wavelength.

In our laboratory, we use essentially the same 4-WSM technique as that described by Hodgson, Sorokin, and Wynne[8] who generated VUV radiation tunable from 200 to 177 nm in Sr vapor. By using Mg, Zn, or Hg vapor as the nonlinear medium, we have extended the tuning range to the transmission limit of LiF windows at $\sim$104 nm. The experimental arrangement is shown in Fig. 2. A N_2 laser (Molectron – UV1000) or excimer KrF/XeCl laser (Lumonics TE-861M) pumps two dye lasers at frequencies ν_1 and ν_2. With the N_2 laser pump, each dye laser produces $\sim$10 kW pulses of $\sim$7 ns duration with linewidths $\sim$0.1 cm^{-1}. Resonant enhancement of $\chi^{(3)}$ is achieved by setting $2\nu_1$ to a 2-photon resonance in the atomic system, and the second laser (ν_2) is scanned over the breadth of a suitable autoionizing level to obtain a broad range of tunability (Fig. 3). The two beams are spatially overlapped in a Glan-Thompson prism, and focused near the exit end of a heat-pipe containing Mg vapor at $\sim$20 torr and He buffer gas at $\sim$200 torr. The resulting coherent radiation is continuously tunable from 174 to 135 nm[15,16], with a flux $>10^8$ photons per pulse. For generating shorter wavelength radiation, the excimer laser is used to pump two dye laser oscillator-amplifier systems with outputs of ∿50 kW. One of the beams is frequency-doubled in a KDP crystal to produce the 2-photon resonance enhancement, $2\nu_1$, of a suitable level in Mg[17], Zn or Hg vapor. With Zn vapor[18], a flux of ∿10^7 photons per pulse is generated from 140 to 106 nm, and with Hg vapor an increase of 10^2 to 10^3 in flux is achieved over the tuning range of 120 to 104 nm[19-22].

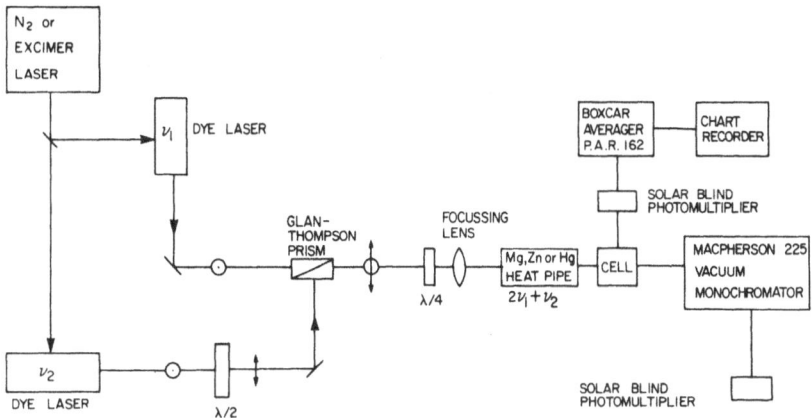

Fig.2. Method of exciting two tunable dye lasers and of combining radiation at ν_1 and ν_2 to generate tunable coherent VUV radiation at $2\nu_1+\nu_2$ by 4-WSM in a metal vapor[15-18,22].

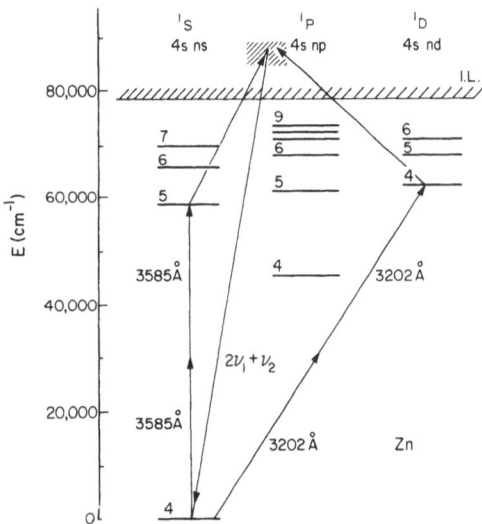

Fig.3. Energy-level diagram of Zn, showing generation of $2\nu_1+\nu_2$ radiation with 2-photon resonance enhancement by states $4s5s\,^1S$ or $4s4d\,^1D$, and with tunability provided by autoionizing level $4snp\,^1P$.

5

Some examples of other experimental arrangements used for generating VUV and XUV radiation are shown in Figs. 4-7. The pioneering work of Harris and his co-workers at Stanford University led to the first announcement of efficient THG and tunable frequency mixing in the VUV and XUV. Their experiments are shown in Figs. 4a,b. The primary laser source was a mode-locked Nd:YAG laser and amplifier producing 1.06 µm radiation in several pulses each of 50 psec duration, and peak power of ∿20 MW. This radiation was doubled to 532 nm in a KDP crystal, and then used directly, or doubled again, or mixed with 1.06 µm radiation in a second KDP crystal, to produce radiation at 532, 266, or 355 nm with ∿10% efficiency. In this way, high-power radiation at several fixed wavelengths was available for nonlinear mixing to shorter wavelengths in metal vapors and rare gases. Initial experiments in a Cd:Ar phase-matched mixture produced summing of 1.064 and 2 x 354.7 to yield 152 nm, and tripling of 532 nm and of

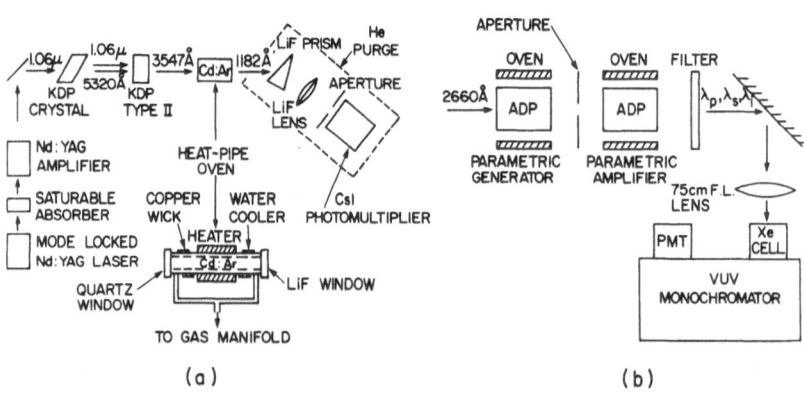

| (a) | (b) |

Fig.4. Experimental methods for generating third harmonic radiation as well as sum- and difference-mixing radiation by nonlinear process (a) in a phase-matched metal vapor[23] at fixed wavelengths, and (b) in xenon gas at tunable wavelengths[24].

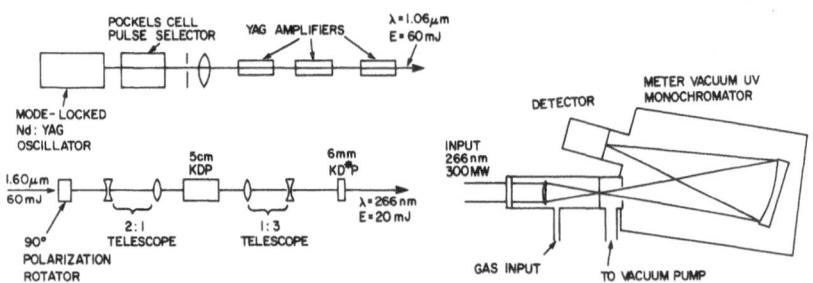

Fig.5. The high-power laser system and amplifiers used to generate XUV radiation at fifth and seventh harmonics of 266 nm radiation at 53.2 and 38.0 nm, respectively, in rare gases[25].

354.7 nm to yield 177.3 and 118.2 nm radiation, respectively[23]. In a phase-matched mixture of Xe:Ar, THG of 354.7 to 118.2 nm was achieved, and in Ar, THG of 266 to 88.7 nm.

Tunability was added to this coherent source[24] by using the 266 nm radiation to pump an ADP parametric generator (Fig. 4b). Variation of the crystal temperature from 50 to 105°C provided a tuning range of 420 to 720 nm, with >100 μJ output in a bandwidth of 0.5 to 2 nm. The pump (266 nm) and resulting signal and idler radiation was then focussed into Xe gas (at pressures up to 1 atm). THG and sum and difference mixing produced VUV radiation tunable over portions of the region from 118 to 147 nm and continuously from 163 to 194 nm at peak powers of $\sim$1 W corresponding to $\sim 10^7$ photons in a 20 psec pulse.

The early work of Harris and co-workers on nonlinear mixing in Xe and Ar was followed by work on Kr and Xe. Recently, extensive wavelength tunability with rare gases was reported by Hilbig and Wallenstein[25]. They used sum-frequency mixing ($\omega_{VUV} = 2\omega_{UV} + \omega_D$) in Kr and Xe to generate VUV radiation of $\sim$20 W power, tunable over most of the range 110 to 130 nm where these gases are negatively dispersive. They also generated VUV radiation of $\sim$50 W power at longer wavelengths in Xe by difference-frequency mixing. The process $\omega_{VUV} = 2\omega_{UV} - \omega_D$ resulted in radiation from 185 to 207 nm, and $\omega_{VUV} = 2\omega_{UV} - \omega_L$ at shorter wavelengths, from 160 to 190 nm. (The frequencies ω_L, ω_D, and ω_{UV} refer to the output of a Nd:YAG laser, a dye laser and harmonic of the dye laser, respectively.) For these experiments, Hilbig and Wallenstein used the second harmonic radiation of a Nd:YAG laser (λ_L) to excite a dye-laser oscillator-amplifier system. This system operated at $\lambda_D = 550$ to 650 nm, with output powers of 3 to 5 MW in pulses of $\sim$6 ns duration, and with bandwidth of $\sim$0.02 cm^{-1}. This tunable visible radiation was then doubled in KDP to produce tunable UV radiation (λ_{UV}) at powers of $\sim$1 MW. Both visible and ultraviolet radiation was focused in the rare gas, and the resulting VUV radiation analyzed with a monochromator and detected by a solar-blind photomultiplier and NO ionization cell. This multi-laser system, with further flexibility in order to fill in the few gaps in tuning range, will be a useful source for spectroscopy in the region 100 to 200 nm.

Higher order frequency conversion has been used to generate radiation at fixed frequencies to wavelengths as short as 38 and 35.5 nm. Reintjes and co-workers[26] at the Naval Research Laboratory in Washington, D.C. have used the fundamental, second, and fourth harmonics of a mode-locked Nd:YAG laser to generate XUV radiation in rare gases in third through fifth and seventh harmonic conversion, and by 6-wave mixing ($\nu_3 = 4\nu_1 \pm \nu_2$). Their powerful laser system and experimental arrangement are shown in Fig. 5. Fifth harmonic of 266.1 nm radiation produced radiation at 53.2 nm in He, Ne, Ar and Kr. The highest conversion efficiency was $\sim 10^{-5}$ in He (at $\sim$50 Torr) and yielded peak pulse powers of $\sim$1 kW. Seventh harmonic at 38 nm was observed in He with peak power of $\sim$100 W.

More recently Bokor et al[13] have used a supersonic He gas jet to produce radiation at 35.5 nm by seventh-harmonic conversion of 248 nm radiation from a KrF excimer laser.

Rhodes and his colleagues[27,28] have developed KrF and ArF laser systems of extremely high brightness for use in THG and 4-WSM below 100 nm. A schematic outline of a typical multi-laser system is shown in Fig. 6. The tunable output of a single-frequency, cw dye laser at $\sim$580 nm was pulse-amplified in a three-stage, XeF-pumped, amplifier. The $\sim$10 ns, 20 mJ pulses were focused into Sr vapor to generate third harmonic radiation at $\sim$193 nm, in 5 ns pulses of 200 mW peak power. These pulses were then amplified in two ArF laser amplfiers to produce $\sim$5 ns pulses of 6 MW peak power and $\sim$0.01 cm^{-1} bandwidth, tunable within the ArF gain profile. Another system, with a final KrF laser amplifier, produced similar output at 248 nm. The KrF radiation was frequency tripled in flowing Xe with the input beam focused at the exit end of a windowless cell, through a 350 nm diameter pinhole. Third harmonic radiation at $\sim$83 nm was dispersed by a monochromator and detected with a windowless photomultiplier. In similar experiments, ArF radiation at $\sim$193.6 nm was focused and tripled to 64 nm in Kr, H$_2$, and Ar. In later work, they used 4-WSM ($\omega_{VUV} = 2\omega_{ArF} + \omega_D$, with radiation of an ArF laser and of a tunable dye laser) in flowing H$_2$ to generate tunable VUV in the region of 79 nm.

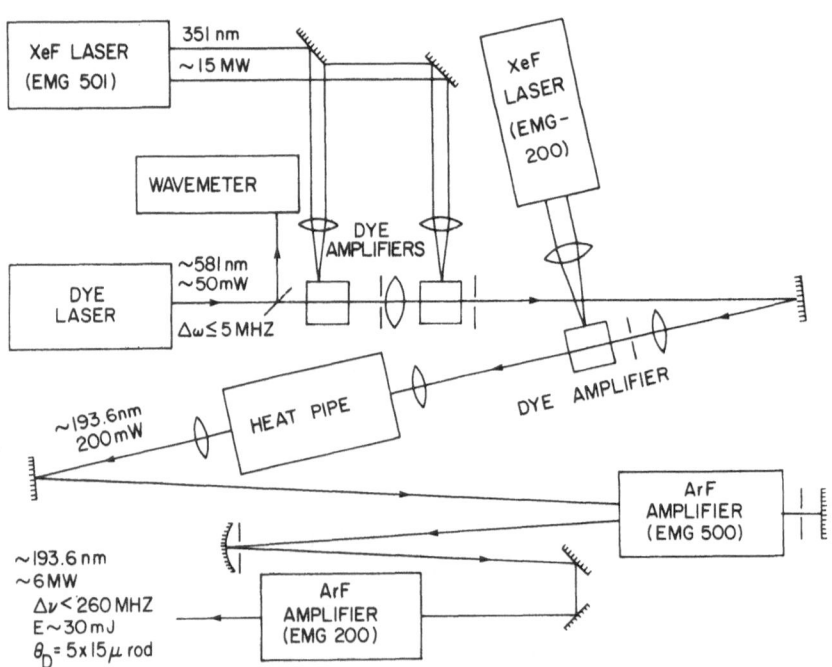

Fig.6. A high spectral brightness laser system for use in generating VUV radiation[27,28].

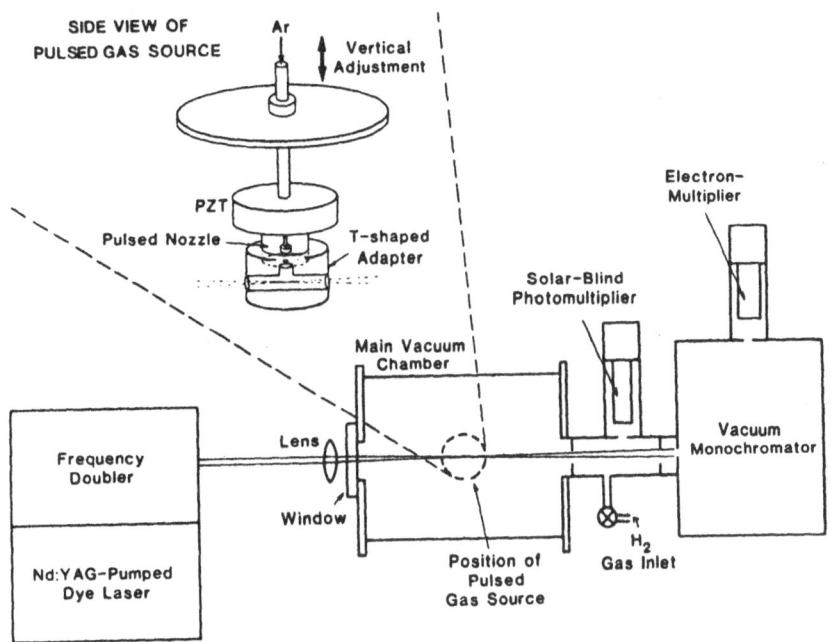

Fig. 7. Laser and pulsed supersonic jet with rare gases used to produce VUV and XUV radiation by third-harmonic generation[14,29].

In Fig. 7 is shown the scheme developed by Kung and his colleagues[14,29] for use of gases at high densities from a pulsed supersonic jet as nonlinear media in VUV and XUV generation. With a
pulsed gas source, the difficulties encountered with the lack of
transmitting materials for use as windows at $\lambda < 100$ nm, are readily
overcome. Moreover, the vacuum system need only be of modest size.
With the design shown in Fig. 7, a Q-switched Nd:YAG laser was used
to provide radiation of 18 MW at 354 nm for frequency tripling to
118 nm in a jet of Xe. Peak power of 260 W was obtained in a 2.9 ns
pulse, corresponding to $\sim 5 \times 10^{11}$ photons per pulse. When second
harmonic radiation from a pulsed dye laser was incident instead,
tunable radiation was generated in Xe, Kr, Ar, Ne, H_2 and CO. Ar
was the most efficient nonlinear medium ($\sim 10^{-7}$) giving 7×10^9
photons per pulse and 10 W peak power, over the range 102.3 to
97.3 nm.

The nonlinear media and laser systems used at the present time
to generate tunable, coherent, VUV and XUV radiation are summarized
in Tables 1 and 2, along with their regions of tunability. Only
systems with demonstrated broad regions of tunability are included
here. Many other systems which make use of rare gases (and even
molecular gases) have been reported, but have limited tunability.
These have been reviewed elsewhere[10].

9

Table 1. Tunable Generation in Rare Gases

λ(nm)	Nonlinear Medium	Processes	Primary Laser	Reference
206 – 160[a]	Xe	$2\lambda_{UV}-\lambda_L$	Nd:YAG–Dye	Hilbig and Wallenstein[25]
195 – 163	Xe	$2\times266\pm\lambda_s,\lambda_i$[b]	Nd:YAG and PO[b]	Kung[24]
147 – 118	Xe	$2\times266\pm\lambda_s,\lambda_i$[b]	Nd:YAG and PO[b]	Kung[24]
147 – 140	Xe:Kr	$3\lambda_{Dye}$	Nd:YAG–Dye	Hilbig and Wallenstein[25]
130 – 110[a]	Kr	$2\lambda_{UV}+\lambda_L$	Nd:YAG–Dye	Hilbig and Wallenstein[25]
123.6–120.3	Kr	$3\lambda_{Dye}$	KrF–Dye	Cotter[30]
123.5–120	Kr:Ar	$3\lambda_{Dye}$	Nd:YAG–Dye	Hilbig and Wallenstein[26]
106.8–105.8	Xe	$3\lambda_{Dye}$	Nd:YAG–Dye	Northrup et al[35]
102.3–97.3	Ar	$3\lambda_{Dye}$	Nd:YAG–Dye	Marinaro et al.[29]
73.6 – 72.0	Ne	$3\lambda_{Dye}$	Nd:YAG–Dye	Hilbig et al[31]

[a] λ_{UV} from a range of laser dyes.
[b] PO $\equiv$ parametric oscillator (signal and idler wavelengths λ_s,λ_i)

Table 2. Tunable Generation in Metal Vapors

Nonlinear Medium (Ioniz. Limit in nm)	λ(nm)	Primary Laser	Reference
Sr(217.8)	195.7–177.8	N_2–Dye	Hodgson et al[8]
Mg(162.1)	174 – 145	N_2–Dye	Banic et al[16]
	160 – 140	N_2–Dye	Wallace and Zdasiuk[15]
	129 – 121	KrF–Dye	McKee et al[17]
Zn(132.0)	140 – 106	XeCl/KrF–Dye	Jamroz et al[18]
Hg(118.0)	125.1–117.4	Nd:YAG–Dye	Mahon and Tomkins[19]
	115.0–93.0	Nd:YAG–Dye	Freeman et al[20]
	196 – 109	Nd:YAG–Dye	Hilbig and Wallenstein[21]
	120 – 104	XeCl–Dye	Herman and Stoicheff[22]

Finally, it may be mentioned that there has been some success in generating continuous-wave VUV radiation by 4-WSM in Sr vapor[32] and by THG in Mg vapor[33].

APPLICATIONS IN SPECTROSCOPY

These VUV sources have been applied to a variety of spectroscopic studies[34], almost always using the highly sensitive method of laser-induced fluorescence. As examples of the diversity of investigations, reported so far, the following may be mentioned: spectra of H_2 and D_2[29,35,36]; reaction of H with HBr and detection of Br[37]; rotationally excited CO from photodissociation of formaldehyde[38]; and rotationally inelastic scattering of CO from LiF surfaces[39]. Several experiments carried out in our laboratory (Fig. 2) on radiative lifetimes, and on excited states of van der Waals molecules are discussed briefly.

The monochromaticity (~ 0.1 cm^{-1}) and short pulse duration (~ 1 to 5 ns) of these sources makes possible the measurement of radiative lifetimes of specific rovibronic levels. Preliminary experiments on CO[40] and on NO[16] have been published. For CO, a more detailed study was recently completed[41]. Initially, the fluorescence excitation spectrum of the (0,0) band of the transition $A^1\Pi \rightarrow X^1\Sigma$ was recorded (154 to 155.5 nm) with an effective resolution of ~ 0.3 cm^{-1}. The observed lines were identified; then the exciting radiation was tuned to each rovibronic line and the decay of fluorescence intensity with time was measured. In this way, lifetimes for levels $J' = 1$ to 29 of the $v' = 0$ level were obtained from the decay rates for transitions in the P, Q, and R branches. The results indicate strong perturbations at $J' = 9$, 16, and 27, as shown by a doubling of lifetimes over those of unperturbed levels ($\tau \sim 10$ ns). Similar investigations were carried out on the levels $v' = 0,1,2...8$ of the $B'^2\Delta$ state of NO[16]. The many perturbations of this state are well-known, and permitted measurements of radiative lifetimes of the $B'^2\Delta$ as well as the nearby Rydberg states $N^2\Delta$ and $F^2\Delta$. Single-exponential decays were found at all pressures for fluorescence from levels $v' = 1,3,5$ of the $B'^2\Delta$ state, and lifetimes $\tau_0 \sim 225$ ns were obtained for these levels. For the levels $v' = 0,2,4,6,7,8$ double-exponential decays were observed, having short-lifetime components (~ 20 to 30 ns) which were ascribed to the perturbing Rydberg states $F^2\Delta$ and $N^2\Delta$.

The electronic spectra of rare gas dimers have been a subject of interest for many years, mainly because these dimers are model systems for studying van der Waals interactions, and because of their potential as media for VUV and XUV lasers. Yet very little is known about the excited states of these dimers. Two experimental techniques were combined in our laboratory for this investigation[42]: four-wave sum-mixing (4-WSM) and a pulsed supersonic jet to produce rotationally and vibrationally cold dimer molecules. In this way it was possible to resolve rovibronic structures in several isotopic band systems of Xe_2, Kr_2 and Ar_2, in the region 150 to 104 nm. Thus, for the first time, the spectroscopic constants and potential energy curves of the lowest three excited states were determined for each of these dimers. The above experimental technique will be applicable to the whole family of rare gas dimers.

REFERENCES

1. J. C. White and D. Henderson, Tunable 178 nm Iodine anti-Stokes Raman laser, Opt. Lett. 7: 204 (1982): Anti-Stokes Raman laser emission at 149 nm in atomic Bromine, Opt. Lett. 8: 520 (1983).

2. H. Egger, T. S. Luk, H. Pummer, T. Srinivasan, and C. K. Rhodes, Stimulated VUV emission following two-photon excitation of H_2, in "Laser Spectroscopy VI", H. P. Weber and W. Lüthy, eds., Springer-Verlag, Berlin (1983) p. 403.

3. T. Srinivasan, H. Egger, T. S. Luk, H. Pummer, and C. K. Rhodes, Stimulated extreme-ultraviolet emission at 93 nm in Krypton, in "Laser Spectroscopy VI", H. P. Weber and W. Lüthy, eds., Springer-Verlag, Berlin (1983) p. 385.

4. W. T. Silfvast and O. R. Wood II, Recombination lasers in the vacuum ultraviolet, in "Laser Techniques for Extreme Ultraviolet Spectroscopy", T. J. McIlrath and R. R. Freeman, eds., American Institute of Physics, New York (1982) p. 128.

5. H. Mahr and U. Roeder, Use of metastable ions for a soft x-ray laser, Opt. Commun. 10: 227 (1974).

6. J. A. Armstrong, N. Bloembergen, J. Ducuing, and P. S. Pershan, Interaction between light waves in a nonlinear dielectric, Phys. Rev. 127: 1918 (1962).

7. G. C. Bjorklund, Effects of focusing on third-order nonlinear processes in isotropic media, IEEE J. Quant. Electr. QE-11: 287 (1975).

8. R. T. Hodgson, P. P. Sorokin, and J. J. Wynne, Tunable coherent vacuum-ultraviolet generation in atomic vapors, Phys. Rev. Lett. 32: 343 (1974).

9. C. R. Vidal, Coherent VUV sources for high resolution spectroscopy, Appl. Opt. 19: 3897 (1980).

10. W. Jamroz and B. P. Stoicheff, Generation of tunable coherent vacuum-ultraviolet radiation, in "Progress in Optics XX", E. Wolf, ed., North-Holland, Amsterdam (1983) p. 325.

11. G. H. C. New and J. F. Ward, Optical third-harmonic generation in gases, Phys. Rev. Lett. 19: 556 (1967).

12. R. B. Miles and S. E. Harris, Proposed third-harmonic generation in phase-matched metal vapors, Appl. Phys. Lett. 19: 385 (1971): Optical third-harmonic generation in alkali metal vapors, IEEE J. Quant. Electr. QE-9: 470 (1973).

13. J. Bokor, P. H. Bucksbaum, and R. R. Freeman, Generation of 35.5 nm coherent radiation, Opt. Lett. 8: 217 (1983).

14. A. H. Kung, Third-harmonic generation in a pulsed supersonic jet of Xenon, Opt. Lett. 8: 24 (1983).

15. S. C. Wallace and G. Zdasiuk, High efficiency four-wave summixing in Mg at 140 nm. Appl. Phys. Lett. 28: 449 (1976).

16. J. R. Banic, R. H. Lipson, T. Efthimiopoulos, and B. P. Stoicheff, Radiative lifetimes of $B'^2\Delta$ state (v' = 0....8) of NO obtained by VUV laser excitation, Opt. Lett. 6: 461 (1981).

17. T. J. McKee, B. P. Stoicheff, and S. C. Wallace, Tunable,

coherent radiation in the Lyman-α region (1210-1290 Å) using Mg vapor, Opt. Lett. 3: 207 (1978).

18. W. Jamroz, P. E. LaRocque, and B. P. Stoicheff, Generation of continuously tunable coherent VUV radiation (140 to 106 nm) in Zn vapor, Opt. Lett. 7: 617 (1982).

19. R. Mahon and F. S. Tomkins, Frequency up-conversion to the VUV in Hg vapor, IEEE J. Quant. Electr. QE-18: 913 (1982).

20. R. R. Freeman, R. M. Jopson, and J. Bokor, Generation of coherent and incoherent radiation below 100 nm in Hg, in "Laser Techniques for Extreme Ultraviolet Spectroscopy", T. J. McIlrath and R. R. Freeman, eds., American Institute of Physics, New York (1982) p. 422.

21. R. Hilbig and R. Wallenstein, Resonant sum and difference frequency mixing in Hg, IEEE J. Quant. Electr. QE-19: 1759 (1983).

22. P. Herman and B. P. Stoicheff, Generation of VUV radiation at 120 to 104 nm by four-wave sum-mixing in Hg vapor, unpublished (1984).

23. A. H. Kung, J. F. Young, G. C. Bjorklund, and S. E. Harris, Generation of vacuum ultraviolet radiation in phase-matched Cd vapor, Phys. Rev. Lett. 29: 985 (1972).

24. A. H. Kung, Generation of tunable picosecond VUV radiation, Appl. Phys. Lett. 25: 653 (1974).

25. R. Hilbig and R. Wallenstein, Narrowband tunable VUV radiation generated by nonresonant sum- and difference-frequency mixing in Xe and Kr, Appl. Opt. 21: 913 (1982); Enhanced production of tunable VUV radiation by phase-matched frequency tripling in Krypton and Xenon, IEEE J. Quant. Electr. QE-17: 1566 (1981).

26. J. Reintjes, Frequency mixing in the extreme ultraviolet, Appl. Opt. 19: 3889 (1980); J. Reintjes, C. Y. She, and R. C. Eckardt, Generation of coherent radiation in the XUV by fifth- and seventh-order frequency conversion in rare gases, IEEE J. Quant. Electr. QE-14: 581 (1978).

27. H. Egger, T. Srinivasan, K. Hohla, H. Scheingraber, C. R. Vidal, H. Pummer, and C. K. Rhodes, A tunable, ultrahigh-spectral-brightness ArF* excimer laser source, Appl. Phys. Lett. 39: 37 (1981).

28. H. Egger, R. T. Hawkins, J. Bokor, H. Pummer, M. Rothschild, and C. K. Rhodes, Generation of high-spectral-brightness tunable XUV radiation at 83 nm, Opt. Lett. 5: 282 (1980).

29. E. E. Marinero, C. T. Rettner, R. N. Zare, and A. H. Kung, Excitation of H_2 using continuously tunable coherent XUV radiation (97.3-102.3 nm), Chem. Phys. Lett. 95: 486 (1983).

30. D. Cotter, Tunable narrow-band coherent VUV source for the Lyman-alpha region, Opt. Commun. 31: 397 (1979).

31. R. Hilbig, A. Lago, and R. Wallenstein, Tunable XUV radiation generated by nonresonant frequency tripling in Neon, Opt. Commun. 49: 297 (1984).

32. R. R. Freeman, G. C. Bjorklund, N. P. Economou, P. F. Liao, and J. E. Bjorkholm, Generation of cw VUV coherent radiation by four-wave sum frequency mixing in Sr vapor, Appl. Phys. Lett. 33: 739 (1978).

33. A. Timmerman and R. Wallenstein, Generation of tunable single-frequency continuous-wave coherent vacuum–ultraviolet-radiation, Opt. Lett. 8: 517 (1983).

34. T. J. McIlrath and R. R. Freeman, "Laser Techniques for Extreme Ultraviolet Spectroscopy", Am. Inst. Physics, New York (1982).

35. M. Rothschild, H. Egger, R. T. Hawkins, J. Bokor, H. Pummer, and C. K. Rhodes, High-resolution spectroscopy of molecular Hydrogen in the extreme ultraviolet region, Phys. Rev. A23: 206 (1981).

36. F. J. Northrup, J. C. Polanyi, S. C. Wallace and J. M. Williamson, VUV laser-induced fluorescence of molecular Hydrogen, Chem. Phys. Lett. 105: 34 (1984).

37. J. W. Hepburn, D. Klimek, K. Liu, R. G. Macdonald, F. J. Northrup, and J. C. Polanyi, Reactive cross section as a function of reagent energy. II. $H(D) + HBr(DBr) \rightarrow H_2(HD,D_2) + Br$, J. Chem. Phys. 74: 6226 (1981).

38. P. Ho and A. V. Smith, Rotationally excited CO from formaldehyde photoionization, Chem. Phys. Lett. 90: 407 (1982).

39. J. W. Hepburn, F. J. Northrup, G. L. Ogram, J. C. Polanyi, and J. M. Williamson, Rotationally inelastic scattering from surfaces. $CO(g) + LiF(001)$, Chem. Phys. Lett. 90: 407 (1982).

40. A. C. Provorov, B. P. Stoicheff, and S. C. Wallace, Fluorescence studies in CO with tunable VUV laser radiation, J. Chem. Phys. 67: 5393 (1977).

41. M. Maeda and B. P. Stoicheff, Measured radiative lifetimes of rovibronic levels in the $A^1\Pi(v = 0)$ state of CO and comparison with theory, in "Laser Techniques in the Extreme Ultraviolet", S. E. Harris and T. B. Lucatorto, eds., Amer. Inst. Physics, New York (1984).

42. R. H. Lipson, P. E. LaRocque, and B. P. Stoicheff, Vacuum-ultraviolet laser-excited spectra of Xe_2, Opt. Lett. 9: 402 (1984).

EXCIMER LASERS : PRACTICAL EXCIMER LASER SOURCES

Pio Burlamacchi

Dipartimento di Energetica, University of Florence
Via S.Marta, 3
50139 Florence, Italy

INTRODUCTION

Since their first demonstration[1] in 1975 rare-gas-halide
excimer lasers have been developed with amazing rapidity when
compared with the development of previous systems. This new generation of lasers have been indeed described as an ultraviolet revolution and a new gas laser workhorse. The basic similarity in design
to CO_2 pulsed TEA lasers played a fundamental role in this development.

Recently high efficiency, high power devices are receiving
increased attention for applications to laser fusion[2], laser chemistry[3,4], remote sensing[5], laser marking of metals[6], and advanced
laser processing of semiconductor materials[7].

The fact that excimer laser can be suitable sources for pumping dye lasers is now well assessed and the combination "excimer
laser pumped high power, narrow-band, tunable dye lasers" is a
practical research tool in many laboratories.

Shortly after the discovery of rare gas halides excited com-
plexes as suitable active materials for UV laser[8] two different
excitation methods were developed. Searles and Hart[1] were the
first to use this class of lasers by obtaining stimulated emission
from XeBr using an electron beam to excite the gaseous species.
The other method for exciting these lasers is the UV-preionized
discharge, first operated by Burnham et al.[9]. A combined method,
the e-beam controlled discharge, was first demonstrated by Mangano
and Jacob[10].

In a beam controlled discharge pumping a discharge voltage is applied across the mixture that is ionized by an e-beam. This method seemed to have the potential of beeing more efficient than pure e-beam pumping, however, to achieve this, key technical issues of discharge stability have to be addressed. Furthermore, the system technology seems to be complex and actually there is a strong tendency to consider alternative ways for large volume preionization as will be illustrated in the following paragraphs.

Electron-beam pumped devices are undergoing a rapid and profitable development in order to produce high energy pulses, essentially for such applications as inertial fusion or weaponery.

The performance characteristics and scalability of discharge excited excimer laser systems are limited by an ionization instability[11,12]. The instability leads to a collapse or filamentation of the discharge usually terminating laser operation in a few tens of nanosecond. Fast transverse discharges have been used very successfully to pump lasers of limited output energy (0.1-1 J) and pulse duration (10-40 ns) and have the advantage of simplicity of construction. High repetition rate is possible with the consequence that high average power can be obtained by using this excitation scheme.

The growth of nonuniformities can be suppressed by preionizing the gas with ultraviolet radiation generated by spark gaps, by corona discharge or X rays. The mechanism of preionization is not very clear since free electrons rapidly undergo dissociative attachment with the halogen donors to form negative ions. However, the ions may act as a secondary source which release electrons when the discharge voltage is applied. A uniform preionization is very important in order to maximize the amount of energy which can be deposited in the gas volume before the plasma collapses.

In the subsequent paragraphs the fundamental aspects for the construction of practical excimer laser devices will be reviewed. The spectroscopy of excimer molecules and their reaction kinetics during the discharge or electron beam excitation is very complex but it has been extensively studied and fairly well understood. As we shall limit our attention to the major practical considerations and recent achievements of excimer laser devices, we address the reader to an introductory bibliography on this subject[13,14,15].

ELECTRON BEAM EXCITED DEVICES

Electron beam devices are mostly used because of their ability to deposit relatively large power densities uniformly into the laser mixture. Fast electrons were in fact used for the direct pumping of rare gas excimers and all exciplex and excimer lasers were first pumped by high energy electron beams.

At the present time the development of e-beam technology for laser pumping seems to be limited to the achievement of high pulse energy[16,17]. One of the scaling limitations of these lasers seemed to be the self magnetic field of the beam that caused the beam to pinch. However a guided magnetic field can eliminate this effect[18] and the achievement of cost effective devices with very high energies seems to be possible. The minimum pump power density required to obtain laser action for an excimer laser is in the order of 100 KW/cm^3. Most devices built up to now have the capability to provide an order of magnitude higher density. Electron beam currents ranging from 1 to 100 KA, with energy up to 3 MeV and pulse duration of 20-200 ns have been used.

A typical laser system is shown schematically in Fig.1. The electron beam source consists of a high voltage generator, such as

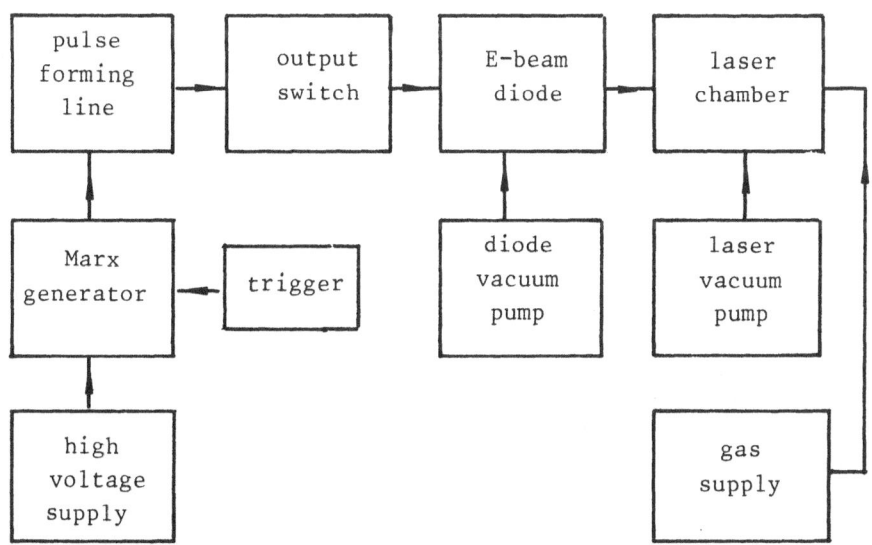

Fig.1.   Scheme of a typical e-beam excited rare gas halide laser system.

a Marx bank or pulse transformer[19],[20], a pulse forming line, usually a Blumeline, to produce ideally square pulses and a vacuum diode. For high voltage pulses (> 100Kv) it is not practical to start with a d.c. voltage because of corona losses, insulation problems and chance of breakdown. The Marx generator is in fact an impulse-voltage multiplicator circuit consisting of a number of capacitors which are charged in parallel and discharged in series through low-inductance spark-gaps and the high voltage is present only during a relatively short pulse. The Marx generator charges the pulse-forming line which is connected to the diode chamber through an output switch. The switch consists in a spark-gap or a rail-gap in which the gas pressure is adjusted in order to reach breakdown (by overvoltage) when the voltage reaches nearly its maximum value. Usually electron emission is from a cold cathode. The diode compartment must be avacuated to a pressure of about 10^{-4} Torr and is separated from the capacitor dielectric by a high voltage insulator and a feed-through. In principle the whole metal part in the vacuum, which is pulse charged to high voltage, can emit electrons, therefore the anode cathode distance is made much larger where no emission may occur, and the cathode is provided with either knife-edges, needles or, more usually, graphite which enhances the field on a microscopic scale. The anode in most cases is a high tensile strength foil, usually of titanium, steinless steel or aluminized dielectric which is sufficiently thin ($\le$ 50 µm) so as to allow efficient penetration by electrons with energies of 200 Kw or greater. An extra foil support, typically consisting of an array of holes or slots with a trasparency of 70-80%,is needed because of the high pressure difference between the laser chamber and the diode chamber. Possible heating of the foil is a major limitation to the possibility of high repetition rate in e-beam pumped excimer lasers.

The maximum current density supplied by a diode is limited by space charge effects. In fact , as the voltage pulse is applied to the cathode, the wiskers or microscopic protrusion explode due to high local electric field, producing heating and evaporation of the material. A plasma is formed, which expands in all directions creating a homogeneous virtual cathode which moves with a velocity of ~ 2.10^4 m/sec to the anode. Closure of the diode is the cause of the pulse length being limited to a few µs. The anode-cathode separation and therefore the impedance of the diode is effectively decreasing with time.

The fast electrons, after traversing the foil, are fairly diffuse in angles. A magnetic field with his direction coincident with the electron flow will restrict the diffusion because the electrons tend to gyrate around the magnetic field[21].

The most efficient way to couple e-beam energy into the laser mixture is by using two beams in a face to face configuration together with a magnetic field. As much as 90% of the fast electron energy can be coupled into the laser mixture with a variation in energy deposition of + 10% across the laser cavity in the beam direction[16,17,18,19,20,21].

AVALANCHE DISCHARGE EXCITED DEVICES

Although in the past the avalanche discharge excitation method appeared to operate at substantially lower efficiency as compared with electron beam devices, the discharge excited laser resulted in a convenient laboratory instrument of small size. A great deal of work mainly induced by a commercial fall-out have been done in order to improve reliability, efficiency and easy of operation. Prototypes and commercial lasers have been built with output energy ranging from 1 mJ to several Joules and repetition frequency up to the KHz range.

Different technical approaches have been choosen depending on the range of output energy and repetition rate of pulses. Common problems are the control of contaminants which may severely limit the lifetime of the gas fill or of the laser itself and the reliability of electronic components which are stressed by high voltage fast risetime pulses. For repetition rates higher than a few pulses per second gas flow between the electrodes is necessary. Actually most of the lasers which have been built are provided with a transverse flow blower. A cross section of a typical laser head is shown in Fig.2 while Fig.3 describes schematically the possible configurations of electrical circuits which are often used in a low energy apparatus.

We will now describe in detail the various technical solutions which at the present time seem to have led to the construction of reliable and efficient UV lasers.

a. Preionization

Preionization is an indispensable means for initiating volume stabilized discharges at high pressure. The simplest and most widely used method to obtain an initial electron density is UV preionization which can be obtained by row of sparks or by corona discharge[22,23,24]. The results of Sze and Loree[25] indicated that

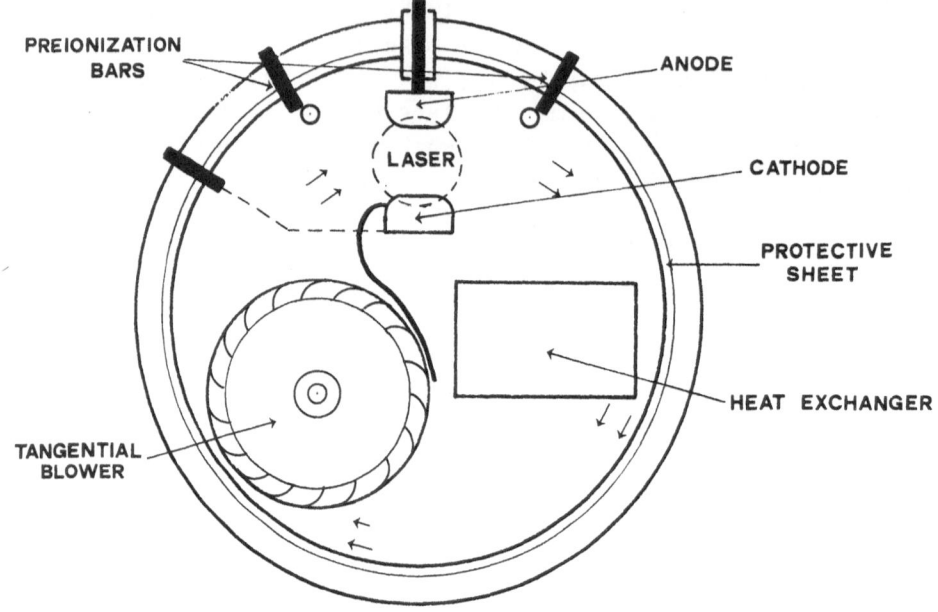

Fig.2. Cross-section of a typical high repetition rate UV preionized avalanche discharge laser.

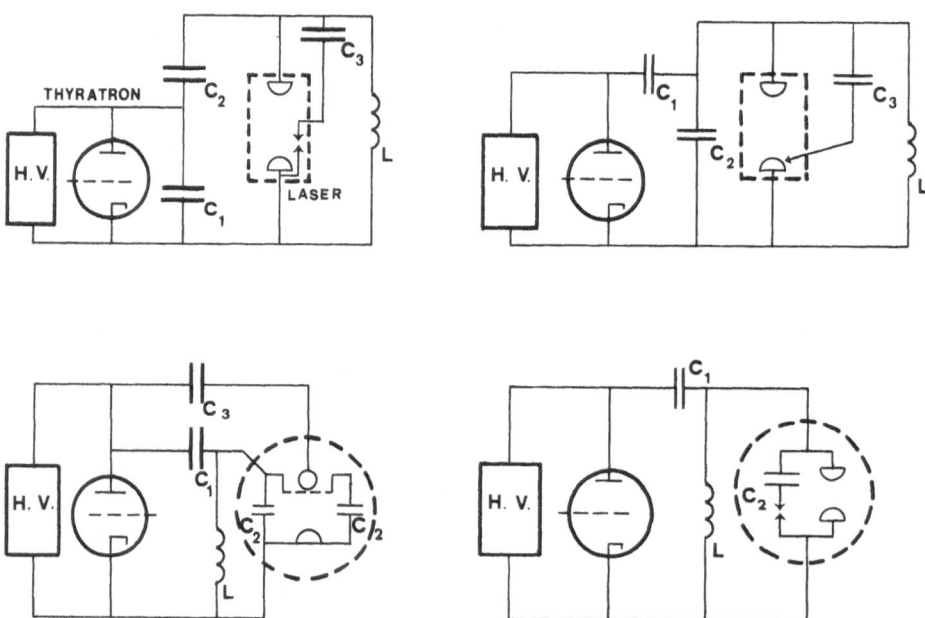

Fig.3. Various possible schemes for electrical excitation of
avalance discharge lasers.

an initial electron number density of 10^8 electrons/cm^3 is necessary. The electron density should be as uniform as possible[26]. For most applications the simplicity, ruggedness and reliability of UV pre-ionization cannot be dismissed, however this method has some limitations, expecially for the preionization of large volume devices or in those cases in which a spatially uniform beam is highly desirable.

X-ray preionization seems to be the most attractive for pre-ionizing large aperture, high pressure discharge lasers[27,28,29,30]. The advantages are as follows:

1) Since the mass penetration depth of X ray is much larger than that of charged particles of UV photons a high-pressure, large volume of laser gas mixture can be preionized.

2) The X-ray source can be separated and controlled outside the chamber. Therefore its design and material selection become more flexible.

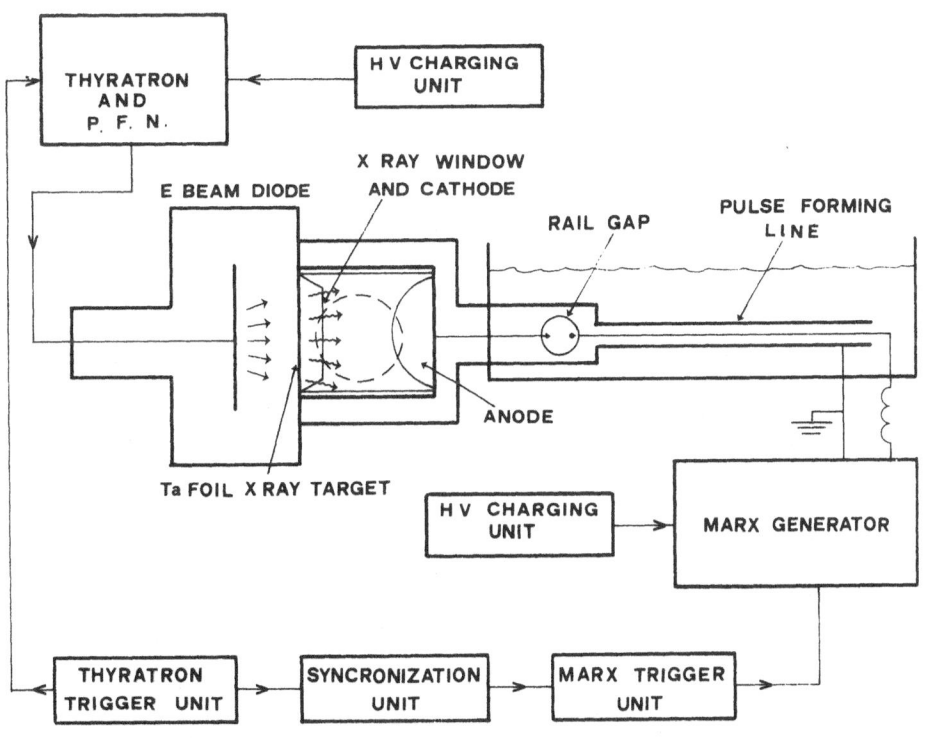

Fig.4. X ray preionized laser. For details of the electrical components see paragraph E.

3) Gas contamination caused by sparks can be eliminated with increase in lifetime of the gas fill.

However, the complexity of the apparatus and the necessity of radiation shielding has, up to now, limited the diffusion of this method. A schematic drawing of a X-ray preionized practical laser system is shown in Fig.4.[30]

b. Magnetic Pulse Compression

The fast risetime needed for excitation of avalanche discharge lasers must be acheaved by designing circuits with very low inductance and impedance parameters. This circuit characteristics along with the high charge voltages and peak currents places stringent requirements on the switch used to connect the primary energy storage capacitors to the excimer gas load. Thyratrons are the common switch choice for high repetition rate devices while in some cases, for repetition rates lower than 50 Hz or for high energy per pulse, spark-gaps are used. Thyratrons were originally developed for radar applications, in which peak current and current rise time are much lower. The more stringent excimer application can be met by thyratrons at the expense of lifetime which is much shorter in terms of number of shots (5.10^7-5.10^8 as compared with 5.10^9 for radar applications).

The commutation time of the thyratron lasts for several tens of ns so that, for fast switching, the current pulse develops before the anode cathode potential has dropped. This means that a consid-

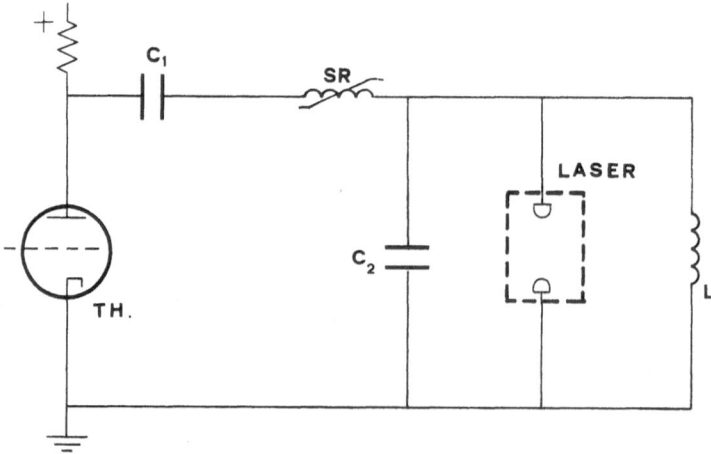

Fig.5.   Scheme of a magnetically assisted discharge circuit. The pulse is delayed but not compressed.

erable amount of power is dissipated inside the tube. The larger
the dissipation the more severe is the stress on the thyratron due
to increased cathode secondary emission and localized heating in
the tube internal structure. A new technique to delay the pulse
without substantial reduction of risetime is to use a magnetic pulse
compression[31,32]. This technique utilizes the nonlinear properties
of various ferromagnetic materials as the basis for magnetic switching elements. A schematic drawing is shown in Fig.5. The saturable
reactor will remain in a high impedance state, behaving as an in-
ductor, only as long as the magnetic core is not saturated. Core
saturation occurs when the current reaches a certain value which
depends on the magnetic material chosen. When the thyratron switches
the core represents a large inductance which slowes down the current
rise-time. As soon as the core saturates the impedance decreases to
near zero and current continues as though the core were not present.
In this case the pulse is delayed but not compressed. Fig.6 represents a pulse compression scheme. Conceptually it may be considered
as two subcircuits; the first circuit consists of L_1, thyratron TH,
C_1 and C_2. The second circuit consists of C_2, the saturable induc-
tor SI and the laser load. The charge stored in C_1 is first transferred to C_2 through the thyratron and C_1 keeping the rate of increase of the current within the thyratron rating. Subsequently C_2
is switched rapidly to the load using the SR, which is designed to
saturate exactly when C_2 is fully charged. Magnetic pulse compression technology appears very attractive, expecially for high average power devices. Butcher and Falhen[32] have demonstrated the pos-

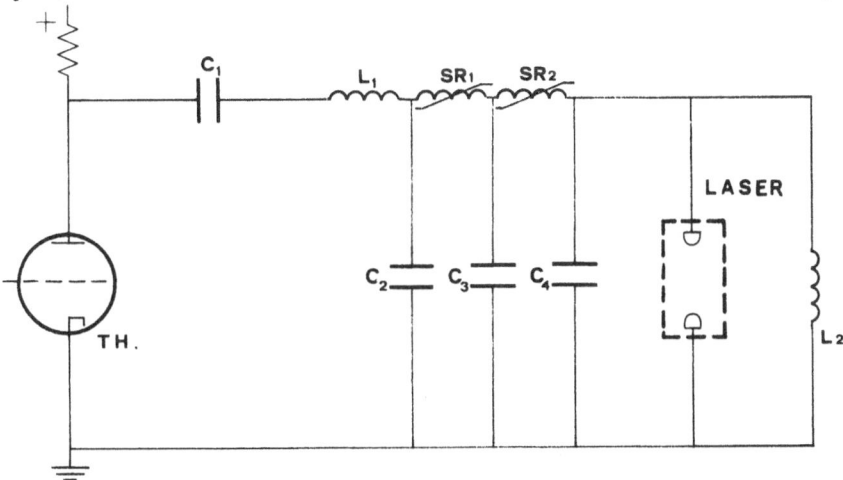

Fig.6.   Scheme for a magnetically assisted discharge circuit with
pulse compression. The scheme shows two compression elements
for a compression ratio of about 12.

sibility of performing with high repetition rate (500 Hz) at an average power level of 150 W. The device could be also operated, at lower average power, by replacing the thyratron with a simple commercially available silicon controlled rectifier and saturable magnetics, resulting in a all-solid-state-driven-laser.

c. High Voltage Prepulse

Where discharge lasers have fallen short in the past is in the transfer of energy from a pulse forming network or line into the gas. Efficient power transfer can in fact only be obtained after the gas is fully broken down. By using a high voltage pre-pulse generated by an interdependent circuit Lang and coworkers[33] have demonstrated the possibility of optimizing impedance matching, reaching a laser efficiency of 4.2% in a 4.2 Joule output pulse. Although some circuit complexity is necessaty to obtain a suitable prepulse and a pulse forming line must be used, this experiment opens the way to a really efficient device which could relax the requirements for high average power operation.

d. High Repetition Rate

The heat released in the gas mixture by the discharge induces a substantial perturbation of the refractive index so that the optical cavity is completely destroyed after each pulse. The recovery time after a pulse depends on the laser geometry and gas pressure and composition and is typically in the order of several hundreds of ms for medium average power systems[34]. In addition to this effect there are problems arising from the formation of atomic and molecular species adversely affecting the excimer kinetics, and a break down voltage drop leading to a reduction in energy deposition. The most effective way to achieve high pulse repetition rate is to "clean" the discharge region by circulating the gas through the electrodes at high speed. Experimental evidence shows that more than 5 gas replacements in the discharge volume between pulses are required to maintain the maximum energy per pulse. Although wind tunnels with centrifugal blowers have been used in the past, to achieve gas velocity of the order of 50 m/sec actually tangential blowers are used, connected to external motors by magnetic shaft seals. The gas mixture has to be cooled by suitable gas to water heat exchangers.

e. Wide Aperture Excimer Lasers

Several applications, as described in the introduction, have motivated the development of wide aperture ($\geq$ 3 x 3 cm^2),high output energy density ($\geq$ 2 J/l),long optical pulse duration ($\geq$ 50 ns) discharge lasers. The technology of such lasers is not very different from the electron beam devices. Fig.4 represents a schematic of a possible arrangement. UV preionization[35,36,37] or X ray preionized[38] systems have been developed. The electrical excitation is derived from a Marx-bank primary energy storage source which is used to pulse charge a pulse forming transmission line (normally a distilled water insulated Blumeline line). A rail gap switch, which is essentially a low inductance multichannel spark gap[39], isolates the pulse forming from the laser electrodes and it switches spontaneously when the line reaches maximum voltage. The X ray unit is triggered independently some time before the switching of the Marx generator. Such systems have given output energies up to tenth of joules[36] with efficiencies in the order of 3%[38].

f. Electrodes Profile

In order to have a stable discharge with the maximum possible energy density it is important to have a very uniform energy loading of the active gas medium. This can be achieved by producing a very uniform field-strength distribution over a certain amount of surface. Specially contoured electrodes have been designed following the work of several authors[40,41,42,43]. However, practical electrode design and construction is always followed by a somewhat empirical retouching, which is done by hand after visual observation of eventual discharge inhomogeneities. In many cases the cathode is a flat or slightly curved net which can transmit UV radiation from the preionization bars. The configuration of the net is normally an empirical choice.

g. Effects and Control of Contaminants

One of the serious problems which people working with excimer lasers have been faced with,is the handling of corrosive gases, such as halogen compounds. The dominant forms of chemical-related degradation include optical damage, corrosion of laser materials and gas contamination[44]. Optics degradation is mostly due to the deposition on the surfaces that are in contact with the gas, of particulate, chlorocarbons and HCl or HF that are present inside the laser. The particulates originate primarily in the discharge and preionized region, while the halogens react with metals. Chorocarbons originate mostly from ultraviolet photons interacting with dielectrics, including O-rings, near the laser beam. Several techniques have been used to prevent coating from accumulating on the optics. Using nickel as construction material greatly reduces the formation of metallic chlorides and fluorides which flow apart in the discharge in form of powder. Plastic has to be avoided and dielectric materials should be limited to ceramics or saturated fluorocarbons (teflon). Also flushing of the optics with clean rare gas alleviates particulate accumulation. Failure to adhere to these guidelines will result in premature performance degradation and short lifetime of the gas fill due to excessive halogen reactivity. As mentioned before, nickel is very desirable due to high resistance to halogen attack. Nickel plating of all basic metals is an accetable technique although its usfulness for main discharge electrodes is limited by a rapid erosion. Despite its high cost the electrodes material should be solid nickel.

Carefull choice of material, proper passivation and initial cleaning of the gas and gas handling devices, in combination with suitable additives such as hydrogen may bring to the construction of devices with a life of 10^{10} pulses[44]. This means that, at least for the XeCl excimer a sealed-off tube should be possible.

CONTROL OF EXCIMER LASER RADIATION

The bandwidth of the output radiation of an excimer laser with a conventional optical cavity, consisting of two plane or slightly curved mirrors, is in the order of 100 cm^{-1} and the divergence is ~`5 mrad. For advanced applications methods for control of the bandwidth, laser frequency, divergence, spatial homogeneity and pulse duration must be devised. The key parameter in this context is the spectral brightness of the radiation, that is the energy per unit time, area, solid angle and bandwidth.

Approaches controlling excimer laser radiation have been implemented using combinations of intracavity gratings, prisms, etalons, apertures and unstable resonators. However, these techniques have limitations when the high energy and efficiency of excimer lasers have to be exploited. From the point of view of practical systems it is now clear that a small, well controlled oscillator followed by single or multipass amplifiers is a convenient choice.

The various techniques for controlling the output radiation can be resumed as follows:

a. Frequency Tuning

The selective elements in the optical resonator is most often constituted by a grating inserted in the optical cavity in various configurations[35,36]. Grazing incidence gratings have provided the best efficiency with reasonable narrow bandwith (0,1 Å)[36]. KrF and ArF have been the most popular candidate probably because of the fact that in both of these cases laser action terminates on an unbound ground state. Hence continuous tuning should be more readily achieved. XeF and XeCl terminate on bound ground states and therefore relative transition probabilities will play a role in determinating their respective tuning ranges.

The bandwidth of a tuned laser oscillator has a lower limit which is imposed by the short duration of the pulse. Long pulse excitation of excimers laser is convenient, although not easy, for the control of spectral brightness and several devices have been designed for this purpose[37]. Longer pulses, up to the order of one microsecond have been achieved by electron-beam pumping.

b. Mode Locking and Short Pulse Generation

If short pulses are to be obtained directly from rare gas halides lasers the system would become considerably simple. As in other well known lasers, active and passive mode locking can be achieved.S.Watanabe and coworkers[48] have used a saturable absorber in form of a suitable dye solution inside the laser cavity in a long pulse laser (150 ns). A modulation of 100% in the output intensity of the pulse train with a sharpening to the duration of ~ 2 ns was achieved. Better results have been obtained, using active mode locking, by modulating the gain with an intracavity Pockel's cell[49]. A train of 100% modulated pulses with ~ 300 ps in duration were measured by the authors.

The minimum achievable pulse duration in direct mode locked pulses is severely limited by the small number of cavity round trips completed during the period of effective gain. However the emission spectrum of rare gas halides lasers is sufficiently broad to permit the amplification of pulses of a few picosencond in duration. Thus by using as an example a XeF laser,the restrinctions placed on the minimum attainable pulse duration by the gain duration are effectively circumvented. Amplification of picosecond pulses have been obtained by a well controlled lasers such as an actively mode locked Nd laser[50] or a dye laser[51] with third harmonic frequency conversion. Tunable output radiation with pulse duration down to ~ 10 ps have been obtained in this way in ArF.

c. Injection Locking

When a high power laser is required to operate with only one longitudinal mode and at a particular frequency the best technique may be injection locking[52,53]. In fact a spectrally pure oscillator output tends to be low power and injection locking involves injecting the output of a low power laser into a broad-band high power laser.

A very high percentage (up to 90%) of the energy available in an amplifier can be extracted in a narrow line by this technique. Transform limited bandwidth (2×10^4) Å has been obtained in a XeF oscillator with 60% efficiency in energy extraction[54]. In some cases, as with XeCl tuned at the peaks of the gain profile, the entire output of a high power laser was confined in a 4.10^{-4} Å line with an injected signal of 20 μJ of UV radiation obtained from an argon ion pumped ring dye laser[55]. Injection locking in very high power devices, with single or multiple beam, seems to become common practice in many applications[17].

REFERENCES

1. S.K.Searles, and G.A.Hart, Stimulated Emission at 281.8 nm from XeBr, Appl.Phys.Lett.27: 243 (1975)
2. J.Goldhar, W.R.Rapapart, and J.R.Murray, An Injection-Locked Unstable Resonator Rare Gas Halide Discharge Laser of Narrow Bandwidth and High Spatial Quality, IEEE J.Quantum Electron. QE16: 235 (1980)
3. G.Eden, R.Burnham, L.F.Champagne, T.Donohue, and N.Djeu, Visible and Ultraviolet Lasers, IEEE Spectrum, p.50 (Apr.1979)
4. V.S.Letokhov, Laser Induced Chemical Processes, Phy.Today, p.34 (Nov.1980)
5. O.Uchino, M.Maeda, M.Hirono, Application of Excimer Lasers to Laser-Radar Observations of the Upper Atmosphere, IEEE J. Quantum Electron. QE15: 1094 (1979)
6. T.J.Mc Kee, Excimer Laser. An Ultraviolet Revolution, Phy.Can. 36: 41 (1980)
7. T.J.Mc Kee, and J.Nilson, Excimer Applications, Laser Focus p.51 (June 1982)
8. J.E.Velazco, and D.W.Setzer, Quenching Studies of $Xe(^3p_2)$ metastable atoms, IEEE J.Quantum Electron. QE11: 708 (1975)

9. R.Burnham, H.W.Harris, and N.Djeu, Xenon Fluoride Laser Excitation by Transverse Electric Discharge, App.Phys.Lett. 28: 86 (1976)

10. J.A.Mangano, and H.J.Jacob, E-Beam Controlled Discharge Pumping of KrF Laser, App.Phys.Lett. 27: 495 (1975)

11. J.D.Daugherty, J.A.Mangano, and J.H.Jacob, Attachment Dominated Electron Beam Ionized Discharges, App.Phys.Lett. 28: 581 (1976)

12. W.L.Nighan,"Principles of Laser Plasmas", G.Bakefi ed., chap.7, p.257, Wiley New York, (1976)

13. C.K.Rhodes Ed., Excimer Lasers, Topics App.Phy. 30, Springer (1979)

14. M.H.R.Hutchinson, Excimers and Excimer Lasers, App.Phys. 21: 95 (1980)

15. M.Rokni, and J.H.Jacob, Rare Gas Halides Lasers, in: "Applied Atomic Collision Physics" vol.3, Academic Press (1982)

16. J.Goldhar, K.S.Jancaitis, and J.R.Murray, 850 J, 1050 ns narrow-band Krypton-Fluoride Laser, CLEO 84 Technical Digest ThB$_2$ p.136 (1984)

17. S.Singer, Recent Advances in KrF Systems Technology, CLEO 84 Technical Digest THB1, p.136 (1984)

18. J.C.Hsia, A Model for UV Preionization in Electric-Discharge-Pumped XeF and KrF Lasers, App.Phys.Lett. 30: 101 (1977)

19. C.B.Edwards, M.H.R.Hitchinson, D.J.Bradley, and M.D.Hutchinson, Repetitive Vacuum Ultraviolet Xenon Excimer Laser, Rev.Sci. Instr. 50: 1201 (1979)

20. E.Fiorentino, T.Letardi, A.Marino, E.Sabia, M.Vannini, Electron Beam Sustained Discharge XeCl Laser.(To be published)

21. J.H.Jacob, Diffusion of Fast Electrons in the Presence of a Magnetic Field, App.Phys.Lett. 31: 252 (1977)

22. H.J.Seguin, and J.Tulip, Photoionization and Photosustained Lasers, App.Phys.Lett. 20: 414 (1972)

23. H.J.Seguin, J.Tulip, and D.Mc Keen, Ultraviolet Photoionization in TEA Lasers, IEEE J.Quantum Elect. QE10: 331 (1974)

24. R.C.Sze, and P.B.Scott, 1/4-J Discharge Pumped KrF Laser, Rev. Sci.Instr. 49: 772 (1978)

25. R.C.Sze, and T.R.Loree, Experimental study of a KrF and ArF Discharge Laser, IEEE J.Quantum Electron.QE14: 944 (1978)

26. S.Sunida, K.Kunitamo, M.Kaburagi, M.Obara, and T.Tujioka, Effect of Preionization Uniformity on a KrF Laser, J.App.Phys. 52: 2682 (1981)

27. S.Sumida, M.Obara, and T.Fujioka, X-Ray-Preionized High Pressure KrF Laser, App.Phys.Lett.33: 913 (1978)

28. S.C.Lin, and J.I.Levotter, X-Ray Preionization of Electric Discharge Laser, App.Phys.Lett. 34: 505 (1979)

29. H.Shields, and A.J.Alcock, Short Pulse X-Ray Preionization of a High Pressure XeCl Gas Discharge Laser, Optics Comm. 42: 128 (1982)

30. K.M.Dorikawa, M.Obara, and T.Fujioka, X-Ray Preionization of Rare-Gas-Halide Lasers, IEEE J.Quantum Electr. QE20: 198 (1984)

31. I.Smilanski, S.R.Byron, and T.R.Burkes, Electrical Excitation of an XeCl Laser Using Magnetic Pulse Compression, App.Phys. Lett. 40 (7): 547 (1982)

32. R.R.Butcher, and T.S.Falhen, Magnetically Switched 150 W XeCl Laser, CLEO 84 Technical Digest THP1, p.202 (1984)

33. W.H.Lang Jr., M.J.Plummer, and E.A.Stappaarts, Efficient Discharge Pumping of an XeCl Laser Using a High Voltage Pre-pulse , App.Phys.Lett. 43 (8): 735 (1983)

34. R.Buffa, P.Burlamacchi, M.Matera, H.F.Ranea Sandoval, and R. Salimbeni, High Repetition Rate Effects in XeCl TEA Lasers, Optic. Comm. 40: 288 (1982)

35. R.S.Taylor, S.Watanabe, A.J.Alcock, K.E.Leopold, and P.B.Carkum, Operating Characteristics of a 5 J (5 J/liter) UV Preionized XeCl Laser, IEEE J. Quantum Electron. QE17: Special Issue, Part.II, 82 (1981)

36. S.Watanabe, and A.Endoh, Wide Aperture Self Sustained Discharge KrF and XeCl Lasers, App.Phys.Lett. 41: 799 (1982)

37. R.S.Taylor, P.B.Carkum, S.Watanabe, K.Leopold, and A.J.Alcock, Tune-Dependent Gain and Absorption in a 5 J UV Preionized XeCl Laser, IEEE J. Quantum.Electr. QE19: 416 (1983)

38. M.R.Osborn, M.H.R. Hutchinson, and P.W.Smith, Improvement in efficiency of X-Ray Preionized XeCl Lasers, CLEO 84 Technical Digest THP4, pag.204 (1984)

39. R.S.Taylor, A.J.Alcock, and K.E.Leopold, Rail Gap Switches for High Output Energy Excimer Lasers, In: "Proc.3rd IEEE Int. Pulsed Power Conference", Albuquerque NM, p.157 (1981)

40. W.Rogowski, Arch.Electrotech. 12: 1 (1923)

41. T.Y.Chang, Improved Uniform Field Electrode Profiles for TEA Laser and High Voltage Applications, Rev.Sci.Instr. 44: 405 (1973)

42. A.E.Stappaerts, A novel Design Method for Discharge Laser Electrode Profiles, App.Phys.Lett. 40 (12): 1018 (1982)

43. G.J.Ernst, Uniform Field Electrodes with Minimum Width, Opt. Comm. 49: 275 (1984)

44. R.Tennant, Control of Contaminants in XeCl Lasers, <u>Laser Focus</u> (Oct. 1981)

45. T.J.Mc Kee, J.Banic, A.Jares, and B.P.Stoicheff, <u>IEEE J.Quantum Electron</u>. QE15: 332 (1979)

46. R.Buffa, P.Burlamacchi, R.Salimbeni, and M.Matera, Efficient Spectral Narrowing of a XeCl TEA Laser, <u>J.Phys.D</u> (App.Phys.) 16: L125 (1983)

47. D.B.Cohn and H.Komine, Long Pulse Excimer Laser Excited by Sequenced Discharges, <u>IEEE J.Quantum Electron</u>. QE19: 786 (1983)

48. S.Watanabe, M.Watanabe, and A.Endoh, Passive Mode Locking of a Long Pulse XeCl Laser, <u>App.Phys.Lett</u>. 43 (6): 533 (1983)

49. G.Reksten, T.Varghese, and W.Margulis, Active Mode Locking of a XeCl Laser, <u>App.Phys.Lett</u>. 39 (2): 129 (1981)

50. I.V.Tomov, R.Fedosejevs, M.C.Richardson, W.J.Sarjeant, A.J. Alcock, andK.L.Lopold, Picosecond XeF Amplified Laser Pulses, <u>App.Phys.Lett</u>. 30: 146 (1977)

51. H.Egger, T.S.Luk, K.Boyer, D.F.Muller, H.Pummer, T.Srinivasan, C.K.Rhodes, Picosecond, Tinable ArF Excimer Laser Source, <u>App.Phys.Lett</u>. 41 (11): 1032 (1982)

52. H.L.Stower, and W.H.Steier, Locking of Laser Oscillations by Light Injection, <u>App.Phys.Lett</u>. 8: 91 (1966) and C.J.Buczek, R.J.Freiberg, and M.L.Skolnick, Laser Injection Locking, <u>Proc. IEEE</u> 16: 15411 (1973)

53. W.W.Chow, Theory of Line Narrowing and Frequency Selection in an Injection Locked Laser, <u>IEEE J.Quantum Electron</u>. QE19: 243 (1983)

54. I.J.Bigio, and M.Slatkine, Transform-Limited-Bandwidth Injection Locking of an XeF Laser with an Ar-Ion at 3511 Å, <u>Opt. Lett</u>. 7: 19 (1982)

55. O.L.Bourne, andA.J.Alcock, A High Power, Narrow Linewidth XeCl Oscillator, <u>App.Phys.Lett</u>. 42(9): 777 (1983)

EXCIMER LASER APPLICATIONS

Dirk Basting and Ulrich Sowada

Lambda Physik GmbH
3400 Göttingen
W.-Germany

A) The Srinivasan Effect

INTRODUCTION

Until recently excimer lasers have been used almost exclusively
in basic research, e.g. to pump dye lasers. However, this is
not the only potentially useful application. The output of
excimer lasers differs in one aspect quite drastically from all
other known uv-light sources, namely fluence, which is number
of photons per cm^2 and sec. The photons are the same, but the
mode of delivery is different. This is basically the reason for
a new effect, which, if it holds up to its promises, will radi-
cally expand excimer laser use in science, industry, and
medicine. This effect is often called "ablative photodecompo-
sition", but it is also known as the "Srinivasan effect".

The Process

If pulsed, uv-light from an excimer laser irradiates an organic
substance (polymer or biological tissue), three consecutive
phenomena add up to the Srinivasan effect (ref. 1): first,
the laser light is absorbed by the organic material; the second
step is breaking of bonds in the large molecules, thus producing
a high concentration of volatile molecular fragments; third,
from the irradiated surface the tiny pieces ablate and carry
away most of the excess energy.

The first step also occurs under low power uv-exposure.
However, the concentration of volatile fragments produced in
the second step requires the high power of the uv-light; with
e.g. a mercury resonance lamp the concentration reached depends

on gain and loss terms, and loss due to recombination or effusion is not negligible on a timescale of seconds. With all the photons in 15 nsec, however, most loss terms can be neglected. So the third step will not happen unless the fluence of uv-light is high enough.

The special feature which makes the Srinivasan-effect different from other types of laser-processing is that practically all energy ends up in the ablated particles. This means that the unirradiated portions of the sample are left unaffected. Thus it is justified to name the Srinivasan-effect "heatless laser etching".

The effect has great potential because it is operative on an unusually wide variety of compounds: polymers like polymethyl methacrylate, polycarbonate, polyimide, poly(ethylene terephtalate) (ref. 2) as well as biological tissue from various parts of the body (refs. 3, 4). This is because the required material properties, namely to strongly absorb uv-light and to exhibit bond breaks after photon absorption, are not special to selected compounds but rather general features. This is particularly true if the excimer laser is operated at 193 nm.

The techniques used to apply the Srinivasan-effect for purposes of materials processing are simple and straightforward. By defining the irradiated area with a mask, edges of unusually high quality are readily produced. With longer wavelength laser processing, (248, 308 nm) the edges show signs of high temperatures: melting and sometimes carbon production.

Several aspects have been studied and detailed reports are available. Computer modelling (ref. 5) has shown that the ablated material has velocities around 1400 m/s, that the ablated matter is ejected into a cone around the surface normal with 25° full angle, and that the material remaining behind is not melting. These computer results have been verified experimentally (ref. 6).

A new tool will be accepted if it is reproducible. This is certainly the case in this new process. The thickness of the layer l which is removed from the exposed surface is a very reproducible quantity. It varies with applied fluence F according to the following relation (see e.g. ref. 7)

$$l = m \cdot \ln (F/\text{threshold fluence})$$

In this equation m and threshold fluence are constants which depend on material and laser wavelength. Typical values for l are around 0.5 μm for irradiation intensities of F = 500 mJ/cm^2. This value of l is quite small, and it would be uneconomical to cut through 1 cm of material. However, the reproducibility of this value gives a degree of depth control which is hard to reach by any other method.

One of the unresolved puzzles about this laser process is "the taper". Figs. 1 a - b show holes drilled in PMMA and photographed through a microscope from the side. The diameter of the cross section decreases steadily with increasing drill depth. Finally, a conical volume has been removed from the sample rather than a cylindrical one. The half angle of the cone depends on laser fluence as evident from the pictures. It has been suggested that the cone angle is related to the fact that the Srinivasan-process is operative only above a threshold value. Then, ablation will stop if the effective fluence is below the threshold:

$$\text{effect. fluence} = F \cdot \cos a \, (1 - R(a)) \leq \text{threshold fluence}$$

Here a is the angle of incidence and R(a) the surface reflectivity. This hypothesis is based on the following idea: if by some mechanism, as e.g. weaker irradiation at the geometrical edge due to diffraction a taper has formed it will be stabilised at an angle under which the effective absorbed fluence on the surface is equal to the threshold fluence.

It turns out that with a constant value for threshold fluence the determined reflectivities R(a) decrease with increasing a. This is against all probability. However, closer inspection of fig. 1 a shows that at low F the hole is not conical, but the tangent angle a is a function of hole diameter! This is rationalised if we let threshold fluence be a function of "free space" above the surface available for ablation. This makes the quantitive understanding of the "taper" a difficult problem.

Applications

Science

The application of this process in scientific studies is hardly limited. It should find use in cutting, trimming or hole drilling whenever high quality results in organic or polymeric materials have to be obtained.

Sometimes observations can be made which are quite pretty. Figs. 2 a to c show cuts on hair. Fig. 2 a is a human hair with attempted cuts at a distance of 40 μm from each other. Fig. 2 b is a slice of a hair from taurotragus euryceros . This easily produced sample shows no sign of deformation due to pressure during the cutting process; therefore, its interior is undisturbed. Since there is no water in hair, the usual deep-freeze cutting procedure will probably not work. Fig. 2 c is a slice of 50 μm thickness of a hair from oreatragus oreatragus . (We are grateful to the Frankfurt Zoo for sending us the animal hair samples.) It is from the middle portion which is white. The hair consists mainly of air, and no scissors could produce a slice without squeezing this delicate structure.

This should demonstrate that excimer laser cutting can be performed to yield high quality cuts on a microscopic scale without pressure or high temperature effects.

Industry

The Srinivasan-effect can easily be utilized in semiconductor technology. Present day large scale integrated circuits are produced by processing certain areas on a silicon wafer. This processing may involve ion implantation, oxydation or coating. Other parts are protected by a photoresist which had been spun or sprayed onto the surface. Photolithography is used to clear the parts to be processed from the photoresist. Up to now this still includes washing out of the illuminated photoresist. With the Srinivasan process a resist image can be obtained without the wet step. The wall profile is determined by the light path and will not be affected by the wet development. It is expected that the achievable resolution depends (as in ordinary wet lithography) on the wavelength of the light, and therefore submicron structures should be within technological reach. The relevant companies are at present investigating whether this process of dry etching will be used to produce the 1 MBit-chip.

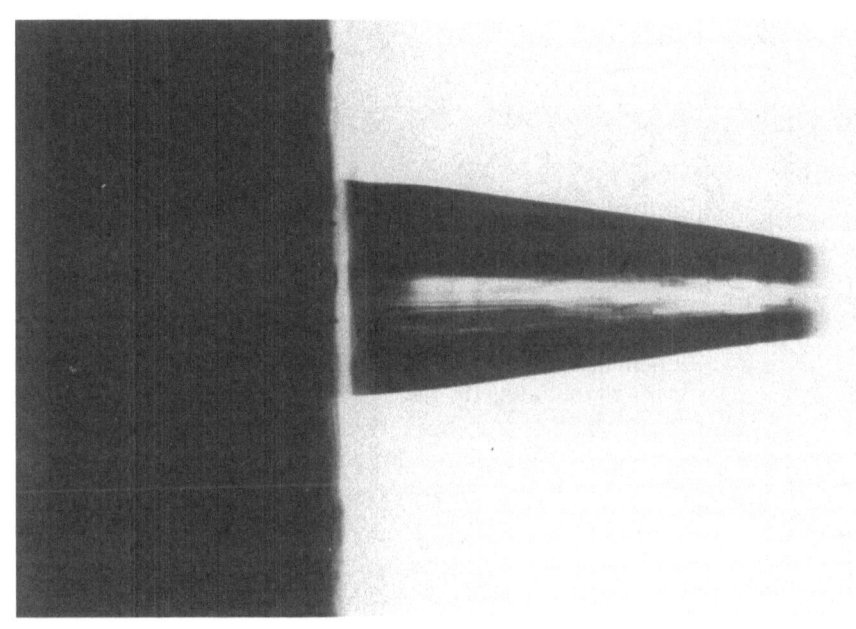

a) 205 mJ/cm^2

b) 520 mJ/cm^2

Fig. 1: Hole drilled in PMMA using the Srinivasan effect; laser at 193 nm.

Fig.2a: Human hair with trial cuts 40 μm apart.
Hair thickness 70 μm.

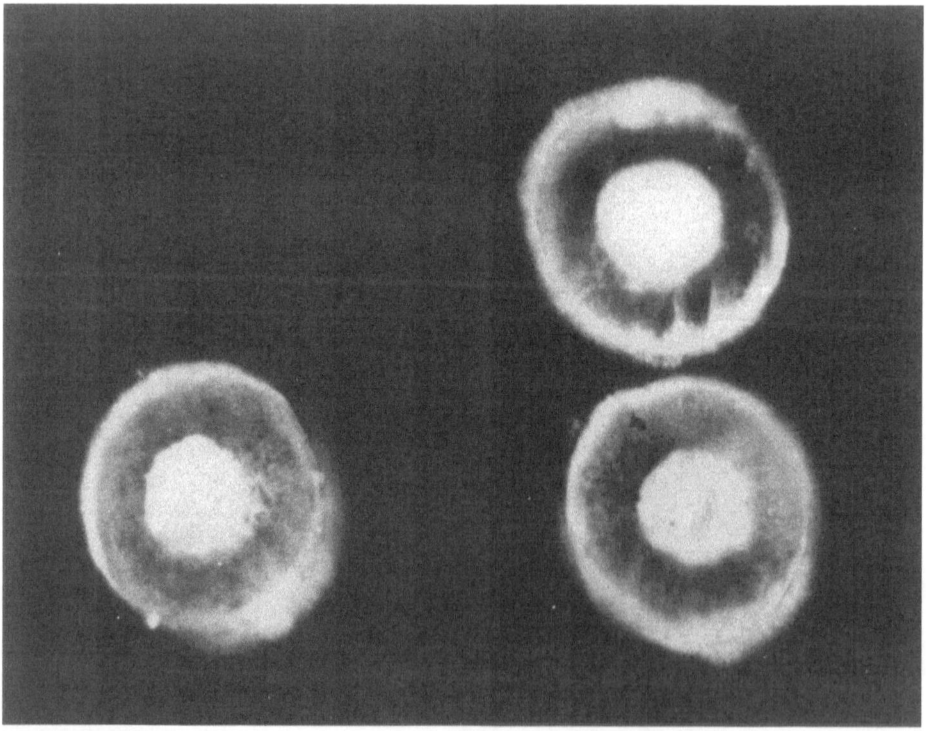

Fig.2b: Hair slice from taurotragus euryceros.
The slice is 50 μm thick; hair diameter is 160 μm.

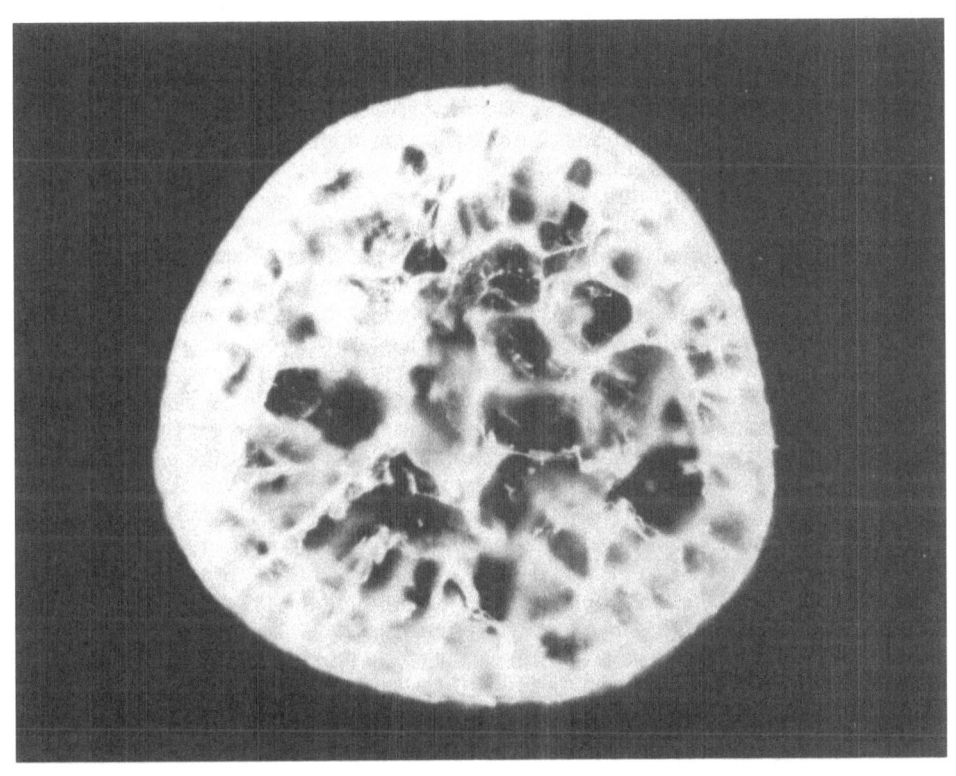

Fig.2c: Hair slice from oreotragus oreotragus. The
slice is 70 μm thick; hair diameter is about 400 μm.

Medicine

The need of an "ideal knife" in many fields of medical surgery
represents another potential application of the Srinivasan effect.
Such an ideal knife should allow to make cuts without heating or
charring of the tissue, and without pulling or tearing. The
quality of cuts which can be achieved with the excimer laser
using Srinivasan's effect has been frequently examined (ref.8).
While the longer excimer wavelenghts (particularly 308 nm) leave
potentially carcinogenic or mutagenic compounds in the walls of
the cut, 193 nm appears to be very close to the ideal knife
with no detectable harmful side effects. The cuts show no signs
of heat on the walls when investigated histologically. The kerf
width is easily made as small as 0.01 mm (see fig. 2 a), which is
the theoretical limit of a CO_2 laser. The depth control (through
the number of applied pulses) of 0.001 mm appears to be beyond
any mechanical tool's capability.

It is investigated these days in several laboratories in Europe
and America, to which surgical problems this new knife can be
applied. Experts in the fields of ophthalmology, neurosurgery,
cardiology, otology, urology, and orthopedy have become interested.
The investigations about this new "knife" may produce results for
the benefit of many patients. Positive results in animals suggest
that in the near future corneal cuts probably will be made with
excimer lasers. The latest models of excimer lasers are reliable
and easy to operate to be a trustworthy surgical tool.

B) Semiconductor Technology

The main difference between excimer lasers and other pulsed
lasers is output wavelength, or photon energy. The conventional
lasers on the market emit either in the infrared spectral region
(CO_2-, Nd:YAG-laser) or in the visible (ruby laser). Even
though there are ways to convert the output to shorter wavelengths, the poor efficiency of these methods are in favor of
the excimer laser. The shorter wavelength implies two conse-
quences: the theoretical resolution in imaging should be higher,
and molecular dissociation after single photon absorption becomes
possible. The semiconductor industry has realized the importance
of these two aspects, and is a good example to describe
industrial applications of excimer lasers.

Theoretically the resolution of an otherwise perfect imaging
system is given by (see e.g. ref. 9)

$$d = 0.6 \cdot \lambda / NA$$

Here NA is the numerical aperture of the imaging system,
the wavelength of the light and d the size of the resultant
diffraction "spot". The importance of this equation for semiconductor production is evident, because the electronic
circuitry on a chip is produced with the help of photolithography, i.e. an imaging of patterns onto the wafer surface
which is then still covered by a photoresist. Spatial resolution
limits the density of the circuits, and therefore illumination
at shorter wavelength offers the possibility of higher integration. Submicron structures can be produced, and a commercial
system for contact printing wafer illumination using an excimer
laser at 193 nm is on the market (ref. 10).

Dopant atoms can be deposited on a wafer by photodissociation of
molecules. A necessary requirement for a reasonable efficiency
is that it should be by a single photon process. Therefore, among
the lasers only the excimer with its high energy photons is
suitable for this task. The conventional technique requires
heating of the substrate to about 1000° C. The photochemical step
induced by the excimer laser makes this risky and difficult
procedure superfluous.

Another task in semiconductor production which may possibly see
our devices on the job is annealing of crystal surfaces or thin
films. The possibility to change output wavelength easily allows
to process different types of workpieces at the respective best
wavelength. Sometimes 351 nm (ref. 11) is preferable to 308 nm,
but also 249 nm is used. The crystallinity is sometimes different for different wavelengths under otherwise similar conditions
(ref. 11). More research is required to determine feasability
and optimum conditions.

References

1   R.Srinivasan, J.Vac.Sci.Technol. B.1 (4) (1983), 923

2   R.Srinivasan and B.Braren, to be published

3   R.Linsker, R.Srinivasan, J.J.Wynne, D.Alonso,
    to be published

4   S.Trokel, R.Srinivasan, B.Braren, Am.J.Ophthal., Dec.83

5   B.Garrison and R.Srinivasan, IBM RC 10181, Sep.26th 1983

6   R.Srinivasan, private communication

7   H.H.G.Jellinek and R.Srinivasan, J.Phys.Chem. July 1984

8   Session on "New Lasers", Symp. on surgical lasers in
    ophthalmology, Paris 28 - 29th June 1984, unpublished

9   M.Born: "Optik"; Springer Verlag 1965; p.183

10 Company Carl Süss, Munich, West Germany

11 Annual Report of the Rutherford Appleton Laboratory,
    Chilton, United Kingdom, 1983; p. 3.14 ff

PROPERTIES OF ALEXANDRITE LASERS

John C. Walling

Allied Corporation
7 Powderhorn Dr.
Mt. Bethel, NJ, 07060

INTRODUCTION

The properties of electrons and electronic states are well suited for storing energy to be emitted as laser radiation. However, as quantum mechanics dictates all bound electronic states have discrete energy. Tunable lasers must be obtained, therefore, by means of either free electron states or by modifying (modulating) the electronic Hamiltonian. In certain classes of lasers, the electronic Hamiltonian is rendered continuously time variant by interactions with vibrations which participate directly in the stimulated emission event, such that part of the stored electronic energy is carried away by vibrational quanta. Dye lasers operate in this way as do a less well known but earlier class, the "phonon terminated," or "vibronic" solid state laser, of which alexandrite is a prime example.

Continuous tunability, a characteristic of such lasers, is achieved by designing the resonator so as to spoil the gain for other than the desired wavelength. Because the vibrational states are closely spaced and their linewidths overlap, their gain is generally broad and continuous with few practical exceptions. Because the (discrete) stored electronic energy is partitioned between photon and vibrational quanta, the broad continuous character of the vibrational spectrum gives rise to broad, continuous tuning of the laser radiation.

The first solid state laser of this type was $Ni:MgF_2$ which was reported in 1963 by L. F. Johnson, R.E. Dietz and H.J. Guggenheim[1]. Although this discovery was of considerable phenomenological interest, the laser required cryogenic cooling and did not lend itself to development as a practical device at that time. During the

43

middle 60's and early 70's several transition metal ions in MgF_2 and related compounds were studied by the Bell Laboratories group[2]. For a few years, the interest in these materials subsided until P. Moulton, A. Moradian, and T.B. Reed at Lincoln Laboratories began to pump many of these materials with a cw Nd:YAG laser[3].

Quite independently, research in beryllium containing oxides as potential laser hosts was initiated by Carl Cline at Allied Corporation in the early 70's. This work gave rise to two successful laser compositions, Nd:BEL[4] and $Cr:BeAl_2O_4$ (alexandrite)[5]. In 1977, alexandrite was found to lase on the vibronic side bands of the Cr^{3+} $^4T_2$ to $^4A_2$ transition and hence was a tunable, vibronic, solid state laser[6,7]. Particularly significant was the fact that vibronic alexandrite not only operated well at room temperature but its performance actually improved at elevated temperature. This discovery eventually led Allied Corporation to seek commercial development of this material and has since stimulated further developments in Cr^{3+} doped garnets[8] and other materials including emerald,[9] that promised similar laser properties to alexandrite. What follows, is a summary of the properties of alexandrite with emphasis on the laser related material properties and the laser performance.

ALEXANDRITE CRYSTALS

Alexandrite is a precious gem crystal of substantial physical beauty. Lightly doped, crystallographically oriented cubes of this orthorhombic crystal, appear blue, red, and green when viewed along with the a, b, and c axes, respectively. An attractive appearance, with hues that depend on the polarization and color balance of the illumination (a phenomenon known as the "alexandrite effect"), plus the rarity of this crystal in nature, have lead to a very high price for the natural stone. The finest quality gems are found in the Ural Mountains of central Asia. They were first discovered there and presented to their namesake Alexander II, Czar of Russia, on his birthday.

Growth and physical properties of Alexandrite

Chrysoberyl, the clear undoped host of alexandrite having the chemical formula $BeAl_2O_4$, is isomorphic with olivine and belongs to the crystallograhpic space group D_{2h}^{16} (Pnma).[10] The synthesis of chrysoberyl was first reported in 1963 by Farrell, Fang, and Newnham[11] who grew the crystal from a $PbO-PbF_2$ flux. These authors analyzed the crystal structure by x-ray diffraction. Concurrently, R. Newnham, et. al. reported using x-ray diffraction to study the Fe^{3+} and Cr^{3+} bearing material[12].

In chrysoberyl, there are two crystallographically distinguished

Al^{3+} sites, each with low point group symmetry, C_s (mirror site) and C_i (inversion site). Both mirror and inversion Al^{3+} ions are octahedrally coordinated with 6 oxygen atoms. The mirror symmetry planes that pass through the mirror sites are perpendicular to the b axis. When Cr^{3+} substitutes for Al^{3+} about 78% of the total Cr^{3+} enters the mirror site (which is slightly larger and a better fit to Cr^{3+}) and plays a dominant, virtually exclusive role in the laser action.

Many of the physical and optical properties of chrysoberyl and alexandrite, are given in Table 1. The hardness, thermal conductivity (2/3 that of sapphire), mechanical strength, and chemical stability are outstanding properties of alexandrite relevant to high-power laser applications.

Crystals of alexandrite, suitable for laser application, are now produced by modified Czochralski growth methods in which the crystals are pulled from melts of the principal components. Because the melt temperature exceeds 1870 °C iridium crucibles are used. Crystals 5 cm in diameter and 15 cm long are produced by this method at Allied Corporation. From such crystals, alexandrite rods are extracted that are up to 1 cm in diameter and 12 cm in length. The Cr^{3+} concentration for laser rods varies over the range from 0.05 at.% to 0.3 at.% depending on rod diameter, pump source, and application.

Research in alexandrite crystal growth has steadily improved the optical quality and purity of the crystal. Typical rods are 0.63 cm in diameter and 11 cm long and display less than one wave optical distortion over the central 90% of the rod cross section when examined by Twyman-Green interferometry.

Electronic Properties of Alexandrite Crystals

The electronic states of transition metal ions in oxide hosts interact strongly with the crystal field, resulting in large energy level shifts, level spittings, significant line broadening, and in many cases, strong vibrational coupling with the lattice. In this respect, transition metal and rare earth metal ions act in a qualitatively different manner. In the latter, the magnetic 4f shell is shielded from the ligand field interaction by the 5d and 5s electrons, and the transitions lines remain comparatively narrow.

For Cr^{3+} the free ion electronic configuration is d^3, indicating three electrons in the d shell. In the presence of the combined action of the spin-orbit coupling and the crystal field, there are formed from these three electrons, several manifolds of states, each with levels having 2-fold (Kramer's) degeneracy.[13]

Optical spectroscopy, both absorption and fluorescence, provide empirical means to determine the energies of several states. The opti-

Table 1. Summary of Alexandrite Properties

PHYSICAL PROPERTIES	SYMBOL	WAVE-LENGTH RANGE	TEMP. RANGE	TYPICAL PARAMETER VALUES	REF
THERMAL CONDUCTIVITY	k		22°C	0.23 W/cm^2	14
BREAKING STRESS	σ_B			4 → 9 x 10^8 P$_a$	7
CR CONCENTRATION	C			0.05 → 0.23 at.%	7
FRACTION CR ON MIRROR SITE	f_m			0.78 ± 0.03	15
OPTICAL PROPERTIES					
REFRACTIVE INDEX	n	250 nm → 2.6 μm	22°C	1.74	7
dn/dT (x10^{-6} k^{-1})		1.15 μm	25°C→ 50°C	E ‖a 9.4 ±0.5 E ‖b 8.3 ±0.5	16
SCATTERING LOSS	ε	1.06 μm	22°C	< 5 db/km	
NONLINEAR REFRACTIVE INDEX COEFFICIENT	γ		22°C	2 ±.03 x 10^{-20} W	17
LASER PROPERTIES					
EMISSION CROSS SECTION	σ_e	700 nm → 825 nm	22°C→ 290°C	7x10^{-21} → 3x10-20 cm^2	
EXCITED STATE ABSORPTION CROSS SECTION	σ_{2a}	700 nm → 825 nm	22°C→ 290°C	2x10^{-21} → 7x10-21 cm^2	
STORAGE TIME	τ		4 K → 500 K	1.5 ms → 70 μs	7
			22°C	262 μs	
GROUND STATE ABSORP. IN PUMP BANDS	σ_{ap}	350 nm → 800 nm	22°C	0.5 x 10^{-19} cm^2	7
EXCITED STATE ABSORP IN PUMP BANDS	σ_{2ap}	425 nm → 670 nm	22°C	0.3 x 10^{-19} cm^2	18
THERMAL LENS		730 nm → 790 nm	90°C	0.33 diopters/kW	27
ABSORBED POWER IN ROD TO FRACTURE				0.6 kW/cm	7
OPTICAL DAMAGE THRES BULK AND SURFACE		750 nm	22°C	> 30 GW/cm^2	19

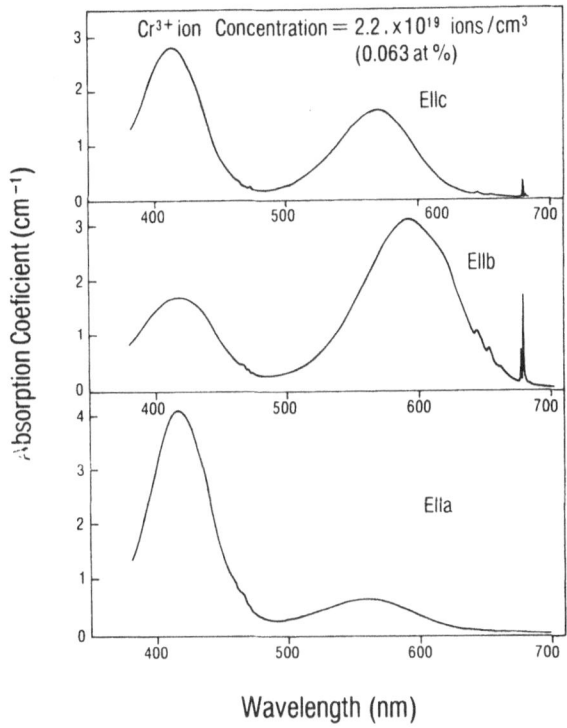

Fig. 1. Absorption spectra of alexandrite at room temperature.[7]

cal absorption and emission spectroscopy data and the inferred energy
level diagram are shown in Figs. 1 through 3, respectively. The
indicated (cross-hatched) broad bands of the energy level diagram in
Fig. 3 correspond to the Stokes' absorption band of the vibronic
transition associated with the indicated multiplet.

Electric dipole transitions are made possible by configurational
mixing with higher lying levels (5s, 5d) because parity prevents
electric dipole, intra-shell transitions. Magnetic dipole, intra-
shell transitions are allowed, but these are characteristically weak.
In alexandrite the magnetic dipole component of transitions is only
about 10% as strong as is the electric dipole component (except where
the electric dipole is explicitly forbidden as is the case for purely
electronic transitions of the inversion site).

The spectroscopic data show a marked polarization dependence,
suggesting that symmetry properties of the mirror-site states and the
crystal field play a significant role. However, group theoretical
analysis indicates that no symmetry restrictions apply, ie., all
transitions are allowed when both crystal field and spin-orbit
interactions are considered. Some understanding of the polarization
dependence is obtained from group theory if the crystal field is

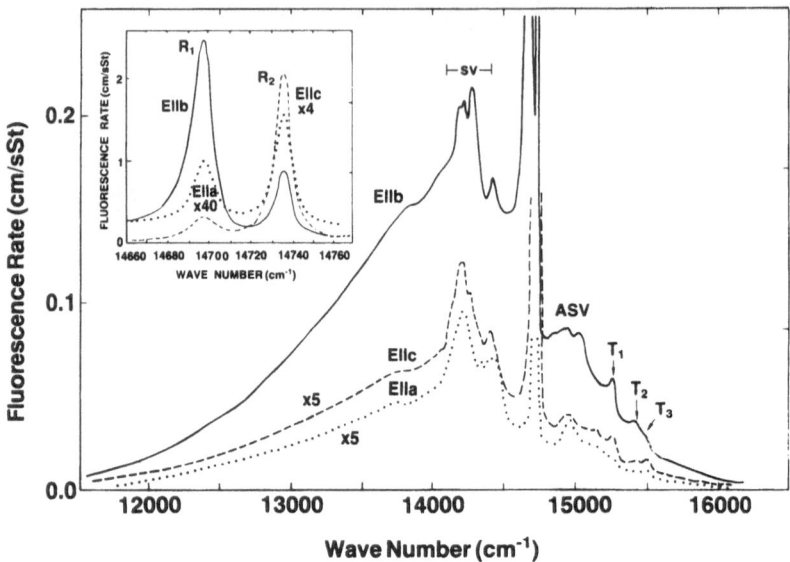

Fig. 2. Fluorescence spectra of alexandrite at room temperature.[7]

assumed to dominate the spin orbit interaction, the latter being a minor perturbation.[20] The symmetry restrictions of the crystal field alone then appear to govern the polarization dependence, but not rigorously, because of the spin-orbit perturbation.

In the optical spectrum, both sharp lines and broad bands are present. The broad bands arise from transitions in which a change in the electronic orbital state configuration occurs and a strong inter-action with the lattice results. By contrast, the sharp lines involve primarily a change in the spin configuration which has a minor impact on the lattice. The broad band transitions are vibronic in charac-ter. Because their breadth exceeds the bandwidth of the entire optical phonon spectrum (as determined from Raman and infra-red scattering), the broad band transitions necessarily involve multiphonon as well as single phonon events. At room temperature and below there are very few optical phonons available to be absorbed. Consequently, both the vibronic emission and absorption processes are dominated by the case where the phonon component is emitted only irrespective of whether the photon component is emitted or absorbed. As a result, the broad vibronic bands appear at different energies in absorption and emission (the Stokes' shift). In particular, the longer wavelength pump band of alexandrite, centered at about 590 nm, has the same electronic component ($^4T_2$) as does the strong fluores-cence band, centered at about 720 nm, responsible for most laser gain.

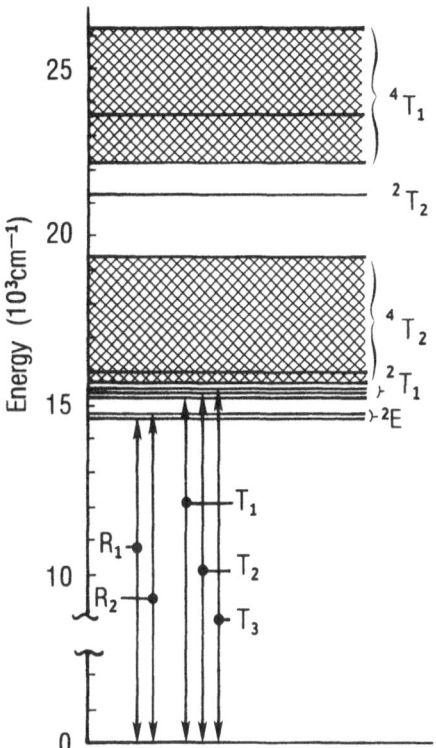

Fig. 3. Cr^{3+} mirror site energy levels of alexandrite.[7]

The lower energy member of the R-lines, indicated in Fig. 3 is the source of gain for 3-level laser action at about 680.4 nm. In this 3 level lasing, alexandrite is directly analogous to ruby.

The relative site occupancy of the mirror and inversion sites by Cr^{3+} is not readily obtained by optical spectroscopy. For an accurate determination of this parameter an EPR analysis was performed.[15] From this analysis, accurate ground state Hamiltonian parameters for the mirror and inversion sites could be determined. With these parameters and the observed signal strength it is possible to arrive at an accurate ratio of mirror to inversion Cr^{3+} ions. It was found that the fraction of total Cr^{3+} ions that reside on the mirror site is 78% ± 3% for both 0.1 at.% and 0.3 at.%. The insensitivity of this parameter to concentrations in this range implies that the value 78% for this parameter is reasonable and appropriate to use for alexandrite laser modeling.

LASER MODELING

Modeling the alexandrite laser is both challenging and necessary because in alexandrite there exists a strong dependence of laser parameters, in particular the emission cross section, with both temperature and wavelength that adds two additional dimensions to the characterization space beyond that required for Nd:YAG.

Modeling Laser Kinetics

There are in fact two models that have arisen in describing alexandrite operation. The first, a 5-level model, is useful for describing qualitatively the dependence of laser gain on temperature and to relate alexandrite vibronic lasing to the classical 4 level mode of laser operation. The second model, a 3-level model, is useful for predicting performance of both the vibronic and R-line laser and is used in conjunction with detailed spectroscopic and single pass gain data to accomplish this task. This approach provides an accurate, experimentally confirmed model, suitable for the purpose of advanced design.

The 5-level model is depicted in Fig. 4. Apart from the extra "storage" level this model is very similar to the classical 4 level

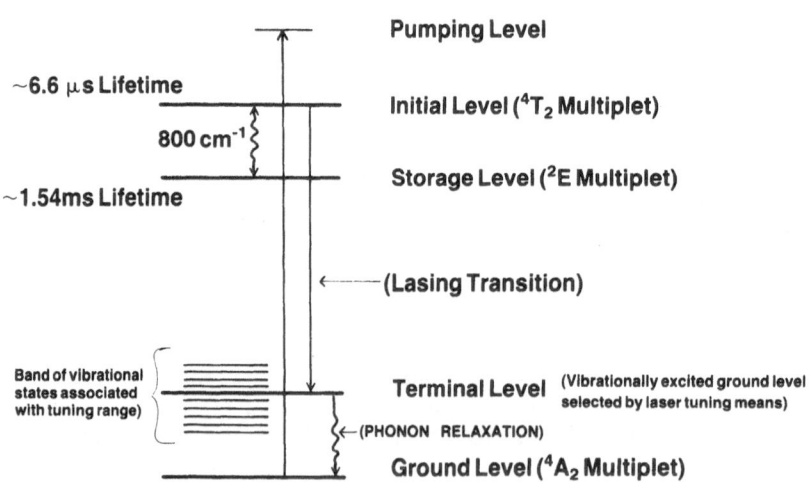

Fig. 4. 5-Level (descriptive) model of alexandrite laser kinetics.

laser model in which the terminal level is formed from the ground state by a phonon excitation that is coherent with the emitted photon. The dephasing and population decay of this phonon component returns the laser to the ground state of the system. The fifth level, a storage level, augments the clasic 4-level picture. In alexandrite this level is the ^2E multiplet. Thermal excitation to the upper laser level, dominantly the $^4T_2$ multiplet, over an 800 cm^{-1} energy gap, gives rise to fluorescence emission and laser gain. Fig. 5 illustrates alexandrite's temperature dependent fluorescence lifetime. The dashed line indicates the theoretical fit to the data (solid line) using the parameters given in Fig. 4 for the two upper levels.

The three level model is based on McCumber's work[21] on the theory of phonon terminated lasers in 1966. This work extends the Einstein relations between the stimulated and spontaneous decay rates of radiative transitions. The McCumber result can be reduced further by certain assumptions[7] to provide a useful relationship between the observed fluorescence decay rate spectrum and the emission cross section that is valid for alexandrite and a number of other vibronic lasers. The key assumption in this picture is that the population distribution within the metastable electronic manifold is in quasi-thermodynamic equilibrium during lasing (as it is during the measurement of fluorescence). The relaxation time for these states is

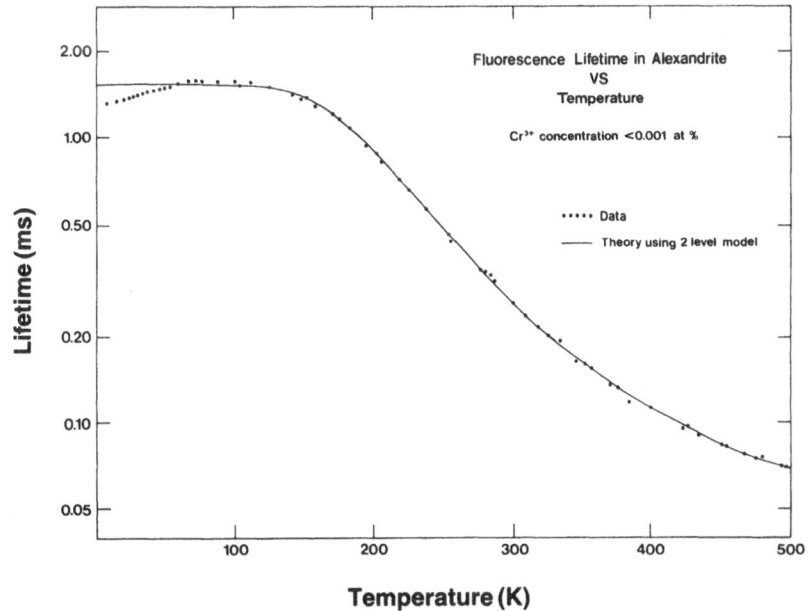

Fig. 5. Alexandrite fluroescence lifetime vs temperature.[7]

expected to be subnanosecond. Consequently, the key assumption is likely valid for laser pulse durations long compared to this time. The mathematical expression for gain so derived is given by[7],

$$g_\lambda(k,E) = \left\{ N^* - (N-N^*)\exp\left[(E-E^*)/k_B T\right] \right\} \cdot \sigma_{e\lambda}^*(k,E)$$

where N and N^* are the total and excited active ion densities, respectively, E is the photon energy, E^* is an effective electronic energy (a weighted average of the electronic energy of the several upper level states), and $\sigma_{e\lambda}^*(k,E)$ is an effective emission cross section at photon energy E. The latter has been shown to be proportional to the measured fluorescence emission, as required by the key assumption. Here the polarization is indexed by λ and the ray trajectory by k. k_B and T are Boltzmann's constant and temperature, respectively.

The above gain expression does not include the effects of excited state absorption, caused by parasitic transitions from the metastable manifold to higher lying states, that competes with stimulated emission for the available photons and thus reduces laser performance. Excited state absorption determines the long wavelength operating limit of alexandrite lasers, about 826 nm.[22]

Modeling Laser Performance

By and large, alexandrite lasers are not limited in Q-switched operation by their energy storage, but rather by their energy extraction capability. Efficient energy extraction is governed by the saturation fluence, which depends inversely on emission cross section, and whether this fluence exceeds that which the laser can tolerate without damage. The calculated performance of generic laser oscillators[23] with emission cross section in the Nd:YAG and alexandrite ranges are shown in Figs. 6 and 7, respectively, for a fixed maximum allowable intra-resonator fluence. From these figures it is apparent that a 0.63 mm diameter Nd:YAG rod, emitting roughly 300 mJ is limited, or nearly so, by superradiance (a phenomenon in which spontaneous emission is amplified by the high gain down the material length to the point where a significant depopulation of the excited state occurs). By contrast, an alexandrite rod of the same size (Fig. 7) is capable of more than 2 J per pulse (but over a longer pulse duration). The lower emission cross section of alexandrite essentially translates into a more energetic pulse, of peak power comparable to that of a Nd:YAG rod of comparable size; this, assumes that the damage threshold is power dependent and comparable for both materials.

By increasing the power density further in alexandrite the full capability of the material to store energy would be utilized. At 2 GW/cm^2 average single pass intracavity power density, several Joules

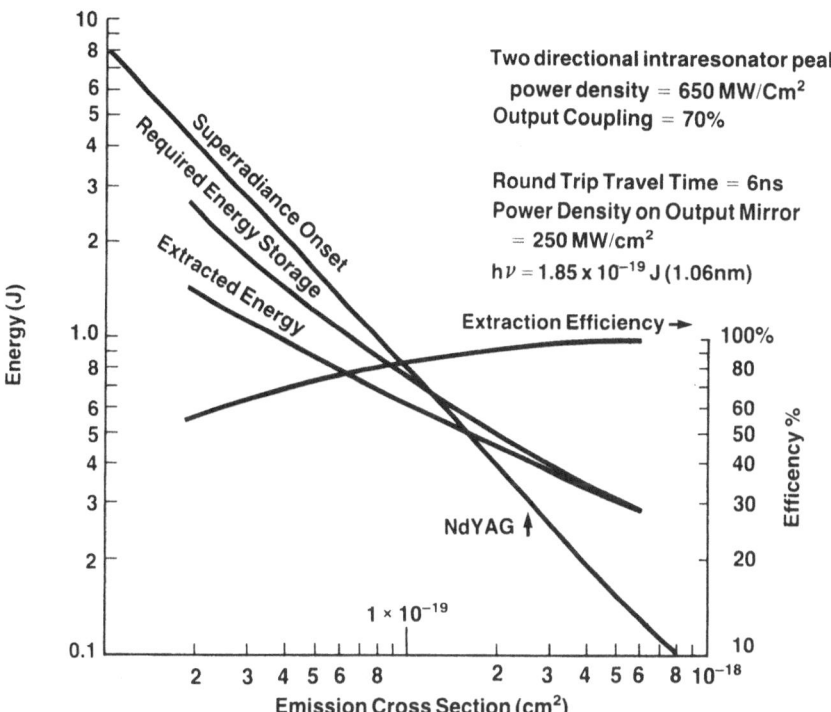

Fig. 6. Calculated oscillator performance vs emission cross section based on Nd:YAG parameters and a 0.634 cm diameter, 10 cm long laser rod.

would be extracted. Oscillator operation at these power densities is problematic because nonlinear interactions within the laser rod cause power instabilities and optical damage. Resonator design measures to suppress these instabilities and the use of oscillator-amplifier configurations to control better the power density are active areas of alexandrite research.

SUITABILITY OF ALEXANDRITE AS A HIGH PERFORMANCE LASER GAIN MEDIUM

All new laser materials undergo a long period of development, characteristic of crystal growth technology. Nevertheless, several excellent high performance crystalline optical materials have been produced and alexandrite now numbers among them. For alexandrite, because of its comparatively high saturation fluence (7 to 25 J/cm^2), excellent crystal quality is required to permit stable, high-performance, efficient operation without damage. The focus is therefore placed on producing material able to handle fluences on this order. Achieving this goal, and by proper resonator and system

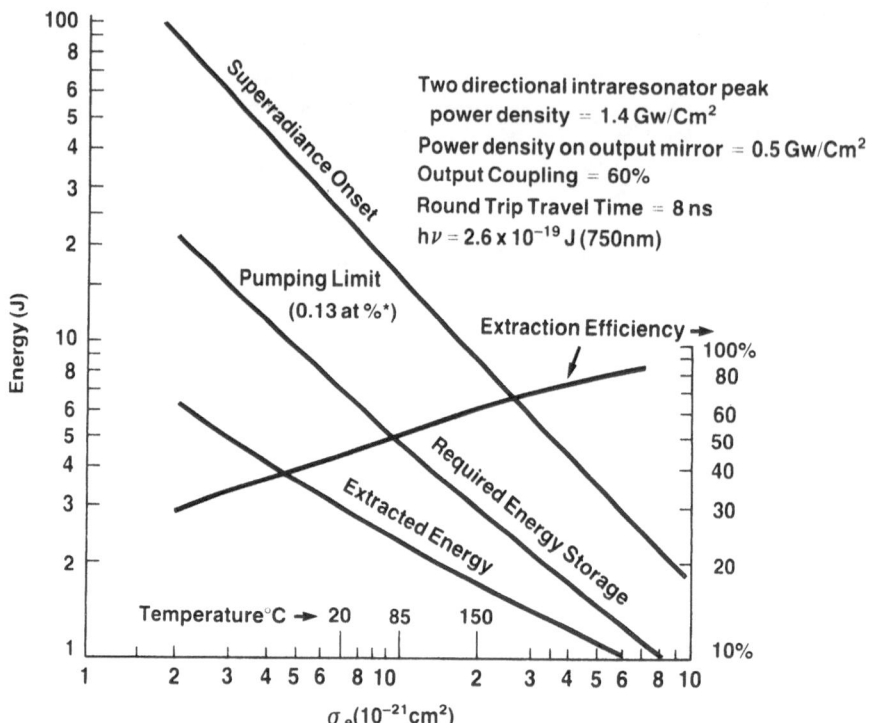

Fig. 7. Calculated alexandrite oscillator performance vs emission cross section based on a 0.634 cm diameter 10 cm long rod.

design, specific energy extraction in the 1 to 3 J/cm³ range would become practical. In fact, operation at the low end of that range is now a practical reality. In this section four topics will be discussed in light of high power applications: optical damage, efficiency, optical quality, and natural birefringence.

Optical Damage

Optical damage results when the localized time integrated power density at a crystalline defect, surface or bulk, exceeds some threshold within the damage response time of the defect. The response time and threshold are characteristic of the defect. A defect may be defined generally as a departure from homogeneity or uniformity within the material. The material surface itself is such a departure, as are dislocations, voids, vacancies, vacancy aggregates, inclusions, surface irregularities, contamination, and surface imperfections caused by fabrication. For high-power laser operation, absolute cleanliness is required to achieve sustained operation at maximum levels of performance.

The intrinsic optical damage threshold of bulk alexandrite and properly prepared alexandrite surfaces has been shown to exceed 23 GW/cm^2 and 270 J/cm^2, respectively, for 12 ns pulses in focused beam testing of local areas.[27] The achievement of comparable performance over large areas is challenging. However, only 10 to 30 J/cm^2 are required to extract power efficiently from alexandrite oscillator rods. In experiments with alexandrite oscillators, optical damage, attributed to optical power density excursions, starts at roughly 1.5 GW/cm^2, double-pass, peak intra-resonator power density. Because this value depends strongly on resonator configuration, there is the possibility that this phenomenon might be controlled or suppressed by improvements in resonator design. In single pass amplifiers, however, more than 1 GW/cm^2 (RMS over the aperture) has been extracted without damage provided the input beam is spatially uniform and well behaved. In these tests, pulses 30 ns in duration at a fluence of 30 J/cm^2 were extracted from alexandrite amplifiers in sustained (damage free) operation[24].

Czochralski growth of alexandrite necessitates an iridium crucible and for a long time iridium particles, in the form of triangular platelets oriented with normals parallel to the a-axis, appeared in the crystal bulk causing a serious problem of bulk damage. The elimination of these inclusions has produced material that exhibits lifetimes greater than 10^6 shots[24] when operating in Q-switched oscillators that generate 20 ns pulses at 10 J/cm^2. Comparable, but early tests on ruby amplifiers produced damage in only 70 shots at this fluence even though amplifiers should be more stable against damage compared to oscillators because the power density is easier to control.[25] More recently, operation at 30 J/cm^2 in alexandrite amplifiers has achieved greater than 100,000 shots without bulk damage. Thus far no bulk damage has been experienced in alexandrite amplifier rods in tests at Allied.

Optical damage in cw alexandrite lasers primarily is limited to the burning of resonator mirrors at high intra-resonator power. The power handling capability of dielectric mirrors of specialized design is as high as several hundred kilowatts per square centimeter, adiquate for the purpose. (The system efficiency depends on being able to maintain a low laser threshold which implies that high-mirror reflectivity, low-loss resonators must be used.)

Efficiency of Alexandrite Lasers

Efficiency in solid state lasers is dependent upon many factors each contributing incrementally to the overall result. There are two main catagories: energy storage efficiency, where electrical energy is converted into energy stored in the laser medium; and extraction efficiency, where the energy stored is extracted in useful laser emission. In long pulsed and cw operation, pumping rates and power efficiency are determinant. In Q-switched operation, energy storage and energy extraction dominate. In either case fluorescence may contribute a significant loss, governed by the fluorescence lifetime (storage time). For alexandrite, the fluorescence lifetime is about 250 µs at ambient operating temperatures, but this depends strongly on temperature. For efficient Q-switched operation, the flash lamp duration must be comparable to or shorter than the fluorescence lifetime or significant fluorescence loss will occur. Thus, the pumping efficiency is highly dependent upon the characteristics of flash lamps.

In practice, Xe flash lamps have proven to be superior to Kr for flash lamp pumping of alexandrite. The emitted spectrum is well matched to the alexandrite absorption bands when the lamps are driven at the pulse energy and pulse duration required for vigorous pumping. Vigorous pumping is more important for alexandrite than for ruby, Nd:YAG, or Nd:Glass. Lamp life can be severly shortened by pumping with to much energy. However, practical lamp life (10^6 to 10^8 shots) has been demonstrated at the levels required by alexandrite.

Maximizing the efficiency of alexandrite lasers is important not only to reduce operating costs but also to reduce the overall size and complexity of practical devices. Thus far the greatest effort has been directed toward improving the energy extraction capability of the material in view of the relatively high saturation fluence. Energy storage efficiency has not yet been optimized. However, operating efficiencies that exceed 1% of the energy stored in the capacitor bank have been achieved with alexandrite in Q-switched operation. For normal mode (free running) operation 2.5% overall efficiency has been achieved. Slope efficiencies as high as 5 or 6% have been generated in long pulse operation with a thermally compensated resonator. Overall efficiency tends to scale with the size of the system. Modeling calculations predict, for example, that efficiencies of Q-switched alexandrite lasers will exceed 3% and possibly be significantly higher for systems of optimum design scale.

Optical Distortion

Strain, from dislocations and Cr^{3+} concentration variations that arise during growth, alter the index and produce optical distortions. It is useful to interpret strain induced optical distortion in terms of an equivalent focal power or focal length. For example, one fringe at the alexandrite wavelength from center to rod edge of a 0.634 cm diameter rod were it produced by a parabolic lens ground into the rod end would correspond to a 5 m focal length. Based on the thermal lensing characteristics of alexandrite, this lens would be created in the same rod by ~ 0.61 kW of average input power applied to the lamps. A typical operating input power for Q-switched operation using a 0.63 cm diameter rod at 20 Hz would be 4kW, which at

0.3% efficiency would provide 12 W of output at 600 mJ per pulse. Clearly, under such operating conditions, the thermally induced lens dwarfs by an order of magnitude the strain distortion, which is less than 1 fringe for the better alexandrite rods. The thermal lens is, to some degree aparabolic and astigmatic, but correctable over the central half of the rod to within a few percent by proper pump chamber design, choice of Cr^{3+} concentration, and lamp operating temperature. The remaining uncorrectable distortion is comparable to the strain distortion, but becomes dominant at very high average pumping powers, for example, when rods are used in tandem to achieve increased gain.

Natural Birefringence

A very positive feature of alexandrite is its natural birefringence, arising from its orthorhombic crystal structure, that overshadows any thermally induced birefringence and prevents depolarization loss during laser operation. In this way alexandrite is vastly superior to Nd:YAG and other cubic materials that suffer greatly from this effect.

PULSED ALEXANDRITE LASERS

In pulsed operation alexandrite lasers are particularly strong in specific energy storage density and in Q-switched energy extraction; at the same time they provide tunability. The emission is continuously tunable from 710 to 826 nm and good efficiency has been achieved over the 730 to 785 nm range. Pulsed alexandrite lasers have been injection locked, mode-locked, PTM and PRM Q-switched, and operated single mode. In this section, the performance of various pulsed alexandrite lasers will be reviewed.

Typical pulsed alexandrite lasers in current use utilize rods 0.63 cm diameter and 11 cm long with a Cr^{3+} concentration of ~ 0.14 at.%. Alexandrite needs to be pumped harder than Nd:YAG on a given pulse, yet it can tolerate higher thermal loads. Consequently, power supples and pumping hardware suitable for Nd:YAG is frequently inadequate to operate alexandrite to the performance levels where optimum performance is obtained. Similarly, hardware built for ruby lasers does not generally have the repetition rate capability required to fully exercise alexandrite rods.

Alexandrite rods are mechanically strong enough to withstand the maximum average load that current Xe flash lamps can deliver. New Xe flash lamps with sapphire jackets will push the material closer to its fracture limit. However, such loads can create severe thermal lensing and stress-optic effects that are difficult to compensate.

Generally, dual lamp elliptical pump chambers are used to pump alexandrite because of the higher and more uniform excitation level that can be achieved with two lamps. In principle, the brightness of the lamp image on the rod is limited to the brightness of the lamp plasma. Hence, in an ideal pumping chamber, one lamp should pump as well as two were its bore of size commensurate with the rod diameter. However, it is found that, if efficiency is not the main issue, the best results are more simply achieved by a dual lamp system.

Early in the development of alexandrite lasers, diffuse ceramic pump chambers were used because they were stable with hot water coolant and because they provided good pumping uniformity. Typical long pulse (non-Q-switched) operation of alexandrite is illustrated in Fig. 8. In this case a 7.6 cm rod was used in a dual lamp diffuse ceramic pump chamber (from J.K. Lasers Ltd.).[23]

Tuning the alexandrite laser has been achieved almost exclusively with birefringent filters largely owing to their convenience, low loss, and high damage resistance. Line widths on the order of 0.2 nm are typically achieved in Q-switched operation with such filters and continous tuning from 720 to 800 nm is possible with a single tuner.

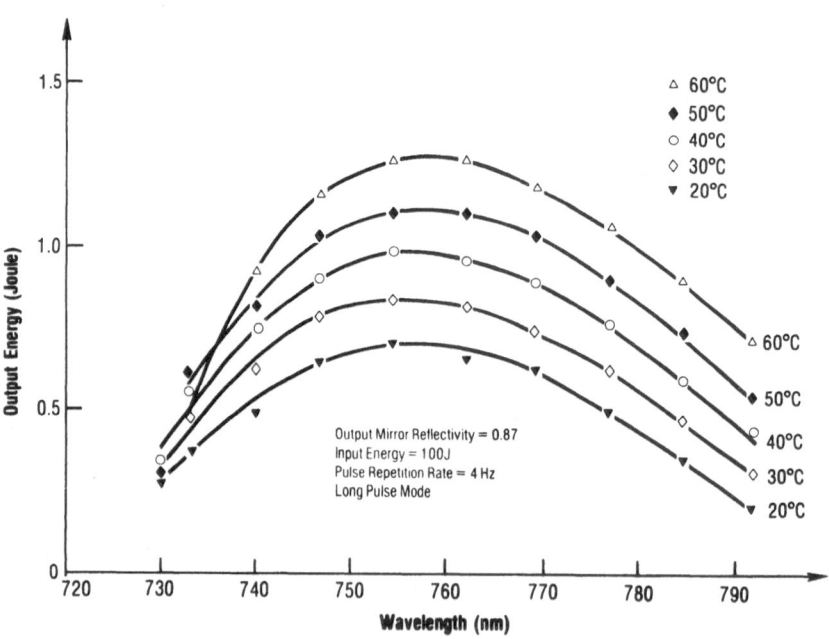

Fig. 8. Temperature and wavelength dependence of alexandrite laser performance.

Q-switching

In Fig. 9 is shown a typical performance curve for an alexandrite laser in pulse reflection mode (PRM) Q-switched operation. In this case, a dual-lamp, specular, elliptical pump chamber was used. The Q-switch is a Brewster angle KD*P crystal in a Brewster angle quartz window enclosure filled with dry nitrogen gas. This cell has some distinct advantages in alexandrite resonators with respect to its low loss, high damage resistance, and long term stability. The Brewster surfaces also provide discrimination against the wrong polarization augmenting the 10/1 ratio in gain from the alexandrite rod and the contribution from the Brewster angle birefringent tuner when used.

In addition to PRM Q-switching, PTM Q-switching (where, by use of a double action Pockel's cell, the power is first transferred from the rod to the resonator before being "dumped" from the resonator) has been explored with satisfactory results. This technique however is highly susceptible to power instabilities and optical damage. To date 60 mJ have been extracted in 4 ns pulses at 10 Hz from specialized resonator configurations.[34] Typical performance is illustrated in Fig. 10.

Oscillator-Amplifier Configuration

In order to achieve high brightness from alexandrite lasers with

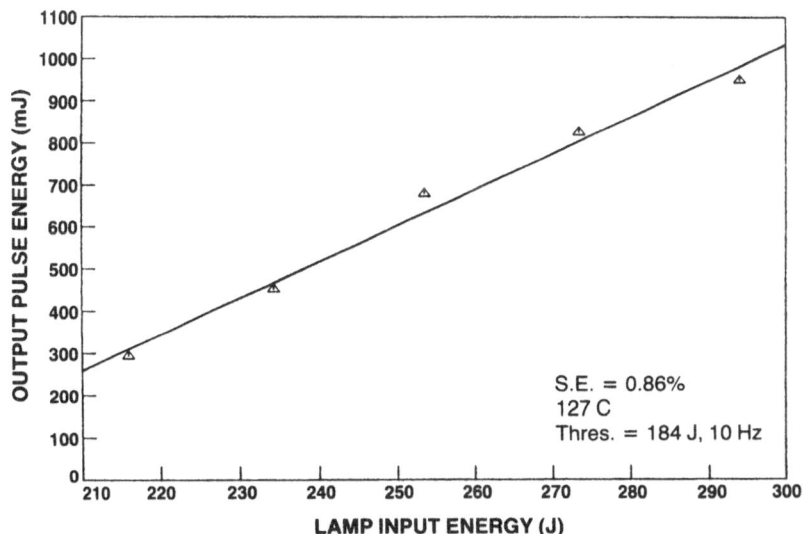

Fig. 9. Typical alexandrite Q-switched performance curve at 750 nm.

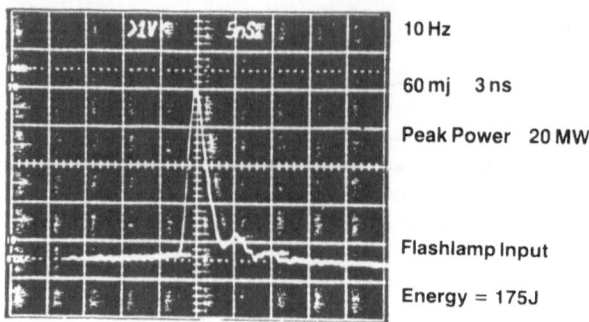

Fig. 10. PTM Q-switched output.

less than perfect optical quality material, the beam over a cross-section that has the fewest fringes of distortion. The contribution to the beam quality (divergence aperture product) of a given pass through a gain medium is proportional to the total number of fringes of distortion sampled. This implies that the beam should pass through small cross sections to maintain optical quality. However, damage thresholds and pumping efficiency considerations provide a practical limit to this approach.

In our experience,[27] a 3 mm aperture with a 6.3 mm diameter rod in a stable, thermally compensated resonator with a Fresnel number of about 100 provides a reasonable compromise for the oscillator. With such resonators tunable, 300 mJ, 15 ns, Q-switched pulses, can be generated at up to 30 Hz PRR with a 4X diffraction-limited beam. Alexandrite single pass amplifiers are capable of amplification factors of 2X to 5X for input beams from this oscillator. Such an oscillator amplifier system has provided ~ 100 MW of peak power in a 5X diffraction limited beam, adequate for reasonably efficient nonlinear frequency conversion by raman or nonlinear mixing processes.

Fig. 11 shows the performance of an alexandrite single pass amplifier at constant pump power as a function of input energy. The curve in Fig. 11 is model predicted performance. The fall off at higher injected energies is the effect of saturation. Even at 3.5 J output, with 2.5 J derived from the 4.5 mm apertured, 11 cm long, amplifier rod, only a small saturation effect is exhibited.

Mode-locking the Pulsed Alexandrite Laser

Interest in mode-locking alexandrite arose as it was recognized that the inherent stability of this broadly tunable laser was better

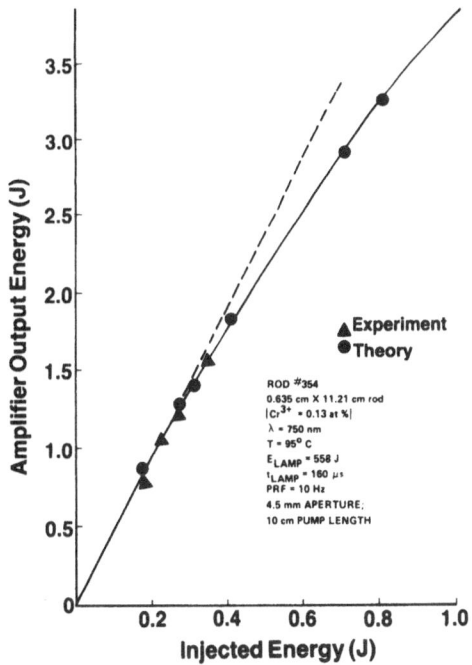

Fig. 11. Alexandrite amplifier performance: comparison with theory.

than dye lasers. Moreover, the ability to store energy implies that high-power mode-locked pulses might also be expected. Another favorable property is the inflection in the refractive index dispersion with wave number that occurs near the center of the laser band, Fig. 12. This inflection indicates a corresponding null in the group velocity dispersion near the center of the band. This property implies that pulse broadening from dispersion is minimized.

Most studies with mode-locked, pulsed alexandrite lasers, at Allied, employed a linear colliding pulse resonator that features two counter-propagating pulses which overlap in the optical center of the resonator. Saturable absorbers are placed at this position to generate passive mode-locked, transform limited, 38 ps pulses. In a similar experiment, 50 ps pulses tunable from 735 to 768 nm were generated.[28] This basic resonator also yields stable 150 ps pulses when the passive dye cell is replaced by an acousto-optic mode-locking device placed near the back reflector.[29] Under Q-switched operation the individual pulse energy from this system has reached 1 to 2 mJ. Passive mode-locking of alexandrite has also been reported by a Soviet group where 8 ps pulses have been achieved.[30]

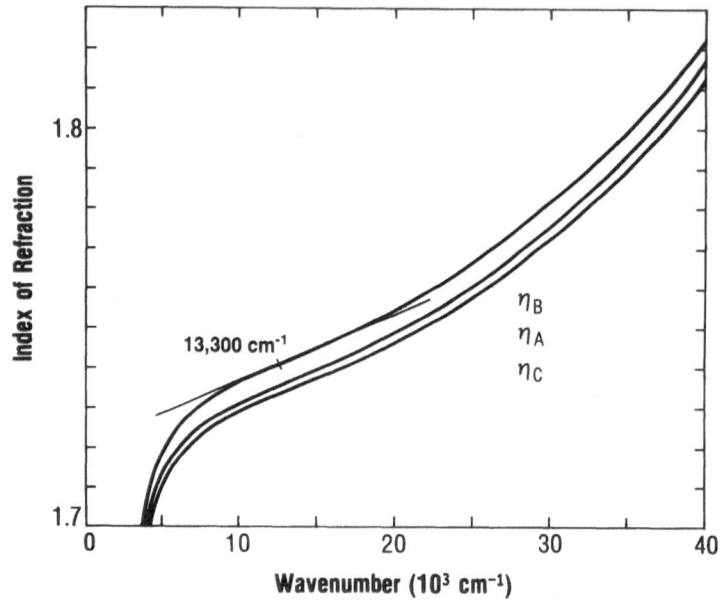

Fig. 12. Refractive index of chrysoberyl vs wave number.

CONTINUOUSLY PUMPED ALEXANDRITE LASERS

Continuous pumping of alexandrite has been achieved both with cw argon and krypton ion lasers in the longitudinal pumping geometry and by Xe and Hg arc lamps in silvered elliptical pump chambers. The ion laser pumping of alexandrite has been used primarily as a means to characterize the laser parameters, but it is also an effective means for generating several watts of cw power tunable from 720 nm to beyond 800 nm with excellent spatial and temporal beam quality. Photon conversion efficiencies (relative to photons absorbed) as high as 85% have been achieved.

The cost and complexity of the ion laser pump source can be obviated by utilizing arc lamp excitation. Much more powerful beams can be produced in this way although it is more difficult to maintain beam stability and mode quality. For the more efficient Hg lamp excitation, multimode performance has reached 50W to 60W, using a single Hg capilary lamp with a 10 cm arc length. Until recently, the emission of both Xe and Hg arc-lamp-pumped alexandrite lasers exhibited strong temporal spikes and extensive modulation. Now, with Xe pumping and 6.5 W of delivered cw power, these oscillations have been reduced to ± 5%. Further improvement in stability at higher power and for Hg pumping as well can be reasonably expected as techniques are perfected.

Early on, the efficiency projections for pumping cw alexandrite lasers by Hg arc lamps were encouraging. However, the high thresholds observed raised questions as to whether alexandrite would be efficacious in this mode. It now is clear that alexandrite is effective as a cw laser. The performance of optimal systems utilizing Hg and Xe arc lamp pumping is compared with cw pumped Nd:YAG lasers, the latter currently represent the baseline for cw solid state laser technology, Fig. 13.[30] This comparison is not made between systems of comparable size, the Nd:YAG lasers are much larger. Accounting for size, the performance of the partially optimized alexandrite lasers compares quite favorably to Nd:YAG.[25]

Continuously driven arc lamps tend to have more line emission than pulsed Xe flash lamps, which are generally operated at a higher current density. Consequently, in selecting a good cw lamp, it is important to determine where in the frequency spectrum the line emission occurs. Compared to Xe cw lamps, the Hg lamp emission is more directly centered in the alexandrite absorption bands and is consequently more readily absorbed by the rod. Moreover, the Hg lamp's

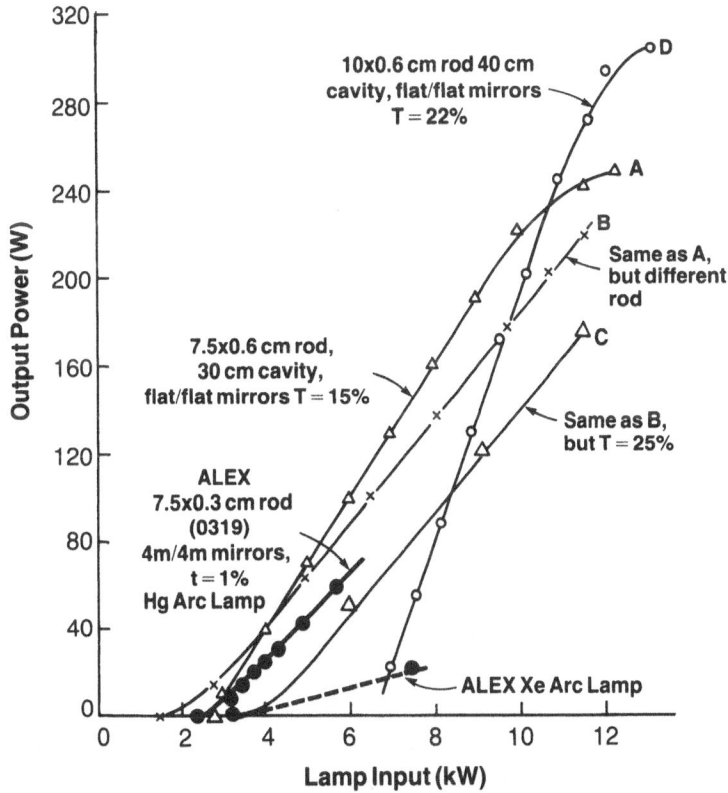

Fig. 13. Comparison of CW alexandrite and Nd:YAG laser performance.

2 mm capillary bore leads to a much brighter emission than obtained from the Xe lamp (with a 4 mm bore) for the same total emission rate. For these reasons the Hg lamp is a more efficacious one. However, Hg lamps run better when ac driven than when driven strictly dc and they result generally in a less stable laser emission than that provided by Xe arc lamps. Consequently, both lamps may find applications.

In Xe pumped, cw-repetitively Q-switched operation, PRR frequencies from 1 Hz to 80 kHz have been achieved. Tunable, repetitive Q-switched operation in alexandrite makes good use of the energy storage and wavelength tuning capability of the material not available jointly in Nd:YAG or dye lasers. The preferred ac operation of Hg lamps may be an advantage when repetitive Q-switched operation up to 1 kHz is desired. For these applications the Q-switch is operated in synchronism with the Hg lamp power peaks. Mode locked operation of arc lamp pumped alexandrite has also been demonstrated.[32]

EXTENDING ALEXANDRITE'S WAVELENGTH RANGE

The fundamental band of alexandrite, 700 nm to 800 nm, is valuable for exciting a wide variety of atomic and molecular species which have narrow line absorption in this range. But also fruitful, is the frequency doubled band from 350 nm to 400 nm where again sharp line interactions occur with a wide variety of target materials. Other bands, accessible by frequency shifted alexandrite, have interesting and valuable applications. It is of interest, therefore, to frequency convert alexandrite to extend the usefulness of this source.

The fact that the drive laser is itself tunable is of inherent advantage in frequency conversion with an alexandrite source. Consequently, even fixed frequency conversion processes, including SHG (second harmonic generation), and SRS (stimulated Raman scattering) provide tunable output. Thus, full spectral coverage of the visible, extending well into the UV and IR can be achieved by various combinations of these processes without resorting to dye laser technology.

Because SHG and SBS are nonlinear processes, high conversion efficiencies depend on high peak power and excellent beam quality from the fundamental source. This is particularly true for SHG where beam divergence is critical. Consequently, much work is being devoted at Allied toward maximizing the optical brightness of the alexandrite source with particular regard to the requirements of the nonlinear conversion processes. Oscillator-amplifier, unstable resonators, and the zig-zag slab geometry are among the approaches being studied. Very encouraging results have been recently achieved in SHG with the oscillator-amplifier where 200 mJ tunable pulses at 380 nm have been generated in KDP at a pulse repetition rate of 10 Hz.[27]

In raman shifting experiments,[27] 30% energy conversion, extracted in the first Stokes, has been achieved from a single pass H_2 gas raman cell.

SUMMARY

Alexandrite lasers offer the potential of exceptional high power tunable laser performance. In order to handle the high fluences and flux levels generated, and in fact required for optimal performance, these lasers must be properly designed and fabricated. Particular attention to creating and maintaining cleanliness for the operational life of the laser is required. Alexandrites proven power handling capability is high by conventional standards with 30 J/cm^2 extracted in 30 ns at 10 Hz from single pass amplifiers in sustained operation.

Alexandrite lasers have been operated in each of the basic configurations commonly used in solid state laser technology, including both pulsed and cw, Q-switched and mode-locked devices.

Applications for alexandrite derive from its 1) tunability, 2) useful fundamental band, 3) high specific energy storage, and 4) inherent simplicity and stability of the overall laser system. Important is the fact that the alexandrite performance is obtained directly from lamp excitation. Currently, alexandrite lasers are being developed for isotope separation, spectroscopy, and LIDAR applications.

As advances in alexandrite laser design continue it can be expected that the inherent advantages of this tunable solid state laser can be realized in a number of specialized market niches. Principal areas of applications include photochemistry, pollutant monitoring, medical diagnostics, medical treatment, remote detection and materials processing.

ACKNOWLEDGEMENTS

The author wishes to thank his many coworkers for allowing portions of their work to be included in this manuscript. Also, he wishes to thank D.F. Heller and H. Samelson for proofreading this manuscript and for their helpful suggestions.

REFERENCES

[1]   L.F. Johnson, R.E. Dietz, and H.J. Guggenheim, "Optical maser oscillations from Ni^{2+} in MgF_2 involving simultaneous emission of phonons," Phys. Rev. Lett., 11, pp 318-320, (1963).

[2]   References to this work can be found in reference #7 below.

[3]   P.F. Moulton, A. Mooradian, and T.B. Reed, in Digest of Technical Papers, Tenth International Quantum Electronics Conference, Optical Society of America, Washington, D.C., 1978, Paper C.2, p 630.

[4]   H.P. Jenssen, R.F. Begley, R. Webb, and R.C. Morris, "Spectroscopic properties and laser performance of Nd^{3+} in lanthanum beryllate", J. Appl. Phys., 47, pp 1496-1500, (1976).

[5]   R.C. Morris and C.F. Cline, U.S. Patent 3,997,853, Dec 14, 1976.

[6]   J.C. Walling, H.P. Jenssen, R.C. Morris, E.W. O'Dell, and O.G. Peterson, Annual Meeting of the Optical Society of America, San Francisco, CA, Oct. 31 - Nov 3, 1978.

[7]   John C. Walling, Otis G. Peterson, Hans P. Jenssen, Robert C. Morris, and E. Wayne O'Dell, "Tunable Alexandrite Lasers", IEEE J. Quant. Elec., QE-16, pp 1302-1315, (1980).

[8]   G. Huber, Paper presented, 1st Annual Conference on Tunable Solid State Lasers, June 13-15, La Jolla Institute, La Jolla, CA, (1984).

[9]   Michael L. Shand and John C. Walling, "A tunable Emerald Laser," IEEE J. Quant. Electr., QE-18, pp 1829-1830, (1982).

[10]   Pnma (D_{2h}^{16}) is described in Internationale Tabellen zur Bestimmung von Kristallstrukturen, 1, Berlin, Germany:Borntrager, (1937).

[11]   E.F. Farrell, J.H. Fang, and R.E. Newnham, "Refinement of the Chrysoberyl Structure", The American Mineralogist, 48, pp 804-810, (1963).

[12]   Robert E. Newnham, "Crystal Structure, Synthesis, and Magnetic Properties of Chrysoberyl," Tech. Rept. #183, Laboratory for Insulation Research, Massachusetts Institute of Technology, (Nov. 1963). Distributed by: National Technical Information Survice, U.S. Dept. of Commerce, 5285 Port Royal Road, Springfield, VA, 22151.

[13]   Yukito Tanabe and Satoru Sugano, "On the Absorption Spectra of Complex Ions II," J. Phys. Soc. Jap., 9, pp 766-779, (1954).

[14]   Measurement performed by R. Taylor, Thermo-Physical Properties Research Center, Perdu University, Lafayette, IN. The value given is an average over a range of orientations.

[15]   C.E. Forbes, "Analysis of the spin-Hamiltonian parameters for Cr^{3+} in mirror and inversion symmetry sites of alexandrite ($Al_{2-x}Cr_xBeO_4$). Determination of the relative site occupancy by EPR.," J. Chem. Phys., 79, pp 2590-2599, (1983).

[16]   R.C. Morris, Materials Laboratory, Corporate Technology, Allied Corporation, (1984).

[17]   M.J. Weber, D. Milam, and W.L. Smith, "Nonlinear Refractive Index of Glasses and Crystals," Opt. Eng., 17, pp 463-469, (1978).

[18]   M.L. Shand, J.C. Walling, and R.C. Morris, "Excited-State Absorption in the Pump Region of Alexandrite," J. Appl. Phys., 52, pp 953-955,1981.

[19]   S.C. Seitel, "Alexandrite Laser Damage Testing," Report to
       Allied Corporation, Michelson Laboratory, Naval Weapons
       Center, China Lake, CA, 93555, (July 2, 1984).
[20]   R.C. Powell, private communications.
[21]   D.E. McCumber, "Theory of phonon terminated optical masers,"
       Phys. Rev., 136, pp A299-A306, (1964); ----,"Einstein rela-
       tions connecting broadband emission and absorption spectra,"
       Phys. Rev., 136, pp A954-A957, (1964).
[22]   Michael L. Shand and John C. Walling, "Excited-State
       Absorption in the Lasing Wavelength Region of Alexandrite,"
       IEEE J. of Quant. Electr., QE-18, pp 1152-1155, (1982).
[23]   C.L. Sam, J.C. Walling, H.P. Jenssen, R.C. Morris, E.W.
       O'Dell, "Characteristics of alexandrite lasers in Q-switched
       and tuned operations," Proceedings of the Society of Photo-
       Optical Instrumentation Engineers (SPIE), 247, pp 130-136,
       (1980).
[24]   D.F. Heller and J.C. Walling, "High-power performance of
       alexandrite lasers," Conference on Lasers and Electro-optics
       (CLEO), Anaheim, CA, Session WI4, June 19-22, (1984).
[25]   W. Koechner, Solid-State Laser Engineering, New York:Springer-
       Verlag, (1976).
[26]   J.J. Barrett, Private commnication., Allied Corporation, Mt.
       Bethel, NJ.
[27]   D.F. Heller, Private communication, Allied Corporation, Mt.
       Bethel, NJ.
[28]   L. Horowitz, P. Papanestor and D.F. Heller, "Mode-locked
       Performance of Tunable Alexandrite lasers," Proceedings
       of the International Conference on Lasers '83 (to be
       published).
[29]   J.C. Walling, and D.F. Heller, "Progress in Alexandrite Laser
       Technology Active Mode-Locked Performance", Proceedings of
       the International Conference on Lasers '82, pp 550-558,
       (1982).
[30]   V.N. Lisitsyn, V.N. Matrosov, V.P. Orekhova, E.V. Pestryakov,
       B.K. Sevast'yanov, V.I. Trunov, V.N. Zenin, and Yu. L
       Renigallo, "Generation of 0.7 - 0.8 μ Picosecond Pulses in an
       Alexandrite Laser With Passive Mode-locking," Sov. J. Quantum
       Electron., pp 368-370, (1982).
[31]   H. Samelson and D.J. Harter, "High-pressure mercury arc lamp
       excited cw alexandrite lasers," Conference on Lasers and
       Electro-Optics (CLEO), Technical Digest, Session W14, Anaheim
       CA, June 19-22, (1984).
[32]   D.J. Harter, Allied Corporation, Mt. Bethel, N.J., private
       communications.

NEW DEVELOPMENTS IN CW DYE LASERS

Bruce Peuse

Coherent, Incorporated, 3210 Porter Drive

P.O. Box 10321, Palo Alto, CA   94303

A dedicated effort at Coherent has resulted in improving the usefulness and performance of the single-frequency dye laser. This work has led to development of a computer-controlled wavemeter dye laser system which expands user control and tunability of a dye laser. The second area of development has led to an increased tuning range of this laser system through intracavity frequency doubling.

Single Frequency Dye Laser

A CW dye laser is an extremely powerful tool, especially for use in laser spectroscopy. By using different organic dyes it is possible to generate single-frequency output from 400 nm to nearly one micron with commercial CW dye laser systems. These lasers are pumped with either CW argon or krytpon ion lasers and are capable of generating several watts of output power. (1,2,3)

The dyes used as the gain medium in these lasers can support lasing action over a very wide part of the spectrum (typically 100 nm). The problem with these lasers is in precisely controlling and tuning the output. To be useful for spectroscopy the output of this laser should be a well-controlled narrow line single-frequency.

The dye laser which in this case is a Coherent Model CR-699-21 (Fig. 1), is forced into single-frequency operation with a stack of intracavity filters, each with increasingly higher selectivity.[4] First, a coarse filter is required which uses the birefringence of crystalling quartz to rotate the polarization of the beam.[5] Twenty-two intracavity Brewster surfaces add loss for any polarization component not in the preferred orientation. This polarization rotation is wavelength-dependent and can be tuned by simply rotating the filter. Two low-loss etalons provide addtional selectivity; a thin etalon with a free spectal range (FSR) of 225 GHz and a thick etalon with a FSR fo 10 GHz. Together, these three filters force the laser to single-frequency

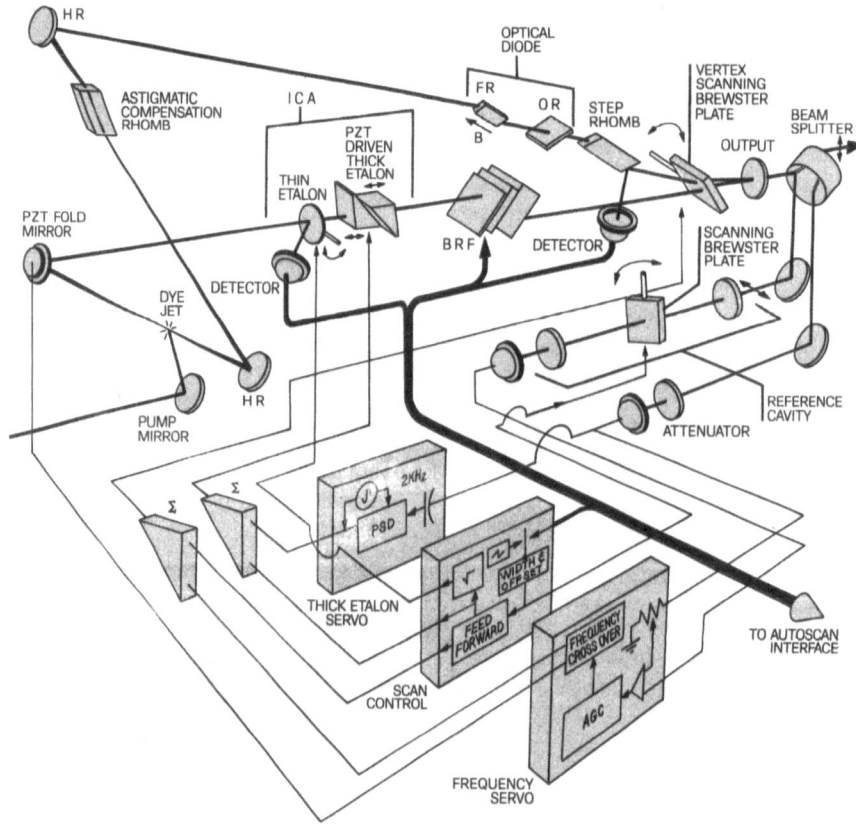

Figure 1.

Schematic diagram of the optical and electronic control systems for the CR-699-21. The additional monitoring and control features required by the Autoscan system are shown.

Now that single-frequency operation is assured, some means of precisely controlling and scanning the frequency of this oscillator must be constructed. In this laser, continuous tuning between the laser cavity modes is accomplished by placing a tipping Brewster plate in the ring resonator cavity. By rotating this plate with a precision galvonometer the laser frequency can be continuously tuned. However, in order to tune more than a few cavity modes, both the thin and thick etalons must be synchronously tuned.

The thick etalon is controlled by a servo system which keeps one of its transmission peaks on the oscillating laser mode. This is done by modulating the length of the thick etalon and monitoring the resulting output laser intensity modulation with a photodetector. The output of this photodetector is demodulated in a phase-sensitive detector and the resulting discriminant is fed back to an etalon length control device, thus enabling the etalon to be continuously peaked on the oscillating longitudinal mode. This thin etalon is controlled by simply feeding forward the same signal that is used to drive the Brewster plate galvo. In this way, laser frequncy scans of up to 30 GHz or 1 cm-1 can be obtained with these commercial systems.

Another major problem with these sytems is that a high level of frequency noise or frequency jitter is present in the output. This is mainly the result of noise generated in the flowing jet-stream of dye. (The high pump power densities necessitate the use of an open flowing jet-stream.) Since this dye jet-stream is located inside the laser resonator, small variations in thickness due to bubbles, acoustic pick-up, pressure variations and others cause the output laser frequency to jitter, so it has an effective linewidth more like 40 MHz instead of the theoretical linewidth of a few Hertz. In order to reduce this linewidth to something which is more useful for high resolution laser spectroscopy, a servo system is used to correct for these laser frequency errors. The laser frequency is locked to a passively stabilized interferometer. An error signal is generated by tuning this reference interferometer so that the laser frequency sits on the side of a transmission peak. The laser intensity is also monitored with a photodetector and the two signals are subtracted so that a zero output occurs when the laser is tuned to the side of the transmission peak. The resulting error signal is fed back to resonator length control devices. These devices are just the galvo-driven tipping Brewster plate and a piezoelectricly-driven cavity mirror. This system reduces the effective linewidth so it is on the order of 1 MHz. Laser frequency drift becomes that of the reference cavity, or in this case less than 50 MHz per hour.

This describes the systems that are commercially available.

These systems will perform precision scans of up to 30 GHz (1 cm⁻¹). After one wavenumber scan it then becomes necessary to manually reset the etalons to obtain another scan in a sequence. This is usually time-consuming and error-prone. Another problem with these systems is locating the laser frequency in the spectrum and tuning to some desired atomic or molecular transition. This usally requires some form of a wavemeter, spectrometer or some other means to locate the laser in frequency space.

Integrated Dye Laser/Wavemeter

These problems were solved by integrating a wavemeter into the dye laser system and then putting the whole system under microprocessor control. The computer uses signals from the wavemeter to measure the laser frequency. The computer controls the dye laser so it can move and scan the laser frequency. To do this, a stepper motor drive was added to the birefringent filter (BRF). The thin etalon and the reference cavity galvo (which controls the laser frequency via the stabilization servo) were put under microprocessor control. Error signals used to peak and move the BRF and thin etalon are generated by monitoring the reflections from the thin etalon and the BRF.

The wavemeter design consists of both a coarse and a fine wavemeter. The coarse wavemeter uses the optical activity of crystalline quartz to measure the wavelength. When polarized light propogates along the optical axis in a piece of crystalling quartz, the polarization vector is rotated by an amount which depends on the wavelength and on the length of the piece of quartz. A signal proportional to the amount of rotation is generated in the wavemeter and then by using the known relationship for optical rotary power in the material, the computer calculates the correct laser frequency to within 120 GHz. The 120 GHz acuracy of the coarse wavemeter is sufficient for use of the fine wavemeter. We call this coarse wavemeter the Optical Activity Meter or OAM.

The high resolution portion of the wavemeter consists of two vacuum-spaced low-finesse etalons which differ in length by 5%. Since the etalon spacers become the secondary length standard by which the wavelength of the laser is compared, they are constructed of Zerodur, a low thermal expansion material. Furthermore, the etalon assembly is housed in an evacuated and thermally stabilized container, so the spacer length is stabilized to better than one part in 10^8.

The free spectral ranges (FSR's) of the two etalons which make up the etalon assembly are about 6.8 GHz and 6.5 GHz. (FSR1 and FSR2 in Figure 2). At some frequency the transmission peaks of these etalons coincide.

As the laser frequency is scanned higher, the interval measured between the peaks of the shorter and longer etalons increases by integer multiples of the differences $\Delta\nu$ between the two FSR's (0.3 GHz).

The two peaks will again coincide after a frequency change of the free-spectral range of the vernier, or about 150 GHz. This frequency interval defines the "order" of this vernier etalon

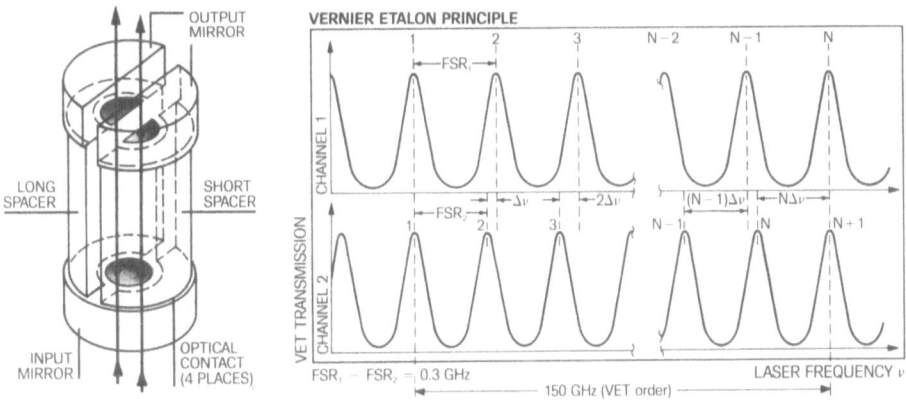

Figure 2.

Schematic diagram of the vernier etalon assembly and display of the vernier etalon principle.

assembly (VET). Within in one VET order, the frequency is uniquely determined by the separation between two adjacent VET peaks--one peak from the longer etalon and the other form the shorter. This is determined by a short (15 GHz) scan of the dye laser. The VET peaks are monitored by two photodetectors and the signals are sent to the computer via an analog to digital interace where the data is stored. Use is made of a digital filter and a statistical averaging algorithm to achieve a readout of the VET channels to better $\pm$ 25 MHz resolution (0.4% of the FSR of one channel). However, a coarse wavelength determination is necessary in order to determine the VET order. This is accomplished with the OAM.

The free-spectral ranges of the two etalons, and the constants for the analytical expression to read out the OAM, are determined by measurements relative to known spectal line references.[6]

Computer Control

The integrating element to this automatic scanning dye laser (Autoscan) system is the computer. Signals from the two photodetectors in the laser head and the five in the wavemeter are fed to an Apple II microcomputer via an interface which converts the analog signal to digital. These signals are analyzed in the computer and, together with the operator commands, are used to generate the appropriate signals to control the CR-699-21 dye laser. These control signals are sent via an interface to the electronic control box and birefringent filter stepping motor.

The interface electronics provide the communication link between the computer and wavemeter and laser. Analog signals from the wavemeter and laser head are multiplexed and converted to digital signals by an 8-bit A/D converter. Similarly, reposnses from the computer are converted to analog control signals by a 16-bit DAC for scanning and a 12-bit DAC for thin etalon control. Timing and input/output functions are provided by the interface electronics. The interface electronics also provide three analog data channels.

Our design philosophy was to do the housekeeping and mechanics of running the Autoscan wavemeter/laser with computer software. This means that the complex sequences of functions necessary to tune the laser and run the wavemeter are performed quickly and automatically by the computer. The software package also included data handling routines for storing, manipulating and displaying up to three channels of data.

The Autoscan program is written in BASIC; however, it makes extensive use of machine language routines. The inherent speed with which machine language programs run allowed powerful signal porcessing routines to be used. The user addresses the program by using a set of simple, easy-to-use keyboard commands.

System Capabilities

By incoporating a computer into the ring laser system, a number of important features can be obtained. Some of these features include stacking of scans, mode-hop detection, scan linearization and automatic etalon peaking.

The complete automation of reading and tuning the dye laser's wavelength enables this system to stack scans. This means that short (10 GHz) scans are butted end-to-end to form a continuous, seamless laser frequency scan, limited in total length only by the dye tuning curve. If the data interval is set to less than 50 MHz, additional data points are taken at the end of each 10 GHz scan.

These extra points are labeled as overlap points and avoid potential discontinuities in the data where the end of one scan is joined to the next.

An important series of routines provided by the Autoscan software is etalon and BRF peaking routines. These peaking routines are used by the computer to interrogate the two sensors in the laser head, and then to maintain both the thin etalon and the BRF at their optimal position.

The laser frequency of the locked dye laser is scanned by changing the peak transmission frequency of the reference cavity. This is done by changing the optical path length of the reference cavity with a galvo-controled scanning Brewster plate. The laser frequency as a function of angle of rotation of the scanning Brewster plate is not quite linear. Furthermore, the angle of rotation, since it is controlled by a galvo, is not necessarily precisely linear with the current drive to the galvo. These nonlinear effects have been found to be a highly reproducible and, therefore, may be compensated for by the microprocessor. In practice, rather than alter the actual scan rate and introduce possible inertial effects, the choice was made to provide a look-up table which adjusts the interval at which data is taken. Thus data is taken at fixed frequency intervals, rather than fixed time intervals during the linear voltage ramp drive to the reference cavity galvo.

To provide relative immunity against mode-hops in the data, in addition to the safeguard of measuring the wavelength of each scan, there is also a fast algorithm which quickly checks for mode-hops and takes corrective action in case one is found. This algorithm detects mode-hops by looking for discontinuities in the data from VET channels during scans. When a mode-hop is detected in a 10 GHz scan segment, the segment is repeated and an error mesage appears on the screen.

The result of this work is a far more useful instrument for performing high resolution, wide scan laser spectroscopy. Measurements that one were considered too time-consuming or too difficult may now be completed using this automatic scanning dye laser system.

Second Harmonic Generation (SHG)

In addition to expanding the capabilities of the CW dye laser, work has also been done to extend the tuning range. Using optical second harmonic generation (SHG), we have expanded the spectral coverage to include the near ultraviolet.[7] Usable single-frequency UV has been generated by intracavity frequency doubling in KDP and LiIO₃, which covers the spectrum from 270 nm to 400 nm.

In (SHG), the output power is dependent on the square of the fundamental intensity. For this reason, frequency doubling is commonly done with pulsed lasers, however pulsed lasers are not very useful for high resolution work. One way of getting high fundamental power is to place the crystal inside an enhancement cavity. Another way is to mount the crystal inside the laser resonator. In this work, the nonlinear crystal was mounted directly in the ring laser resonator of a CR-699-21. The standard output coupler was replaced with a special low transmission output coupler to further enhance the fundamental intracavity intensity.

The SHG power output P_{SHG} from a nonlinear crystal can be expressed by the following relationship:

$$P_{SHG} \sim P_F{}^2 L^2 \left[\frac{\sin \Delta KL/2}{\Delta KL/2} \right]^{\frac{1}{2}}$$

where

$$\Delta K = 2\pi(n_w - n_{2w})/\lambda_F$$

The fundamental power is given by P_F and the interaction length by L. The frequency dependent index ($n_{w,2w}$) results in a phase mismatch for the fundamental and the second harmonic wave. The two waves must be in phase for efficient SHG.

The technique that is widely used to satisfy the phase matching requirement, $\Delta K = 0$, takes advantage of the natural birefringence of anisotropic crystals. It is possible to satisfy the phase matching condition in a uniaxial crystal, by selecting an ordinary wave and extraordinary wave so that the indexes n_w and n_{2w} will be equal. For KDP and LiIO₃, which are negative uniaxial crystals, a phase matching angle can be found between the propagation direction and the crystal optic axis. For this propogation direction the ordinary fundamental wave will phase match with the extraordinary second harmonic wave. To tune the UV frequency requires tuning the fundamental dye laser frequency and then rotating the crystal to obtain a new phase matching angle. This technique of angle tuning frequency doubling makes for an easy to operate system and can be used over a broad tuning range. It allows the UV output to be continuously scanned over 30 GHz for a fixed phase match angle with only a moderate power modulation ($\leq$ 5%).

The doubling cyrstals are mounted in the upper collimated beam of the ring laser (Fig. 1). This helps to maintain good UV output power and mode while minimizing UV crystal damage and thermal lensing. Each crystal is cut at Brewster's angle to further minimize losses. A total of four crystals have been used, each cut so that Brewster's angle and the phase match angle are both optimized for given dyes. (See Figure 3.) The UV is brought out of the ring cavity with a special beam splitter, coated to reflect the UV and pass the fundamental. The beam splitter is mounted at Brewster's angle to minimize intracavity losses. SHG output of up to 25 mW has been demonstrated with spectral coverage from 267 nm to 400 nm.

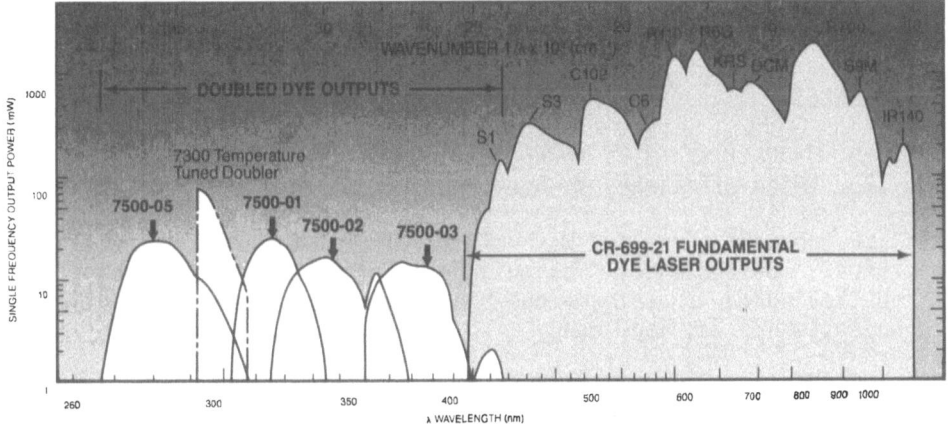

Figure 3.

Plot of the UV single frequency tuning curves generated with a Coherent Model 7500 angle tuned frequency doubler and a CR-699-21 ring dye laser.

Conclusion

The development work that has been carried out to increase the capabilities and usefulness of the ring dye laser as a spectroscopic tool has lead to several important improvements. By integrating a precision wavemeter together with a stabilized ring dye laser system and putting this system under microprocessor control, this laser has been made into a far more powerful tool. The time required to perform high resolution, wide scan, laser spectroscopy can be dramatically reduced. Measurements that once were considered too time consuming or too difficult can now be completed using this automatic scanning (Autoscan) dye laser system.

A second area of work has lead to an increased tuning range for the CW ring dye laser. By intracavity frequency doubling, usable single-frequency ultraviolet in the spectral region from 270 nm to 400 nm has been obtained. This development adds an important spectral region for the researcher to perform high resolution studies.

A continuing effort is being made at Coherent in these and other areas to further extend the performance and usefulness of this powerful research tool.

REFERENCES

1. S.M. Jarrett and J.F. Young, Optics Letters 4 (June '79) 176-178, "High efficiency single-frequency CW ring dye laser."

2. T.F. Johnston, Jr. and W. Proffitt, IEEE J Quant. Elec. QE-16 (April '80) 483-488, "Design and performance of a broad-band optical diode to enforce one-direction travelling-wave operation of a ring laser."

3. T.F. Johnston, Jr., R.H. Brady and W. Proffitt, Appl. Optics 21 (1 July '82) 2307-2316, "Powerful Single-frequency ring dye laser spanning the visible spectrum"

4. T.F. Johnston, Jr. J.L. Hobart. R.C. Rempel, and G.H. Williams, U.S. Patent No. 4,1150,342 (1979).

5. A. Bloom, J. Opt. Soc. Am. 64 (April '74) 447-452, "Modes of a laser resonator containing tilted birefringent plates".

6. S. Gerstenkorn and P. Luc "Atlas du Spectre d' Absorption de la Molecule d' Liode" (Editions du C.M.R.S., Paris 1978), 550 pages.

7. G.D. Boyd and D.A. Kleinman, J. Appl. Phys. 39, 3597-3639 "Parametric Interaction of Focussed Gaussian Light Beams".

SELECTED REPRINTS ON PULSE COMPRESSION

I. TEMPORAL COMPRESSION OF LIGHT

John K. Wigmore and Daniel R. Grischkowsky

IBM Thomas J. Watson Research Center
Yorktown Heights
New York

Abstract

The 5890 Å output from a CW dye laser was converted into a train
of 0.5 ns pulses by frequency modulation and passage through a near-
resonant atomic vapor delay line of Na. The theory of the process is
discussed in both the time and frequency domains. Using a modulation
index of 120 at a frequency of 17.8 MHz, we obtained values for the
temporal compression ratio and intensity enhancement of 112 and 14,
easily the largest that have been reported.

INTRODUCTION

Most optical pulse compression schemes are based on the idea
originally developed for chirp radar whereby a frequency swept pulse
is sent through a dispersive delay line[1-16]. In simple terms, the
group velocity of the light is determined by its instantaneous frequency so that different portions of the pulse travel at different
speeds through the delay line. The length of the line is adjusted so
that the leading edge of the pulse is delayed by just the right
amount to overlap the trailing edge at the output of the delay line.
Under optimum conditions the output pulsewidth is given by the reciprocal of the bandwidth of the frequency sweep.

Optical pulse compression using this scheme was achieved relatively easily for ps pulse[4,5,7,9,16]. For such short pulsewidths

Reprinted with permission from IEEE Journal of Quantum Electronics,
QE-14(4):310-315, (1978). Copyright © 1978 IEEE.

the nonresonant delay lines have sufficient dispersion to allow good compression with reasonable path lengths. The most utilized delay line has been the diffraction grating pair[4,6-9,11,16]. Cells filled with Kerr liquids have also been proposed[3], but the required path lengths are very long. Another important approach is the Gires-Tournois interferometer[2,5,12].

The recent introduction by Grischkowsky[13,14] of the near-resonant atomic vapor delay line has allowed pulse compression techniques to be extended into the ns regime. Such a delay line utilizes the large dispersion near an atomic resonance of an alkali metal vapor, and may be as much as 1000 times more dispersive[10,17] than the nonresonant lines, allowing compression of many ns for delay lines typically 100 cm in length. In Grischkowsky's experiments, sinusoidally frequency modulated dye laser pulses nearly resonant with the $5S_{1/2} \leftrightarrow 5P_{1/2}$ transition (7948 Å) of rubidium were propagated through a 100 cm Rb vapor cell[13,14]. Depending on the experimental condition, either a single 10 ns pulse was compressed to 1 ns, or an initially smooth 30 ns pulse was changed to a series of 1 ns spikes separated by the period of the modulation. Using a modulation frequency of 150 MHz and a modulation index of $\phi_0 \approx 1.5$ rad, where $2\phi_0$ is the peak-to-peak phase modulation, he obtained a temporal compression ratio of about 7 (one modulation cycle is taken as the duration of the input "pulse") and an intensity enhancement of 1.5.

A closely related device is Loy's dispersive modulator[18,19], in which the resonant frequency of the transition itself is modulated, rather than the light frequency. Grischkowsky and Loy[20] showed that the technique is mathematically equivalent to that described above, and Loy obtained similar numbers for the temporal compression of 10 μm light using Stark modulation of an NH_2 transition at a frequency of 80 MHz.

Grischkowsky's technique was extended to CW light by Bjorkholm, Turner and Pearson[15]. Using a Na vapor delay line and a beam almost resonant with the $3S_{1/2} \leftrightarrow 3P_{3/2}$ transition at 5890 Å, they obtained an output train of subnanosecond pulses separated by the 5 ns period of the frequency modulation. For a modulation index ϕ_0 8 rad at 197 MHz, they were able to compress the spikes on the output pulse train to 240 ps, thereby achieving a compression ratio of 21. The intensity of the spikes were six times larger than that of the input CW beam.

In this paper we wish to report on further experiments with a 5890 Å light beam and a sodium vapor delay line. We were able to produce a much greater modulation index of 120 rad at 17.8 MHz resulting in a frequency modulation bandwidth of 4.3 GHz and a large improvement in compression over previous work. The observed output pulse train was composed of a series of 500 ps pulses separated by the 56 ns period of the modulation implying a compression ratio of 112. The peak intensity was 14 times that of the input CW beam. These figures are easily the largest that have been obtained in an optical temporal compression experiment and start to approach the values typical of radar[1].

The arrangement of the paper is as follows. We begin by discussing the theory of the compression process in the inituitive time domain picture. We derive a useful analytic solution and introduce a singularity distance which is the optimum length for the delay line. We then summarize the theory in the frequency domain obtained using linear dispersion theory (LDT), which is necessary in order to calculate the pulse shape at the compression singularity. Following a description of the experimental arrangements, we present our results and compare them with the computer LDT solution. Finally, we discuss the utility of this approach for obtaining a source of stable and repetitive short pulses.

THEORY

A. Time Domain

The simple picture of the compression process outlined in the previous section will now be restated somewhat more precisely. Since the energy and the instantaneous frequency are both assumed to travel with the group velocity, every point on the input envelope can be distinguished uniquely by its instantaneous frequency ω'. Furthermore, as the envelope propagates through the vapor and the frequency dispersion of the group velocity causes the separation of these conceptual points to change, energy does not flow past any of the points. Consequently, if the points move closer together, the intensity increases, and in principle this process may continue until the points overlap and an intersection is reached.

The above argument can be stated in mathematical terms to derive an expression for the optical intensity during the compression process. Consider two points on the pulse envelope which at the entrance to the delay line are separated in time by the infinitesimal δt_i and in instantaneous frequency by $\delta \omega'$. Later in the discussion we will consider the limit as $\delta t_i \to 0$ and $\delta \omega' \to 0$. As these points propagate through the delay line, their separation in time will change according to their difference in group velocity. The intensity associated with each point will also be reduced by simple linear absorption. If we designate the separation in time between the two points at the output of the delay line by δt_f, we obtain the following result for the output intensity

$$\mathcal{E}^2(\tau,z) = \mathcal{E}_0^2(\tau)\exp(-\alpha z)\delta t_i/\delta t_f. \tag{1}$$

Since we are dealing with the propagation of a pulse (1) is stated in terms of the reduced time τ, which is defined as

$$\tau = t - z/\nu_g \tag{2}$$

where ν_g is the group velocity, determined by the instantaneous frequency; ω' is equal to $\omega + \partial\phi/\partial t$, where ω is the carrier frequency of the laser, and ϕ is the time dependent phase angle (for sinusoidal modulation $\phi = \phi_0 \sin \Lambda t$); α is the linear absorption coefficient; $\mathcal{E}_0^2(\tau)$ is the input pulse envelope at $z = 0$ (for a CW input beam this term is a constant). The output time separation δt_f is given from our previous argument, by

$$\delta t_f = \delta t_i + z[1/\nu_g(\omega' + \delta\omega') - 1/\nu_g(\omega')], \tag{3}$$

which in the limit $\delta\omega' \to 0$ is equivalent to

$$\frac{\delta t_f}{\delta t_i} = 1 + z \frac{\delta\omega'}{\delta t_i} \frac{\partial}{\partial\omega'} \frac{1}{\nu_g} . \tag{4}$$

Since in this limit $\delta t_i \to 0$ also, (4) can be reduced to

$$\frac{\delta t_f}{\delta t_i} = 1 + z \frac{\partial\omega'}{\partial r} \frac{\partial}{\partial\omega'} \frac{1}{\nu_g} . \tag{5}$$

Equation (1) then becomes

$$\mathcal{E}^2(\tau,z) = \mathcal{E}_0^2(\tau)\exp(-\alpha z) \left[1 + z \frac{\partial\omega'}{\partial\tau} \frac{\partial}{\partial\omega'} \frac{1}{\nu_g} \right]^{-1} . \tag{6}$$

In order to convert this result into real time, $\mathcal{E}^2(t,z)$, t must be calculated from the given values of τ and z, using (2).

Equation (6) is an analytic description of optical pulse compression in the time-domain regime. In order for this solution to apply, as previously stated, both the energy and instantaneous frequency must propagate at the group velocity. The theoretical prediction of (6) for a CW input beam with a sinusoidally modulated instantaneous frequency is showed in Figure 1. The simple wave propagation of the instantaneous frequency is illustrated in Figure 1(a). The dots represent the conceptual points referred to above, and the lines are characteristics of constant group velocity. Clearly, as the characteristics come closer together, the intensity increases, and a singularity occurs if two characteristics intersect. The behavior of the intensity envelope is shown in Figure 1(b).

It is important to know how close to the singularity this solution is valid. Although in the initial stages of the compression process the output envelope predicted by (6) agrees extremely well

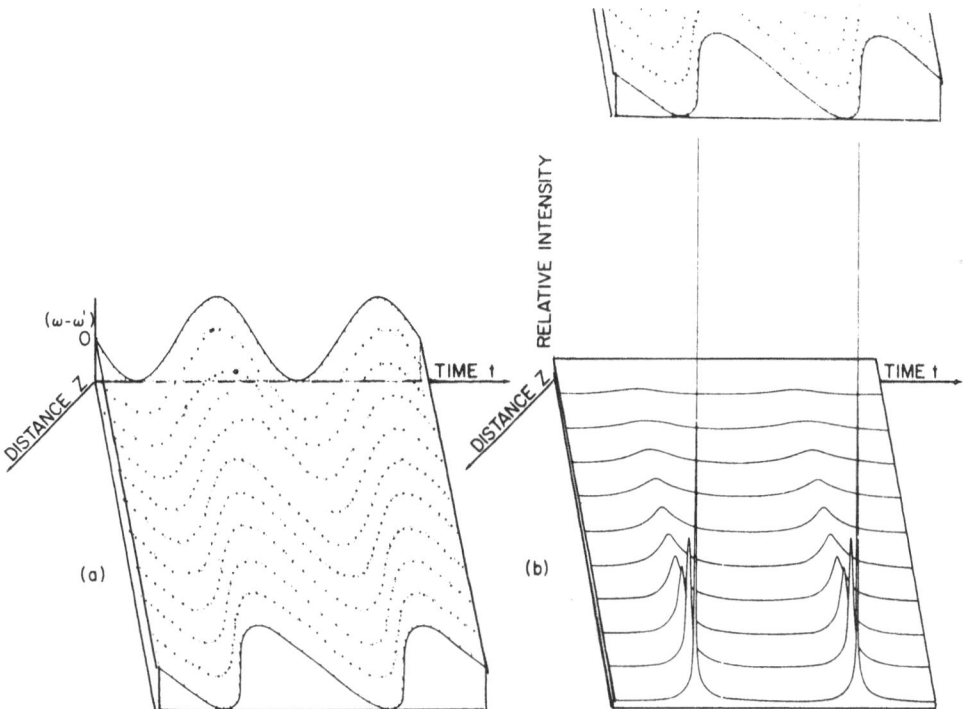

Fig. 1. (a) Propagation of the instantaneous frequency ω' through the vapor of two-level atoms. The constant carrier frequency is denoted by ω, and we plot the frequency difference (ω−ω') as a function of distance z and time t. Initially ω' changes sinusoidally, but after traveling a sufficient distance through the vapor, the time dependence changes to the sawtooth variation. The characteristics are represented by the solid lines, which are the trajectories of the conceptual points. (b) The changes in intensity resulting from the distortion of the sinusoidal distribution of ω'. These results are independent of the system parameters, which serve only to determine the scale factors of the frequency, time, and distance scales.

with both experiment and linear dispersion theory (LDT), when pulse-widths are reached which are the order of that allowed by the input bandwidth, LDT gives better agreement with experiment. Consequently, the main advantage of considering pulse compression in the time domain is that it provides an easily visualized picture of the re-shaping process. However, accurate calculations can be made as long as the pulse compression is well within that allowed by the input frequency bandwidth[14]. In addition, the input pulse shape and the form of the frequency modulation often do not readily allow Fourier decomposition, and for these situations (6) provides a valuable and easier alternative to LDT.

Finally, the conditions required for the singularity allow one to estimate the length z_s of the dispersive delay line for optimum pulse compression, that is, when $\delta t_f = 0$. The denominator of (6) vanishes when

$$\frac{1}{z_s} = - \frac{\partial \omega'}{\partial \tau} \frac{\partial}{\partial \omega'} \frac{1}{\nu_g} .$$ (7)

From our experimental observation, we found that the best pulse compression was indeed obtained for propagation distances approximately equal to z_s. Consequently, knowing the frequency modulation of the input light, we were able to choose the experimental parameters so that the length of the cell was a reasonable value.

In order to calculate z_s, we needed the following expression for the group velocity in sodium vapor.

$$\frac{1}{\nu_g} = \frac{1}{c} \left\{ 1 + \frac{2\pi\omega P_{12}^2}{\hbar} \left[\frac{N_1}{(\Delta\omega'_1)^2} + \frac{N_2}{(\Delta\omega'_2)^2} \right] \right\} .$$ (8)

In (8), the subscripts 1 and 2 refer to the two hyperfine components $F = 1$ and $F = 2$ of the $3S_{1/2}$ ground state of sodium. These two components are separated by 0.06 cm^{-1}. N_1 is the atomic number density in the $F = 1$ ground state; $\Delta\omega'_1 = (\omega_1 - \omega')$ where ω_1 is the resonant frequency from the $F = 1$ components of the $3S_{1/2}$ ground state to the $3P_{3/2}$ state. The hyperfine splitting of the $3P_{3/2}$ state can be neglected. $P_{12} = 5.23 \times 10^{-18}$ ESU is the matrix element of the electric dipole moment operator between the $3S_{1/2} \leftrightarrow 3P_{3/2}$ states of Na for linearly polarized light.

B. Frequency Domain

As we mentioned in Section A, LDT, which treats the pulse compression problem in the so-called "frequency domain", is an alternative picture of the low intensity pulse compression process. The input wave to the dispersive delay line is Fourier analyzed. The different frequencies then all propagate through the line at different phase velocities thus changing their relative-phase angles. They are also attenuated according to their frequency by passage through the line. At the end of the line the components are recombined to give the amplitude of the output pulse. For our experimental situation, this theory has already been discussed by Bjorkholm et al.[15] and in the context of a similar device, the dispersive modulator by Grischkowsky and Loy[20]. Here, we shall use the formalism and notation of[20] to calculate the output pulse train.

In particular, from[20, Equation (8)] the output field at the end of a delay line of length z is given by the following sum of Fourier components:

$$E(z,t) = \mathcal{E}_0 \hat{x} \left\{ \exp[-\alpha_0 x/2] J_0(\phi_0) \cos(k_0 z - \omega t) \right.$$

$$+ \sum_{j=1}^{\infty} J_j(\phi_0) [\exp[-\alpha_j^+ z/2] \cos [k_j^+ z - (\omega + j\Lambda)t]$$

$$\left. + (-1)^j \exp[-\alpha_j^- z/2] \cos [k_j^- z - (\omega - j\Lambda)t]] \right\}. \qquad (9)$$

In (9) α_0, α_j^+, and α_j^- are the absorption coefficients for light with angular frequencies ω, $(\omega + j\Lambda)$, and $(\omega - j\Lambda)$, respectively, similarly, k_0, k_j^+, and k_j^- are the propagation vectors; $\mathcal{E}_0$ is a constant; ω is the angular frequency of the laser; ϕ_0 is the modulation index and Λ is the angular modulation frequency; $J_j(\phi_0)$ designates the Bessel function of the first kind of order j with argument ϕ_0.

Because, compared to the situation of [20], our frequency offset was always considerably larger than the Doppler width of 0.06 cm^{-1}, we were able to use the absorption coefficient appropriate for the Lorentzian wings of the two hyperfine components of the resonance line, i.e.

$$\alpha_0 = \frac{2\pi\omega P_{12}^2}{c\tau_s \hbar} \left[\frac{N_1}{(\Delta\omega_1)^2} + \frac{N_2}{(\Delta\omega_2)^2} \right], \qquad (10)$$

where $\tau_s = 16$ ns is the spontaneous radiative lifetime for the $3P_{1/2}$ state. Equation (10) is stated for our case where collisions between the Na atoms are negligible. Similarly, the index of refraction n_0 at frequency ω was equal to

$$n_0 = 1 + \frac{2\pi P_{12}^2}{\hbar} \left[\frac{N_1}{\Delta\omega_1} + \frac{N_2}{\Delta\omega_2} \right]. \qquad (11)$$

The evaluation of (9) was carried out by computer, and the summation terminated at $j = 150$, further terms contributing negligibly.

EXPERIMENTAL DETAILS

Our laser was a Spectra-Physics model 580A CW dye laser, delivering up to 50 mW at the selected frequency. The modulator was a 20 x 5 x 1 mm rectangular parallelopiped of single crystal lithium tantalate, mounted between capacitor plates in a resonant circuit. Light propagated along the x axis (20 mm) of the material with the electric field vector parallel to the applied RF field along the z axis (1 mm). One of our objectives was to seek as high a value of compression ratio, and hence of intensity enhancement, as possible, so in order to obtain the largest possible modulation index, the RF voltage was applied in a pulse mode to the modulator crystal. Using a Matec model 6600 generator, electric fields up to 2×10^4 V/cm could be applied before the air broke down and shorted out the capacitor. Since the final pulsewidth should not be compressed to narrower than the 300 ps limit of the detecting system, the use of such a high modulation index meant that we had to start with a much lower modulation frequency than that of the previous workers[13-15]. The magnitude of the modulation index could be estimated from the strength of the electric field across the $LiTaO_3$ but because of the acoustic effects described later, there was doubt as to the exact value of the electrooptic constant that should be used. We therefore measured ϕ_0 directly by observing the frequency modulation sidebands using a piezoelectrically scanned Fabry-Perot interferometer, Spectra-Physics model 470.

It proved convenient also to pulse the light beam itself, using a Lasermetrics LM-1 pockels cell driven by a 5 μs, 1 kV pulse from a Velonex 350 pulse generator, and followed by a crossed polarizer. In this way, the possibility of thermal damage of the $LiTaO_3$ was lessened and, in addition, the sensitivity of the Spectra-Physics model 508 avalanche photodetector was increased by a factor of four. The initial trigger was supplied by the Velonex generation, and the 1 μs RF pulse containing some 18 cycles of modulation was delivered to the modulation during the time that light was passing through. In summary, the light was repetitively switched through the Na vapor cell in 5 μs pulses, during the 5 μs pulses, 1 μs RF modulation pulses were applied; temporal compression occurred only during the 1 μs when the light was frequency modulated. No transient effects due to the pulsing were observed. The output from the detector was displayed on a Tektronix 7904 oscilloscope either directly using a 7A19 preamplifier, or with a 7S11 sampling unit and an S-4 sampling head. The response time limit, about 300 ps, was due to the detector itself.

The other free variable in the experiment was the product Nz where N was the number density of sodium atoms in the delay line, and z was the length of the line. It was not obvious from a superficial inspection of the equations whether greater compression and enhancement would be best achieved by higher or lower Nz. However, calculations showed that for a delay line approximately 100 cm long, if the singularity was to occur for values of the offset frequency reasonably larger than the Doppler width and hyperfine splitting (approximately 0.06 cm^{-1}), then N would have to be in the range approximately 10^{12}-10^{13}/cm^3. Over this range of N, the optimum intensity enhancement ratio varied by only about 20 percent. Above a temperature of about 250°C, corresponding to N of the order of 3×10^{13}/cm^3 absorption of light due to collision broadening became an important loss mechanism, and in addition the cell windows became degraded by the hot Na.

The actual delay line was a Pyrex cell 79 cm long and heated by resistive tape to between 220 and 250°C. Because of the likelihood of temperature gradients in the cell, we did not use the value of N calculated directly from the measured temperature. Instead, through (10), we inferred values from measurements of the absorption of the cell and of the offset frequency.

The greatest experimental problem that we encountered was that of acoustic standing waves in the $LiTaO_3$, induced by reason of the strong piezoelectricity of the material. In addition to modifying the effective electrooptic coefficients, the standing waves were particularly troublesome in producing an inhomogeneous modulation of the laser beam. That is, different portions of the beam contained different frequency components which could result in spatial and amplitude modulation in addition to the intended frequency modulation. Since the detector element area was only 0.04 mm^2, very careful focusing and positioning of the beam was required. In an attempt to damp the acoustic resonances, we experimented with the design of the modulator and the manner in which the $LiTaO_3$ crystal was mounted between the capacitor plates. In addition, since the acoustic resonances had a Q considerably higher that that of the RF resonance, some improvement was achieved by carefully tuning the modulation frequency halfway between two acoustic modes, but we were never able to eliminate their effects completely.

EXPERIMENTAL RESULTS

The results that could be achieved using this modulation scheme are illustrated in Figures 2-4. Figure 2 shows a train of light pulses emerging from the Na delay line after modulation and compression. It is seen that a "packet" of constant intensity light lasting 56 ns was compressed down to a pulse less than 1 ns wide. The tail following each pulse is thought to be due partly to the acoustic resonances that were discussed earlier, and partly to the fact that our detector's response to a short pulse has a low level tail with a recovery time of many nanoseconds. We estimated that only about one fifth of the 56 ns light "packet" was actually going into the sharp spike. Nevertheless, in order for a packet even 3.4 m in length to be compressed to a pulse less than 30 cm in length in a cell only 79 cm long, the front of the packet had to travel approximately four times more slowly than the rear. Both velocities were considerably less than c; such large dispersion is easily achieved by the use of a near-resonant atomic vapor delay line[17].

In order to display the full train of pulses and to illustrate the large compression ratios that we obtained, the data of Figure 2 were taken using an oscilloscope 7A19 preamplifier having a bandwidth of 500 MHz. The true shape and magnitude of the pulses reaching the detector were thereby somewhat distorted, and in order to observe the

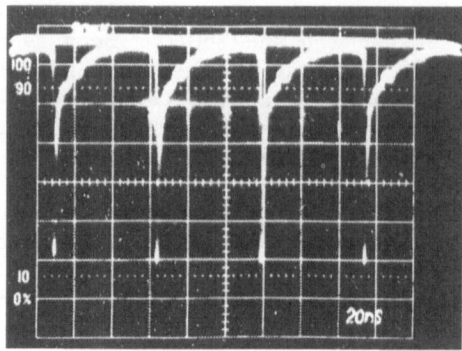

Fig. 2. The train of pulses emerging from the dispersive delay line, observed directly on a 500 MHz oscilloscope. The baseline gives the detector response for zero light input. Scales (each large division): vertical 20 mV, horizontal 20 ns.

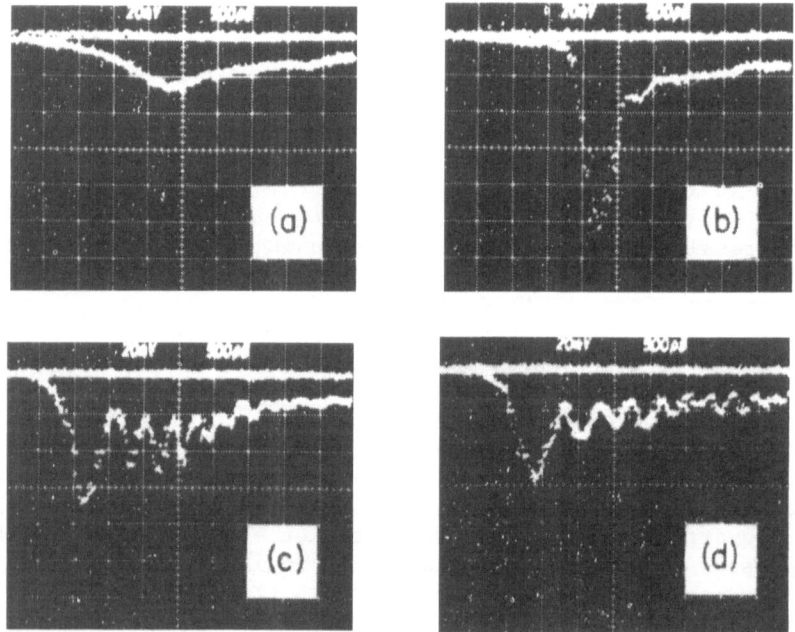

Fig. 3. The compressed pulse observed by a sampling unit for different laser frequencies, separated by one laser mode: (a) f_c + 390 MHz (f_c is the laser frequency for optimum compression); (b) f_c; (c) f_c – 390 MHz; (d) f_c – 780 MHz. Decreasing the laser frequency decreases the frequency offset from the resonance line. Scales (each large division): vertical 20 mV, horizontal 500 ps.

pulse shape in detail, the amplitude was replaced with a sampling unit. The signal-to-noise of the sampled traces in Figures 3 and 4 demonstrates the stability and reproducibility of the compression process.

Figure 3 illustrates the extreme sensitivity of the compression singularity to frequency that was described earlier and shown in Figure 1. The different traces were taken for the laser tuned to adjacent cavity modes separated by only 390 MHz. A change of laser frequency of very much less than this amount was sufficient to cause measurable dephasing of the different frequency components. The asymmetry of the singularity with respect to frequency offset should be noted. As the dispersion was increased (frequency offset decreased) the compressed pulse built up smoothly, following the prediction of (6), until at the maximum all the Fourier components were in phase. With a further increase in dispersion beyond the singularity, the dephasing components beat together to produce the observed oscillations of intensity.

By means of a continuously variable etalon in the laser cavity, the laser frequency was adjusted for the maximum intensity enhancement (Figure 4), which occurred at a measured offset of $\Delta\omega_2/2\pi c = 0.29 \pm 0.03$ cm^{-1}, on the high frequency side of the line. The experimental parameters were deliberately set so that the pulse shape was not detector limited, and for a frequency of 17.8 MHz and $\phi_0 = 120$, the minimum width [full width at half maximum (FWHM)] was 500 ps. Starting from higher modulation frequencies we could never obtain output pulses narrower than 300 ps, which we believe to be the limit of our detector. Figure 4 also displays (upper trace) the light intensity observed in the absence of the RF modulation. From the average of many such photographs, an apparent intensity enhancement ratio of 23 could be calculated. In order to determine the true intensity enhancement, a knowledge of the optical attenuation coefficient at the laser frequency was also required. This was obtained at the end of the run by measuring the light output from the delay line, normalized against the unattenuated beam, as the cell cooled down to room temperature. The ratio of the attenuated to the unattenuated intensities for the conditions appropriate to Figure 4 was 0.6 ± 0.06, resulting in a true intensity enhancement of 14. A value of $N = 1.5 \times 10^{13}/\text{cm}^3$ was calculated from this absorption. The measured pulsewidth of 500 ps implied a compression ratio of 112. These figures, for the temporal compression ratio and for the intensity enhancement, are easily the highest that have been obtained in optical pulse compression experiments. Nonetheless, they are still below the theoretical expectations.

For comparison, Figure 5 shows the computed pulse at peak compression obtained using the LDT frequency domain argument. With the values of the experimental parameters, $\phi_0 = 120$ $\Lambda/2\pi = 17.8$ MHz and $N = 1.5 \times 10^{13}$ cm^{-3}, the computed compression occurred at an offset

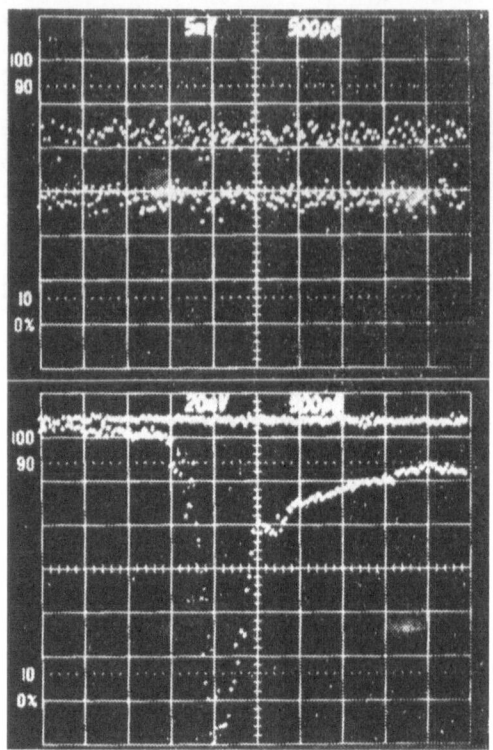

Fig. 4. The compressed pulse at optimum compression, for the experimental parameters $\Delta\omega_2/2\pi c = -0.29 \pm 0.03$ cm^{-1}, $\phi_0 = 120$, $\Lambda/2\pi = 17.8$ MHz, $N = (1.5 \pm 0.2) \times 10^{13}$ cm^{-3}. The upper traces measure the light input to the detector without modulation ($\phi_U = 0$). A value of 14 for the intensity enhancement was inferred from this and the absorption data. Scales (each large division): upper vertical 5 mV, lower vertical 20 mV, horizontal 500 ps.

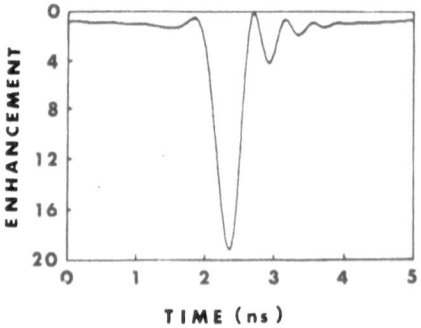

Fig. 5. Calculated pulse shape for the parameters $\Delta\omega_2 = -0.263$ cm^{-1}, $\phi_0 = 120$, $\Lambda = 17.8$ MHz, $N = 1.5 \times 10^{13}$ cm^{-3}.

$\Delta\omega_2/2\pi c = -0.263 cm^{-1}$. From Figure 5 we see than an intensity enhancement of 19 and a final pulsewidth of 390 ps might be expected under ideal conditions. Both these numbers are rather better than the measured values, and in addition the slight oscillations predicted to follow the main pulse are largely absent from the experimental trace. For reasons explained in the previous section, we believe that these discrepancies are mainly due to the acoustic standing waves, which are particularly troublesome at frequencies of a few MHz. However, compression and modulation experiments in the ps regime would be comparatively free from such effects since they would involve much higher harmonics of the resonances with correspondingly lower excitation. In addition, the nature of the pulse compression process is such that, as the compressions ratio increases, the pulsewidth and intensity enhancement become correspondingly more sensitive to fluctuations in the experimental parameters, particularly the frequency offset.

In the initial stages of the experiment, we used calculations of the singularity distance z_s as a guide in our choice of parameters. First we obtained (experimentally) as high a modulation index as possible. From (7), we were then able to show the feasibility of our desired objectives using atomic number densities of the order of $10^{13}/cm^3$, frequency offset large compared to the Doppler width of $0.06 cm^{-1}$, and cell lengths of 100 cm. As an example of the accuracy of these calculations, for the same parameters as used in the computer LDT fit for Figure 5, the calculated value of z_s was 105 cm, to be compared with the actual cell length of 79 cm.

We believe the compression process can be further extended and that pulsewidths in the ps range should be possible. In this regard it should be noted that the output pulse from one delay line could be further modulated and used as the input pulse to another line. Overall intensity enhancements of 100 should be possible using these techniques. As an example of this approach, consider the further compression of 0.5 ns pulses as obtained in our experiment. For this case the 0.5 ns pulses would again be sinusoidally frequency modulated, but with the higher and commensurate modulation frequency of 500 MHz. Assume also a modulation index of 100 rad and a frequency offset of 4 cm^{-1}. This relatively large offset could be obtained by using longitudinal magnetic fields of approximately 30 kG in both the initial Na cell (used to obtain the 0.5 ns pulses) and the second Na cell. The light would be circularly polarized in both cells, but its sense of rotation would be changed between the first and second cell. Thus, the light could be close (0.3 cm^{-1}) to the resonance line ($3S_{1/2} \leftrightarrow 3P_{1/2}$ at 5896 Å) in the first cell and relatively far away (4 cm^{-1}) from the resonance line in the second cell. From our experience with sinusoidal modulation we would estimate at optimum compression a pulsewidth of the order of 20 ps. If we require the length of the second delay line to be 100 cm, then by using the above parameter and (7) for z_s which we require to be 100 cm, we calculate the number density of Na vapor in the second cell to be $N = 2.5 \times 10^{14}/cm^3$. This relatively low value of N shows that the proposed scheme is a reasonable one.

In conclusion, our results demonstrate the utility of the near-resonant delay line for achieving a source of stable, repetitive short pulses. The pulsewidth is variable and can be in the ps range. We have demonstrated output pulse intensities 14 times larger than the intensity of the input light, and intensity enhancements of 100 are thought to be possible. The pulse separation is locked to the modulation frequency which can be controlled very precisely thereby allowing the pulse train to be used as an optical clock. Furthermore, the modulation frequency can be changed easily, compared to the fixed pulse separation of mode-locked lasers.

Acknowledgements

The authors would like to thank M. M. T. Loy and J. J. Wynne for their careful readings of the manuscript. R. L. Melcher graciously loaned us the Matec Pulsed RF generator. The skilled technical assistance of R. J. Bennett was helpful for this work.

REFERENCES

1. J. R. Klauder, A. C. Price, S. Darlington, and W. J. Albersheim, The theory and design of chirp radars, Bell.Syst.Tech.J., 39:745-808 (1960).
2. F. Gires and P. Tournois, Interféromètre utilisable pour la compression d'impulsions lumineuses moduleés en frequence, Compt.Rend.Acad.Sci.(Paris), 258:6112-6115 (1964).
3. J. A. Giordmaine, M. A. Duguay, and J. W. Hansen, Compression of optical pulses, IEEE J.Quantum Electron, QE-4:252-255 (1968).
4. E. B. Treacy, Compression of picosecond light pulses, Phys.Lett., 28A:34-35 (1968).
5. M. A. Duguay and J. W. Hansen, Compression of pulses from a mode-locked He-Ne laser, Appl.Phys.Lett., 14:14-15 (1969).
6. R. A. Fisher, P. L. Kelley, and T. K. Gustafson, Subpicosecond pulse generation using the optical Kerr effect, Appl.Phys.Lett., 14:140-143 (1969).
7. A. Laubereau, External frequency modulation and compression of picosecond pulses, Phys.Lett., 29A:539-540 (1969).
8. E. B. Treacy, Optical pulse compression with diffraction gratings, IEEE J.Quantum Electron, QE-5:454-458 (1969).
9. A. Laubereau and D. von der Linde, Frequenzmodulation and Kompression ultrakurzer Lichtimpulse, Z.Naturforsch, 25A:1626-1642 (1970).
10. B. Ya. Zel'dovich and I. I. Sobel'man, Possibility of shortening light pulses in alkali-metal vapor, ZhETF Pis'ma Red, 13:182-185 (1971); JETP Lett., 13:129-131 (1971).

11. R. A. Fisher and W. Bischel, The role of linear dispersion in plane-wave self-phase modulation, Appl.Phys.Lett., 23:661-663 (1973).

12. Pulse compression for more efficient operation of solid-state laser amplifier chains, Appl.Phys.Lett., 24:468-470 (1974).

13. D. Grischkowsky, Compression of low-intensity, phase modulated light pulses, IEEE J.Quantum Electron, QE-10:723 (1974).

14. Optical pulse compression, Appl.Phys.Lett., 25:566-568 (1974).

15. J. E. Bjorkholm, E. H. Turner, and D. B. Pearson, Conversion of c.w. light into a train of subnanosecond pulses using frequency modulation and the dispersion of a near-resonant atomic vapor, Appl.Phys.Lett., 26:564-566 (1975).

16. R. H. Lehmberg and J. M. McMahon, Compression of 100 psec laser pulses, Appl.Phys.Lett., 28:204-206 (1976).

17. D. Grischkowsky, Adiabatic following and slow optical pulse propagation in rubidium vapor, Phys.Rev.A, 7:2096-2102 (1973).

18. M. M. T. Loy, A dispersive modulator, Appl.Phys.Lett., 26:99-101 (1975).

19. M. M. T. Loy, The dispersive modulator - A new concept in optical pulse compression, IEEE J.Quantum Electron, QE-13:388-392 (1977).

20. D. Grischkowsky and M. M. T. Loy, Theory of the dispersive modulator, Appl.Phys.Lett., 26:156-158 (1975).

II. RECOMPRESSION OF OPTICAL PULSES BROADENED

BY PASSAGE THROUGH OPTICAL FIBERS

Hiroki Nakatsuka and D. Grischkowsky

IBM Thomas J. Watson Research Center
Yorktown Heights
New York

Abstract

A new technique for achieving distortion-free pulse propagation through single-mode optical fibers is demonstrated. Mode-locked dye-laser pulses with 3.3 psec pulse widths and a wavelength of 5878Å were propagated through a 325-m single-mode optical fiber and emerged with 13-psec pulse widths. These output pulses were recompressed to their original 3.3-psec pulse widths by passage through a 50-cm near-resonant atomic sodium-vapor delay line.

INTRODUCTION

Two key parameters describing the performance of single-mode optical fibers are the group-velocity dispersion (GVD) and the attenuation of the optical signal as functions of wavelength. The bandwidth of data transmission is determined by GVD, whereas the practical distance between repeaters is determined by the attenuation. Distortion-free pulse propagation has been shown to occur at the wavelength (typically 1.3 µm for fused silica fibers) where GVD vanishes[1,2]. However, for many applications it is desirable to propagate distortion-free pulses at other wavelengths where GVD is strong.

Reprinted with permission from Optics Letters, 6:13 (1981).

In this Letter we introduce, for the first reported time, a new technique for achieving distortion-free pulse propagation through optical fibers at essentially any wavelength including those at which the fiber has a large GVD. As an example, we present experimental results in Figure 1 showing the distortion-free propagation of 3.3-psec (FWHM), 5878-Å optical pulses through a 325-m single-mode optical fiber, followed by a 50-cm near-resonant atomic sodium-vapor delay line[3-5]. For this combination of the optical fiber and the dispersive delay line, the large GVD of the fiber multiplied by its length is equal to and opposite that of the delay line. The key feature of this technique is that passage through the single-mode optical fiber changes the 3.3-psec input pulse into a frequency-swept (chirped) 13.0-psec output pulse. As is well known, frequency-swept optical pulses can be compressed by passage through a suitable dispersive delay line[3-11], where, roughly speaking the dispersion is such that, at the output of the delay line, the frequency components on the trailing edge of the pulse have caught up with the components at the leading edge. Passage of the chirped 13-psec pulses through the sodium-vapor dispersive delay line recompresses the pulses to their original 3.3-psec pulse width. In addition, we have reversed the order by passing the optical pulses first through the delay line and then through the optical fiber. For this reversed case, the passage through the delay line changes the 3.3-psec input pulses into chirped (the chirp is opposite that produced by the fiber) 13-psec output pulses. The subsequent passage through the optical fiber recompress the pulses to their original 3.3-psec pulse width[10,11].

We used a synchronously pumped mode-locked dye laser to perform our measurements. The pumping source is a Spectra-Physics Model 171 argon-ion laser, mode locked by a Model 342S mode locker. This source pumps a Coherent Model 490 tunable dye laser with an extended cavity. As can be seen in Figure 1(a) with this combination we obtain 3.3-psec dye-laser pulses. The average linewidth, as measured with a Fabry-Perot, interferometer is approximately 6 cm^{-1} (2.1 Å)*.

The arrangement of the experiment is indicated in the schematic diagram of Figure 2. The recompressed pulse width is measured with

---

*This measured linewidth is larger than the linewidth deduced from the output pulse width from the fiber, knowing the temporal dispersion of the fiber to be 12 psec/Å. The explanation for this discrepancy, we believe, is that the laser frequency jitters from pulse to pulse during the exposure time of the interferogram. This jitter leads to an apparently larger linewidth. However, the output of the fiber is not affected by this jitter, and the width of the output pulses provides a measure of the actual bandwidth of the individual pulses.

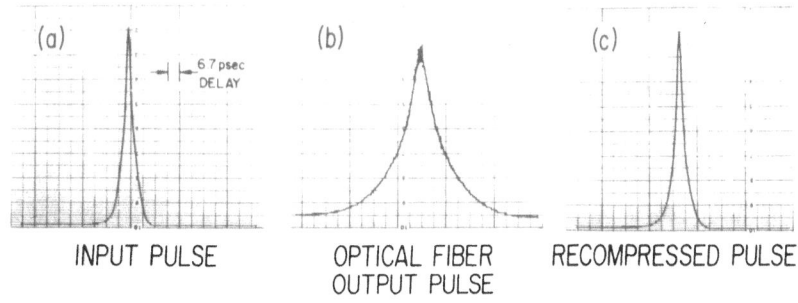

Fig. 1. Autocorrelation pulse-width measurements. Each large
         division corresponds to 1 mm of travel (6.7-psec delay) of
         the retroreflector shown in Figure 3. (a) Mode-locked
         dye-laser 3.3-psec output pulses, (b) 13.0-psec output
         pulses after passage through the 325-m single-mode optical
         fiber, and (c) recompressed 3.3-psec output pulses after
         passage through the optical fiber and the near-resonant
         atomic sodium dispersive delay line.

the autocorrelation scheme shown**. The background-free measurement
is made using noncollinear mixing in a temperature-tuned ADA crystal.
By simple changes in the optical train, we have also measured the
pulse widths of the input dye-laser pulses and the delayed output
pulses from the 325-m single-mode optical fiber. The fiber has a
core diameter of 4 μm and is manufactured in commercial quantities by
ITT as Type-T-1601. Typically the output power from our fiber is 25%
of the value of the input power. Because of the beam pattern of the
dye laser, only about one half of the input power is actually coupled
into the fiber, and absorption that is due to passage through the
fiber reduces the output power by another factor of 2. An important
experimental consideration is that the intensity in the fiber must be
kept low enough so that significant self-phase modulation[12] does
not occur. For our situation, self-phase modulation was not a problem if the average power coupled into the fiber was less that 0.1 mW.
As is shown in Figure 1(b), passage through the fiber broadens the
initial 3.3-psec pulse to 13.0 psec. The output pulses from the
fiber then propagate through the near-resonant atomic delay line, a
double-pass sodium-vapor cell with a hot zone of 25-cm length. By an

---

**It is of interest to point out that we initially tried a cross-
correlation scheme similar to that of Reference 2. For our measurement we used noncollinear mixing of relatively strong undelayed
probing pulses with the relatively weak delayed pulses. Passage
through the fiber introduced a delay of approximately 1.5 μsec. We
were forced to abandon this approach because the relative jitter
(typically ±2 psec) between the two pulse trains prevented accurate
measurement of the recompressed output pulse widths.

absorption measurement, we determined the sodium number density to be $3 \times 10^{15}/\text{cm}^3$. The heat-pipe-type cell also contained 5 Torr of He buffer gas.

As is shown in Figure 1(c), the output pulses from the delay line have been recompressed to their original 3.3-psec pulse width. The essentially perfect recompression is striking. The recompressed output pulse has no additional structure and appears to be identical with the input pulse.

In order to achieve this precise cancellation of the pulse broadening caused by the GVD of the fiber, the product of GVD and the fiber length, here designated by the term temporal dispersion, must be equal to and opposite that of the delay line, or, expressed in other words, the temporal dispersion of the combination of the fiber and the delay line must be zero for undistorted pulse propagation. In Figure 3, we show the measured delay of the fiber delay-line combination as a function of wavelength. The temporal dispersion is the slope of this curve. These measurements were made using a Tektronix sampling scope with an S-4 sampling head (25-psec rise time) and a Spectra-Physics Model 403B ultrafast photodiode. This combination gave instrument-limited pulse widths (FWHM) of 80 psec. In the figure the temporal dispersion is seen to be zero at 5878 Å, where the first derivative of the curve vanishes. This wavelength 12 Å on the short-wavelength side from the D_2 line at 5890 Å, is the same as that corresponding to the minimum recompressed pulse width, shown in Figure 1(c). The width of the minimum is roughly 4 Å, which is significantly larger that the measured 2-Å frequency jitter* of our pulses. An important point to note is that we achieve zero temporal dispersion at a wavelength where the absorption of the delay line is less than 20%***.

Comparing our total temporal delay of Figure 3 with the equivalent curve shown in References 1 and 2, one can see how our technique can introduce an additional zero in the temporal dispersion at almost any desired wavelength. This wavelength can be tuned by changing the atomic number density in the delay line and by changing the length of the delay line. With higher number densities and/or longer optical paths, the position of zero temporal dispersion will move further away from the atomic resonance and the width of the minimum will increase. For example, if, by using a multiple-pass optical delay line, we increase the effective length of our delay line by 64 times,

---

***This value of absorption can be significantly reduced by reducing the number density N and increasing the path length. This is because we are operating in the resonant collision-broadening regime, where the absorption coefficient on the wings of the line that is due to these resonant collisions is proportional to the square of N.

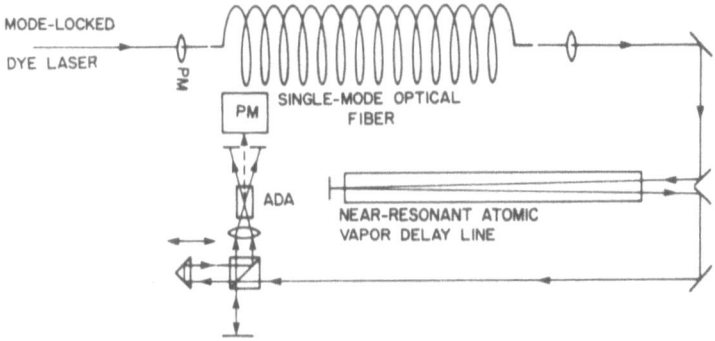

Fig. 2. Schematic diagram of the experiment.

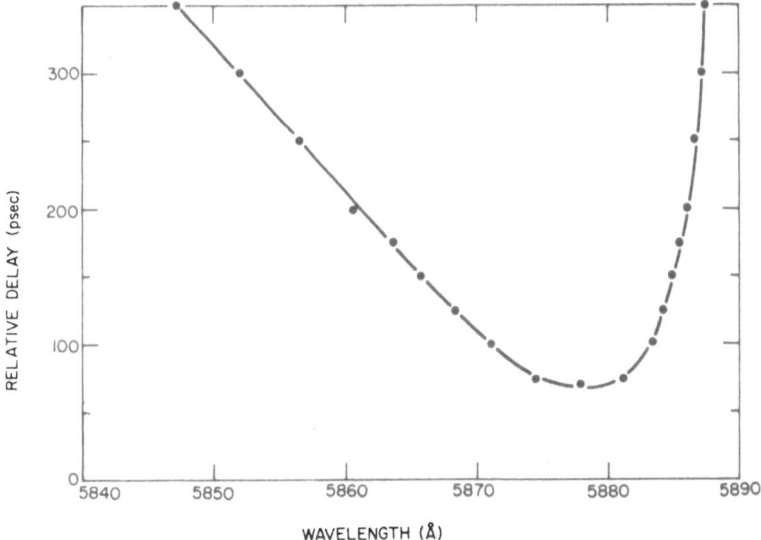

Fig. 3. Temporal delay versus wavelength of the combination of the 325-m single-mode optical fiber and the 50-cm near-resonant atomic sodium-vapor dispersive delay line.

the zero point of the temporal dispersion will shift from 12 to 48 Å from the D_2 resonance line.

A final comment is that our type of fiber has a relative minimum in the absorption coefficient at approximately 8500 Å and at this wavelength the GVD is less than one fourth of the value at 5900 Å. These features, together with the fact that atomic cesium has a strong resonance line at 8521 Å, could lead to an important practical application of our technique. The combination of our type of single-mode optical fiber and a near-resonant atomic cesium delay line with

parameters similar to those used in our experiment could propagate
mode-locked injection-semiconductor laser pulses[13] without distortion for lengths of several kilometers and with minimum loss.

Acknowledgements

We would like to acknowledge the expert technical assistance of
A. C. Balant. The informed advice concerning glass-fiber techniques
of J. S. Harper and M. Johnson was important to this work. The
careful reading of this manuscript by A. C. Balant and J. J. Wynne
was most helpful.

The research was partially supported by the U.S. Office of Naval
Research.

REFERENCES

1. A. Kawana et al., Pulse broadening in long-span single-mode
   fibers around a material-dispersion-free wave-length,
   Opt.Lett., 2:106-108 (1978).
2. D. M. Bloom et al., Direct demonstration of distortionless
   picosecond-pulse propagation in kilometer-length optical
   fibers, Opt.Lett., 4:297-299 (1979).
3. D. Grischkowsky, Optical pulse compression, Appl.Phys.Lett.,
   25:566-568 (1974).
4. J. E. Bjorkholm, E. H. Turner, and D. B. Pearson, Conversion of
   c.w. light into a train of subnanosecond pulses using
   frequency modulation and the dispersion of a near-resonant
   atomic vapor, Appl.Phys.Lett., 26:564-566 (1975).
5. J. K. Wigmore and D. Grischkowsky, Temporal compression of
   light, IEEE J.Quantum Electron, QE-14:310-315 (1978).
6. J. R. Klauder et al., The theory and design of chirp radars,
   Bell Syst.Tech.J., 39:745-808 (1960).
7. J. A. Giordamine, M. A. Duguay, and J. W. Hansen, Compression of
   optical pulses, IEEE J.Quantum Electron, QE-4:252-255 (1968).
8. E. B. Treacy, Compression of picosecond light pulses,
   Phys.Lett., 28:34-35 (1968).
9. M. A. Duguay and J. W. Hansen, Compression of pulses from a
   mode-locked He-Ne laser, Appl.Phys.Lett., 14:14-15 (1969).
10. T. Suzuki and T. Fukumoto, Use of chirp pulses to improve the
    pulse transmission characteristics in a dielectric optical
    waveguide, Electron.Commun.Jpn., 59-C:117-125 (1976).
11. J. V. Wright and B. P. Nelson, Pulse compression in optical
    fibers, Electron.Lett., 13:361-363 (1977).
12. R. H. Stolen and C. Lin, Self-phase-modulation in silica optical
    fibers, Phys.Rev., A17:1448-1453 (1978).
13. E. P. Ippen, D. J. Eilenberger, and R. W. Dixon, Picosecond
    pulse generation by passive mode-locking of diode lasers,
    Appl.Phys.Lett., 37:267-269 (1980).

III. NONLINEAR PICOSECOND-PULSE PROPAGATION THROUGH OPTICAL

FIBERS WITH POSITIVE GROUP VELOCITY DISPERSION

Hiroki Nakatsuka*, D. Grischkowsky and A. C. Balant

IBM Thomas J. Watson Research Center
Yorktown Heights, New York
*Permanent address: Department of Physics
Faculty of Science, Kyoto University, Kyoto 606, Japan

Abstract

The predictions of the nonlinear Schrodinger equation have been tested by passing 5.5-psec optical pulses through a 70-m single-mode optical fiber. With use of a precise cross correlation technique based on pulse compressions, dramatic reshaping of the input pulses into flat-topped, frequency-broadened, and positively chirped 20-psec output pulses with self-steepened fall times of less than 2 psec was observed. The observations are in good agreement with theory.

PACS numbers: 42.65.-k

The recent availability of single-mode optical fibers has opened a new era in nonlinear pulse propagation studies[1,2]. This is due mainly to (1) the small core size (typically 4 μm diam) of the fibers, which allows strong nonlinear effects to occur with modest input powers; (2) the well characterized transverse profile of the beam; and (3) the long optical paths. The nonlinear pulse problem breaks into two different aspects according to whether the group velocity dispersion (GVD) of the optical fiber is positive or negative (for typical fused silica fibers, GVD is positive for wavelengths shorter than 1.3 μm and negative for longer wavelengths). For negative GVD, the frequency sweep due to nonlinear self-phase-modulation causes pulse narrowing, while for positive GVD, pulse broadening occurs.

Reprinted by permission of Physical Review Letters, 47(13):910-913, (1981). Copyright © 1981 The American Physical Society.

There have been a number of theoretical studies directly relevant to nonlinear pulse propagation in single-mode fibers under a
variety of conditions[3-15] but there have been few experimental
studies[1,2]. The only previous nonlinear pulse-reshaping experiment
in single-mode fibers was for the case of negative GVD. This was the
recent, well characterized work of Mollenauer, Stolen, and Gordon[2]
in which the pulse narrowing and pulse splitting, characteristic of
solitons were observed.

In this letter we report the first experimental observations of
pulse reshaping and frequency modulation effects for nonlinear pulse
propagation in single-mode optical fibers with large positive GVD.
We observed that smooth 5.5-psec [full width at half-maximum (FWHM)]
input pulses with peak powers of 10 W were changed by their nonlinear
passage through a 70-m fiber into 20-psec flat-topped pulses with
self-steepened trailing edges falling off faster than 2 psec, our
resolution limit. The output pulses had a broadened frequency spectrum characteristic of linearly chirped, square pulses. We verified
that the 20-psec pulses were linearly chirped by compressing them to
1.5 psec, more than 3 times shorter than the input pulse widths. A
most important feature of our experimental technique was that we used
these 1.5-psec recompressed pulses as probing pulses to measure the
output pulse shape from the fiber. The obtained resolution of 2 psec
allowed us to compare our observations in detail with a numerical
integration of the nonlinear Schrödinger equation.

Our experimental setup is indicated in Figure 1. The synchronously pumped, mode-locked dye laser and the atomic-sodium-vapor delay
line have already been described[16]. Here, we used 5.5-psec (FWHM)
pulses with linewidths less than 2.5 cm^{-1}, close to the transform
limit. The input pulse widths to the single-mode optical fiber
(manufactured by ITT as Type T-1601 with a core diameter of 4 μm) and

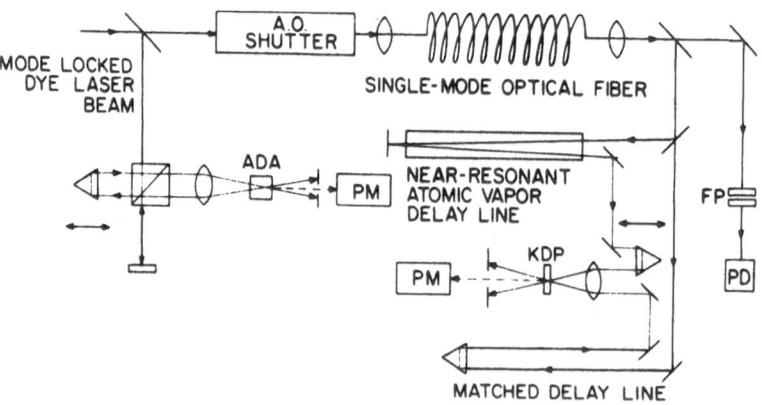

Fig. 1. Schematic diagram of the experiment.

the subsequent output pulse widths were measured simultaneously by noncollinear second-harmonic generation in a temperature-tuned ammonium dihydrogen arsenate (ADA) crystal and an angle-tuned potassium dihydrogen phosphate (KDP) crystal. The backward Rayleigh scattering from the fiber was decoupled from the laser by using an acousto-optic shutter to pass single pulses at 4 MHz. The optical arrangement for the cross-correlation measurements is shown. We also measured, by autocorrelation, the fiber output pulse widths and the recompressed pulse widths. The peak input power to the fiber was determined from the duty cycle and from the measured average output power from the fiber. To obtain the input and output spectra, we scanned the tilt angle of the Fabry-Perot interferometer while monitoring the transmitted beam.

Our observations are depicted in Figure 2 for 5874-Å, 10-W, 5.5-psec input pulses to the 70-m optical fiber. In Figure 2(a) the autocorrelated output pulse from the fiber is displayed. The triangular shape is indicative of the square pulse shape, while the sharpness of the peak illustrates that the pulse has detail on $\sim$1-psec time scale. The positive chirp of this output pulse enabled us to obtain the exceptional compression shown in Figure 2(b), by passing this pulse through the sodium-vapor recompression cell[16]. The 20-psec pulse was compressed to approximately 1.5 psec with a corresponding increase in intensity. This recompressed pulse served as an excellent probing pulse for the cross-correlation measurement of the output pulse, as shown in Figure 2(c). In addition, the jitter in the repetition rate of the laser was eliminated from the measurement because we cross correlated with the same pulse from the mode-locked train. However, we were still sensitive to the ± 0.5-cm^{-1} laser-frequency jitter because of the group velocity dispersion of the recompression cell of approximately 2 psec/cm^{-1}. Therefore we had a timing jitter in our cross correlation of $\pm$ 1 psec. The main features of the cross-correlation measurement of the output pulse are the flat top and the asymmetry between the leading and trailing edges of the pulse. The measured 2-psec drop-off of the trailing edge is our resolution limit. In Figure 2(d), the input spectrum is seen to be quite clean with a linewidth (FWHM), including frequency jitter, of 2.5 cm^{-1}. The structure of the spectral output from the fiber is characteristic of a linearly chirped, square pulse.

The reduced wave equations describing the propagation of the electric field amplitude E and the phase ϕ through the single-mode fiber are

$$\frac{\partial E}{\partial z} + k_1 \frac{\partial E}{\partial t} = - \frac{k_2}{2} \left[2 \frac{\partial \phi}{\partial t} \frac{\partial E}{\partial t} + E \frac{\partial^2 \phi}{\partial t^2} \right] , \qquad (1a)$$

$$\frac{\partial \phi}{\partial z} + k_1 \frac{\partial \phi}{\partial t} = + \frac{k_2}{2} \left[\frac{1}{E} \frac{\partial^2 E}{\partial t^2} - \left(\frac{\partial \phi}{\partial t} \right)^2 \right] - \kappa E^2, \qquad (1b)$$

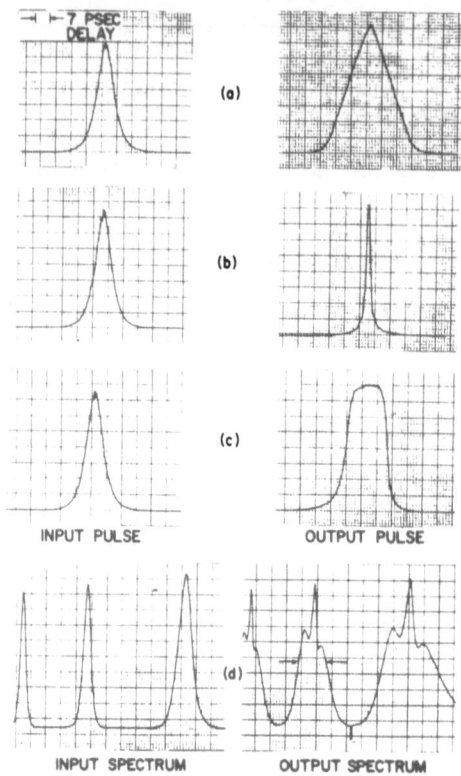

INPUT PULSE   OUTPUT PULSE

INPUT SPECTRUM   OUTPUT SPECTRUM

Fig. 2.   Input and output pulses and spectra from the 70-m single-
          mode optical fiber. For all the pulses each large division
          corresponds to a 7-psec delay. (a) Autocorrelated input and
          output pulses. (b) Autocorrelated input and recompressed
          output pulses. (c) Autocorrelated input and cross-
          correlated output pulses. (d) Input and output Fabry-Perot
          inteferograms. Spectral range is 25 cm^{-1}; finesse is 100;
          the linewidth indicated by the arrows is 10 cm^{-1}.

where the propagating electric field is given by

$$E = E \cos(k_0 z - \omega_0 t - \phi). \tag{2}$$

In Equations (1) and (2) $k_1 = \partial k/\partial \omega = 1/v_g$ is the reciprocal of the
group velocity v_g at the carrier frequency ω_0; $k_2 = \partial^2 k/\partial \omega^2$ is the
group velocity dispersion at ω_0; $\kappa = 0.5\alpha k_0 n_2/n_0$ (see Reference 16);
$k_0 = 2\pi n_0/\lambda_0$ is the propagation vector for the carrier wave with the
index of refraction n_0 and vacuum wavelength λ_0; and n_2 is the nonlinear index of refraction. With the substitution $u = E \exp(i\phi)$,
these equations are equivalent to the nonlinear Schrödinger
Equation[2-4,7].

$$i \left(\frac{\partial u}{\partial z} + k_1 \frac{\partial u}{\partial t} \right) = -\frac{1}{2} k_2 \frac{\partial^2 u}{\partial t^2} + \kappa |u|^2 u, \qquad (3)$$

which also appears in many other branches of physics, e.g. plasma physics, superconductivity, low-temperature physics, water waves, and vortex motion[3,4]. The derivation of Equation (3) is based mainly on the slowly varying envelope approximation, where the amplitude E and phase ϕ changes slowly compared with the oscillation period. Because of the high optical carrier frequency, the physical problem best described by Equation (3) is that of nonlinear optical pulse propagation in single-mode fibers.

In obtaining the numerical solutions to Equation (1) we used the following parameters. The peak input power was 10 W, and n_2 was 1.1×10^{-13} esu[1]. For convenience, α was set equal to unity*. The group velocity dispersion for our 70-m fiber was measured to be 2.5 psec/Å. The input pulse shape (5.3 psec FWHM) in Figure 3 is the square of a hyperbolic secant with the addition of the asymmetric wings shown, and agrees well with the measured autocorrelated input pulse. We assumed the pulse to be transform limited.

The results of our calculation for these input conditions are shown in Figure 3. The most important feature of the pulse reshaping are the rapid pulse broadening, the square pulse shape with self-steepened edges, and the positive chirp (not shown) of 13 cm^{-1} across the 20-psec output pulse**.

The calculation of Figure 3 is compared with the experimental results in Figure 4. The major features are well explained. The calculated peak of the autocorrelation is somewhat rounder than the measurement shown in Figure 2(a), indicating that the experiment may have a sharper falling edge than the calculation. We believe that the observed wings on the pulses are simply the wings of the input pulses which pass through the fiber undistorted until they are swept up by the expanding edges of the square pulse. Considering the nonlinear nature of this propagation problem, the agreement between theory and experiment is quite good.

---

*The constant $\alpha \sim 1$ adapts these equations to describe propagation in single-mode fibers [References 6, 7, 10, 11, and 14].
**The general trend of both the data and the calculations as functions of increasing input power is the following. For input power levels below approximately 0.1 W, the pulse is undistorted by passage through the 70-m fiber. As the power is increased above this value, the pulse broadening, chirping, and self-steepening increase monotonically up to our maximum available input power of 10 W.

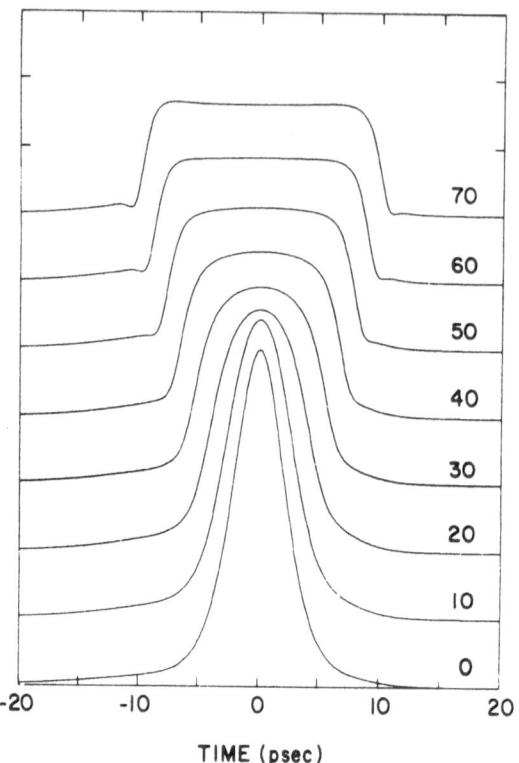

Fig. 3. Calculated pulse shapes vs distance (in meters) propagated
         in the fiber.

A small but perhaps fundamental disagreement is that experimentally the output pulse has a much slower rising edge than the
resolution-limited falling edge. We have tried many calculations
with asymmetric input pulses having different rise and fall times,
and with and without an initial frequency chirp, but we cannot calculate this feature of our data. Some possible explanation could be
the following. (1) an unexpectedly large nonlinear group velocity
effect; (2) the nonlinear index may have a small but finite relaxation time; (3) the slowly varying envelope approximation, used to
derive the nonlinear Schrödinger equation, may not have sufficient
accuracy for long-path, picosecond-pulse propagation; and (4) some
effect if the probing pulse were asymmetric.

In conclusion, our work has shown strong and well characterized
pulse-reshaping effects in single-mode optical fibers with positive
GVD. By using a precise cross-correlation scheme, we measured the
output pulse shape from the fiber with 2-psec precision and thereby
tested both the predictions and the validity of the nonlinear
Schrödinger equation.

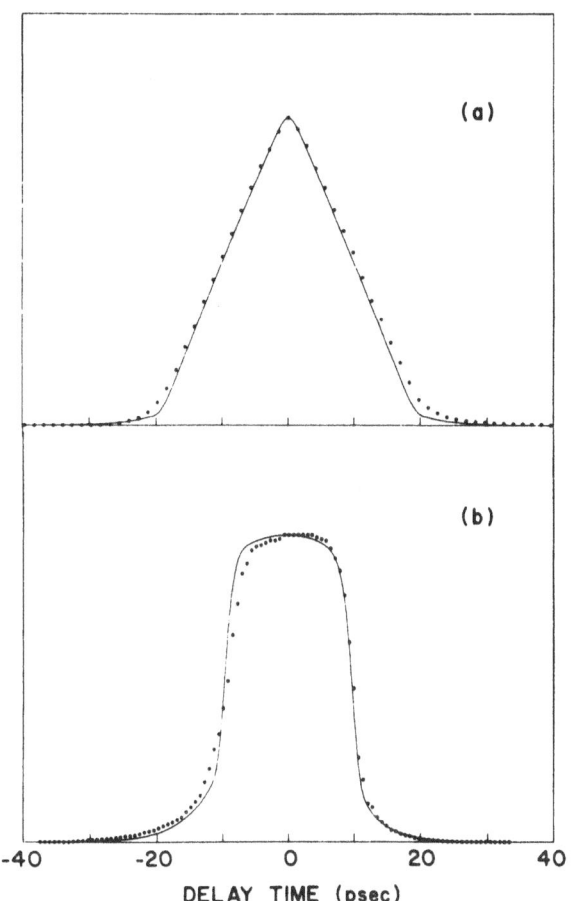

Fig. 4. Comparison between theory (solid line) and experiment (dots). (a) Autocorrelation of the output pulses. (b) Cross correlation.

Acknowledgements

We would like to acknowledge the careful readings of this manuscript and helpful comments of M. M. T. Loy, Aa. S. Sudbo, and J. J. Wynne. This work was partially supported by the U.S. Office of Naval Research.

REFERENCES

1. R. H. Stolen and C. Lin, Phys.Rev., A17:1448 (1978).
2. L. F. Mollenauer, R. H. Stolen, and J. P. Gordon, Phys.Rev.Lett., 45:1095 (1980).

3. G. B. Whitham, Linear and Nonlinear Waves, Wiley, New York (1974).

4. Solitons in Action, K. Lonngren and A. Scott, eds., Academic, New York (1978).

5. V. E. Zakarov and A. B. Shabat, Zh.Eksp.Teor.Fiz., 61:118 (1972) [Sov.Phys.JETP, 34:62 (1972)], and 64:1627 (1973) [37:823 (1973)].

6. R. H. Stolen and A. Ashkin, Appl.Phys.Lett., 22:294 (1973).

7. A. Hasegawa and F. Tappert, Appl.Phys.Lett., 23:142, 171 (1973).

8. J. Satsuma and N. Yajima, Prog.Theor.Phys.Suppl., 55:284 (1974).

9. R. A. Fisher and W. K. Bischel, J.Appl.Phys., 46:4921 (1975).

10. M. Jain and N. Tzoar, J.Appl.Phys., 49:4649 (1978), and Opt.Lett., 3:202 (1978).

11. B. Bendow, P. D. Gianino, N. Tzoar, and J. Jain, J.Opt.Soc.Am., 70:539 (1980).

12. B. Bendow and P. D. Gianino, Opt.Lett., 4:164 (1979).

13. R. H. Stolen, Proc.IEEE, 68:1232 (1980).

14. B. Crosignani, C. H. Papas, and P. DiPorto, Opt.Lett., 6:61 (1981).

15. N. Tzoar and M. Jain, Phys.Rev., A23:1266 (1981).

16. H. Nakatsuka and D. Grischkowsky, Opt.Lett., 6:13 (1981).

IV. OPTICAL PULSE COMPRESSION BASED

ON ENHANCED FREQUENCY CHIRPING

D. Grischkowsky and A. C. Balant

IBM Thomas J. Watson Research Center
Yorktown Heights
New York

Abstract

Through numerical simulations, we show that, under relatively general conditions, passage of an intense picosecond pulse through a single-mode optical fiber can cause the pulse to become strongly frequency broadened with a positive chirp (linear frequency sweep) describing essentially all of the energy of the output pulse. Also, because the optical fiber supports only a single transverse mode, the entire output beam profile has the same frequency modulation. These two features allow for unprecedented optical pulse compression.

PACS numbers: 42.65 – k, 42.80.Mv

Most optical pulse compression schemes are based on the idea originally developed for chirp radar whereby a frequency swept pulse is sent through a linearly dispersive delay line[1-4]. Since the group velocity of the light is determined by its instantaneous frequency, different portions of the pulse travel at different speeds through the delay line. If the length of the line is adjusted so that the leading edge of the pulse is delayed by just the right amount to overlap the trailing edge at the output of the delay line, the output pulse can be as short as the reciprocal of the bandwidth of the frequency sweep.

The use of this scheme requires a method for producing a linearly frequency swept (chirped) optical pulse. For picosecond

Reprinted with permission from Applied Physics Letters, 41(1):1-3 (1982). Copyright © 1982 American Institute of Physics.

optical pulses, it is quite difficult, using electro-optic techniques, to obtain the required magnitude of frequency sweep. Consequently, applications to date have used either the chirp produced by the mode-locked laser itself[2], or the chirp produced by the nonlinear process of self-phase modulation[3,4,7] (SPM) in a nonlinear optical material. For the simplest SPM case, the instantaneous frequency is proportional to the time derivative of the optical pulse shape. Thus, as illustrated in Figure 1, only the central region of the pulse has the proper chirp for subsequent pulse compression, whereas the opposite chirp on the wings of the pulse will lead to pulse expansion (temporal broadening). Another problem with SPM is that, because of the laser beam's varying spatial intensity profile, the magnitude of the chirp will vary with distance from the center of the beam. This prevents optimal compression of the entire beam profile at any given dispersive delay setting.

In this letter we present a new approach to chirping picosecond pulses, namely by propagating them through a single-mode optical fiber. During passage through the fiber, both the pulse shape and the frequency bandwidth are broadened by the combined action of self-phase modulation (with positive n_2) and positive group velocity dispersion. Under easily achieved conditions of pulse intensity and fiber length, essentially the entire output pulse can be positively, linearly chirped (enhanced frequency chirping). Moreover, because of the single-mode propagation in the fiber[9-11], the entire output beam has the same chirp i.e. the chirp is independent of transverse position on the output beam. This approach made possible the measurement technique introduced in a recent nonlinear pulse propagation experiment[11] involving optical fibers. Our purpose here is to illustrate by both a simple discussion and a detailed calculation the usefulness and importance of this method to pulse compression applications. We will show that the enhanced frequency chirp enables the output pulse to be subsequently compressed to the frequency transform limit with greatly reduced wings on the compressed pulse, while the lack of any spatial effects in the frequency modulation allows the entire pulse to compress as a spatial unit.

To estimate the operating conditions required to obtain the enhanced chirping of the entire output pulse from the fiber, we present the following simple discussion. Firstly, to be able to obtain significant pulse compression, the bandwidth of the pulse must be substantially increased. Secondly, in order for the output pulse to be described by a single linear chirp, the pulse shape must be markedly changed by the combined action of self-phase modulation and positive group velocity dispersion [GVD]. We call this combined process "dispersive self-phase modulation" (DSPM). Enhanced frequency chirping will be possible if DSPM is strong enough to produce significant frequency broadening before the power is reduced to the linear range by the well-known pulsewidth broadening action of GVD. In order to clarify these conditions, we introduce here (and will

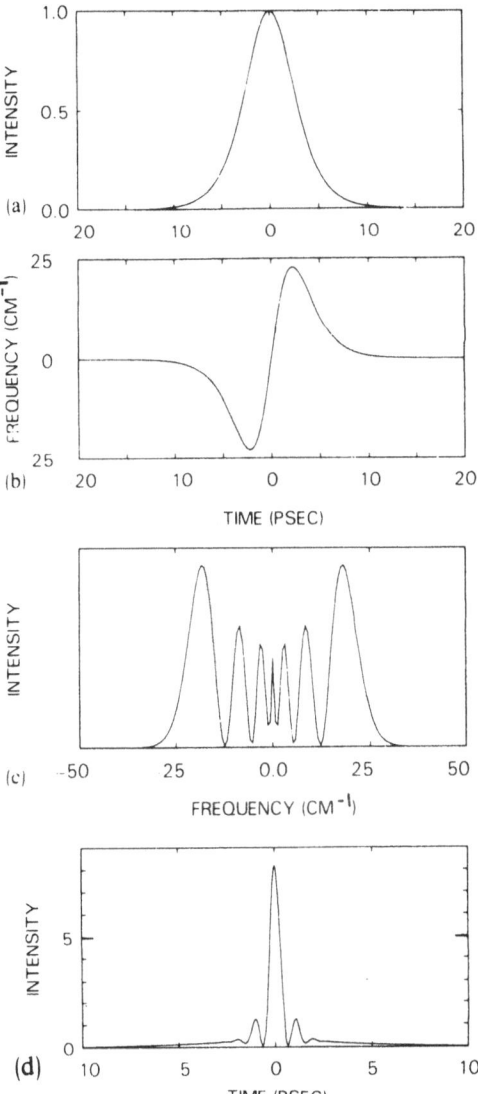

Fig. 1. (a) (Hyperbolic secant)^2 pulse shape with 6-ps pulsewidth.
(b) Self-phase modulation proportional to the derivative of
the pulse shape. Frequency in cm^{-1} (1 cm^{-1} = 30 GHz). (c)
Spectral intensity of the Fourier transform of the pulse.
(d) Calculated compressed pulse.

calculate below) two propagation lengths Z_L and Z_s, where Z_L is
the length required for the input pulsewidth to double due only to
GVD and Z_s is the length required for the pulse edges to become
arbitrarily sharp due to DSPM. For very short propagation lengths

($Z \ll Z_s \ll Z_L$), significant changes can occur in the pulse spectrum due to SPM with no changes in the pulse shape. If the path length is increased, so that $Z \sim Z_s \ll Z_L$, DSPM can produce square, self-steepened output pulses. In this case the frequency chirped portion of the pulse contains most of the pulse energy, and optimum compression can be achieved. Much longer propagation paths ($Z \gg Z$) merely broaden the pulse without significantly increasing the chirp.

From the work of Marcuse[12] and Fisher and Bishcel[6] we obtain Z_L for unmodulated pulses as

$$Z_L = \sqrt{3}\Delta t_0 / (k_2 \Delta\omega_0) , \tag{1}$$

Δt_0 is the full width at half-maximum (FWHM) pulsewidth, $\Delta\omega_0$ is the FWHM bandwidth, $k_2 = \partial^2 k/\partial\omega^2 = \partial(1/v_g)/\partial\omega$ is the fiber GVD. To calculate Z_s we use a simple picture introduced by Fisher and Bishcel[6] where SPM and GVD are considered separately. The pulse is first considered to pass undistorted through a region of strong nonlinearity where only SPM occurs. The resulting instantaneous frequency ω' is given by[13].

$$\omega' = \omega_0 - \frac{\omega_0}{c} Z \frac{n_2}{2} \frac{\partial E^2}{\partial t^2} . \tag{2}$$

Here, ω_0 the carrier frequency, n_2 is the nonlinear index of refraction, and E is the electric field amplitude. Then, the SPM pulse propagates through a strictly linear material with positive GVD, where the edges of the pulse sharpen and reach a singularity at Z_s,[5,8].

$$1/Z_s = - k_2 \frac{\partial\omega'}{\partial t} , \tag{3}$$

where the time derivative is evaluated at the input to the linear section[14].

We will now consider the example of a 100-W, 6-ps (FWHM), transform-limited (hyperbolic secant)[2] input pulse to a single-mode optical fiber with a 4-μm core diameter, a GVD of 2.5 ps/Å for a 70-m length[11] and an n_2 of 1.1×10^{-13} esu.[10,11,13]. From Equation (1), Z_L is evaluated to be Z_L = 480 m. Enhanced frequency chirping with substantial bandwidth broadening will be possible if Z_s is much less than Z_L. In order to estimate the SPM, we use Equation (2) with $Z = Z_s/2$. Then, using this SPM pulse as the input to the linear section of the fiber, we calculate Z_s from Equation (3) to be Z_s = 14 m. Clearly, this is an acceptable value. We would expect from our experience with both experiments and numerical calculations that the optimal chirp would be produced at about $2Z_s$.

At a more precise description of nonlinear pulse propagation in optical fibers we performed a numerical integration of the reduced wave equations for the input pulse of the above example. The equations describing the propagation of the electric field amplitude E and the phase ϕ through the single mode fiber are[9-11].

$$\frac{\partial E}{\partial Z} + k_1 \frac{\partial E}{\partial t} = - \frac{k_2}{2} \left(2 \frac{\partial \phi}{\partial t} \frac{\partial E}{\partial t} + E \frac{\partial^2 \phi}{\partial t^2} \right) , \tag{4a}$$

$$\frac{\partial \phi}{\partial Z} + k_1 \frac{\partial \phi}{\partial t} = + \frac{k_2}{2} \left(\frac{1}{E} \frac{\partial^2 E}{\partial t^2} - \left(\frac{\partial \phi}{\partial t} \right)^2 \right) - \kappa E^2 , \tag{4b}$$

where the propagating electric field is given by

$$E = E \cos(k_0 z - \omega_0 t - \phi) . \tag{5}$$

In Equations (4) and (5), $k_1 = \partial k/\partial \omega = 1/\nu_g$, $\kappa = 0.5 k_0 n_2/n_0$, $k_0 = 2\pi n_0/\lambda_0$ is the propagation vector for the carrier wave with the index of refaction n_0 and the vacuum wavelength $\lambda_0 = 5870\text{Å}$. The fact that the plane-wave Equation (4) have been experimentally shown to describe nonlinear propagation in a single-mode fiber[9-11] demonstrates that SPM is independent of transverse position on the profile of the propagating beam.

As the pulse propagates through the fiber the calculated reshaping proceeds smoothly from the transform-limited input pulse shown in Figure 1(a) to the characteristic square pulse shape shown in Figure 2(a), which occurs at 30 m. Simultaneously, the developing chirp grows in magnitude and describes more and more of the total pulse extent. For this example, the chirp shown in Figure 2(b) describes more than 95% of the total energy, compared to only 59% for the self-phase modulated pulse shown in Figure 1. In agreement with our simple discussion, enhanced frequency chirping occurs for the propagation distance $Z \sim 2Z_s$, where for this example $Z = 30$ m and $Z_s = 14$ m. Our numerical calculations also show that enhanced chirping is obtained for fiber lengths greater than Z_s. This demonstrates the practical feasibility of the method. However, for distances longer than $2Z_s$ the pulse continues to broaden but the magnitude of the chirp does not increase significantly and little is gained by using longer fibers. Thus $Z \sim 2Z_s$ is a reasonable estimate for the optimal fiber length.

These relationships were initially demonstrated experimentally in Reference 11, where enhanced chirping was obtained for 10-W, 5.5ps, (hyperbolic secant)^2 pulses propagated through a 70-m fiber for which Z_s is calculated to be 40 m. These pulses were subsequently compressed to 1.5 ps by passage through a sodium vapor cell. More recently, Shank et al.[15] used this technique to produce the shortest optical pulses (30 fs) to date. For their experiment 7-kW,

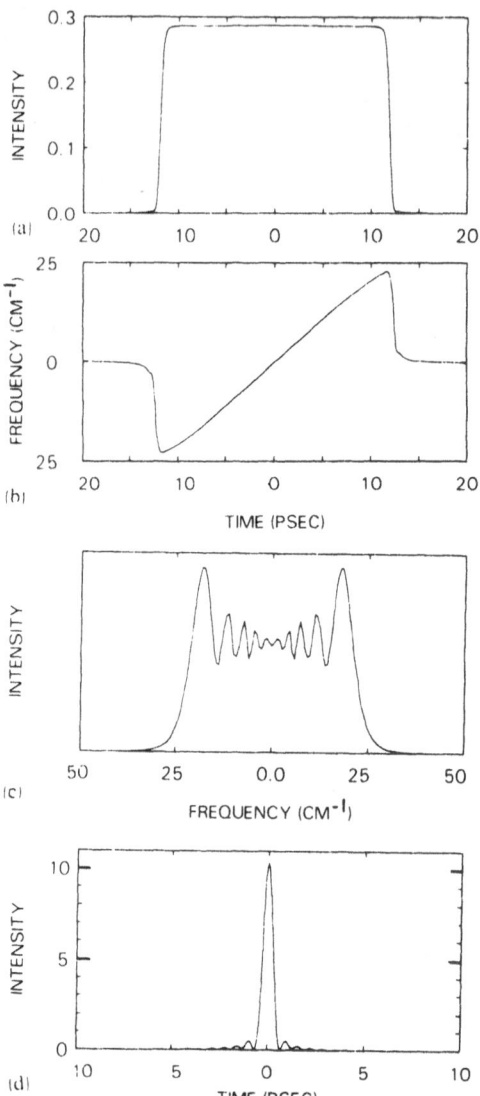

Fig. 2.   (a) Calculated DSPM output pulse from 30 m of optical fiber.
           (b) Frequency modulation of the output pulse.   (c) Spectral
           intensity of the Fourier transform of the output pulse.   (d)
           Calculated compressed pulse.

90-fs (hyperbolic secant)[2] pulses were propagated through a 15-cm
fiber for which Z_s is calculated to be 2.3 cm.   These pulses were
subsequently compressed to 30 fs by passage through a grating pair.
In agreement with their calculation, good compression would also have
been obtained with $Z \sim 2Z$.

The pulse compression obtainable with the fiber output pulse (DSPM pulse) of Figure 2(a) will now be compared to that for the pure SPM case of Figure 1. For this comparison the (hyperbolic secant)² input pulses and the magnitude of the frequency modulation Figures 1(b) and 2(b) are the same. These pulses are Fourier analyzed (Figures 1(c) and 2(c) and considered to pass through dispersive delay lines with group velocity dispersion opposite to that of the optical fiber. At the output of the delay lines the Fourier components are summed to give the compressed pulses. The lengths of the delay lines are adjusted to obtain the shortest pulses shown in Figures 1(d) and 2(d). For both cases the compressed pulsewidths are approximately 0.6 ps compared with the input pulsewidths of 6 ps, and the intensities of the compressed pulses have correspondingly increased by about 10 times. The intensity of the compressed DSPM pulse is higher than the SPM pulse because the central peak of the compressed DSPM pulse contains more the total energy. Furthermore,

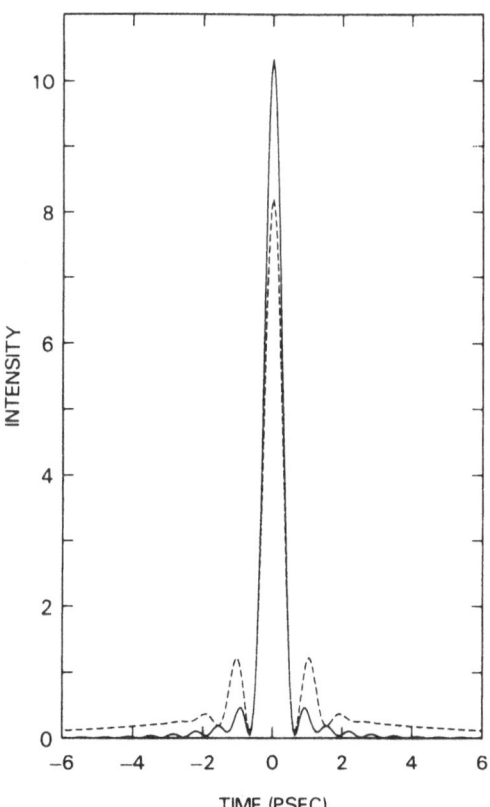

Fig. 3. Comparison of the compressed pulses of Figures 1(d) and 2(d). Solid line is the DSPM pulse; dashed line is the SPM pulse.

the compressed DSPM pulse has much less extensive wings than the SPM pulse, as illustrated by the expanded scale of Figure 3. The relative energies is the wings of the two compressed pulses, as measured from the null closest to the central peaks, are 42% for the SPM pulses and only 10.6% for the DSPM pulse. The ratio of the intensity of the main pulse to the secondary maxima pulse lobes is 22 for the DSPM pulse and 7 for the SPM pulse. For the extended wings measured from ±2 ps the relative energies are 25% for the SPM pulse compared to only 3% for the DSPM pulse. The significantly reduced wings resulting from enhanced frequency chirping should prove important to the applications of these compressed DSPM pulses.

Acknowledgement

We would like to acknowledge helpful comments by J. J. Wynne concerning this manuscript. This work was partially supported by the U.S. Office of Naval Research.

REFERENCES

1. J. R. Klauder, A. C. Price, S. Darlington, and W. J. Albersheim, Bell.Syst.Tech.J., 39:745 (1960).
2. E. B. Treacy, Phys.Lett., 28A:34 (1968).
3. R. A. Fisher, P. L. Kelley, and T. K. Gustafson, Appl.Phys.Lett., 14:140 (1969).
4. A. Laubereau, Phys.Lett., 29A:539 (1969).
5. D. Grischkowsky, Appl.Phys.Lett., 25:566 (1974).
6. R. A. Fisher and W. K. Bischel, J.Appl.Phys., 46:4921 (1975).
7. R. H. Lehmberg and J. M. McMahon, Appl.Phys.Lett., 28:204 (1976); R. H. Lehmberg, J. Reintjes, and R. C. Eckardt, Opt.Commun., 22:95 (1977).
8. J. K. Wigmore and D. Grischkowsky, IEEE J.Quantum Electron, QE-14:310 (1978). This paper contains an extensive listing of the pulse compression literature.
9. A. Hasegawa and F. Tappert, Appl.Phys.Lett., 23:142 (1973).
10. L. F. Mollenauer, R. H. Stolen, and J. P. Gordon, Phys.Rev.Lett., 45:1095 (1980).
11. H. Nakatsuka, D. Grischkowsky, and A. C. Balant, Phys.Rev.Lett., 47:910 (1981).
12. D. Marcuse, Appl.Opt., 19:1653 (1980).
13. R. H. Stolen and C. Lin, Phys.Rev., A17:1448 (1978).
14. For a Gaussian pulse our $\overline{Z}_s \sim Z_{shock}/3$ of Reference 6.
15. C. V. Shank, R. L. Fork, R. Yen, R. H. Stolen, and W. J. Tomlinson, Appl.Phys.Lett., 40:761 (1982).

V. PULSE COMPRESSION USING OPTICAL FIBERS

B. Nikolaus* and D. Grischkowsky

IBM Thomas J. Watson Research Center
Yorktown Heights, New York
*Institut fur Angewandte Physik, Universitat Heidelberg
Heidelberg Federal Republic of Germany

Abstract

We report a factor of 12 compression of the 5.4-ps, 1-kW pulse
from a mode-locked dye laser. The pulses were frequency broadened
and linearly chirped by the combined action of self-phase modulation
and group velocity dispersion during passage through a 30-m single-
mode optical fiber. The fiber output pulses were then compressed to
450-fs, 3-kW pulses by passage through a diffraction grating based
dispersive delay line. These short pulses were tunable over the
300-Å range of the laser dye.

PACS numbers: 42.65 - k, 42.80.Mv

A powerful method of optical pulse compression utilizing nonlinear propagation in single-mode fibers has recently been introduced[1-4]. The propagation in the fiber broadens and chirps the
pulse due to the combined action of group velocity dispersion (GVD)
and self-phase modulation (SPM)[5]. The pulse can then be compressed
to the Fourier limit by passage through a suitable dispersive delay
line.

This technique solves the three main problems which have hin-
dered the practical use of optical pulse compression. Firstly, the
single-mode propagation eliminates the transverse instability of
self-focusing, which otherwise causes large shot to shot variations

in the self-phase modulation and the output beam smoothness and divergence. Secondly, again due to the single-mode propagation, the entire output beam profile has the same frequency modulation. This feature allows the output pulse to be compressed as a single spatial unit. Thirdly, the combination of strong SPM with GVD in the fiber causes the linear frequency sweep of the output pulse to describe essentially all of the energy (enhanced frequency chirping), thereby enabling subsequent pulse compression with much reduced wings[4]. In addition, the fiber method makes pulse compression feasible at modest powers due to the high intensity inside the approximately 4-μm-diameter fiber core and the long optical paths.

In the letter reported here, we have demonstrated the above features by compressing, by a factor of 12, the 5.4 ps [full width at half-maximum (FWHM)] frequency-tunable pulses from a commercial synchronously pumped, mode-locked, cavity-dumped dye laser. The measured peak power of the tunable 450-fs compressed pulses was about three times higher than the peak output power of the mode-locked laser. The recompressed pulses had the predicted[4] exceptionally clean pulse shapes with much reduced wings.

We will now describe the laser system, the measurement tech- niques used, and the optical pulse compressor. The pumping source was a Spectra-Physics model 171 argon-ion laser, mode locked by a model 342S mode locker. This source pumped a Coherent model 490 tunable dye laser (rhodamine 6G) with an extended cavity including a Spectra-Physics model 3448 cavity dumper. With this combination we obtained 5.4-ps (FWHM), 1-kW dye-laser pulses. The linewidth was typically 2.7 cm^{-1} indicating that the pulses were close to being tranform limited. In addition to increasing the peak output power, the cavity dumper decoupled the laser from the backward Rayleigh scattering from the fiber. The pulse widths were measured by the standard technique of noncollinear (background-free) second-harmonic generation in an angle tuned 1-mm-thick potassium dihydrogen phosphate (KDP) crystal.

The optical pulse compressor, consisting of the fiber and a dispersive delay line, is shown schematically in Figure 1. The laser pulses were coupled into a 30-m-long single-mode optical fiber (ITT type T-1601), with a core diameter of 4 μm. By careful focusing with a 10X microscope objective lens, we coupled more than 60% of the incident light into the fiber. The absorption in traversing the 30-m fiber was only about 6%, while the reflection from the end face was 4%. The output light from the fiber was recollimated by another 10 x objective lens. This 3-mm-diam beam was then incident at 60° onto the dispersive delay line consisting of a Zeiss holographic grating with 2400 lines/mm and a right angle prism[6]. This combination is equivalent to the more usual grating pair[7], but for a fixed dispersion requires only 1/2 the separation of the grating pair. By monitoring the compressed pulse, we adjusted the laser

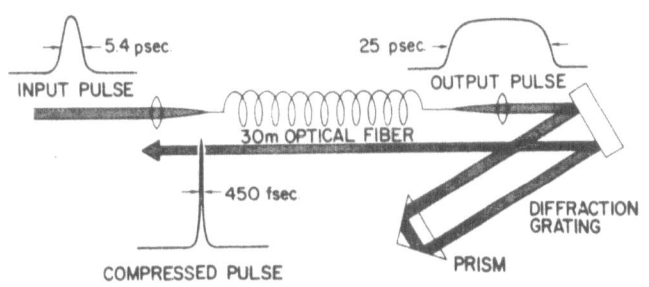

Fig. 1. Optical pulse compressor.

system and the separation between the grating and the prism to obtain
the minimum pulse width. Typically, the separation was 62 cm and was
set to a precision of ±5 mm. The diffraction efficiency of the
grating is 80% for the optical polarization. Our measured throughput
for the optical compressor was 25%.

Our results are illustrated in Figure 2, where the time scale is
the same for both pulses. A comparison of the input and compressed
pulses illustrates our excellent compression ratio of 12. Assuming a
hyperbolic secant squared pulse shape for the laser pulse (the input
to the optical pulse compressor), the pulse width obtained from the
autocorrelation measurement shown in Figure 2(a) is 5.4 ps. The peak
power of 1 kW is determined from a measurement of the average power
of 22 mW and the pulse repetition rate of 4 MHz. The output from the
30-m fiber was typically 12 mW indicating that more than 1/2 of the
incident power was coupled into the fiber. As discussed previously[4], the combined effects of self-phase modulation and group
velocity dispersion caused the output pulse from the fiber to be
broadened in both time and frequency. Typically, the pulse broadened
to 25 ps and the bandwidth increased 37 cm^{-1}. Most importantly,
passage through the fiber resulted in a linear frequency sweep which
described essentially all of the output pulse. Because of this frequency sweep, the fiber output pulse was easily compressed to the
pulse shown in Figure 2(b) by passage through the dispersive delay
line illustrated in Figure 1. Because of its sharpness and small
wings, the clean, compressed pulse shape is ideal for measurement
application. The shape is assumed to be similar to a hyperbolic-
secant squared for which this autocorrelation measurement gives a
450-fs pulse width (FWHM). Knowing this pulse width, the repetition
rate, and the average output power of 5 mW from the optical pulse
compressor, we obtain a peak power of about 3 kW for the compressed
pulses.

An important characteristic of the optical pulse compressor is
its broadband operation. We have tuned the dye laser wavelength from
5700 to 6000 Å and with only slight angular adjustments to the dis-

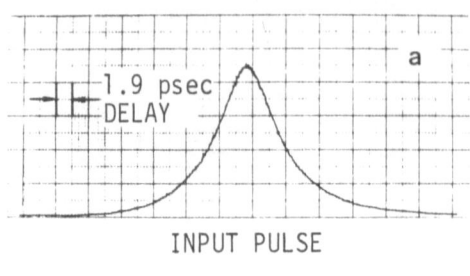

INPUT PULSE

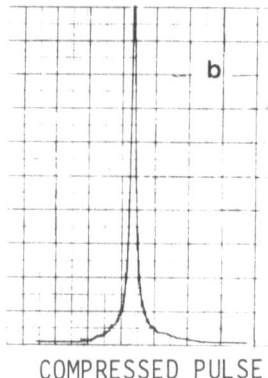

COMPRESSED PULSE

Fig. 2.   (a) Input pulse to the optical pulse compressor. (b)
Compressed pulse from the optical pulse compressor. For
both pulses each large division corresponds to the indicated
delay time of 1.9 ps.

persive delay line have maintained pulse widths of less than 500 fs
throughout this range.

In summary, we have experimentally demonstrated that this pulse
compression method can be readily adapted to the standard synchronously pumped, cavity-dumped, mode-locked dye laser systems to reduce
the pulse widths by 12 times, while simultaneously increasing the
peak powers by about 3 times. Since the tunability is maintained,
this technique allows the study of ultrashort processes in the subpicosecond domain, where frequency scanning is required.

Acknowledgements

We would like to acknowledge the many helpful discussions with
A. C. Balant and the careful readings of the manuscript by A. C.
Balant and J. J. Wynne. This research was partially supported by the
U.S. Office of Naval Research.

REFERENCES

1. H. Nakatsuka and D. Grischkowsky, Opt.Lett., 6:13 (1981).
2. H. Nakatsuka, D. Grischkowsky, and A. C. Balant, Phys.Rev.Lett.,
   47:910 (1981).
3. C. V. Shank, R. L. Fork, R. Yen, R. H. Stolen, and W. J.
   Tomlinson, Appl.Phys.Lett., 40:761 (1982).
4. D. Grischkowsky and A. C. Balant, Appl.Phys.Lett., 41:1 (1982).
5. R. R. Alfano, L. L. Hope, and S. L. Shapiro, Phys.Rev., A6:433
   (1972).
6. A. Laubereau and D. von der Linde, Z.Naturforsch., 24A:1626
   (1970).
7. E. B. Treacy, IEEE J.Quantum Electron, QE5:454 (1969).

SPECTRAL CHARACTERISTICS OF SEMICONDUCTOR DIODE LASERS*

A. Mooradian

Lincoln Laboratory, Massachusetts Institute of Technology
Lexington, Massachusetts 02173-0073

INTRODUCTION

The spectral linewidth and phase noise characteristics of single-frequency semiconductor diode lasers have become quite important for application to such areas as frequency standards, fiber optical sensors, laser gyros, and optical communications. This work describes the spectral broadening mechanisms which have been observed for monolithic (GaAl)As and lead-salt diode lasers as well as the spectral characteristics of external-cavity controlled (GaAl)As diode lasers.

LINEWIDTH BROADENING MECHANISMS

Power Dependent Broadening

Many of the fundamental broadening mechanisms for a semiconductor laser can be conveniently described with the help of Fig. 1. The laser field amplitude is represented in the complex plane by

$$E = CI^{1/2} \exp(i\phi) \quad , \tag{1}$$

where I and ϕ are the time dependent intensity and phase of the laser field, and C is a dimensional constant. When a spontaneous emission photon is radiated into the mode of the laser field, it adds as a small vector with a random phase θ_i to produce a new field intensity $I + \Delta I$ with a phase change of $\Delta\phi$. Because these spontaneous emission events are random, the phase ϕ executes Brownian motion and has a Gaussian probability distribution. Lax[1] has performed a very extensive analysis of noise in laser oscillators and has shown that the

*This work was sponsored by the Department of the Air Force.

autocorrelation function of a laser field can be written approximately as

$$\langle E(t)*E(0)\rangle \approx E(0)^2\exp(-\langle\Delta\phi^2\rangle/2)\exp(i\omega t) \qquad (2)$$

where

$$\Delta\phi = \phi(t)-\phi(0) \qquad (3)$$

and ω is the laser angular frequency. Equation (2) has neglected amplitude flucuations. The power spectrum is then obtained by taking the Fourier transform of the autocorrelation function. The laser linewidth originally described by the Schawlow-Townes theory[2] in which only the quantum phase fluctuations were considered produced an expression for the linewidth given by

$$2\Gamma = (\pi h\nu\Gamma_c^2/P_c)n_{sp} \qquad , \qquad (4)$$

where 2Γ is the full width at half-maximum of the laser line at frequency ν, P_c is the intracavity power emitted by the stimulated emission, and Γ_c is the passive resonator linewidth given by

$$\Gamma_c = c(\beta L-\ell n R)/2\pi n L \qquad , \qquad (5)$$

where n is an effective refractive index including dispersion, β is the mode loss coefficient, L is the cavity length, n_{sp} is a population factor described below, and R is the facet reflectivity. Equation (4) applies to the case where the driving noise source, spontaneous emission, is homogeneous and band limited. The factor n_{sp} in Eq. (4) is the ratio of spontaneous emission rate per mode to the stimulated emission rate per laser photon and for a semiconductor laser is given by

$$n_{sp} = \frac{f_c(E_2)[1-f_v(E_1)]}{f_c(E_2)-f_v(E_1)} \qquad , \qquad (6)$$

where $f_c(E_2)$ and $f_v(E_1)$ are the occupancy factors for the upper and lower laser levels, respectively. Equation (6) becomes for Fermi or Maxwellian statistics

$$n_{sp} = \{1-\exp[(h\nu+E_{FV}-E_{FC})/kT]\}^{-1} \qquad , \qquad (7)$$

where E_{FC} and E_{FV} are the conduction-band and valence-band quasi-Fermi levels, k is Bolzmann's constant, and T is the temperature. In most lasers n_{sp} is nearly unity because for four-level devices the terminal state is usually empty. For semiconductor lasers with non-degenerate carrier distribution functions n_{sp} is greater than unity at room temperature. For (GaAl)As diode lasers, $(h\nu+E_{FV}-E_{FC}) \approx$ 15 meV at room temperature[3] and n_{sp} is about 2.3, approaching unity at 77 K. This spontaneous emission factor can be physically understood by noting that a finite terminal state population can cause absorption of laser photons which in turn will produce more spontaneous photons to be radiated into the laser mode with random phase.

The Schawlow-Townes description of laser linewidth has been shown to be inadequate to describe the results of measurements for (GaAl)As diode lasers.[4] An additional broadening mechanism which was considered but neglected as sizable effect by Lax[1,5] and Haug and Haken[6] has been reexamined recently by Henry,[7] and later by Vahala and Yariv.[8] This additional broadening mechanism can be understood by again referring to Fig. 1. In addition to the usual broadening caused by phase fluctuations arising from spontaneous emission events, there is an additional contribution to the line broadening which comes from a phase change associated with the laser field intensity change induced by spontaneous emission. The carrier density and hence gain will fluctuate to restore the laser field amplitude to the steady-state value. The change in the gain or imaginary part of the refractive index arising from this carrier density change is accompanied by a change in the real part of the refractive index. The time it takes to restore equilibrium is the relaxation time of about one nanosecond, and this is long enough for an observable broadening of the laser linewidth to occur. This additional broadening mechanism modifies the original Schawlow-Townes linewidth to give[7,8]

$$2\Gamma = (\pi h \nu \Gamma_c^2 / P_c) n_{sp} (1+\alpha^2) \quad , \tag{8}$$

where α is the ratio of change in the real part of refractive index to the change in the imaginary part of refractive index due to spontaneous emission events over the same period of time. The real part of refractive index change can be related to laser mode shifts with change in carrier density using

$$\Delta n_r = (n/\lambda)\Delta\lambda \quad , \tag{9}$$

while the imaginary part of the refractive index change can be related to gain changes as a function of carrier density using

$$\Delta n_i = -(c/2\omega)\Delta g \quad . \tag{10}$$

All of the above contributions to laser linewidth can be seen to be inversely dependent upon laser power from Fig. 1 since the amplitude of the spontaneous emission photons remains fixed as the laser intensity grows. This additional line broadening mechanism is similar to the problem of a detuned gas laser treated by Lax[1,5] where the cavity resonance and the optical transition frequencies do not coincide. Haug and Haken[6] also discussed a $1+\alpha^2$ contribution to the laser linewidth but like Lax did not consider it as significant.

The linewidth problem has been discussed in various ways includ-

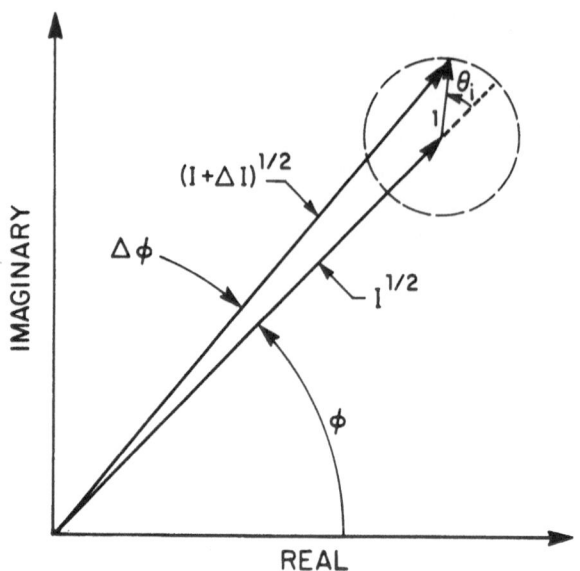

Fig. 1. Representation of the laser field in the complex plane with the proportionality constant in Eq. (1) set equal to one.

ing the Van der Pol oscillator model[1,9,10] in which the frequency width of an LC triode oscillator was expressed as a function of the circuit parameters. This treatment can be extended[8] to account for the additional broadening mechanisms present in (GaAl)As type lasers. The differential equation describing the response of a noise driven oscillator is given by[10]

$$\frac{\partial^2 E}{\partial t^2} + \frac{1}{\tau_c} \frac{\partial E}{\partial t} + \omega^2 E + \frac{1}{n_0^2} \frac{\partial^2 P}{\partial t^2} = N(t) \quad , \tag{11}$$

where $N(t)$ is driving noise term, ω is laser frequency, τ_c is the cavity lifetime, and n_0 is the refractive index which does not include contributions from interband transitions. The induced polarization, P, is given by

$$P = [\chi^{(1)} + \chi^{(3)} E^2] E \quad , \tag{12}$$

where

$$\chi^{(3)} = \chi_r^{(3)} + i \chi_i^{(3)} \quad . \tag{13}$$

The real part of $\chi^{(3)}$ is due to index saturation while the imaginary part of $\chi^{(3)}$ is due to gain saturation. The results of this approach are essentially the same as those of Henry.

Phase Noise Sidebands

The spectral lineshape has, until recently, been considered to be Lorentzian. Osterwalder and Rickett[11] first reported the observation of sidebands associated with each longitudinal mode of a multi-mode (GaAl)As diode laser with an intensity of a few percent of the central peak and a frequency shift of 1-2 GHz. Recently, Daino et al.[12] and later Vahala et al.[13] have made more careful studies of this effect on single-frequency devices and explained the occurrence of these sidebands as due to a peak in the phase noise spectrum at the relaxation frequency. Henry[14] has presented a comprehensive theory of these sidebands which is derived from his initial treatment of the enhanced laser linewidth due to intensity fluctuations.

By referring to Fig. 1 it is easy to see that the perturbed laser field, $I+\Delta I$, will return to its average value I after undergoing damped relaxation oscillations. A delayed phase change will occur during these damped relaxation oscillations. The modulation of both the intensity and phase will therefore produce sidebands at the relaxation frequency. These sidebands have been observed[13] to increase in frequency and decrease in amplitude as the power is increased.

Power Independent Broadening

An additional line broadening has been observed[15,16] which has not been included in previous treatments and is discussed below. This is a power-independent contribution to the linewidth which has been described phenomenologically as due to refractive index fluctuaations resulting from the statititical fluctuations in the number of conduction electrons in the small active gain volumes of (GaAl)As diode lasers. Typical gain volumes for such lasers are 200 x 2 x 0.2 μm.

The frequency of laser oscillation can be written as

$$\nu = (\nu_c \Gamma_g + \nu_g \Gamma_c)/(\Gamma_c + \Gamma_g) \quad , \tag{14}$$

where ν_c is the cavity mode frequency, ν_g is the gain peak frequency, and Γ_c and Γ_g are the spectral half-widths of the passive cavity mode and the gain spectrum, respectively. The cavity mode frequency fluctuations are related to changes in the refractive index via the phenomenological relation

$$\delta\nu_c = (\nu_c/n)(dn/dN)_{N_t \nu} \, \delta N \quad , \tag{15}$$

where δN is the mean square fluctuation in the number N of the electrons in the gain volume and dn/dN is evaluated at the laser frequency and at N equal to the number of electrons at laser threshold, N_t. For most cases of interest, the mean square fluctuation in electron number is assumed to be

$$\delta N = \sqrt{N_t} \quad . \tag{16}$$

The resulting expression for the laser frequency fluctuations is given by

$$\delta \nu = (\nu_c M \sqrt{N_t}/n)(dn/dN)_{N_t} \nu \, [\Gamma_g/(\Gamma_g + \Gamma_c)] \quad , \tag{17}$$

where M is a mode confinement factor which determines the fraction of the optical cavity frequency affected by index changes and is given by the ratio of cross-sectional optical mode area to carrier confinement area. The presence of the laser field will strongly damp the gain at high power levels and no carrier density fluctuations at frequencies less than the damped relaxation rate should be expected.

Fluctuations in the peak of the gain spectrum center frequency with electron number can also contribute to the linewidth when the carrier distribution becomes degenerate at low temperatures. Assuming that the terminal states for the lasing process are empty, the worst case is given by

$$\delta \nu = (2RE_{FC}/3 \sqrt{N}) \, [\Gamma_c/(\Gamma_c + \Gamma_g)] \quad , \tag{18}$$

where R is a factor which accounts for the effects of band gap renormalization and is about 0.6 for (GaAl)As devices. The frequency change associated with this effect is of the same sign as the effects of Eq. (17).

An alternative model of the power-independent broadening has been presented by Vahala and Yariv[17] involving electron occupation fluctuations. This model does not require a change in electron number but depends upon the thermal fluctuations of electrons of order kT around the quasi-Fermi surface to produce a refractive index fluctuation at the laser frequency. The expression for the power-independent linewidth can be calculated using the formulation of Eq. (2) to get

$$2\Gamma = \frac{kT}{V} \left(\frac{\pi \omega A M}{n_0^2} \right)^2 \int_{-\infty}^{\infty} D(\Omega) \left(\frac{\omega - \Omega}{(\omega - \Omega)^2 + 1/T_2^2} \right)^2$$

$$\times \, [T_c(\partial f_c/\partial E_{FC}) + T_v(\partial f_v/\partial E_{FV})] d\Omega \quad , \tag{19}$$

where V is the gain volume, A is proportional to the matrix element, M is the mode confinement factor, $D(\Omega)$ is the reduced density of states, T_2 is the dephasing time, T_c and T_v are the intraband relaxation times, f_c and f_v are the conduction and valence band occupation probabilities, and T is the absolute temperature. Many of the parameters in this model are not determined to a high degree of accuracy including the use of the Kane band structure matrix element. The intraband scattering times T_c and T_v are determined from mobility measurements[18] on different device structures and may not be the exact values operative in the semiconductor lasers. In any event, T_c and T_v would be expected to increase only slowly below 80 K resulting in Eq. (19) approaching zero as the temperature went to zero. This is in contrast with the experimentally measured[19] power independent broadening for TJS and CSP lasers at liquid helium temperature described below.

EXPERIMENTAL RESULTS AND COMPARISON WITH THEORY

Power-Dependent Broadening

The power-dependent linewidth of various types of (GaAl)As semiconductor lasers have been measured experimentally and compared with theory.[20] Experiments were carried out as a function of power and temperature on single-frequency channel-substrate planar (CSP) Hitachi and transverse-junction stripe (TJS) Mitsubishi diode lasers which operated continuously at room temperature. All devices which were measured operated with their output in a single frequency with "kink-free" power versus current at all temperatures. An example of the power versus current curves are shown in Fig. 2 for a Mitsubishi TJS device. In addition, measurements were made with the laser mode frequency tuned near the center of its tuning range to prevent mode hopping instabilities which would have a tendency to broaden the output line.

The devices were thermally isolated in a Dewar to reduce temperature instabilities to an insignificant level and to operate at reduced temperatures. Precautions were taken to prevent optical feedback by optically isolating the diodes. Injection current noise was also minimized by using a shielded lead-acid battery together with input filters. Later experiments were carried out with a special temperature controller and stable electronic power supply to eliminate long term drift effects.

Initial lineshape measurements were made using an external cavity, single-frequency (GaAl)As diode laser of < 15 kHz linewidth(21) heterodyned with the monolithic diode lasers. The same external cavity device was also used to determine the resolution of a scanning Fabry-Perot interferometer used for most other measurements. Frequency calibration of these instruments were also made by producing modulation sidebands on the monolithic devices with a precision frequency oscillator. The measurements produced similar results for all the devices studied, independent of the fabrication technique.

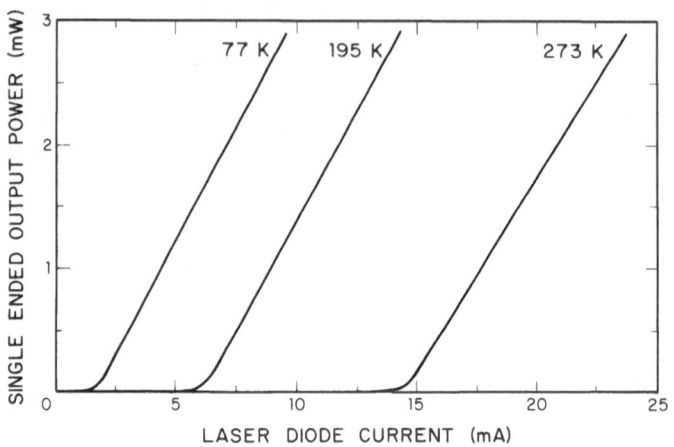

Fig. 2. Single-ended output power versus current for a Mitsubishi TJS diode laser at 273, 195, and 77 K.

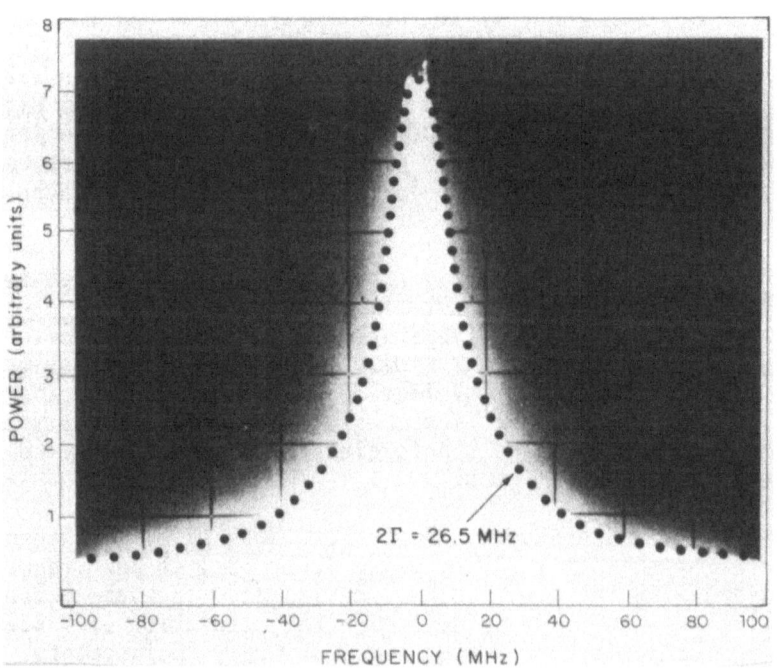

Fig. 3. Heterodyne beat spectrum between an external cavity (GaAl)As cw diode laser and a monolithic Hitachi CSP diode laser at 20 C. Fit is to a Lorentzian.

Figure 3 shows the spectrum of a monolithic diode laser heterodyned with a stable external cavity laser. The lineshape of the monolithic device was determined to be approximately Lorentzian and the linewidth was equal to that measured by the Fabry-Perot interferometer.

The laser linewidth as a function of reciprocal output power is shown in Figs. 4a, b and c at temperatures of 273, 195 and 77 K, respectively for a Mitsubishi TJS device. The linewidth is seen to decrease linearly with reciprocal output power as predicted by Eq. (4). In addition, there is seen to be a finite intercept which increases as the temperature decreases. This power independent contribution to the linewidth is attributed to electron number fluctuations described by Eq. (17). The linewidth versus reciprocal power for a Hitachi CSP laser is shown in Fig. 5.

In order to compare experiment with theory, the linewidth must be expressed in terms of the experimental parameters to give

$$2\Gamma = (h\nu/8\pi P_0)(c/nL)^2(\ell nR-\beta L)(\ell nR)n_{sp}(1+\alpha^2) \qquad (20)$$

where P_0 is the single-ended output power. The loss coefficient β was determined from the slope efficiency for each device measured. The parameter α was determined from measurements of $\Delta\lambda$ and Δg with current. Because of the approximate linearity of $\Delta\lambda$ and Δg with electron density for the experimental condition of a near-degenerate conduction band and a highly degenerate valence band, as is the case for the devices discussed here, changes of laser mode wavelength and gain integrated over a change in carrier density from zero injection current to threshold should give a reasonable estimate for α, as only the ratio of these parameters enters. The spectral shift of the Fabry-Perot laser mode center frequency as a function of injection current up to threshold was used to estimate $\Delta\lambda$. The thermal contribution to the spectral shift is approximately an order of magnitude smaller than that due to carrier density variation, in contrast to lead-salt lasers, and can be neglected in the estimation of $\Delta n'$. The laser output power versus injection current was used to estimate the change in gain for a change of carrier density from zero injection current up to threshold.

At 273 K, for example, the calculated change in gain up to threshold is 133 cm^{-1} and the spectral shift of the laser mode extrapolated from threshold to zero injection current gives an observed $\Delta\lambda$ of -8.1 Å. Using Eqs. (9) and (10) gives a value of 15.5 for $(1+\alpha^2)$. The inverse power dependence ofthe linewidth is then calculated to be 67.2 MHz mW which compares to the least square fit to the data of 74.7 MHz mW. Table I shows the parameters which were measured and the related theoretical estimates. It is important to note that parameter values for different devices fabricated the same way can vary sufficiently to require their determination for each comparison between experiment and theory. Such variations can change

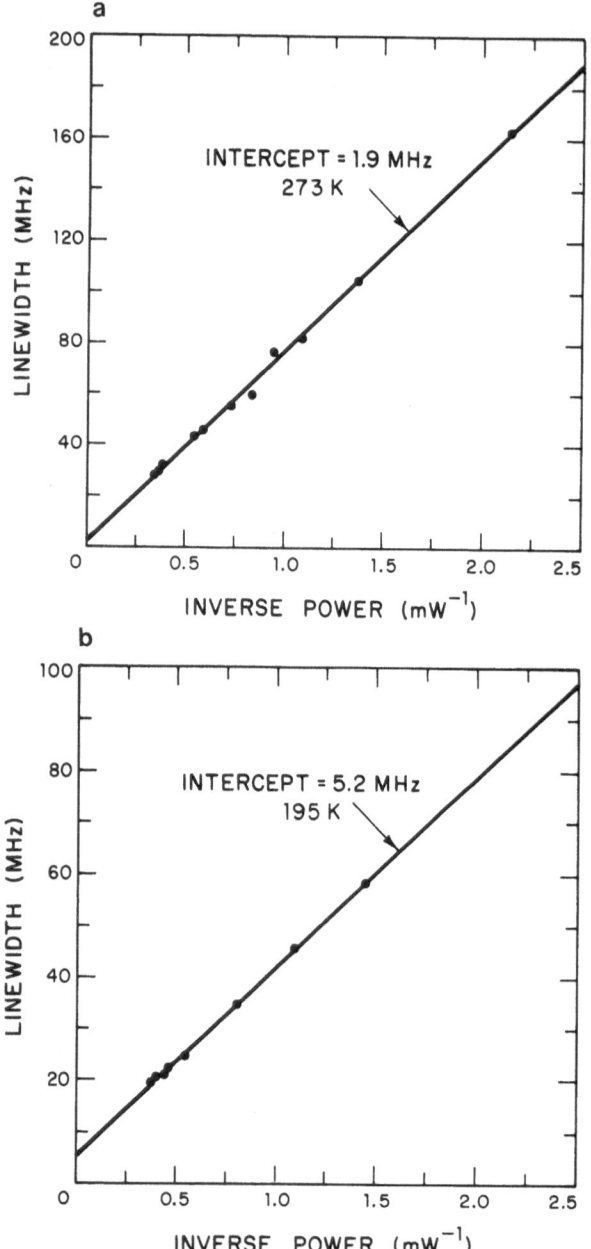

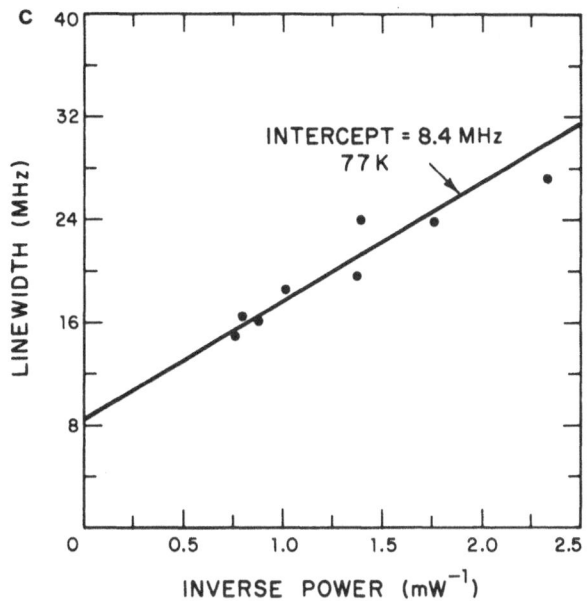

Fig. 4. Linewidth of a Mitsubishi TJS diode laser as a function of the inverse of the single-ended output power at three temperatures: (a) 273 K, (b) 195 K, and (c) 77 K.

the observed linewidth for a given power level by as much as 10-20 percent. A significantly larger variation between devices made by different technique can occur as is seen in the comparison between Figs. 4 and 5.

Power Independent Broadening

For the power independent contribution to the laser linewidth, the change of refractive index at N_t with electron number was determined by monitoring the spectral shift of the Fabry-Perot modes as a function of injection current up to laser threshold. Typical data are shown in Fig. 6 for the spectral mode shift at 195 K. The rate of change of refractive index is expected to be linear with carrier density. The relation between carrier density and injection current, however, is parabolic[22] up to threshold which causes the parabolic shift of mode frequency with injection current. The parameter M in Eq. (17) is estimated to be ~ 0.4 for TJS devices, and the factor $\Gamma_g/(\Gamma_c+\Gamma_g)$ is approximately unity. Table II

Table I. A summary of measured device data and theoretical estimates of the power-dependent linewidth contribution for Mitsubishi TJS diode lasers.

	77 K	195 K	273 K
λ(nm)	803	824	842
$\Delta\lambda$(Å)	-2.6	-6.0	-8.1
W_g(cm^{-1})	88.3	100	133
n_{sp} (calculated)	1.12	1.69	2.12
$(1 + \beta^2)$	5.1	16.4	15.5
Theoretical estimate of $2\Gamma P_0$ (MHz mW)	8.2	43.7	67.2
Experimental observation of $2\Gamma P_0$ (MHz mW)	9.28	36.7	74.7

Table II. A summary of measured device data and theoretical estimates of the power-independent linewidth contribution for Mitsubishi TJS diode lasers.

	77 K	195 K	273 K
λ(nm)	803	824	843
I_t (mA)	1.3	6.3	13.5
N_t	1.7×10^7	8.4×10^7	1.8×10^8
$(\partial n/\partial N)_{N_t \nu}$	8.4×10^{-11}	2.2×10^{-11}	5.2×10^{-12}
$2\Gamma_{theory}$ (MHz)	11.5	6.5	2.2
$2\Gamma_{exp}$ (MHz)	8.4	5.2	1.9

summarizes the experimental parameters and the comparison between experiment and theory at three temperatures.

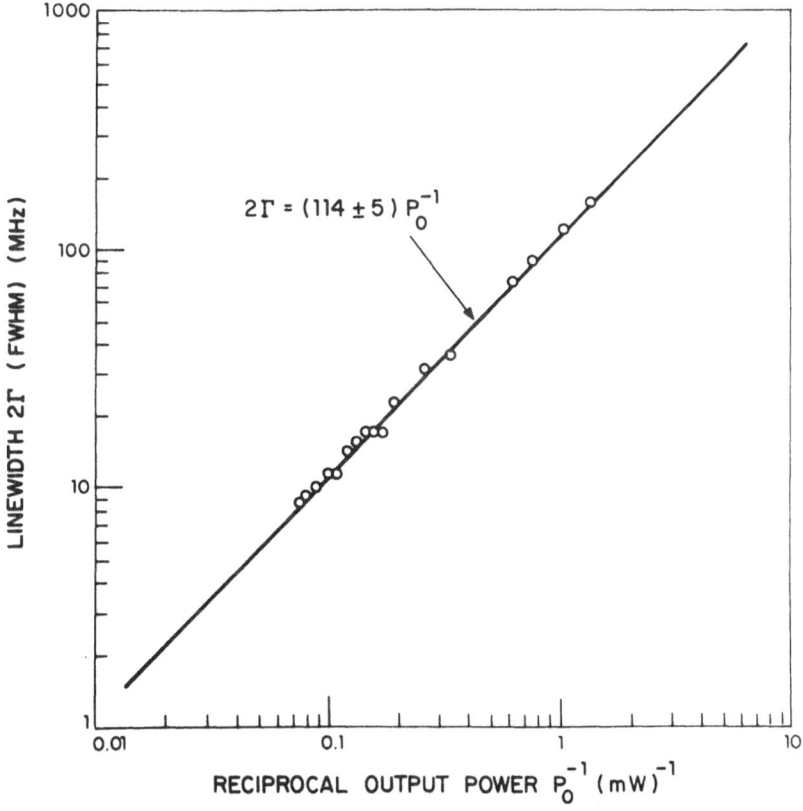

Fig. 5. Linewidth of a Hitachi CSP diode laser at 0 C as a function of the inverse of the single-ended output power. The slope is 114 ± 5 MHz with an infinite power intercept of 2.7 ± 0.8 MHz. Mean variance of data is ± 1.5 MHz for output powers greater than 2 mW.

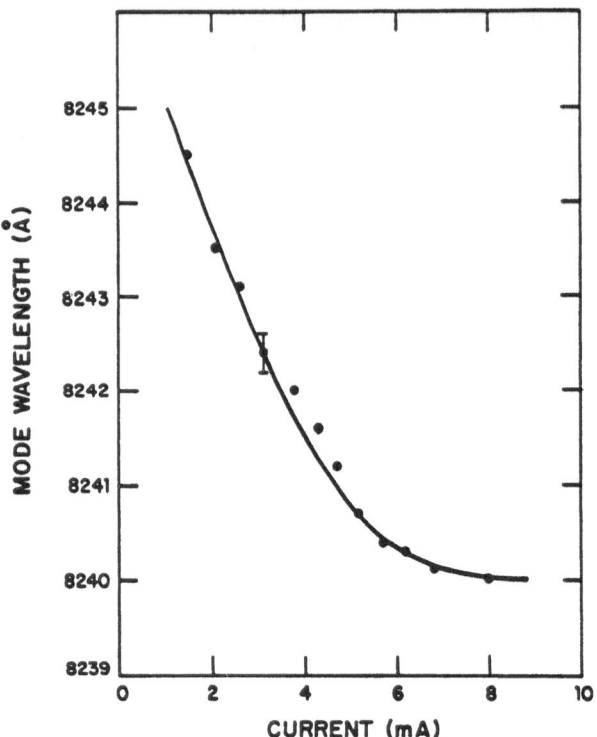

Fig. 6. Shift of the Fabry-Perot laser cavity peak wavelength at the laser frequency for a Mitsubishi TJS laser at 195 K.

The contribution to the power independent linewidth due to fluctuations in the peak of the gain spectrum center frequency with electron number fluctuation becomes significant only at low temperature and would be no more than a few MHz for temperatures below about 20 K. In addition to the data shown in Table II, measurements[23] made at liquid helium temperature (Fig. 7) indicates a power independent broadening value of about 30 MHz which is also consistent with the greatly reduced threshold currents and increased value of resonant dispersion. This behavior is in agreement with the phenomenological power-independent broadening model of ref. (16) and the power-dependent theory of ref. (8). This is in apparent contrast with the predictions of the Vahala and Yariv[17] theory in which the linewidth of Eq. (19) approaches 0 as T → 0. The exact origin of the fluctuations is not clear. However, it appears that this broadening may be related to the presence of 1/f noise[24] in the device.

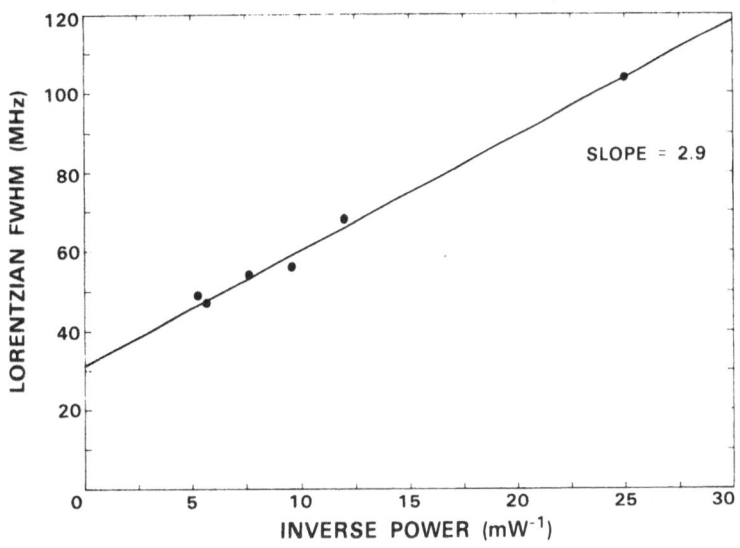

Fig. 7. Linewidth of Mitsubishi TJS laser at 1.7 K.

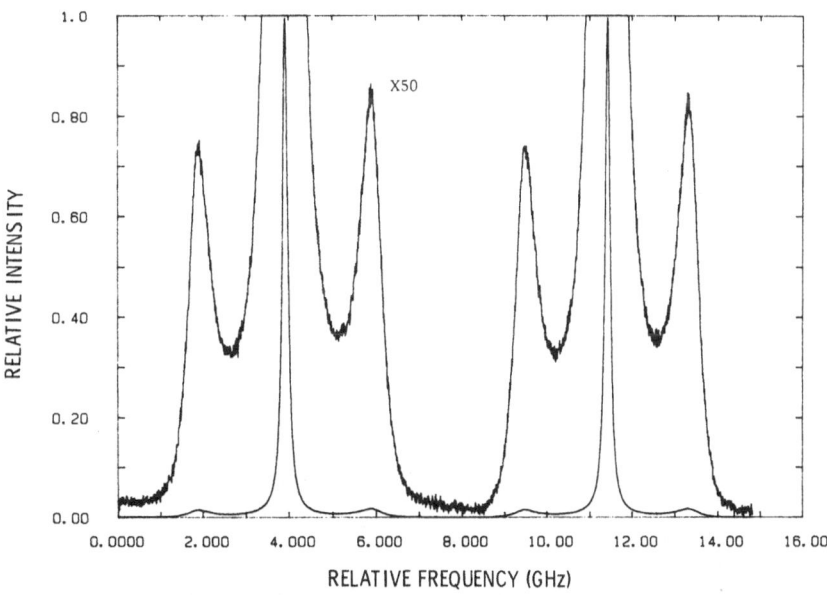

Fig. 8. Fabry-Perot trace of the output of a Hitachi CSP diode laser at 20 C showing relaxation resonance sidebands. Resolution of Fabry-Perot was 35 MHz. Single-ended output power of diode laser was between one and two milliwatts.

It is interesting to note that the power independent contribution to the laser linewidth is a dominant broadening mechanism at high output power levels at room temperature and at most power levels at low temperature. This effect limits the usefulness of these devices for many high resolution applications but it can be overcome by use of an external cavity as described in the next section. The implications of this power independent broadening mechanism are most significant as device structures become even smaller.

Phase Noise Sidebands

The sidebands at the relaxation frequency described above are shown experimentally[25] in Fig. 8. The magnitude of the frequency shift has been observed to vary as the square root of the power in the main laser mode in agreement with the dependence of the relaxation oscillation frequency with stimulated emission rate. The intensity of these sidebands linearly decrease in magnitude with increasing laser mode intensity.

EXTERNAL CAVITY (GaAl)As LASERS

This section describes the spectral characteristics of (GaAl)As diode lasers operated in external cavity structures. Use of an external cavity can overcome many of the spectral limitations described in the preceding sections and provide a device with a very narrow linewidth and controlled tuning characteristics. A disadvantage of an external cavity structure is the greater complexity of a hybrid structure over a monolithic device. With advances in fabrication of small optical components, this does not turn out to be a significant disadvantage. In fact, considering the hybrid technology of fiber optical connectors and transmission systems in general, the new generation miniature external cavity devices are quite reasonable.

The CSP and TJS lasers described in the previous sections operate with nearly all of their output power in a single frequency in contrast to earlier generation devices which had a multi-longitudinal mode output even when they operated in what appeared to be a single spatial mode. These early lasers were of the wide stripe design (i.e., greater than 10 μm junction width) which for a typical cavity length of 200 to 300 μm meant a Fresnel number of greater than one. The narrow stripe width (~ 1-2 μm) of the CSP and TJS lasers has led to single-frequency operation in part because these lasers have a Fresnel number near unity; however, the actual device fabrication technique strongly contributes to the mode structure. The subject of multi-longitudinal mode operation of diode lasers which presumably operate in a single spatial mode has been discussed but not completely understood for some time. Recently, Liu et al.[26] have studied the intensity statistics of diode lasers in which the output occurred in a predominantly single mode. It was found that a weaker adjacent longitudinal mode could for a short period of time increase in power

a predominantly single mode. It was found that a weaker adjacent
longitudinal mode could for a short period of time increase in power
to the power of the main mode, while that main mode power decreased
in intensity so that the total power remained constant.

Use of an external cavity to force oscillation in a single fundamental spatial mode[21] for a diode laser which would normally operate with many longitudinal modes had the effect of producing a
single-frequency output with nearly all of the double-ended multimode
power of the diode laser before it was operated in the external cavity. Figure 9 shows a typical multimode output spectrum of a
(GaAl)As diode laser operating continuously at room temperature.
These lasers were antireflection coated on both ends so that no laser
action would occur at the maximum injection current levels used before coating. Figure 10 shows the output spectrum from this device
when operated within an external cavity as shown schematically in
Fig. 11. Similar results were repeatedly obtained with several dozen
such lasers. By use of very wide junction lasers it would be possible to extract a few hundred milliwatts of continuous output power
at room temperature from one diode laser in a single, stable frequency and one or more watts of power at low temperature.

It is important to note that when a normally multimode device is
forced to oscillate in a truly single spatial mode, nearly all of the
multimode output can be extracted in a single frequency. This has
been demonstrated in an external cavity structure even when operated
without a tuning element in the cavity.[19] This implies that there
exists neither spectral nor spatial hole burning in these lasers.

Spatial hole burning, a common source of multi-longitudinal mode
operation in many lasers, occurs when the local field intensity of
the standing wave in the laser cavity saturates the local population
inversion. The adjacent cavity mode of next order will not overlap
in the center of the cavity and can be sustained because the population inversion is localized. In the case of a semiconductor laser,
the inversion is from mobile carriers and any hole burning will be
washed out due to thermal diffusion. The time for thermal diffusion
over a distance of about one quarter of the wavelength of the standing wave in the material (several hundred Å) is comparable or faster
than the stimulated emission rate.

Spectral hole burning is a second source of multi-longitudinal
mode operation in lasers with inhomogeneous gain broadening. For
interband transitions in materials such as (GaAl)As, most of the
transitions are direct with only removal of the k selection rule
near the band edge for the doping levels ($\sim 10^{18}$ cm^{-3}) typically encountered in these devices. At room temperature, phonon
scattering as well as carrier-carrier scattering produce an intraband scattering time on the order of 10^{-13} sec as determined from
traditional methods. Figure 12 shows the interband laser transition

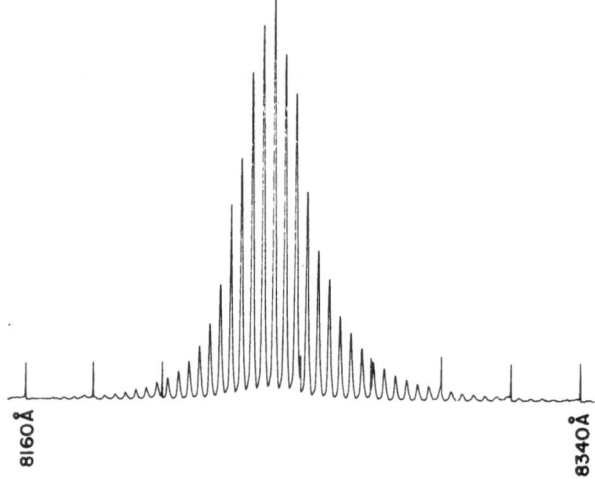

Fig. 9.   Emission spectrum of cw room temperature (GaAl)As diode
laser before operation in external cavity.

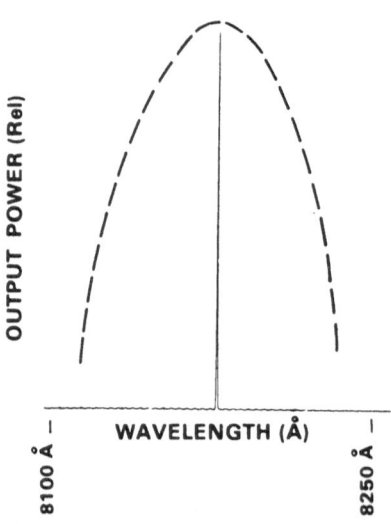

Fig. 10. Spectrometer trace and tuning range of a grating controlled external cavity (GaAl)As diode laser operating
cw at room temperature.   Maximum output power was several
milliwatts.

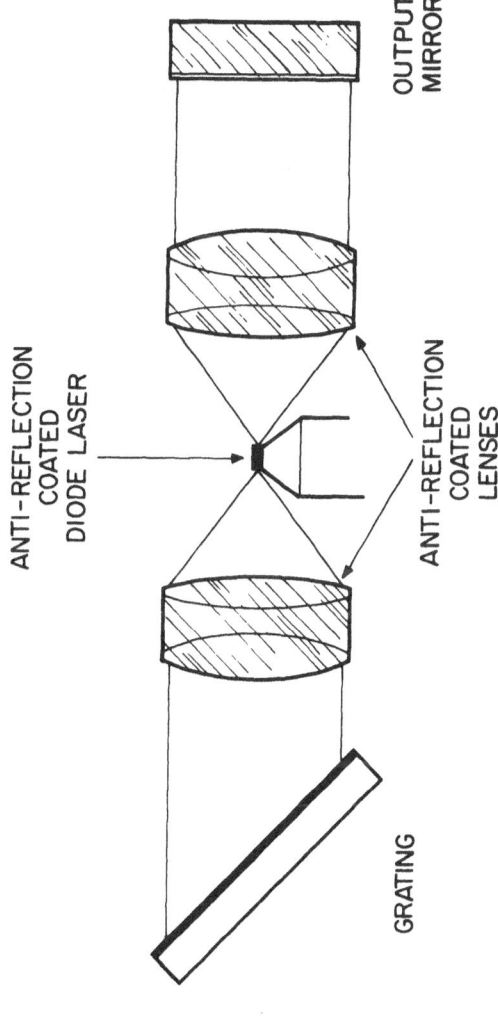

ANTI-REFLECTION
COATED
DIODE LASER

OUTPUT
MIRROR

ANTI-REFLECTION
COATED
LENSES

GRATING

Fig. 11. Schematic diagram of a double-ended external cavity diode laser.

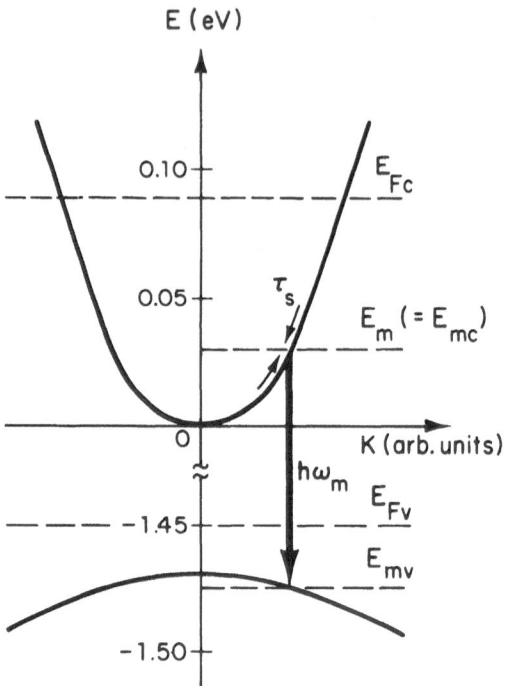

Fig. 12. Schematic diagram of the simplified band structure of
(GaAl)As showing the direct interband laser transition.

while Fig. 13 shows schematically the hole which would be formed in
the electron distribution function and the corresponding effect on
the spontaneous emission emitted from the recombination of carriers
in the gain region. Because of the large hole effective mass, valence band hole burning effects can be neglected. Calculated hole
widths[12] for an intraband scattering time of 10^{-12} and 10^{-13} s
are 1.3 and 13 meV, respectively. Corresponding hole depths are
4×10^{-3} to 4×10^{-5} of the peak of the luminescent intensity at
the laser frequency.

An attempt has been made to measure the hole burning on a
(GaAl)As diode laser[21] using an external cavity device. By observing the luminescent emission along the axis of the laser it was possible to externally aperture the beam and observe only the emission
coming from the mode volume of the laser where hole burning would be
expected to occur. Figure 14 shows the results of experiments performed on an external cavity device operating with a one milliwatt
output power compared with theoretical calculations of the effects
of a hole burned with a one percent depth. No holes were observed

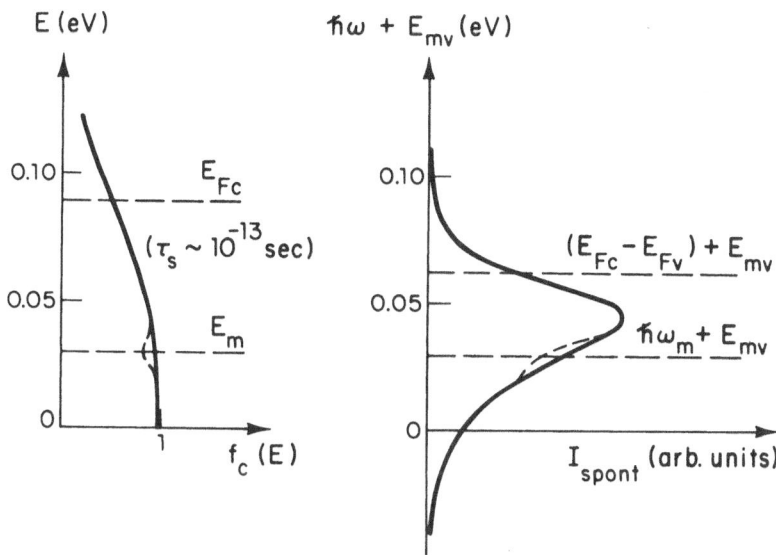

Fig. 13.   Representation of hole burned into electron distribution
           function and corresponding hole as it would appear on the
           electron-hole recombination spectrum.

to the sensitivity limit of about 0.5% of the intensity of the luminescent intensity indicating that the intraband scattering time
would be faster than about 5×10^{-13} sec according to existing
models.[21] This is consistent with the fact that the external cavity as well as monolithic CSP and TJS type lasers operate in a single
axial mode.

The limitations imposed by the fundamental linewidth broadening
mechanisms described in Section II can be overcome by the use of an
external cavity. The linewidth will be narrowed by the square of the
ratio of the relative cavity lengths when the unloaded cavity length
is significantly larger than the length of the diode. Figure 15
demonstrates the narrow linewidth that can be achieved[15] using an
external cavity structure heterodyned together. The observed linewidth of 30 kHz implies that the linewidth for each of the lasers is
less than 15 kHz. This is for the case of two Lorentzian lines. The
expected fundamental linewidth for each of the lasers in this experiment at their output power levels would be about 400 Hz. The measurement limitation was set by the RF spectrum analyzer. With present
technology, it should be possible to construct a small external cavity device with a fundamental width of about one hertz and an un-
locked jitter width of a few kilohertz.

The amplitude noise spectrum is shown in Fig. 16 from essentially dc to 2 GHz. Only slight traces of 120 Hz and an external

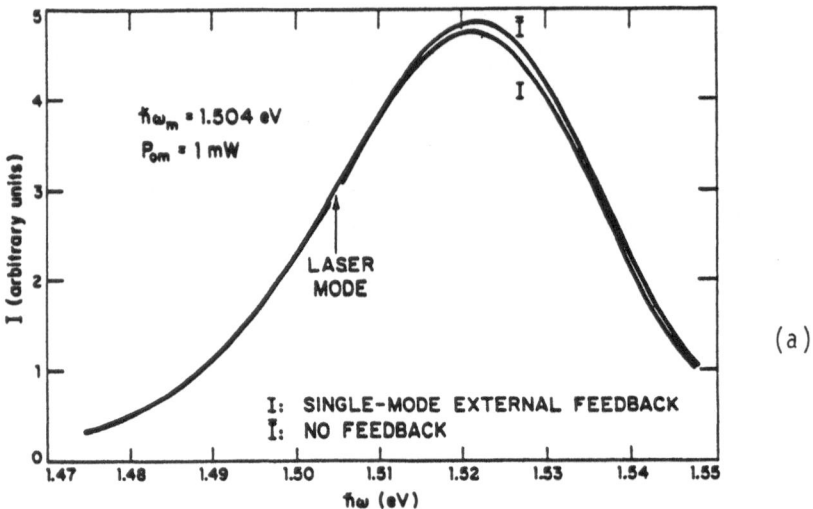

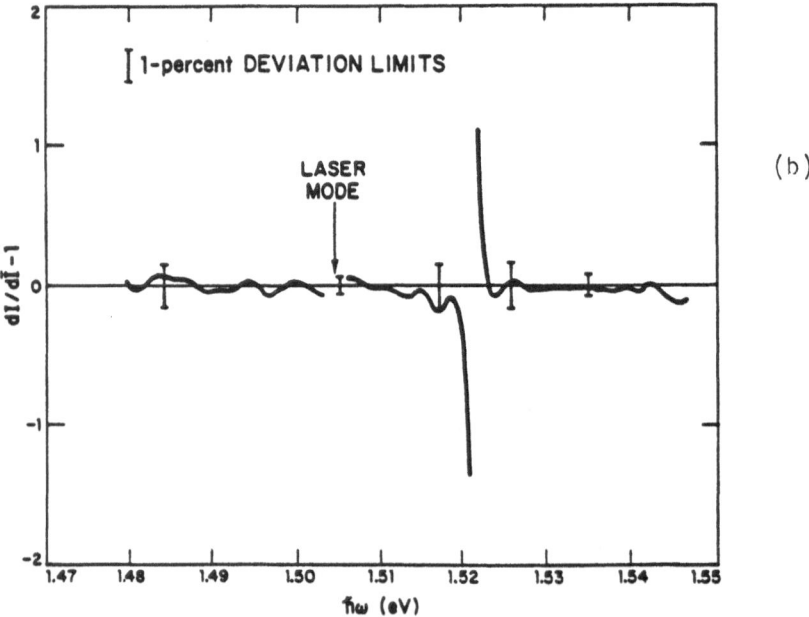

Fig. 14. (a) Spontaneous emission spectrum from (GaAl)As room temperature external cavity laser with and without feedback.
(b) Spectrum of the ratio of the energy derivatives of
the lasing and nonlasing spectra in (a).

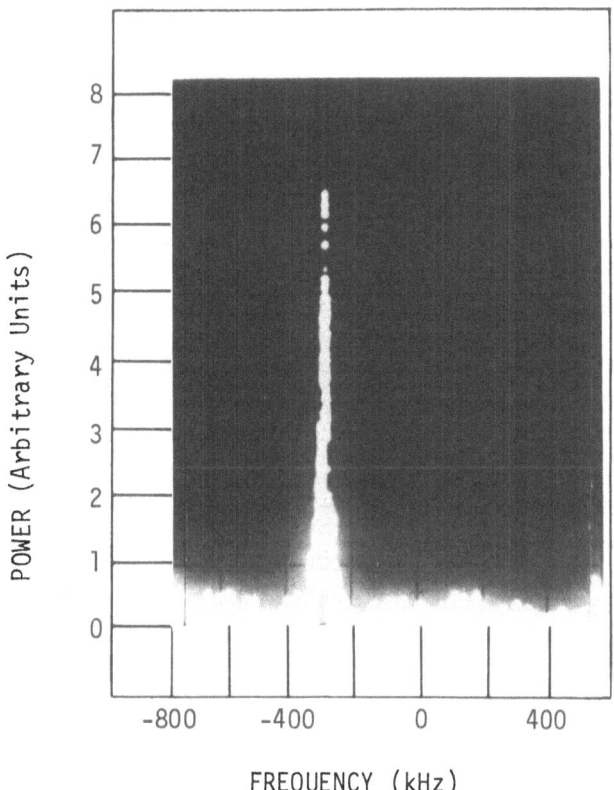

Fig. 15. Heterodyne beat spectrum between two spectral cavity (GaAl)As cw diode lasers at 850 nm with output powers of 0.4 and 0.34 mW, respectively.

cavity relaxation resonance at 25-40 MHz can be seen. A smaller and simpler external cavity structure would eventually have very high amplitude and phase stability making it an ideal source for metrological applications.

External cavity lasers have gone through various stages of evolution and like many other types of laser devices, they must be in a reasonably stable cavity structure in order to examine many of their interesting physical properties. An intermediate design is shown in Fig. 19 in which a Hitachi CSP diode laser having both ends antireflection coated is in a stable cavity employing a microscope objective lenses. Most of this structure is fabricated from Invar for stability. With the very low (< 0.3%) reflectivity coatings presently possible for (GaAl)As diode lasers, a greatly simplified cavity design using only one microscope objective lens with the diode

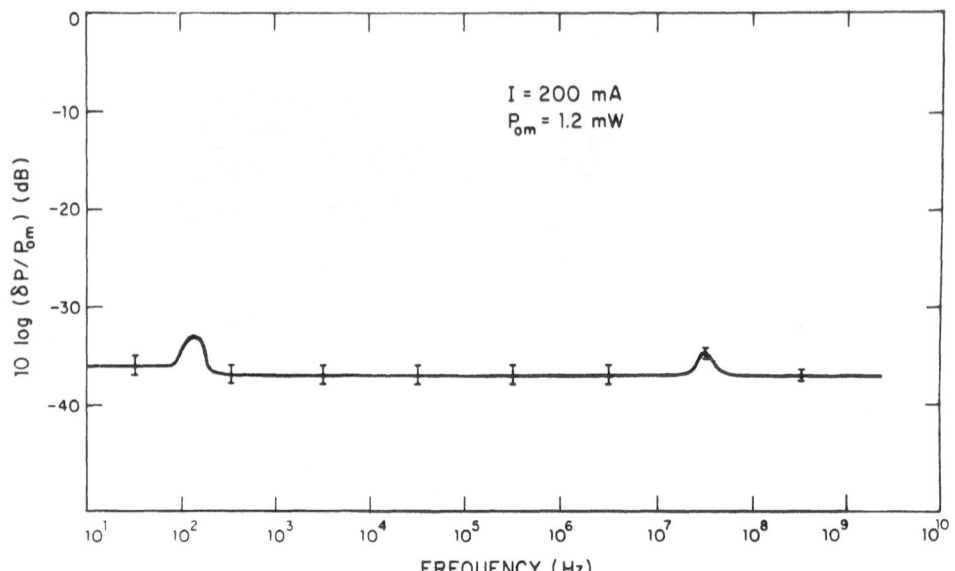

Fig. 16. Laser intensity noise spectrum detected in a bandwidth of
10 Hz over the frequency range 10 Hz to 100 kHz, 100 Hz
from 100 kHz to 1 MHz, 10 kHz from 1 MHz to 10 MHz, and
100 kHz from 10 MHz to 2 GHz. The spectra are essentially
flat in each frequency interval and are therefore joined
continuously for this complilation. The magnitudes of the
noise peaks must be divided by the corresponding bandwidths for quantitative comparison. The power scale is
normalized to the average mode intensity for a 10-Hz
detection bandwidth. Error bars indicate the average
scatter in the recorded spectra over the component frequency ranges. The peak at 120 Hz is due to power supply
ripple while the peak in the range 25-40 MHz is due to
relaxation oscillations of the external cavity.

antireflection coated on one side and 100% reflectivity coated on the
other side allows the use of simple, commercially available diode
lasers. Commercially available microscope objective lenses do not
have significant loss at typical (GaAl)As diode laser wavelengths and
can be used in a simple rigid cavity design with a significantly larger volume-to-surface ratio than previous designs. A very compact,
stable, external-cavity device which is also relatively inexpensive

can be fabricated for various applications in high resolution spectroscopy, short-pulse generation, and optically pumped frequency standards.

Fig. 17. Photograph of a stable external cavity (GaAl)As diode laser.

REFERENCES

1. M. Lax, Phys. Rev. 160, 290 (1967).
2. A. L. Schawlow and C. H. Townes, Phys. Rev. 112, 1940 (1958).
3. C. H. Henry, R. A. Logan, and K. A. Bertness, J. Appl. Phys. 52, 4453 (1981).
4. M. W. Fleming and A. Mooradian, Appl. Phys. Lett. 38, 511 (1981).
5. M. Lax, Phys. Rev. 157, 213 (1967).
6. H. Haug and H. Haken, Z. Physik 204, 262 (1967).
7. C. H. Henry, IEEE J. Quantum Electron. 18, 259 (1982).
8. K. Vahala and A. Yariv, IEEE J. Quantum Electron. 19, 1096 (1983).
9. M. Sargent, M. O. Scully, and W. Lamb, Jr., in Laser Physics (Addison-Wesley Publishing Company, Reading, Massachusetts, 1974).
10. Y. Yamamato, IEEE J. Quantum Electron. 19, 34 (1983).
11. J. M. Osterwalder and B. J. Rickett, Proc. IEEE 67, 1671 (1979).
12. B. Daino, P. Spano, M. Tamburrini, and S. Piazzola, IEEE J. Quantum Electron. 19, 266 (1983).
13. K. Vahala, Ch. Harder, and A. Yariv, Appl. Phys. Lett. 42, 211 (1983).
14. C. H. Henry, IEEE J. Quantum Electron. (to be published).
15. A. Mooradian, D. Welford, and M. W. Fleming, in Proceedings of Fifth-International Conference on Laser Spectroscopy, Jasper, Canada, 1981, edited by A. R. W. McKellar, T. Oka, and B. P. Stoicheff (Springer-Verlag, 1981), p. 67.
16. D. Welford and A. Mooradian, Appl. Phys. Lett. 40, 560 (1982).
17. K. Vahala and A. Yariv, Appl. Phys. Lett. 43, 140 (1983).
18. H. L. Stoumer, A. Pinczuk, A. C. Gossard, and W. Wiegmann, Appl. Phys. Lett. 38, 691 (1981).
19. Measurements by J. Harrison and A. Mooradian.
20. D. Welford and A. Mooradian, Appl. Phys. Lett. 40, 865 (1982).
21. M. W. Fleming and A. Mooradian, IEEE J. Quantum Electron. 17, 44 (1981).
22. S. E. H. Turley, G. H. B. Thompson, and D. F. Lovelace, Electron. Lett. 15, 256 (1979).
23. J. Harrison and A. Mooradian, Appl. Phys. Lett. 45, 318 (1984).
24. K. Kikuchi and T. Okoshi, Electron. Lett. 19, 813 (1983).
25. Measurements by J. Harrison and A. Mooradian.
26. P. L. Liu, L. E. Fencil, I. P. Kaminow, T. P. Lee, and C. A. Burrus, to be published in IEEE J. Quantum Electron.

EXTENSIONS OF SEMICONDUCTOR LASERS

TO HIGHER POWER AND LONGER WAVELENGTH

Roy Lang

Opto-Electronics Research Laboratories
NEC Corporation
Kawasaki 213, Japan

1. INTRODUCTION

Semiconductor Injection Lasers or Laser Diodes (LDs) constitute a unique class of lasers because of their practical advantages, which include compact size, low voltage (1-2V) and small current (10^1-10^2 mA) required for operation, high efficiency, ability to be modulated directly with up to Gb/s current signal, and broad tunability. These features, together with the fact that they can be mass produced with semiconductor electronics technology, have made LDs most significant laser sources for industrial applications, such as optical fiber communication and optical information processing.

The following is a brief introductory review on the current status of the LD research and development. Emphasis is placed on the activities towards high power operation and towards longer wavelength oscillation in 1 μm region. For more general introduction, readers are referred to textbooks.[1,2]

Basic LD Structures

The initial version of the diode lasers, which were first realized in 1962, had a simple pn diode structure formed in either GaAs [3,4] or GaP_xAs_{1-x} [5] crystal, and were called homojunction lasers. The optical cavity is composed of a pair of flat, parallel crystal facets formed by cleavage. By simply forward biasing the diode, carriers overflowing across the pn junction (mostly electrons into the p region) create a minority carrier population in excess of the thermal equilibrium, which gives rise to optical amplification as the carriers recombine with stimulated emission. However, the homojunction LDs could not operate continuously at room temperature. The reason became gradually apparent as being that the diffusion of electrons injected across the pn junction into the p region, with the typical diffusion length of a few μm, necessitated an excessively high current density in achieving the high injected carrier concentration of 10^{18} cm^{-3} required for the room temperature lasing.

The continuous operation at room temperature, which was considered imperative for practical application of these lasers, was realized in 1970, when so called double heterostructure (DH) was introduced [6,7]. Depicted in Fig. 1 is a double heterostructure (DH) LD. It differs from the simple homojunction structure in that the thin light emitting (GaAs) active layer is sandwiched between wider energy gap p and n ($Al_xGa_{1-x}As$) cladding layers. The energy gap differences between the active and the cladding layers, typically a few hundred meV, serve as potential barriers and confine the injected carriers into the active layer. By making the active layer as thin as 0.1-0.2 μm, i.e., much thinner than the carrier diffusion length, the room temperature threshold current density could be reduced to less than several KA/cm^2, which was small enough to permit CW operation.

In order to further reduce the thermal loading and maintain homogeneous lasing across the injected region, "stripe geometry" was introduced, in which the current injection was confined to a narrow stripe region a few to a few tens of μm wide.[8] This stripe geometry, combined with double heterostructure, is the simplest practical structure for room temperature CW operation of GaAs and InGaAsP LDs.

Material and Emission Wavelength

The emission wavelength of a LD is determined essentially by the active layer energy gap. In order to realize an LD emitting at a suitable wavelength, direct gap material with appropriate energy gap must be found. Furthermore, a material combination for the active layer and the cladding layers with sufficient energy gap difference and closely matched lattice constants is needed to

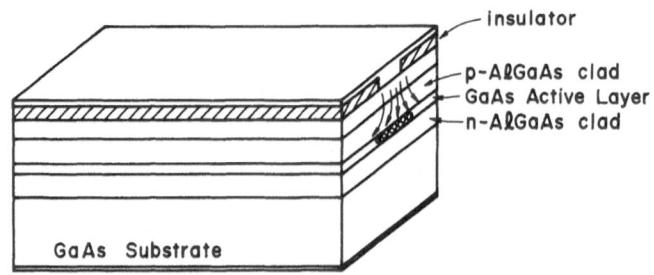

Fig.1 Stripe geometry double heterostructure for GaAs laser diode.

form the double heterostructure. $Al_yGa_{1-y}As$ active layer-$Al_xGa_{1-x}As$ cladding layer combination (y < x) covering 0.7-0.9 μm range and $In_xGa_{-x}As_yP_{1-y}$ active layer-InP cladding layer combination covering 1.2-1.6 μm range have been most highly developed, but other material combinations are also being studied to extend the wavelength coverage of laser diodes, as indicated in Fig. 2.

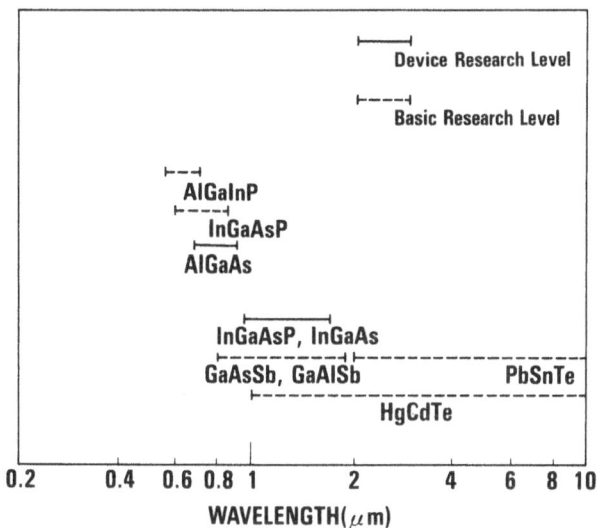

Fig. 2 Material for diode lasers.

2. QUANTUM ELECTRONICS OF INJECTION LASERS

Lasing Transitions

While lasing transitions in most other lasers take place between discrete electronic states localized in an individual atom or molecule, those in injection lasers occur between energy bands, that is, between sets of continuously distributed electronic states which spread over many atoms in crystal. From a theoretical stand point, the difference is not essential, and the laser equations can be formulated quite analogously to those for an inhomogeneously broadened two level system.[9] However, in practical aspects, this difference renders many unique features in injection laser lasing properties.

An electronic state in a band can be labeled, apart from the spin quantum number, with wave vector **k**. Energy vs. **k** relations, that is the band structures, near the conduction and valence band edges for a direct gap III-V semiconductor, is shown in Fig. 3.

In an ideal band-to-band optical transition, the **k** vector is conserved. That is, the transition connects a conduction band state with wave vector **k** to the valence band state having an identical **k** vector. If this "k-selection" rule is obeyed, the band-to-band transitions can be viewed simply as a collection of transitions in an inhomogeneously broadened two level system. Therefore, the real and imaginary part of the complex susceptibility, χ_1 and χ_2, associated with the transitions between the conduction and the valence bands, at frequency Ω, can be expressed as

$$\chi_1 + i\chi_2 = \sum_k \frac{|M_{kq}|^2}{\varepsilon_o} \frac{P_{c,k} - P_{v,k}}{(\hbar\Omega - \varepsilon_{c,k} + \varepsilon_{v,k}) + i\hbar\gamma_t} \qquad (1)$$

where M_{kq} is the momentum matrix element, and t is the inverse transverse relaxation time. P_{ck} and P_{vk} are occupation probabilities, where c and v respectively denote conduction and valence bands.

Relaxation Processes and the Occupation Probabilities

At thermal equilibrium, the occupation probability of a state, having energy $\varepsilon_{\alpha,k}$, is given by Fermi distribution function $f(\varepsilon_{\alpha,k})$, where α =c or v and

$$f(\varepsilon_{\alpha,k}) = [1 + exp \frac{\varepsilon_{\alpha,k} - \mu}{k_B T}]^{-1} \qquad (2)$$

Here, k_B is the Boltzman constant, T is the carrier temperature and μ is the Fermi level. With current injection through the pn junction, electrons are excited from valence band to the

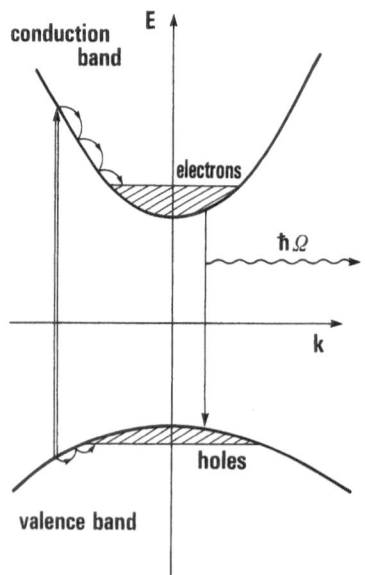

Fig.3 Qualitative energy band structure near the direct gap semiconductor band edge. Optical pumping and lasing transitions are shown with arrows.

conduction band, creating electron and hole population in excess of the thermal equilibrium in respective bands. The excited carriers exchange energy with the crystal lattice primarily through LO phonon scattering at picosecond rates. They also redistribute energy among like carriers through carrier-carrier scattering at a rate around 0.1 ps. Since these relaxation processes are much faster than the inter-band optical recombinations occuring at nanosecond rates, the electrons and holes can be regarded as being separately in equilibrium with the crystal lattice.

The state occupation probability under such a quasi-equilibrium condition can be described with quasi-Fermi distribution function $f_\alpha(\varepsilon)$ for $\alpha = c$ or v, where

$$f_\alpha(\varepsilon) = [1 + exp \frac{\varepsilon - \mu_\alpha}{k_B T}]^{-1} \qquad (3)$$

which is just the Fermi function (2) with the Fermi level replaced with parameters μ_α, called quasi-Fermi levels, defined separately for each energy band. These parameter values are uniquely related, respectively, to conduction band electron concentration n and valence band hole concentration p.

Because the intraband relaxation is so much faster than the interband transitions, the quasi-Fermi distributions in respective bands are expected to be maintained even above the lasing threshold, where the stimulated emission deplete selectively the population in those electronic states interacting directly with the monochromatic laser field. That is, the rapid spectral

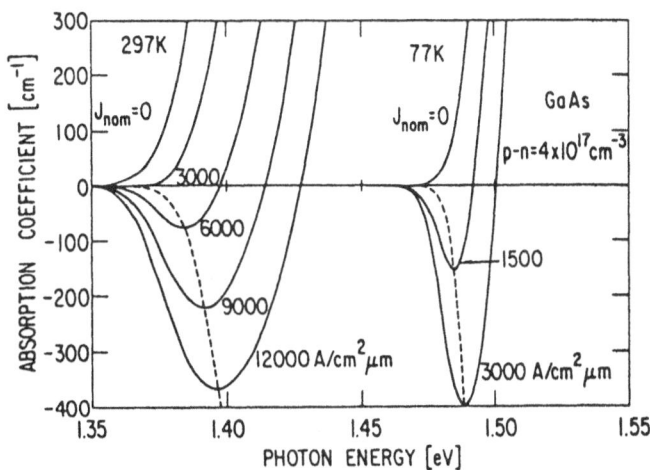

Fig. 4 Theoretical absorption spectra for GaAs at various pumping levels expressed with nominal current density (current density per 1 μm thick active layer). After ref. 14.

diffusion due to the intraband relaxation processes make the gain saturate effectively homogeneously over the spectrum, and so called spectral hole-burning is negligible at ordinary driving levels. Absence of appreciable spectral hole-burning has been confirmed experimentally.[10] However, deviation from the quasi-Fermi distributions may take place in the form of population pulsations in those **k** states interacting directly with multimode lasing fields. Such population pulsations seem to explain [11,12] observed spectral behaviors[13] suggestive of strong mode coupling.

Gain and Refractive Index Spectra

The imaginary and real parts of the complex susceptibility respectively yeild the gain and the refractive index spectra. However, quantitative agreement with experimental observations requires incorporation of many-body and impurity effects. F. Stern[14] derived theoretical gain curves for GaAs shown in Fig. 4, where these effects are taken into account in phenomenological fashion.

Large gain values, on the order of 10^2 cm^{-1}, broad spectrum width and the peak position and width varying with excitation are among the characteristic features of LD gain spectrum. The gain peak shifts towards a longer wavelength with increasing temperature, with the temperature coefficient amounting, near room temperature to about 3 A/K.

Associated with the conspicuous asymmetry in the gain spectrum is that in the refractive index spectrum[15]. As a result of the asymmetry, the dispersion curves do not zero-cross at the gain spectrum peak, and, consequently, the refractive index changes appreciably with the excited carrier density, even near the gain peak. The free carrier plasma effect also has contribution to the refractive index, which varies with the carrier density.[16] The strong carrier density dependence yields unique features to many aspects of diode laser properties. Examples are self-focusing of laser radiation into the spatially burned holes, which creates mode instability[17,18] and dynamic instability[19], linewidth broadening through coupling of AM noise with FM noise[20], and asymmetric tuning characteristics at injection locking[21].

3. OUTPUT POWER LIMITING FACTORS

In many applications, high output power is desired. At present, in most commercially available GaAs LDs, maximum coherent CW output from a single emission stripe is limited to a modest level of a few tens of mW. The major power limiting factors are, (1) catastrophic optical damage to the output facets, (2) instability of the spatial modes and (3) thermal effects or heating.

Catastrophic Optical Damage (COD)

A GaAs LD is permanently destroyed with mechanical damage at the emission spot on the output facet, when it is driven at an output level greater than a certain threshold value. This is called catastrophic optical damage (COD). The threshold power density for COD is difficult to determine accurately, but it is estimated to be around a few MW/cm^2 at CW operation. M. Wada et al[22] recently reported the 4 MW/cm^2 critical power density for their twin ridge terraced substrate lasers, which amounted approximately to 25 mW output per 1 μm width of the emission stripe. Critical power level P_c is greater at pulsed operation, and it increases to 4-8 MW/cm^2 as the exciting current pulse width is reduced to 100 ns.

The mechanism leading to COD has been studied by C.H. Henry et al[23], and by T. Kamejima et al[24]. According to their findings, COD is triggered by absorption of the laser radiation in a less excited region in the active layer, which results in localized heating of the absorbing region. Because the energy gap decreases with temperature rise, absorption increases with heating. When the laser intensity is above a certain critical level, this absorption and heating process instigates into a thermal runaway, resulting in crystal melting at around 1500 K and consequent mechanical damage in the absorbing region. The optical damage usually starts at regions near the output mirror facets, which tend to be absorptive, because the surface states mediate non-radiative recombination, which deplete the excited carriers.

For GaAs and (AlGa)As LDs COD is the most serious factor which limits coherent CW output. In case of InGaAsP LDs, COD has never been observed at device operation, and the critical power level is believed to be at least an order of magnitude greater than that for GaAs LDs.

Transverse Mode Instability

Since COD limits the output power density, the maximum output can be increased by enlarging the beam cross section. An obvious way to achieve this is to widen the stripe width. However, stripe geometry DH structure is not suitable for high power operation, because the transverse modes become unstable as the output increases, and the widening of the stripe makes the situation worse.

At an early stage of GaAs DH laser development, it was realized that the stripe geometry lasers had many peculiarities in their laser performance. For example, the output frequently saturated at a few mW level, showing a kink in the output vs.

current curve. Experimental examination of the near-field pattern revealed that the emission spot moves rapidly along the junction plane at the kink. It was also found that other laser characteristics, such as the spectral properties, tend to degrade at or above the kink. These peculiar behaviors result from lack of stable optical guiding in stripe geometry lasers.

In stripe geometry DH lasers, the active layer, having a narrower band gap, has a greater refractive index than the cladding layers. The relative difference in refractive indices, amounting to a few %, makes the double heterostructure an efficient slab dielectric waveguide, providing optical confinement for the laser radiation into the thin active layer. However, there is no such waveguiding mechanism built along the heterojunction planes into the crystal structure. The lasing radiation in stripe geometry DH LDs is laterally confined in the stripe region, primarily because of the gain (and loss) spatial variation around the stripe, where the carriers are injected.[26] This type of waveguiding, called gain-guiding, is not quite ideal in obtaining a highly coherent, diffraction limited output with stable modal pattern, because of the enhanced spontaneous emission contribution[27] and astigmatism[26] inherent in it.

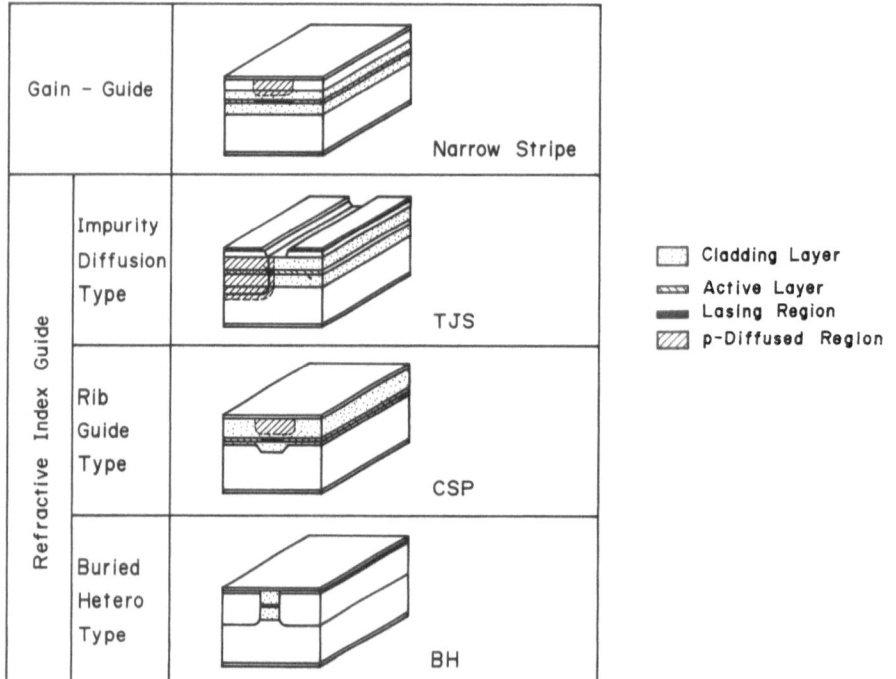

Fig. 5 Lateral transverse mode stabilized GaAs laser diode structures.

Moreover, the lasing lateral mode in a stripe geometry laser tends to be unstable as the excitation increases. The mechanism leading to the mode instability has been explained as the self-focusing of the laser beam into the spatially burned hole in the lateral profile of the excited carrier density.[17,18]

Various modifications to the double heterostructure, which are compatible with epitaxial crystal growth technique, have been devised to overcome this problem. Listed in Fig. 5 are some of the structures proved effective in stabilizing the transverse modes.

One structure is represented by a stripe geometry laser, with the stripe width decreased to less than 10 m to reduce the spatial hole-burning. LDs of the other type have built-in lateral refractive index variation, either real or equivalent, along the active layer. Among them, the impurity diffusion type, such as Transverse Junction Stripe (TJS) lasers[28], utilizes the refractive index variation with impurity type and concentration. In rib guide type lasers, the active layer is laterally homogeneous, but equivalent refractive index variation is introduced by the rib guide type structure fabricated, for example, using a substrate with channels. A Channeled Substrate Planar (CSP) laser[29] is one of the examples. The Buried Heterostructure (BH)[30] type has an active region surrounded entirely by a wider gap cladding region. With this structure, very effective confinement of both radiation and injected carriers can be attained, although the fabrication requires more complex processes than the other structures.

Thermal Effects

The injected current generates heat at and around the active region through non-radiative recombination, ohmic dissipation, etc. Thermal resistance for LDs depends on the structure, but it is typically a few to several tens of K/Watt with μs relaxation time. Therefore, when driven with dc current or a long pulse, the active region temperature is raised by tens of degrees.

Threshold current I_{th} and differential efficiency η_d for LDs depend rather sensitively on temperature. Empirically, I_{th} values vary as $\exp(T/T_o)$, where T_o is called characteristic temperature. The T_o values vary somewhat with the device structure, but it is around 150 K in GaAs LDs, which reflects the spreading in energy space of the quasi-Fermi distribution, which increases with temperature. InGaAsP LDs are more sensitive to the temperature and T_o values range from 50 to 70 K near room temperature. η_d tends to decrease with temperature approximately as $\exp(-T/T'_o)$. The T'_o value is around 200 K for both GaAs and InGaAsP LDs, and scatters considerably depending on the device structure.

The active region temperature rise, combined with the temperature sensitivity of I_{th} and η_d, have so far limited the maximum output at CW operation to 100-200 mW in GaAs LDs and to around 100 mW in InGaAsP LDs. Reduction in heat generation and thermal resistance are among the most important factors in designing CW high power LDs.

4. HIGH POWER INJECTION LASER STRUCTURES

Various modifications to the stripe geometry DH structure have been introduced to increase the obtainable coherent CW output power from GaAs and AlGaAs lasers. Described in the following are some representative examples.

Window Structures

The first type aims to increase the maximum output by increasing the COD threshold power itself. "Window Structure (WS)" has been developed for this purpose, in which no carriers are injected into the parts of an active region near the output facet, but these parts are made transparent to the laser radiation. The transparent "windows" eliminate the optical absorption which triggers the COD.

The original WS[31,32] was based on the planar stripe DH structure, with the stripe forming Zn diffusion made deep enough to reach the active region and the stripe ends being indented from the crystal facets. In this way, the near facets regions were unexcited, while they remained transparent to laser radiation because the Zn difused active region had a smaller effective band gap compared with the undiffused area. Several times increase in the COD level has been achieved with the WS structure[28].

Spot Size Expansion

An alternative approach to increase the COD output power

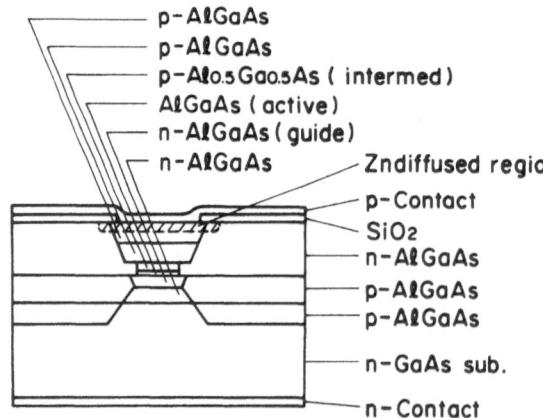

p-AlGaAs
p-AlGaAs
p-Al$_{0.5}$Ga$_{0.5}$As (intermed)
AlGaAs (active)
n-AlGaAs(guide)
n-AlGaAs

Zndiffused region
p-Contact
SiO$_2$
n-AlGaAs
p-AlGaAs
p-AlGaAs

n-GaAs sub.

n-Contact

Fig. 6 Buried Coarctate Mesa (BCM) structure for GaAs laser diode.

level is to increase the emission spot size at the output facet. Generally speaking, the spot size can be expanded by weakening the waveguiding around the active region. Thus, there is a delicate trade-off with the lateral mode stabilization. Constricted DH (CDH) structure[33] and Twin Ridge Substrate (TRS) structure[22] are examples in this category. Buried Coarctate Mesa (BCM) structure[34], shown in Fig. 6, is another example, which is based on BH structure.

Phase-Locked Array Lasers

The available output of an LD can be multiplied by adopting an array or stack of emission stripes, although there is, in general, no phase relationship or coherence between emissions from different stripes. Attempts are being made to construct LD arrays with sufficiently small separation between stripes to induce optical coupling and, consequently, phase-locking among the stripes. The 2.6 W maximum cw output has been achieved with 40 stripe array with 3 μm separation between stripes.[35] A problem in this approach is that, because the effective width of the wave guide, composed of the arrayed stripes, is large enough to sustain lateral transverse modes to very high orders, it is, in principle, very difficult to ensure a single lateral mode oscillation required for truly coherent emission.[36]

7. InGaAsP/InP LDS

GaAs LDs development was strongly motivated by the expectation of their application as sources for optical fiber communication. One of the most important reasons was that their emission wavelength matched closely with lowest transmission loss band of optical fibers in the early stages of their technology development. By the time practical GaAs/AlGaAs DH LDs were realized, the advances in the optical fiber technology had opened a lower loss and low dispersion window, around 1.3 μm, and an even lower loss window, around 1.55 μm. The lowest loss attained at 1.55 μm is 0.2 dB/km, which essentially equals the theoretical limit expected for silica based fibers.[37]

A search for a new LD material resulted in finding that quarternary compound $In_xGa_{1-x}As_yP_{1-y}$ lattice matched to InP substrate could cover these wavelength ranges. Intensive effort, started in the late 1970's realized practical LDs by the early 1980's with this material system. Presently, further improvement in the LD performance is very actively pursued.

Features of InGaAsP LDs

There are no essential differences in the laser physics between quaternary and (AlGa)As LDs, but there are some differences in the LD characteristics which are tied to differences in the semiconductor material properties. Marked advantages which the quaternary LDs have, in comparison with GaAs LDs, are (i) at least an order of magnitude greater COD level and (ii) much less tendency toward defect growth in the active region during device operation which, in GaAs LDs, was a major cause of device failure. The causes for such differences are not yet fully understood.

A major disadvantage is that quaternary LDs are much more temperature sensitive than GaAs LDs. Characteristic temperature T_0 for I_{th} is 50-80 K for 1.3 µm LDs, and somewhat smaller for 1.55 µm LDs. Non-radiative Auger recombinations, having a much higher rate in InGaAsP, are suspected to cause the enhanced temperature sensitivity.[38,39] Carrier leakage over the hetero-barrier[40] and absorption through intervalence band transitions[41] are other possible causes. While these are material properties unlikely to be tailored, improvements in the crystal quality and device structure have been found to considerably ease the problem.

Buried Heterostructure LDs

Because of the strongly temperature sensitive I_{th}, it is particularly important for a quaternary LD to reduce heat generation within the device. To this end, variations in buried heterostructure (BH) are frequently adopted for high quality LDs. Depicted in Fig. 7 is an example of such a structure, called Double Channel-Planar Buried Heterostructure (DC-PBH).[42] Devices with its 1.3 µm version have low threshold current of around 20 mA, large T_0 value ranging between 70 and 80 K, and maximum CW operable temperature up to 115 C. Moreover, high differential efficiency of nearly 80 % with electrical to optical conversion efficiency exceeding 40 % at maximum have been attained.

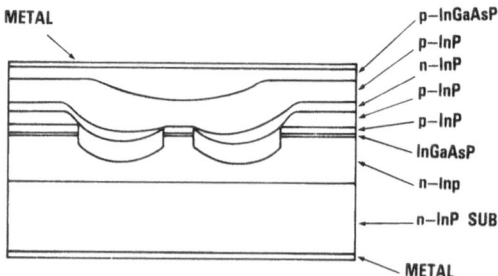

Fig. 7 Double Channel-Planar Buried Heterostructure (DC-PBH) for InGaAsP/InP laser diode.

Distributed Feedback Lasers

As stated earlier, LD gain saturation is effectively homogeneous spectrally, and, as far as the light propagation direction is concerned, it is homogeneous spatially as well. As a result, LDs with refractive index guiding structure oscillate in a single mode or at most a few longitudinal modes, when driven with constant current. However, many longitudinal modes within the broad gain spectrum width are excited, when the injection current is modulated at frequency greater than a few hundred MHz, or frequency close to that of the relaxation oscillation in a GHz range. This happens because the carrier density and, consequently, the level of the gain spectrum are modulated with the high frequency current modulation. Such spectrum broadening of optical pulses is quite undesirable in high bit rate long distance fiber communication, because the chromatic dispersion in fibers, typically 15-20 ps/A/km at 1.55 µm, causes broadening of transmitted pulses and, hence, inter-symbol interference.

Much effort has been made to realize LDs which can oscillate in single mode, even at high bit rate direct modulation, by increasing the frequency selectivity of the optical cavity. There have been two different approaches along this direction. One is to employ a coupled Fabry-Perot cavity. The Cleaved Coupled Cavity (C³) LD [43] is an example of this type. Another approach is to use so called Distributed Feedback (DFB) cavity.

Depicted in Fig. 8 is an example of DFB LD structure, which is named DFB DC-PBH LD.[44] In DFB LDs, the optical feedback is attained with Bragg reflection by the grating, formed very closely along the active layer. In this example it is formed on a quaternary waveguide layer on top of the active layer, which has a wider band gap than the active layer. A DFB cavity with finite grating length can support a multitude of longitudinal modes, like

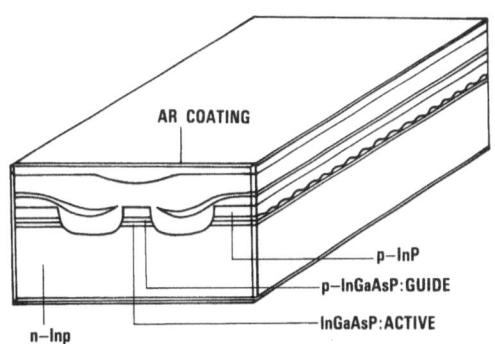

Fig. 8 Distributed Feedback (DFB) laser based on DC-PBH structure. Facet AR coating is applied to suppress Fabry-Perot cavity modes.

a Fabry-Perot cavity, but the feedback ratio is strongly wavelength dependent. As a result, stable single longitudinal mode oscillation can be maintained in a wide output range, and even under high bit rate modulation. Other characteristics, such as the threshold current and efficiency, are comparable to those for the Fabry-Perot cavity LDs. Therefore, DFB buried hetero-structure LDs appear to be very promising as light sources for long distance, high bit rate optical fiber communication systems.

8. SUMMARY

Present status of semiconductor laser development has been reviewed with emphasis on the high output power capability of GaAs/AlGaAs LDs and the recent activities in the long wavelength InGaAsP/InP lasers. The improvements in lasing characteristics and reliability have made these lasers very attractive, both for engineering applications and for scientific studies. There are also some unique aspects in the LD lasing processes itself, which awaits studies from a quantum electronics viewpoint. An example is the chaotic output behavior of an LD with external feedback.[45,46] Possiblities offered by so called quantum well structure[47], namely, a DH structure with the active layer thickness around 100 A, which is comparable to the electron de Broglie wavelength, are among other examples.

The author is grateful to T. Uchida, F. Saito, A. Ueki and S. Matsushita for their kind support and encouragement. He is also grateful to K. Kobayashi, T. Furuse, T. Suzuki, M. Ueno and I. Mito of Opti-Electronics Res. Labs., NEC, for helpful comments and cooperation in preparing this manuscript.

REFERENCES

1. H. C. Casey and M. B. Panish, "Heterostucture Lasers (parts A & B)," Academic Press (New York, 1980).
2. G. H. B. Thompson, "Physics of Semiconductor Laser Devices," John Willey (1980).
3. R. N. Hall et al, Phys. Rev. Lett. $\underline{9}$, 366 (1962).
4. M. I. Nathan et al, Appl. Phys. Lett. $\underline{1}$, 62 (1962).
5. N. Holonyak, Jr., and S. F. Bevacqa, Appl. Phys. Lett. $\underline{1}$, 82 (1962).
6. I. Hayashi et al, Appl. Phys. Lett. $\underline{17}$, 109 (1970).
7. Zh. I. Alferov et al, Sov. Phys. Semicond. $\underline{4}$, 1573 (1971). (Translated from Fiz. Tekh. Poluprovodn. $\underline{4}$, 1826 (1970))
8. J. C. Dyment Appl. Phys. Lett. $\underline{10}$, 84 (1967).
9. H. Haug, Phys, Rev. $\underline{184}$, 338 (1969). See also refernces therein.
10. M. W. Fleming and A. Mooradian, IEEE I. Quantum Electron., $\underline{QE-17}$,44 (1981).
11. R. Lang, in Digest of Jpn. Phys. Soc. Spring Meet., laGL-13(1979) (in Japanese).

12. R. F. Kazarinov and C. H. Henry, J. Appl. Phys., $\underline{53}$, 463 (1982).
13. M. Nakamura et al, J. Appl. Phys., $\underline{49}$, 4644 (1978).
14. F. Stern, J. Appl. Phys. $\underline{47}$, 5382 (1976).
15. C. H. Henry et al, J. Appl. Phys. $\underline{52}$, 4457 (1981).
16. G. H. B. Thompson, Opto-Electron., $\underline{4}$, 257 (1972).
17. G. H. B. Thomson et al, IEE J. Solid State Electron. Div., $\underline{2}$, 12 (1978).
18. R. Lang, IEEE J. Quantum Electron., $\underline{QE-15}$, 718 (1979).
19. R. Lang, Jpn. J. Appl. Phys., $\underline{19}$, L93 (1980)
20. C. H. Henry, IEEE J. Quantum Electron., $\underline{QE-18}$, 259 (1982).
21. R. Lang, IEEE J. Quantum Electron., $\underline{QE-18}$, 976 (1982).
22. M. Wada et al, Appl. Phys. Lett. $\underline{42}$, 853 (1983).
23. C. H. Henry et al, J. Appl. Phys. $\underline{50}$, 3721 (1979).
24. T. Kamejima et al, Jpn. J. Appl. Phys. Suppl. $\underline{19-1}$, 425 (1980).
25. T. Furuse, unpublished.
26. D. D. Cook and F. R. Nash, J. Appl. Phys., $\underline{46}$, 1660 (1975).
27. K. Peterman, IEEE J. Quantum Electron., $\underline{QE-18}$, 976 (1982).
28. H. Namizaki et al, J. Appl. Phys., $\underline{45}$, 2785 (1974).
29. K. Aiki et al, IEEE J. Quantum Electron., $\underline{QE-14}$, 89 (1979)
30. T. Tsukada, J. Appl. Phys., $\underline{45}$, 4899 (1974).
31. H. Yonezu et al, IEEE J. Quantum Electron., $\underline{QE-15}$, 775 (1979).
32. M. Ueno, IEEE J. Appl. Phys., $\underline{45}$, 4899 (1974).
33. D. Botez et al, Electron. Lett., $\underline{19}$, 882 (1983).
34. K. Endo et al, Electron. Lett., $\underline{20}$, 728 (1984).
35. D. R. Scifres et al, Electron. Lett., $\underline{19}$, 169 (1983).
36. K. Nishi and R. Lang, in preparation.
37. T. Miya et al, Electron. Lett., $\underline{15}$, 106 (1979).
38. Y. Horikoshi and Y. Furukawa, Jpn. J. Appl. Phys., $\underline{18}$, 809 (1979).
39. T. Uji et al, IEEE Trans. Electron. Devices, $\underline{ED-30}$, 316 (1983).
40. S. Yamakoshi et al, Appl. Phys. Lett., $\underline{40}$, 144 (1982).
41. A. R. Adams et al, Jpn. J. Appl. Phys. $\underline{19}$, L621 (1980).
42. I. Mito et al, IEEE J. Light Wave Tech., $\underline{LT-1}$, 195 (1983).
43. W. T. Tsang and N. A. Olsson, Appl. Phys. Lett., $\underline{43}$, 527 (1983).
44. M. Kitamura et al, IEEE J. Light Wave Tech., $\underline{LT-2}$, 363 (1984).
45. H. Kawaguchi and K. Ohtsuka, Appl. Phys. Lett., $\underline{45}$, 934 91984).
46. R. Dingle in "Festkorper Probleme XV, Advances in Solid-State Physics," p.21, Pergamon-Vieweg (Berlin, 1975).

PROPERTIES AND APPLICATIONS OF C3 LASERS

N. A. Olsson

AT&T Bell Laboratories

Murray Hill, NJ   07974

Abstract

We review the basic principles and operation characteristics of the Cleaved Coupled Cavity (C^3) semiconductor laser. Based on the unique features of the C^3 laser, single frequency operation and tunability, we also show some of the possible applications of the C^3 laser.

INTRODUCTION

The Cleaved-Coupled-Cavity or C^3 [1,2] laser derives its name both from the way it is made, cleaved, and from its basic principle of operation, coupled-cavity. Basically it is a semiconductor laser with two optically coupled but electrically isolated cavities formed by re-cleaving an ordinary laser chip in two. In this paper we will review the device characteristics of the C^3 laser along with some of its applications. These applications include fiber optic communication, electro-optical signal processing, optical bistability, and fiber measurements. All of the above are based on the unique features of the C^3 laser, namely: single frequency operation and tunability.

C^3 LASER CHARACTERISTICS

The C^3 laser shown in Fig. 1 consists of two standard Fabry-Perot (F-P) cavity laser diodes of approximately equal lengths which are self-aligned and very closely coupled to form a two-cavity resonator. The active stripe of each diode is precisely aligned with respect to the other and are separated by a distance

< 5 µm. All the reflecting facets are formed by cleaving along crystallographic planes and are hence perfectly mirror flat and parallel to each other. Complete electrical isolation (> 50 Kohm) between the individual diodes also results. The total length of a C^3 laser can be as short as 100 µm with a typical length of 200-400 µm.

The basic working principle is illustrated in Fig. 2. The propagating mode in each active stripe can have a different effective refractive index, Neff, even though they have the same geometrical shape, size, and material composition. This is because Neff is a function of the carrier density in the active stripe. This can be varied by varying the injection current below threshold when the junction voltage is not saturated. Thus the F-P mode spacing for stripes 1 and 2 will be different and are represented schematically with solid lines in Fig. 2. Because the two cavities are coupled, only those F-P modes from each cavity that coincide spectrally will interfere constructively and become the enforced modes of the coupled cavity. The spectral separation of these enhanced F-P modes will be significantly larger than that of the original individual F-P modes and depends on the relative lengths of the two cavities. For a 9:10 length ratio of the two cavities, the spacing of the enforced modes will be 10X the mode spacing of the individual laser sections. For the enforced mode near the gain maximum, the normal gain roll off of the laser material is sufficient to suppress the lasing of the adjacent enforced modes even under high speed modulation [3]. Thus, the first feature of the C^3 laser, single longitudinal mode operation, is based on the interaction of two coupled cavities. (A rigorus

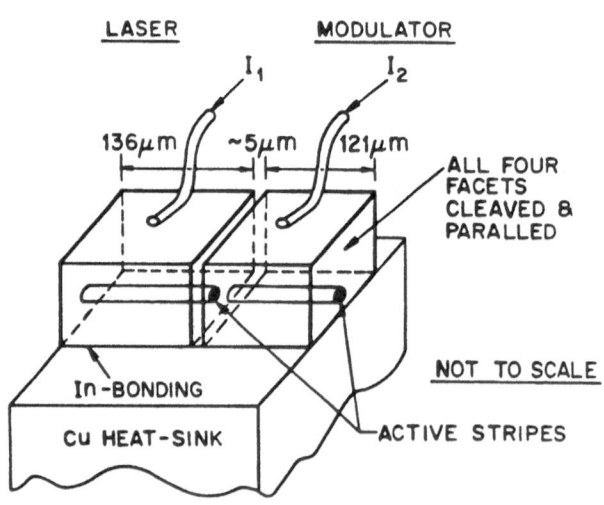

Figure 1. Schematic of a C^3 laser.

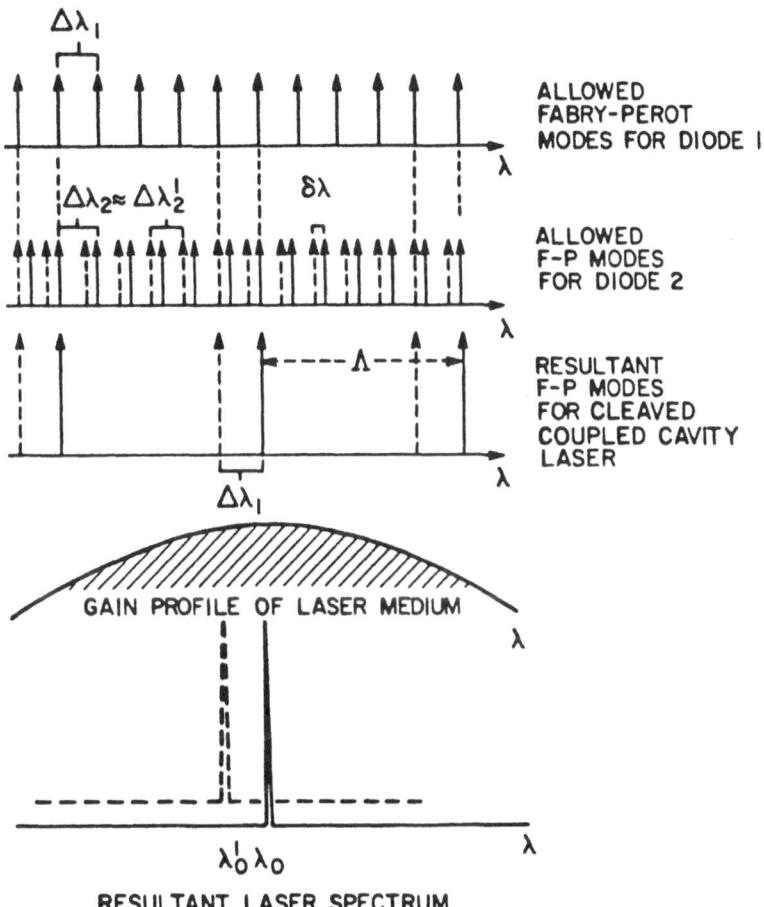

Figure 2. Basic working principle of the C^3 laser. The dashed lines indicate the changes occuring when the modulator current is changed and the laser is tuned.

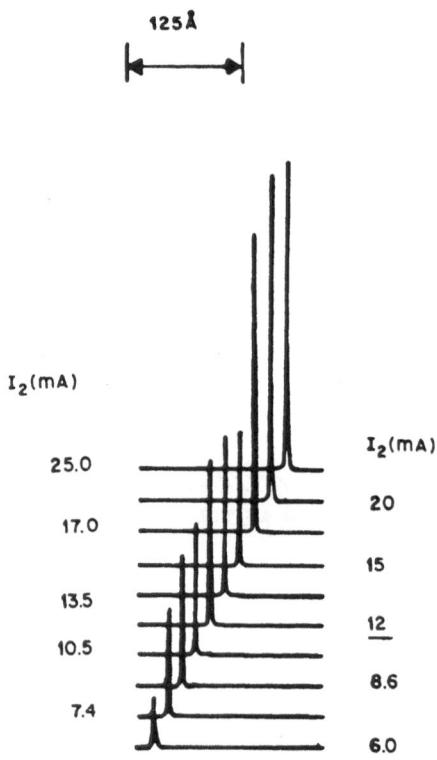

Figure 3. Example of the discrete tunability of a 1.5 μm C^3 laser.

analysis of the operation of the C^3 laser must also take into account the effect of a third cavity, that formed by the gap between the two lasers [4]).

With the above understanding, one can easily show how tunability can be achieved in C^3 lasers. Let laser 1 be biased with an injection current I1 above lasting threshold, thus it acts as a laser. Let laser 2 be biased with some current I2 below threshold, thus it acts as an etalon. Under these conditions, the situation is described by the solid lines in Fig. 2. Now, if one increases current I2, a change in the carrier density in the etalon will change its refractive index. This results in a shift in the position of the modes of laser 2 towards shorter wavelength as shown by the dashed line in Fig. 2. As a result, the F-P modes from laser 1 and the etalon that originally coincide become misaligned and the adjacent mode on the shorter wavelength side comes into coincidence and becomes the new inforced mode. This results in a shift of the lasing wavelength of the C^3 laser one mode spacing. Since the change in the refractive index necessary to shift to the next mode is small, only a small change in I2 is sufficient. Examples of the tunability of a 1.5 μm C^3 laser is shown in Fig. 3. We note here that the frequency tuning of the C^3 laser can be very fast, limited by the carrier lifetime of the modulator section. The switching time between two modes can be as fast as 1 ns. We have here discussed only the discrete tuning feature of the C^3 laser, however, there is also a regime of continous tuning both with enhanced and supressed tuning rate. Further details can be found in ref. [5].

APPLICATIONS

In discussing the applications of C^3 lasers, it is convenient to think of the C^3 laser as a device having a number of possible states. A regular laser has two states, it is either on or off. A C^3 laser, on the other hand, has many possible states. From the unique features of both single frequency operation and tunability, we can regard each possible emission wavelength as a state of the laser. This is schematically shown in Fig. 4. Most of the applications of the C^3 laser are found by assigning appropriate names to the states and then controlling the device currents to access the different states in accordance with the operation of the particular application. For example, the states can be called "channel number" or "analog level" for operation of the C^3 laser as an optical switch or analog-to-digital converter respectively. In what follows we will discuss some of the possible applications of the C^3 laser.

OPTICAL COMMUNICATION

In one of the first applications of the C^3 laser, long haul high data rate optical fiber transmission, only the single frequency characteristics of of the C^3 laser were used [6,7]. To obtain the longest possible span between repeaters in fiber optic communication systems, one must operate the laser at the wavelength where the fiber has the lowest loss. For silica based fibers currently used, this minimum loss wavelength lies in the 1.5 μm region. However, in this wavelength region, the fibers also have chromatic dispersion, that is, light pulses of different color travel down the fiber with different speeds. At 1.5 μm wavelength, the dispersion is typically 15 ps/nmKm or, for a 100 Km long fiber, two light pulses separate 5 nm (50A) in wavelength launched at the same time, will arrive at the other end 7.5 ns apart. Obviously this dispersion severely limits the data rate for long haul systems using regular semiconductor lasers with spectral widths of typically, 5 nm. By using a C^3 laser operating in a single longitudinal mode, the spectral width is reduced to approximately 0.1 nm and the effect of chromatic dispersion is dramatically reduced. Several record breaking fiber transmission experiments have demonstrated the usefulness of the C^3 laser in this respect [6,7]. We can also use the tunability of the C^3 laser to enhance the information capacity of fiber communication systems. In this case, the states, or wavelengths, of the laser represents a multilevel code. Because the information content of a signal is proportional to log2(N) where N is the number of possible states, going from a traditional binary 2 state (on-off) coding to a C^3 laser based 4 or 8 level coding, the data rate is doubled or tripled. A 4-level 2-channel system was recently demonstrated [8].

OPTICAL SWITCHING

In the optical switching application [9] of the C^3 laser we designate the possible states in Fig. 4 as different channels. In this switching system, the route or outport port an optical signal will follow is determined by the wavelength of the optical signal. By switching the lasing wavelength of the C^3 laser transmitter, different routes or output ports can be addressed. The key parameters of a switching system – the access time and the number of addressable ports – is determined in this case by how fast the C^3 laser can be tuned and by how many discrete frequencies are available from the C^3 laser. Switching times as short as 1 ns and up to 13 discrete frequencies are available, indicating that high performance switching systems can be built with this new scheme.

A general NXN switch is outlined in Fig. 5. It consists of N C^3 lasers that can be tuned to N discrete wavelengths λ_1 to λ_N. The optical outputs from the lasers are combined and enters a wavelength division multiplexer (WDD). In the WDD, the different wavelengths are separated and each wavelength is detected with a separate detector. If desirable, the separated wavelengths can be

	TOTAL POWER	λ_1	λ_2	λ_3	λ_4	λ_N
ON	REGULAR LASER	c^3	c^3	c^3	c^3	c^3
OFF	REGULAR LASER	c^3	c^3	c^3	c^3	c^3

Figure 4. The possible "states" of a regular Fabry-Perot laser and a c^3 laser.

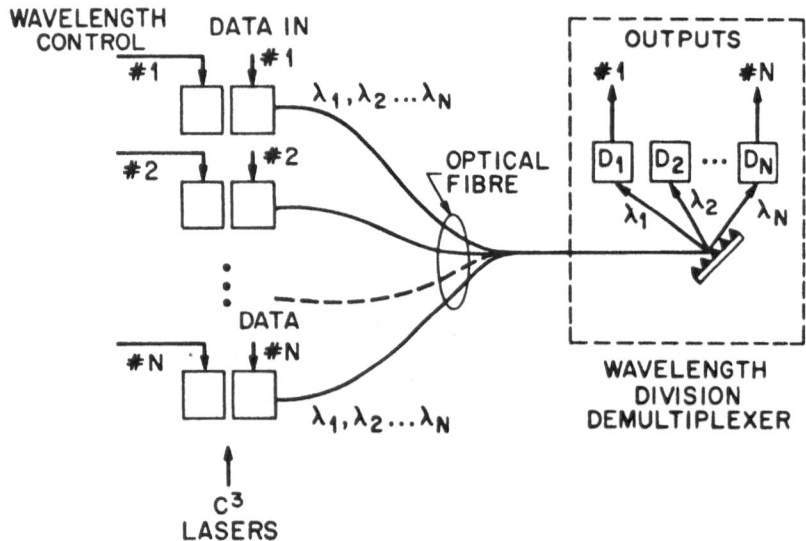

Figure 5. A general NXN optical switch based on C^3 lasers.

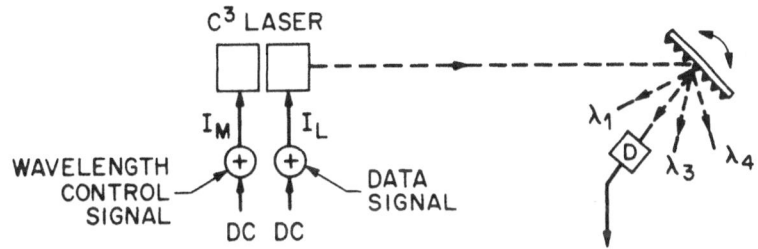

Figure 6. Experimental setup for a 1 x 4 optical switch.

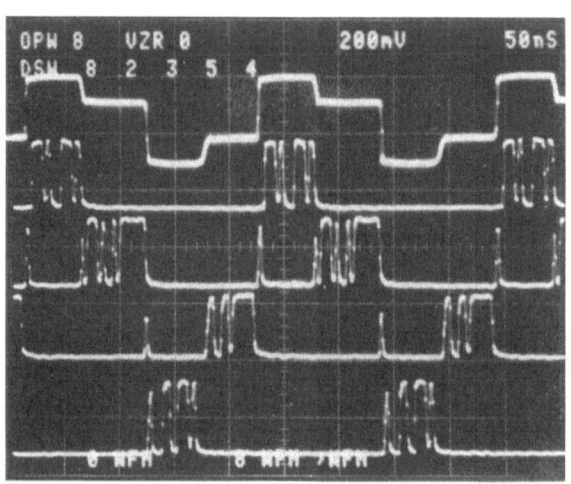

Figure 7. Waveforms for the 1 x 4 switch. Top trace: Wavelength control signal. Four lower traces: Optical outputs at the four wavelengths.

launched into lightguides for further transmission before detection.

A practical realization of a 1X4 switch using a C³ laser operating at 1.3 μm is shown in Fig. 6. The laser current consists of a DC bias and a 300 Mbit/s, 64 bit, NRZ word. The wavelength control current applied to the modulator section of the C³ laser is synchronized with the data signal and is a four level step waveform (see the top trace of Fig. 7). The step amplitudes have been adjusted such that each of the currents correspond to a different wavelength. As a result, the 64 bit data word will be transmitted in 4 subwords, each subword transmitted at a different wavelength. Using a diffraction grating, the wavelengths are separated and detected. The top trace in Fig. 7 shows the wavelength control signal and the lower traces show the optical outputs at wavelengths 1.313, 1.315, 1.317, and 1.319 μm respectively. When the laser is switched between channels that are not adjacent to each other a short pulse or "glitch" will appear in the in-between channels. This is clearly seen in Fig. 7 for the outputs at 1.315 and 1.317 μm. The glitch, however, is very short, approximately 3 ns, and does not seriously degrade the system performance. Among the advantages of this C³ laser based switching scheme is the extremely short access time, the low crosstalk and that the routing of the optical signal is determined at the transmitter.

ANALOG-to-DIGITAL CONVERSION

In the previous application, the states of the C³ laser were designated as "channel number". In this section we will describe an analog-to-digital converter [10] and the states will represent digitized analog intervals. This device is an electro-optical interface that both performs A/D conversion and, since the outputs are optical, can also simultaneously act as an optical transmitter. The scheme is based on the discrete frequency tunability of the C³ laser.

A typical lasing wavelength versus modulator current behavior is shown in Fig. 8 for a 1.3 μm GaInAsP C³ laser. The laser step tunes between the longitudinal modes of the laser section of the C³ laser and the lasing wavelength is a staircase function of the modulator current. Hence, the C³ laser acts as an level discriminator of the analog modulator current and the digitized optical output is coded in terms of the optical wavelength. If desired, the optical output can be transmitted over optical fibers for further processing at a distant location. When the laser section of the C³ laser is operated CW, the device continuously samples the analog input, however, the device can be strobed simply by pulsing the laser section on and off. To convert the digitized optical output to electrical signals, a combination of a

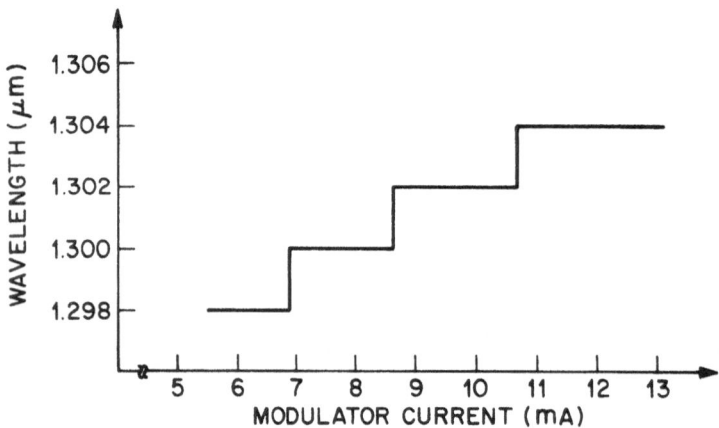

Figure 8.   Lasing wavelength vs modulator current for a 1.3 μm C^3 laser.

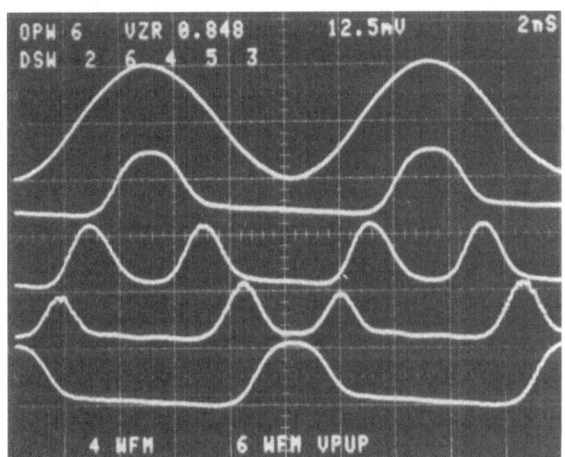

Figure 9. The 100 MHz input signal (top trace) and the digitized outputs at wavelengths 1.304, 1.302, 1.300 and 1.298 μm respectively for the lower traces.

dispersive grating and a detector array is used as shown in Fig. 9. The grating deflects each possible wavelength into a separate detector and the digitized output is available in electrical form.

To demonstrate the high speed capability of this device, a 100 MHz signal was applied to the modulator. The modulator was also pre-biased with a DC current and the laser section was operated at 2.5X the threshold current. The 100 MHz signal is shown in top trace in Fig. 9 and the resulting digitized optical outputs at wavelengths 1.304, 1.302, 1.300, and 1.298 μm are shown in four lower traces respectively. The rise time of the optical signals are detection limited and the real crossover time between two levels is less than 1 ns as obtained with a sampling oscilloscope. The less than 1 ns switching time shows that sampling rates as high as 1 Gsample/s can be used.

By using a C^3 laser which can be tuned over a wider range (the tuning range is determined by the length ratio of the two sections of the C^3 laser), the precision and dynamic range of the A/D conversion can be increased. However, there is a trade off between speed and the number of digitized levels. The resolution of the A/D conversion can also be increased by the "wrap-around" characteristics of the tuning of the C^3 laser. That is, after the C^3 laser has been tuned a certain number of longitudinal modes, the lasing wavelength jumps back to the starting position and the tuning sequence is repeated. This behavior follows from the periodic nature of the mode selection on C^3 lasers. A second C^3 laser which receives an appropriate fraction of the signal, digitizes the next decade in the same fashion as conventional A/D converters.

To summarize, we have demonstrated a new kind of electro-optical interface which acts both as an A/D converter and optical transmitter. The simple design and ultra-high-speed character-istics of the device indicates important applications in electro-optical interfacing and data processing.

SPECTRAL BISTABILITY

Bistable devices are important in several different areas such as optical computers, signal processing, and memory devices. One class of bistable devices are based on a hysteresis observed in the light current characteristics of some regular laser diodes. This bistability arises from inhomogeneous current injection along the active stripe and/or saturable absorption effects. This bistability always occurs at threshold for lasing of the device which implies two main drawbacks. First, because the "off" state is below threshold, the turn-on delay will be limited by the spontaneous lifetime leading to long delay times of several tens of nanoseconds. Second, because the "on" state is by necessity close to threshold, the available output power is limited. In this section we describe a new kind of optical bistability observed in C^3 lasers [11]. The bistability is in the spectral domain and is observed for drive currents substantially above threshold, thus leading to very short switching times and high output powers.

It was found that for some C^3 lasers when both sections are operated above threshold, the laser was spectrally bistable. That is, in the bistable region the C^3 laser can lase in either of two longitudinal modes but never in both simultaneously. Which mode is lasing depends on how the operating point was approached. It has been shown by analyzing the coupled rate equations that such bistable regions can exist and do not depend on saturable absorption [4].

The bistable region of an 1.5 µm C^3 laser is shown in Fig. 10. Spectral bistability is observed inside the wedge shaped region. As seen in Fig. 10, the bistable region is large and extends to at least to 3X the threshold current (maximum tested) and is 10 mA wide at 60 mA of laser current. For bias points inside the bistable region, lasing occurs either at 1.5102 µm when approached with decreasing modulator current or at 1.5122 µm when approached with decreasing modulator current. The variation in total output power between the two states is less than 5%. The spectrally resolved output power is shown in Fig. 11, (a) for the output at 1.5122 µm and (b) at 1.5102 µm.

To measure the switching speed between the bistable states, the modulator was biased inside the hysteresis loop and a train of short (200 ps) alternating positive and negative current pulses was superimposed on the DC bias to switch the device between the two states. The top trace of Fig. 12 shows the electrical switching pulses and the bottom trace shows the optical output at 1.5122 µm. The switching time is less than 1 ns and should be compared to typical switching times of 10-50 ns in bistable semiconductor lasers having the "off"-state below threshold.

One possible application of this bistable device would be as an ultrafast transmitter for optical frequency shift keying (FSK) communicaton systems [12]. This is shown in Fig. 13 where a bistable C^3 laser have been biased inside the hysteresis loop and a 300 Mbit/s signal is modulating the laser between its two states. The two top traces show the optical outputs at the two wavelengths and the bottom trace the electrical data signal. The advantages of this system as compared to regular on-off keying are threefold; First, the laser operates at full power all the time, thus giving a 3dB power advantage. Second, the extinction ratio is automatically extremely high. Third, the amplitude of the modulation signal is non-critical. The only requirement is that

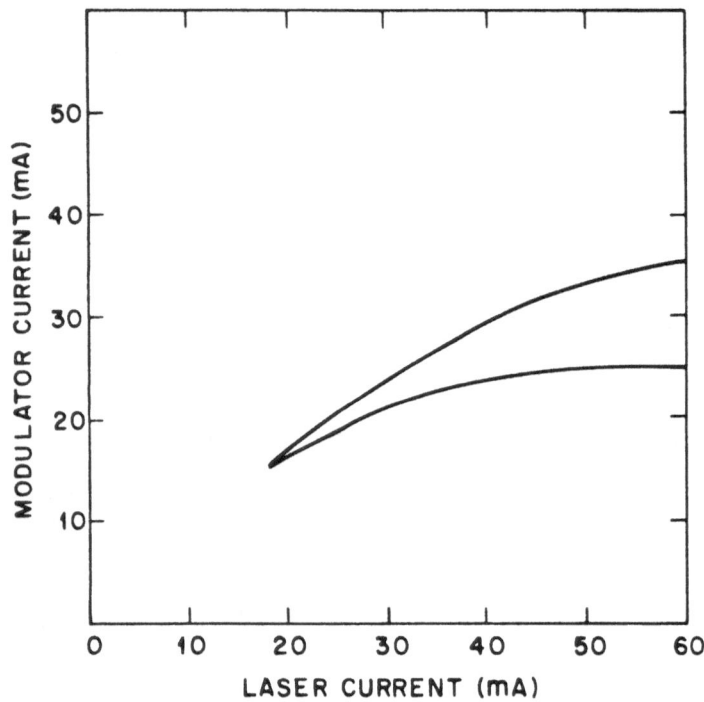

Figure 10. The spectrally bistable region of a 1.5 μm C³ laser. Spectral bistability is observed inside the wedge shaped region.

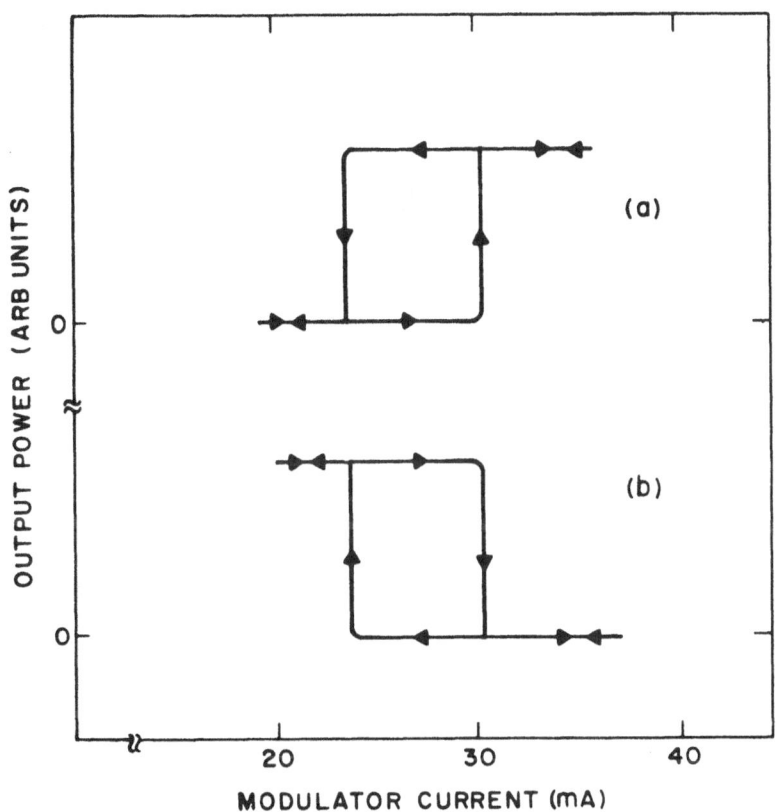

Figure 11. Spectrally resolved output power. (a) at 1.5122 μm and (b) at 1.5102 μm.

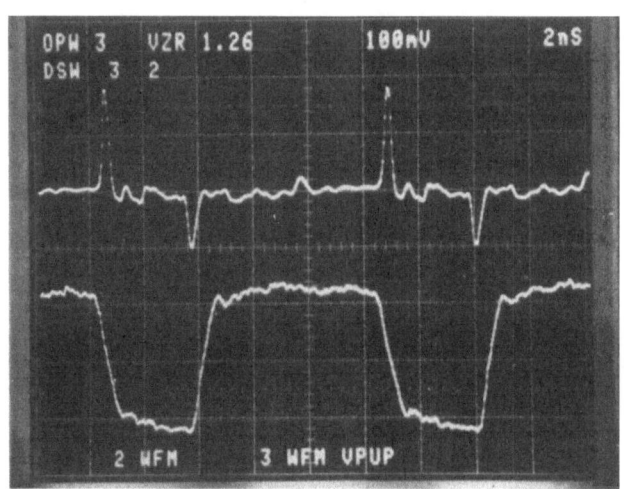

Figure 12. High speed switching between the bistable states. Top
trace: the electrical   switching   pulses.   Bottom trace: the
optical output at 1.5122 µm.   The time scale is 2 ns/division.

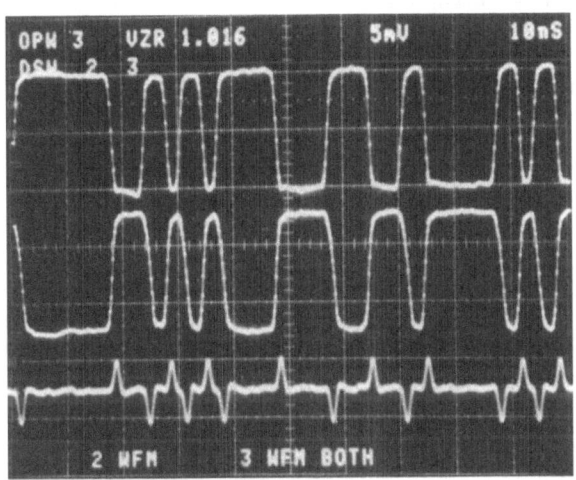

Figure 13. Frequency shift keying with a spectrally bistable laser. The two upper traces are the complementary outputs at the two wavelengths. The bottom trace shows the electrical data signal.

they are large enough to bring the modulator momentarily outside the bistable region.

FIBER MEASUREMENTS

As a last example of applications of C^3 lasers, we describe a technique for measurements of the chromatic dispersion of optical fibers [13]. It is a phase-shift technique based on the wavelength dependence of the phase shift an amplitude modulated light wave experience after propagation through the test fiber. The experimental set-up for this measurement is shown in Fig. 14. A C^3 laser is amplitude modulated at some frequency f and the light is coupled into the fiber to be measured. A RF vector voltmeter measures the phase shift between the amplitude modulated light at the beginning and end of the fiber. During the measurement, the C^3 laser is tuned between its longitudinal modes and the corresponding phase shifts are recorded. The dispersion, D, of the fiber can be expressed as:

$$D = \frac{1}{2\pi f \ell} \frac{d\Phi}{d\lambda}$$

where f is the modulation frequency, ℓ the fiber length and $d\Phi$ the phase shift measured for a wavelength shift $d\lambda$. Fig. 15 shows such a measurement for a 7.5 km long fiber at 1.52 μm wavelength. In Fig. 15, the phase shifts have been recalculated to an equivalent time delay and the slope of the curve gives a dispersion of 15.6 ps/nmkm. This is a highly accurate method for dispersion measurements, for example, for a 1 km length of fiber and a modulation frequency of 1 GHz, a chromatic dispersion as small as 0.03 ps/nmkm can be measured. With a small modification to the measurement routine, measuring the phase shift as function of the modulation frequency, the propagation delay and hence the fiber length can be determined with a 10-7 accuracy.

SUMMARY

This paper has given a condensed overview of the operational characteristics and some of the possible applications of the C^3 laser. The interested reader will find the details of the experiments in the cited references. Reference 4, for example, gives a thorough account of the theory of the C^3 laser. Obviously, we have not covered all the applications of the C^3 laser. With its unique features of single frequency operation, ultra high speed amplitude and frequency modulation, both discrete and continuous, bistable operation, compactness, and ease of manufacture, most applications are probably yet to be seen.

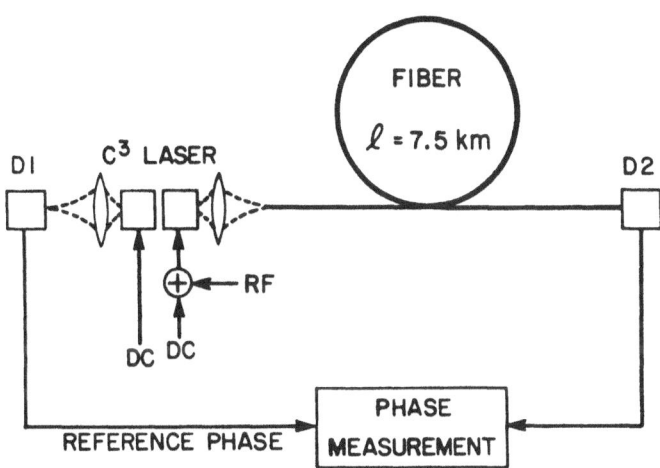

Figure 14. Experimental setup for fiber dispersion measurements.

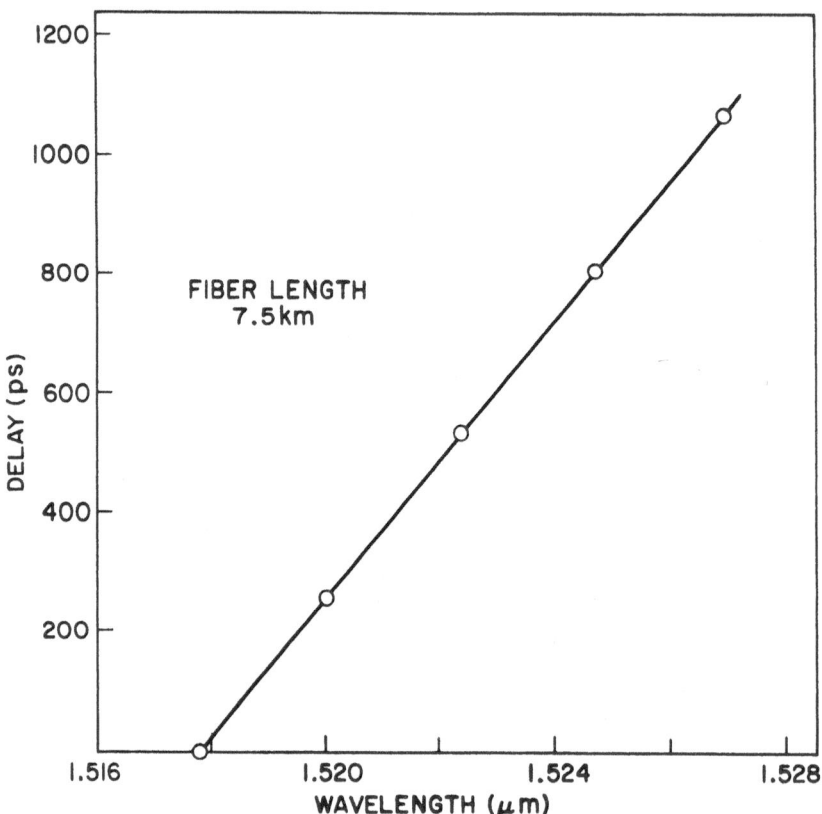

Figure 15. Phase shifts (re-calculated as equivalent time delays) versus wavelength of the C^3 laser around 1.5 μm and for a 7.5 km long fiber.

REFERENCES

1. W. T. Tsang, N. A. Olsson and R. A. Logan, "High speed direct single frequency modulation with large tuning rate and frequuency excursion in cleaved coupled cavity semiconductor lasers", Appl. Phys. Lett., 42, p 650, 1983.
2. W. T. Tsang, "The C^3 laser", Scientific American, 251, p 148, 1984.
3. N. A. Olsson, N. K. Dutta, P. Besomi, R. J. Nelson, R. A. Linke and R. S. Tucker, "2Gbit/s operation of single longitudinal mode 1.5 μm DCPBH C^3 lasers", Electron. Lett., 20, p 395, 1984.
4. C. H. Henry and R. F. Kazarinov, "Stabilization of single frequency operation of coupled cavity lasers", IEEE J. of Quantum Electronics, QE-20, p 733, 1984.
5. W. T. Tsang and N. A. Olsson, "Enhanced frequency modulation in cleaved coupled cavity semiconductor lasers with reduced spurious intensity modulation", Appl. Phys. Lett., 43, p 527, 1983.
6. W. T. Tsang, R. A. Logan, N. A. Olsson, et al, "119 km, 420 Mbit/s transmission with a 1.55 μm single frequency laser", Proc. OSA conference OFC-83, New Orleans, 1983.
7. R. A. Linke, et al, "1 Gbits/s transmission experiment over 101 km of single mode fiber using a 1.55 μm ridge guide C^3 laser" Electron Lett., 19 p 775 1983.
8. W. T. Tsang, N. A. Olsson, and R. A. Logan, "A demonstration of multilevel multi channel optical frequency shift keying with cleaved coupled cavity lasers", Electron. Lett., 19, p 341, 1983.
9. N. A. Olsson and W. T. Tsang, "An optical switching and routing system using frequency tunable cleaved coupled cavity semiconductor lasers", IEEE J. of Quantum Elec., QE-20, p 332, 1984.
10. N. A. Olsson, W. T. Tsang, R. A. Logan and C. K. N. Patel, "Gigahertz bit rate analog to digital conversion with optical outputs using cleaved coupled cavity semiconductor lasers", Appl. Phys. Lett., 43, p 1091, 1983.
11. N. A. Olsson, W. T. Tsang and R. A. Logan, "Spectral bistability in coupled cavity lasers", Appl. Phys. Lett., 44, p 375, 1984.
12. N. A. Olsson and W. T. Tsang, "Wideband frequency shift keying with a spectrally bistable cleaved coupled cavity semiconductor laser", Electron. Lett., 19, p 808, 1983.
13. N. A. Olsson, N. K. Dutta, R. A. Logan and P. Besomi, "Fiber dispersion and propagation delay measurements with frequency and amplitude modulated cleaved coupled cavity semiconductor lasers", Optics Lett., 9, p 180, 1984.

HIGH POWER, TUNABLE WAVEGUIDE CO_2 LASERS

Franco Strumia and Nadia Ioli

Dipartimento di Fisica - Università di Pisa and
GNSM - CNR
Piazza Torricelli 2,   56100 Pisa - Italy

INTRODUCTION

The CO_2 laser is the most popular source of coherent mid-infrared radiation and is widely used both for scientific and technological applications. High efficiency and high power can be obtained either in CW and in pulsed regime. The emitted wavelength can be tuned over many different lines in the interval 9-11 μm by using a diffraction grating as an intracavity dispersive element. The CO_2 laser can also be frequency stabilized with high reproducibility. As a secondary frequency standard it plays a fundamental role in the measurement of the speed of light and in the new definition of the unit of length. Other important scientific applications are plasma generation, molecular multiphoton dissociation, isotopic separation, LIDAR, molecular spectroscopy, and generation of medium and far-infrared (MIR,FIR) coherent radiation either by stimulated Raman scattering or resonant optical pumping of molecular transitions. By means of the molecular FIR laser optically pumped by the CO_2 laser this spectral region was covered for the first time with thousands of CW laser lines of relatively high power so that the laser spectroscopy has been extended down to the microwave region [1].
Many different designs have been developed for the CO_2 laser in order to match such a broad range of applications. The use of waveguide resonators is dictated by the need to increase the frequency tunability around each CO_2 lasing line in the CW operation.

Another advantage of the waveguide laser is its smaller and compact size.

The feasibility of hollow waveguide gas discharge lasers was demonstrated in the case of He-Ne [2] and CO_2 [3] lasers. Reviews on hollow waveguide laser have been already published in the past[4,5]. In the present paper we will be more concerned with the description of relatively high power CO_2 WG laser with full tunability around a large number of individual CO_2 laser lines.

The best results were obtained in our laboratory where the possibility of optically pumping FIR laser lines with a large pump offset was demonstrated[6,7].

THE CO_2 LASER

As is well known the active medium is a gas mixture of CO_2, N_2 and He. The laser emission takes place between the first asymmetric rotovibrational state (0,0,1) and the symmetric states (1,0,0) and (0,2,0) that are strongly coupled by Fermi resonance. For symmetry rules only the even J levels exist in the (1,0,0) and (0,2,0) states and P and R laser lines with even J can be obtained for $2 \leq J \leq 50$. A first band with P and R branches is located in the 10-11 μm region, a second band in the 9-10 μm region.

The (0,0,1) level is selectively populated by resonant collisional energy transfer from the vibrationally excited N_2 molecules. The depopulation of the lower laser levels (1,0,0) and (0,2,0) is greatly enhanced by the fast thermalization of the He gas [8,9].

THE HOLLOW WAVEGUIDE

In 1964 Mercatili and Schmeltzer demonstrated theoretically [8,10] that a hollow dielectric tube can transmit the E.M. radiation with a very small attenuation if the inner diameter $d \gtrsim 10^2 \lambda$. It is not necessary for the guide material to be transparent to the E.M. radiation, but on the other hand even a small bending of the WG increase dramatically the transmission losses.

The E.M. mode transmitted with the lowest attenuation is the EH_{11}. The field is linearly polarized and the intensity along the radius r is proportional to the Bessel function J_0 (fig.1)

$$E \propto J_0(2.405 \, x/\tau) \qquad \qquad 1)$$

The attenuation constant is given by:

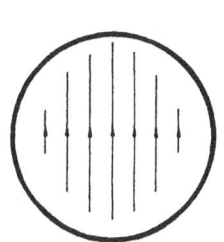 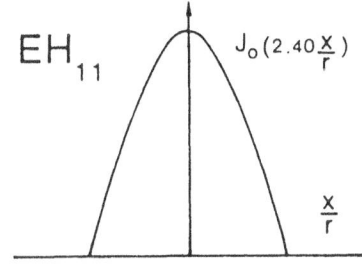

$$EH_{11} \qquad J_0\left(2.40\frac{x}{r}\right)$$

FIG. 1

$$a_{11} = \left(\frac{1.2\lambda}{\pi}\right)^2 \frac{1}{r^3} Re\left[\frac{\nu^2+1}{2\sqrt{\nu^2-1}}\right] \qquad 2)$$

where $\nu = \sqrt{\varepsilon_1}$ is the complex refraction index of the waveguide material. For $\lambda = 10\ \mu m$ a low attenuation is obtained for $r \geq 0.5$ mm. At the waveguide end the guided EM modes are coupled to the free space beam modes. It was demonstrated [11] that more than 98% of the EH_{11} mode power is coupled into a TEM_{00} gaussian mode with a waist

$$w_0 = 0.643\ r \qquad 3)$$

The WG resonator can be terminated with flat mirrors. The coupling losses between the WG and a flat mirror at a distance z are given by [12]

$$\delta' \simeq 0.57(z/b)^{3/2} \qquad 4)$$

where $z \ll b$ and $b = \pi w_0^2/\lambda$ is the confocal parameter of the gaussian beam.

The materials generally used for the hollow WG are Pyrex glass, fused SiO_2, and ceramics like Al_2O_3, BeO, BN that have a better heat conductivity. For these reasons ceramic is preferred in commercial sealed-off laser. However the hole must be mecanically drilled. On the contrary Pyrex or fused SiO_2 WG are easily found in any length and diameter with high quality holes. Since the ceramic materials do not offer any other advantage, glass WG are to be preferred in laboratory lasers. In fact it was found[13] that the attenuation constant was $a_{11} \simeq 10^{-3}$ cm^{-1} for all the above materials when $r \approx 0.1$ cm. Moreover it is experimentally demonstrated that the best output power per unit length of WG is the same for the above materials.

The interest for WG CO_2 lasers is motivated by the following points:

-laser emission on the fundamental TEM_{00} mode can be easily obtained since the waveguide selective attenuation acts as a filter for the high order modes;

-the small signal gain α_o is 2 - 3 times larger than in conventional laser allowing a very short cavity length L;

-the pressure of the active medium can be larger also increasing the homogeneous linewidth of the gain curve well above the Doppler width. Since the cavity free spectral range is given by $\Delta \nu_c = c/2nL$ a combination of a short cavity and a relatively high pressure allows an increase of the frequency tunability around each CO_2 laser line up to an order of magnitude. However the resolving power of the grating is proportional to the number of lines shined by the laser beam. For a diameter less than 2.5-3 mm the resolving power is to small and the unwanted line hopping effect cannot be avoided expecially in the case of the 9R branches. It is worth noting that the use of an intracavity beam expander is not convenient since both cavity length and cavity losses are increased.

DESIGN CRITERIA FOR CO_2 LASER

The Doppler width of the CO_2 laser lines is $\Delta \nu_D$ = 66 MHz at the average temperature of the gas discharge ($\simeq$ 420 °K), while the pressure broadening is

$$\Delta \nu_H = \Gamma = 3.3 \text{ MHz / mbar} \qquad 5)$$

for a gas mixture He:N_2:CO_2 = 4:2:1 [4,5,14]. As a consequence the CO_2 laser has a homogeneous gain for a pressure p > 15-20 mbar. In this case the output intensity , in the limit of a small gain is given by

$$I_{ou} = \frac{T I_s}{2} \left(\frac{2 \alpha_o l}{\delta + T} - 1 \right) \qquad 6)$$

where α_o is the unsaturated gain, l the discharge length, T the transmission of the output mirror, δ the roundtrip cavity losses, and I_s the saturation intensity. From eq. 6 the optimum transmission T can be computed and the maximum power is given by

$$P_M = I_{ou}^M A = \frac{A I_s}{2} \left(\sqrt{2 \alpha_o l} - \sqrt{\delta} \right)^2 \simeq A I_s \alpha_o l \qquad 7)$$

192

where A is the beam cross-section that for a gaussian beam is conventionally $A = \pi w_0^2$. In the case of a laser with a large single pass gain a more general equation can be written

$$I_{ou} = \frac{T I_s \sqrt{R}}{(\sqrt{R} + \sqrt{R_1})(1 - \sqrt{RR_1})} \left[\alpha_0 \ell + \ln \sqrt{RR_1} \right] \qquad 8)$$

where R is the reflectivity of the output mirror and R_1 that of the other mirror or the diffraction grating. The coupling losses defined in eq. 4 are included in the actual reflectivity

$$R = 1 - T - A' - \delta' \qquad 9)$$

where A' is the absorption of the mirror or grating. The small signal gain at the center of the laser lines $\alpha_0(\nu_0)$ increases proportionally to the gas pressure ($\alpha_0 \propto p$) for a Doppler broadened line and becomes independent of p when $\Gamma \gtrsim \Delta\nu_D$. By increasing further the pressure the heath diffusion rate toward the WG walls decreases reducing the population inversion and $\alpha_0 \propto D/p$, where D is the diffusion coefficient of the mixture. In conclusion α_0 has a maximum for a pressure between 20 or 100 mbar depending on the WG diameter and gas mixture.
The saturation intensity I_s for molecular transitions is given by

$$I_s = 4\pi^2 h \nu \Gamma \tau_{sp} \left[\lambda^2 \left(\frac{1}{K_u} + \frac{1}{K_\ell} \frac{g_u}{g_\ell} \right) \right]^{-1} \qquad 10)$$

where u and l refer respectively to the upper and lower level, τ_{sp} is the lifetime of the upper level for spontaneous emission, and K_u and K_ℓ are the relaxation rates. As a consequence we expect $I_s \propto p^2$. This result holds only for a low pressure since the diffusion rate of the molecules in and out of the laser beam is an important effect [16]. Thus for higher pressure we have $I \propto p$. By increasing further the pressure the optimum electron temperature cannot be longer obtained in the glow discharge and I_s is expected to reach a maximum and than to decrease [17]. The optimum electron temperature is obtained for $E/N \approx 2 - 3 \ 10^{-16}$ v cm^2.

EXPERIMENTAL RESULTS

We have built several CW CO_2 lasers by using either Pyrex or fused SiO_2 commercial tubes, obtaining an output power per unit length of discharge of 40-45 W/m with a grating tuned cavity and

DC excitation. We obtained also a full free spectral range tunability and a TEM$_{OO}$ mode for over 85 CO_2 lines. These results are superior to those reported in the literature and are a demonstration of the convenience in using Pyrex tubes for laboratory lasers.

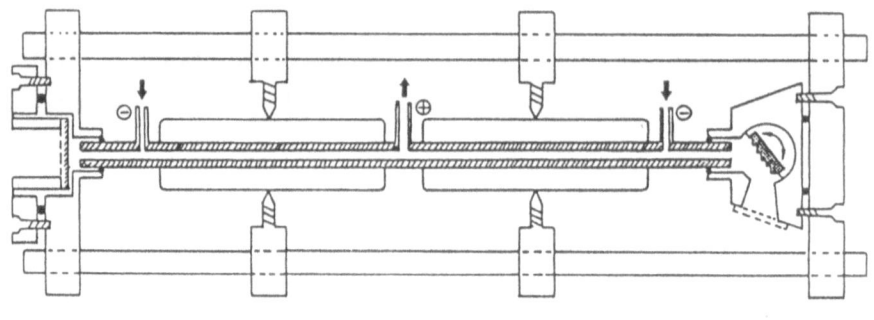

FIG. 2

In fig 2 is shown schematically the structure of the WG laser. The WG is fixed to the end plates of the frame by O-rings. The gas in and outputs are obtained by drilling 2-3 mm holes in the WG and in the sidearms are also inserted the stainless steel electrodes 6 mm in diameter. The waveguide is also supported by screws that are used for the final adjustment. The adjustment is made while the laser is running by looking for the maximum output power which coincides with a perfectly straigth WG and a TEM$_{OO}$ mode.

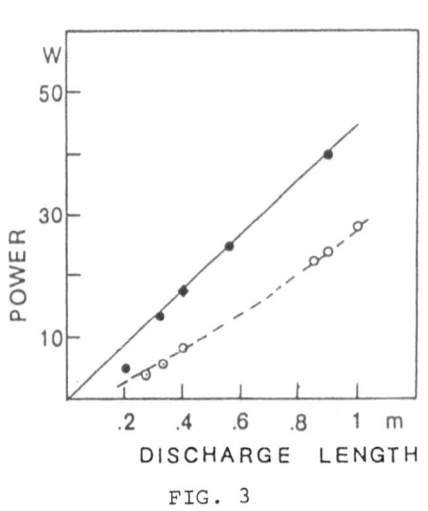

FIG. 3

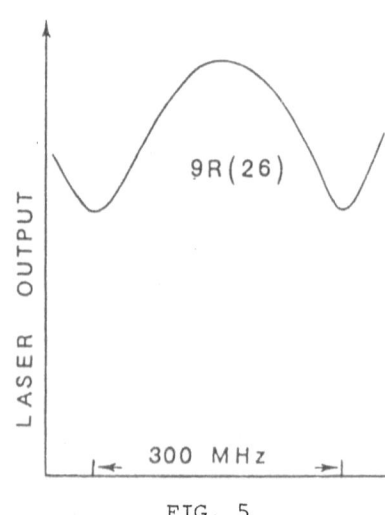

FIG. 5

The flexibility of the Pyrex WG is one of the most important advantages over the ceramic WG where this final adjustement is impossible. The grating is contained in a vacuum tight box and can be rotated independently of its alignment with the WG. A ZnSe window allows the extraction of the zeroth order beam. The output mirror is mounted on a PZT for laser frequency tuning. The cooling is obtained by circulating a water solution of ethylene-glycol. The most convenient temperature was found to be between -10 and 0 °C for WG with an external diameter of 8 mm. In fig 3 is shown the output power obtained as a function of the discharge length. The inner diameter of the WG was between 3 and 5 mm. It is remarkable the increase in the output power obtained by substituting conventional commercial gratings with high efficiency gratings (reflectivity about 98 %). In the second case the output power is the same obtained for a laser without grating. (O Conventional gratings, ● R = 98% gratings, ◆ without grating). In fig 4 is shown the power obtained for different CO_2 lines and TEM_{OO} mode (Pyrex WG r = 1.6 mm, l = 32 cm). For every line a frequency tuning corresponding to the cavity free spectral range was observed as shown in fig 5.

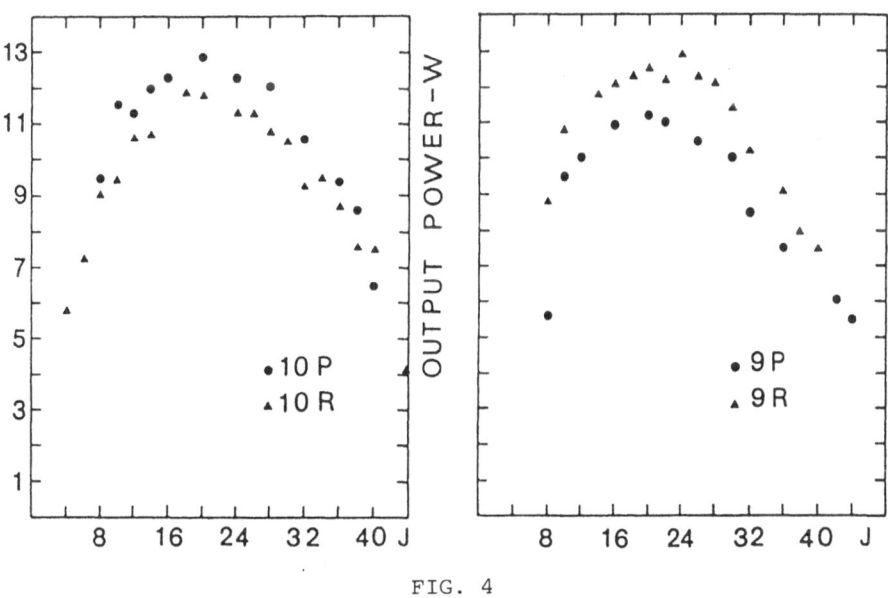

FIG. 4

The discharge current can also be chopped and the laser emits pulses of 50-100 µs with a peak power 5-10 times larger than that observed in CW regime and a repetition frequency of 500-1500 Hz.

This pulsed regime is particularly useful for optically pumped far infrared lasers.

Since the gain in WG laser is large, also a waveguide amplifier can be a useful and practical device. It is then possible to use very short WG laser with a large tunability and to increase the low power by the amplifier. By using a laser amplifier it is also possible to obtain accurate measurements of α_o and I_S .

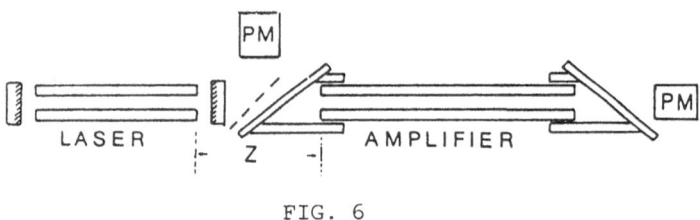

FIG. 6

The experimental apparatus is shown in fig 6. The coupling between the laser and the waveguide amplifier is obtained by simply aligning the amplifier on the laser beam and using a maximum distance z = 10-15 cm. In this case about 90% of the laser power is transmitted by the WG. We have also observed that the amplified beam conserves the gaussian TEM $_{oo}$ mode as shown in fig 7 (the continuous line is the best-fit of a gaussian to the experimental points).

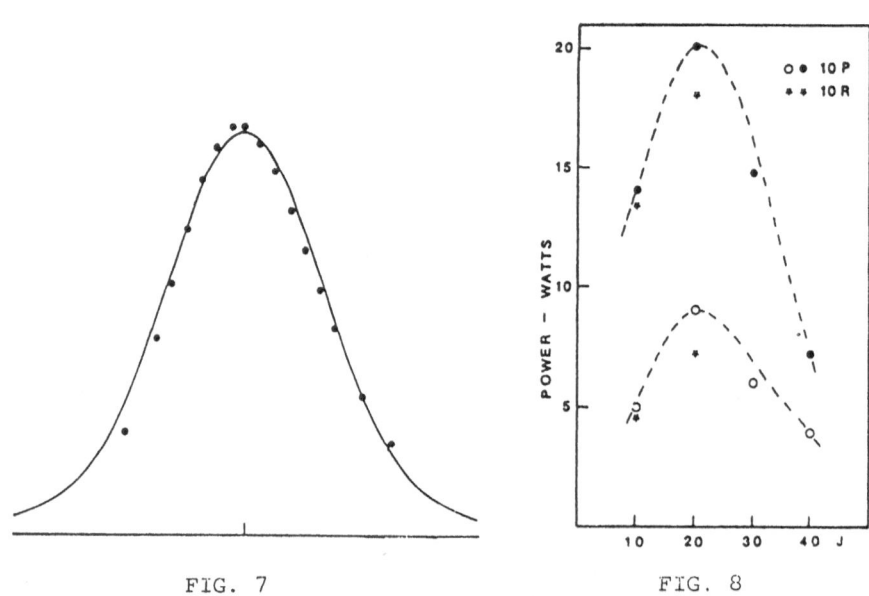

FIG. 7 FIG. 8

This result demonstrates that the WG acts as a filter against a saturation distortion also in a single pass regime ad that the measurements of α_o and I_S are greatly simplified since the input and outpt powers may be directly compared.

In fig 8 is shown an example of the power increase of the laser beam obtained with a relatively short (l = 86 cm, r = 1.75 mm, 7 mA, 45 Torr.) WG amplifier. The laser power is more than doubled and the power extracted from the amplifier is about 30-40 % of that obtainable if the same WG would be.used as oscillator. The efficiency should improve by increasing the amplifier length or the input power. The amplification is given by the equation:

$$\ln (I_2 / I_1) + (I_2 - I_1) / I_S = \alpha_o l \qquad\qquad 11)$$

where I_2 and I_1 are the output and input intensities respectively and $P = I \pi w_o^2$.

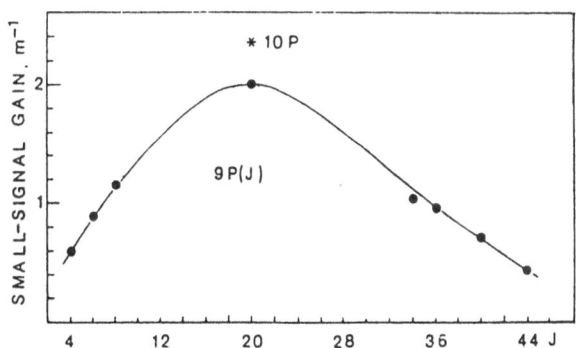

FIG. 9 - Values of α_o for a WG of l = 90 cm, r = 1.9 mm, i=3.5 mA, p = 35 Torr , He:N$_2$:CO$_2$ = 2:1.5:1.

In fig 9 are shown the α_o values obtained for several lines of the P branches. In fig 10 are shown the values of α_o and I_S as a function of pressure and for two different discharge currents. The qualitative dependence on the pressure is in agreement with the prediction of the previous section. We have also found that, while α_o decreases for high and low J values (fig 9), the correspondig I_S values increase. The product $\alpha_o I_S$ was found to be independent of J within the experimental errors ($\sim$ 20%) for all the CO$_2$ laser lines. This result is in agreement with the theory since $\alpha_o \propto |\langle J|R|J'\rangle|^2$ and $I_S \propto (|\langle J|R|J'\rangle|^2)^{-1}$. The above measurements were obtained with the laser frequency set at the center of the

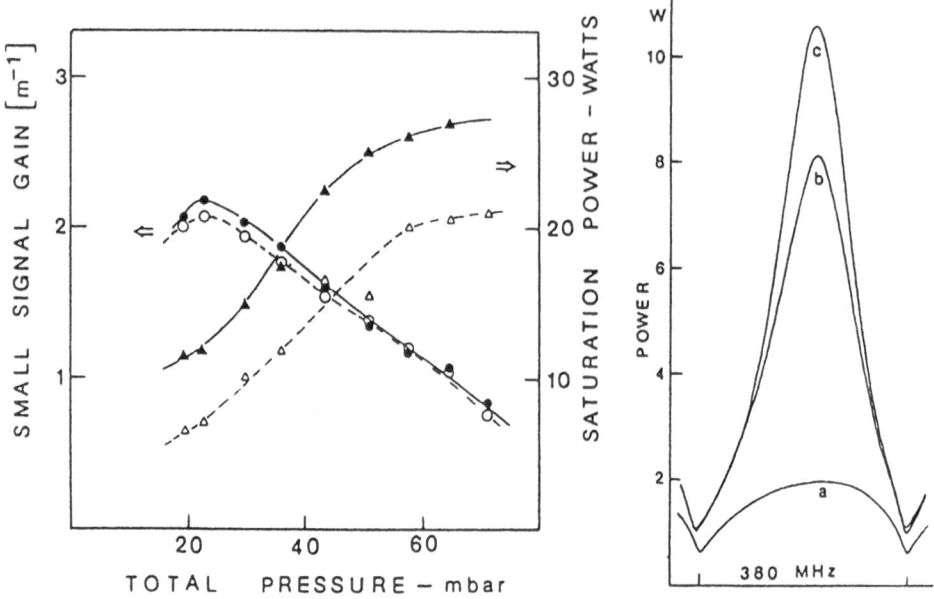

FIG. 10 -Values of α_0 and $P_s = \pi w_0^2 I_s$ for the 10P(20) line and a SiO$_2$ WG as a function of gas pressure: r=1.9 mm, l=90 cm He:N$_2$:CO$_2$=4.3:1.3:1, continous line i=5.5mA, broken line i=3 mA.

FIG. 11 -Amplification of the laser beam as a function of the frequency tuning: l=133 cm, r=2.5 mm, i=5 mA, mixture and line as in fig 10, a) amplifier off, b) amplifier on 40 mbar, c) 30 mbar.

CO$_2$ lines. The amplification of a WG as a function of the laser frequency is shown in fig 11. The best amplification at the end of the free spectral range is obtained for a relatively low pressure. In fact the broadening of the gain curve obtained by increasing the pressure cannot compensate the global decrease of the gain as shown in fig 10.

CONCLUSION

We have demonstrated that Pyrex glass WG can be conveniently used for CO$_2$ laser. The output power of about 40-45 W/m can be obtained for a gas pressure of about 50 mbar. The power is independent of the WG diameter that however must be at least 3 mm for obtaining a full tunability around each laser line also in the case of R branches. WG amplifier can be conveniently used for doubling or tripling the laser output power without degradation of the TEM_{oo} gaussian mode.

REFERENCES

1. F. Strumia, " High resolution laser spectroscopy in the Far-Infrared" in " Advances in laser spectroscopy ", F.T. Arecchi, F. Strumia, H. Walther eds, Plenum Press- NATO-ASI Series vol. B95, New York 1983, pag. 267
2. P.W. Smith, Appl. Phys. Lett. $\underline{19}$, 132 (1971)
3. T.J. Bridges, E.G. Burkhart, P.W. Smith, Appl Phys. Lett. $\underline{20}$, 508 (1972)
4. J.J. Degnan, " The waveguide laser: a review" Appl. Phys. $\underline{11}$, 1 (1976)
5. R.L. Abrams, " Waveguide gas lasers " in " Laser Handbook vol. 3 ", M.L. Stitch ed., North-Holland, 1979 pag. 41
6. N. Ioli, G. Moruzzi and F. Strumia, Lett. Nuovo Cim. $\underline{28}$, 257 (1980)
7. M. Inguscio, N. Ioli, A. Moretti, G. Moruzzi and F. Strumia Opt. Comm. $\underline{37}$, 211 (1981)
8. P.K. Cheo, " CO_2 lasers " in " Lasers vol. 3 ", A.K. Levine and A.J. De Maria eds., Dekker N.Y. 1971 pag. 111
9. D.C. Tyte, " Carbon dioxide lasers " in " Advances in quantum electronics vol. 1 ", D.W. Godwin ed. Academic Press NJ 1970 pag. 129
10. E.A. Marcatili and Schmeltzer, Bell System Tech. J. $\underline{43}$, 1783 (1964)
11. R.L. Abrams, IEEE J. Quant. Electron. $\underline{QE\ 8}$, 838 (1972)
12. J.J. Degnan and D.R. Hall, IEEE J. Quant. Electron $\underline{QE\ 9}$, 901 (1973)
13. D.R. Hall, E.K. Gorton, R.M. Jenkins, J. Appl. Phys. $\underline{48}$,1212 (1977)
14. N. Ioli and F. Strumia, " Laser a CO_2 a guida d'onda" in Final Report of "Progetto Finalizzato Laser di potenza" CNR in press.
15. W.W. Rigrod, J. Appl. Phys. $\underline{36}$, 2487 (1965)
16. C.P. Christensen, C. Freed and H.A. Haus, IEEE J. Quant. Elec. $\underline{QE\ 5}$, 276 (1969)
17. V.V. Grigoryants, B.A. Kuzyakov and A.M. Simitsyn, Sov. J. Quantum Electr. $\underline{9}$, 158 and 452 (1979)

MID INFRARED OPTICALLY PUMPED MOLECULAR GAS LASERS

R.G. Harrison and H.N. Rutt*

Department of Physics, Heriot-Watt University
Riccarton, Edinburgh, U.K.
*UKAEA Culham Laboratory, Abingdon, Oxon, U.K.

I INTRODUCTION

Optically pumped molecular lasers (OPML) based on rotational
transitions have been long recognised as sources of efficient and
powerful far infrared (FIR) emission (50 μ - 3 mm). More recently
the spectral coverage of these systems has been extended to the
mid infrared (5-50 μm); motivated by the urgent need for such
sources in photochemistry and spectroscopy. Based on the resonant
excitation of vibrational rotational transitions, for which
moderately high gains of 10^{-2} to 10^{-1} cm^{-1} may be readily obtained,
many emissions in this spectral region have already been obtained,
with powers in favourable cases being in excess of 1 MW and photon
conversion efficiencies approaching 100%. The striking success of
this approach ,providing an efficient, non destructable and simple
method for infra-red generation owes much to the development of
several powerful pulsed discharge excited molecular lasers. Of
these the TEA CO_2 system has emerged as the dominant pumping source
generating an efficient multi megawatt single line emission step
tunable ($\sim$ 2 cm^{-1} step interval) over a fairly wide range (9-11μm)
for which numerous molecules have vibrational absorption bands.
The emissions shown in fig. (1), obtained using this pump source,
contribute most of those reported over the last eight years; the
predominance of emissions around 16 μm being in response to the
demand for a source at this wavelength for the uranium enrichment
programme. This chapter briefly reviews understanding and
development of these systems over the last decade with some
emphasis on recent progress in the field since 1982; comprehensive
reviews on work prior to this date are to be found in refs (1-4).

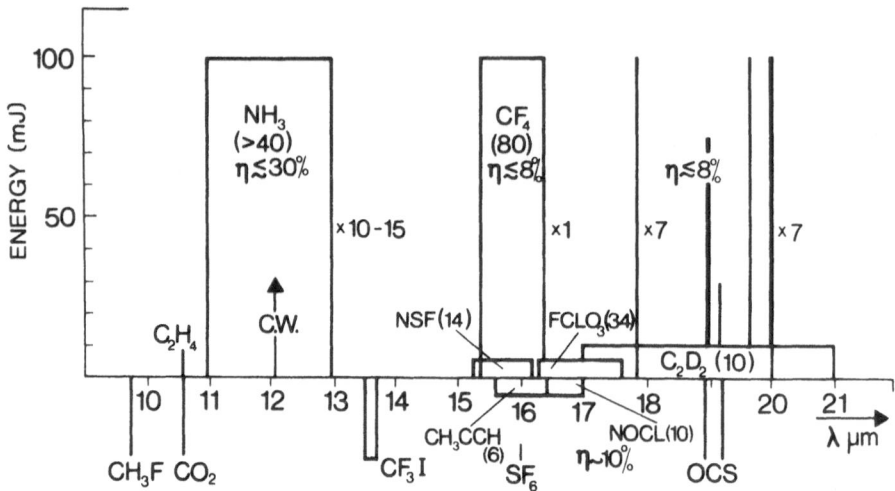

Fig. 1.   OPML emissions obtained with TEA CO_2 laser.
Those below axis have unspecified output energies.
Number of emissions (in brackets) together with
efficiency η are given for each gas.

II SPECIFIC OPTICALLY PUMPED MOLECULAR SYSTEMS

Depending on the nature of pumping and lasing transitions
OPMLs can be categorised in 3 groups as shown in fig. 2, each of
which has its own characteristic features.

a)   Lasers based on inversion in a hot band

For the first subgroup, type A_1, a combination level is excited
and lasing occurs on a hot band.   Type A_2, where an overtone level
is excited followed by lasing on a single vibrational quantum, can
be considered to be a special case of type A_1.   In both of these,
because the pump transition involves two or more vibrational quanta
absorption of the pump beam is often small, but gain on laser
transitions can be large since a single quantum transition occurs.

These lasers are far the most promising as regards a search for new MIR sources, most of the emissions shown in fig. (1) being of this type. However pressure scaling of these systems, desirable for continuously tunable emission is limited since collisional depopulation of the excited levels can be fast and also because of the presence of ground state absorption close to the frequency of the lasing band.

Many criteria affect the suitability of a molecule for use in OPMLs of this type. Evidently the possession of an appropriate pair of energy levels whose combination band is symmetry allowed is essential. Other simple guidelines are given below[51] some of which are also relevant to other types of OPML discussed here.

i. Laser transition strength. The inversion is effectively destroyed at approximately the gas kinetic rate by R-R collisional transfer. For pressures $\sim$ 1 Torr, and assuming a short pump pulse this lifetime is 100 ns. To obtain build up from nosie we require $\alpha c \tau \geq 30$ where α is the laser gain, τ the inversion lifetime and c the speed of light. Hence we require $\alpha > 1\% \text{ cm}^{-1}$, corresponding to a moderate strength infrared band.

ii. Pump transition strength. To just achieve inversion we require

$$\frac{\Delta\nu_A}{\Delta\nu_p} \frac{I\alpha_p}{h\nu} = \exp(-E/kT) \, Nd/Q$$

where I is the pump fluence (J cm^{-2}), α the pump absorption coefficient, ν the pump frequency, E the lower laser level energy, Q the total partition function, d the lower laser level degeneracy (we assume d $\gg$ 1) and $\Delta\nu_p$, $\Delta\nu_A$ the pump laser and absorption line widths respectively. The other symbols have their usual meaning. Taking typical conditions of 1 Torr, 200 K and a 400 cm^{-1} lower level we have

$$I \, \alpha_p \geq 4 \times 10^{-5} \, (\Delta\nu_p/\Delta\nu_A) \cdot (d/Q)$$

Since d/Q is typically $\ll$ 0.1 and the ratio of line widths 25, and I $\sim$ 1 is easily achieved with a CO_2 TEA laser, even extremely weak pump absorptions will suffice. Such weak combination bands are often not reported in the literature, and are easily obscured by overlapping line wings, impurities, etc. Weak pump bands clearly demand long path lengths for efficient operation, placing constraints on the pump laser divergence and making OPML cryogenic cell design cumbersome.

iii. Overlapping absorption. Lasers of this type are automatically overlapped by the corresponding absorbing transition from the ground state. The rotational structure of the transition must be resolved at Doppler limited resolution for laser action to be possible. The total overlapping line density is greatly increased if other low lying modes (typically distortional or torsional) give rise to numerous overlapping hot bands.

iv. Isotopic composition. Mono-isotopic species have the best chance of OPML action. In general only one isotopic species is pumped at a time, the others giving rise to overlapping absorptions and rapid, resonant relaxation processes. For elements with minor isotopes with abundances of a few percent such as C, O, N and S the effect is negligible. However for elements such as Cl, Br where two or more isotopes are present in comparable abundance laser action is unlikely in unenriched material.

v. Partition function. Under collision free conditions the extractable energy density depends directly on the total partition function, since this partly determines the number of molecules available in the lower level of the pump transition. The vibrational partition function is dominated by the lowest lying modes and their degeneracy, whilst the rotational partition function depends on the molecular constants and symmetry. Low lying, degenerate modes again adversely affect the OPML.

The very low value of the rotational partition function in the light, linear molecule C_2D_2 ($\sim$ 200) as compared to most other OPML's (e.g. $\sim 3 \times 10^4$ for $FClO_3$) partly accounts for the fact that this molecule is the most powerful OPML of this type.

vi. Relaxation rates. Aside from inducing overlapping hot band absorptions and increasing the vibrational partition function, low lying modes correlate with rapid V-T relaxation rates (the Slater-Lambert law) which in extreme cases can be as fast as R-R rates. The R-R rate is usually within a factor of 2-3 of gas kinetic for most molecules, and so cannot be strongly influenced.

vii. Chemical and physical properties. Virtually all the adverse effects are reduced at low temperatures, and only the C_2D_2 OPML works well at 300K. The vapour pressure curve for the materials is thus of paramount importance. There are only a limited number of high damage threshold, mid-infrared transparent materials, and compatibility with these is a problem in some cases (e.g. SiF_4, NSF).

viii. Competing transitions. If both components of the combination band are infrared active it may be necessary to suppress the component which is not required. Since many of the molecules have large permanent dipole moments and are known FIR lasers (e.g. OCS, CH_3CCH) interaction with FIR laser transitions can occur in both beneficial and deleterious manner (see below).

The qualitative criteria discussed above agree well with the properties of the molecular systems briefly reviewed below. It is noted that the performance figures quoted below (see also fig. 1) for some of these systems notably CF_4, C_2D_2 and propyne have since been considerably improved using a tunable single mode TEA CO_2 pump source (see section (III)).

Of the large number of molecules that have been made to lase in the 16 μm region, namely CF_4[5], $NOCl$[5], $FClO_3$[6], NSF[7], CH_3CCH[8], NH_3[9-11] and CO_2[12] the most extensively developed is CF_4 for which outputs as high as 100 mJ with photon conversion efficiencies of ∿ 5% have been reported by several authors[13,14]. More than 80 emissions around 615 cm^{-1} have been obtained using different isotope forms of the pump and lasing molecule[15] and the development of high pulse repetition rate systems (80 Hz) have yielded average

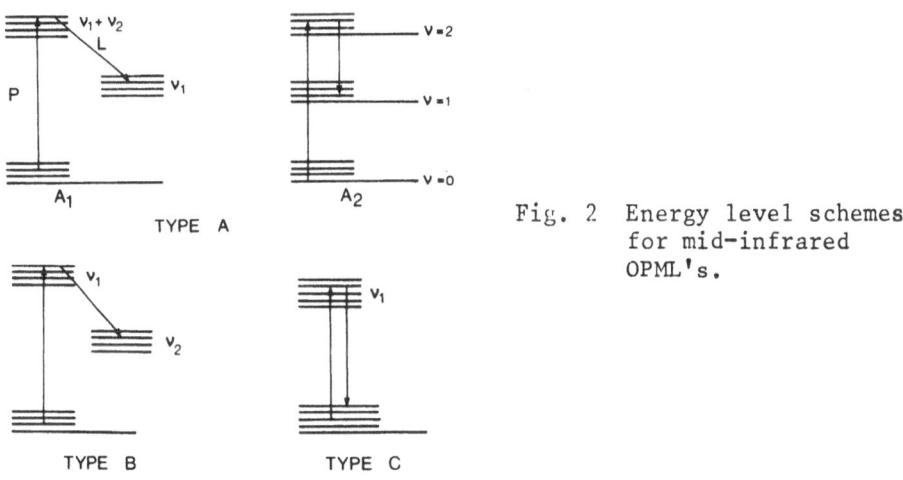

TYPE A

TYPE B   TYPE C

Fig. 2 Energy level schemes for mid-infrared OPML's.

powers of 2.5W [16]. A common feature of lasers of this type is the significant improvement in the maximum operating pressure (for CF_4 from 5 to 23 torr) and energy output achieved on cooling the gas[17] thereby reducing the population of the terminal level of the lasing transition and also enabling optimization of the population distribution of the lower pump level. Multiphoton dissociation of UF_6 has been studied using the CF_4 system[18,19]. Emissions from $FClO_3$[6] in 565-613 cm^{-1} region, from NSF[7] in 618-650 cm^{-1}, and from propyne[8] in 609-637 cm^{-1} range compare favourably in performance to that of CF_4 and have the added advantage of spectral features in better match with those desirable for laser isotope separation of Uranium.

Notable at longer wavelengths is the impressive energy extractions (as high as 750 mJ) from some of the fifteen emission lines in the 17.4 - 20.5 μm region from C_2D_2[20,21] and recently HCCD[73].

Hot band lasing has been obtained over a number of lines in 4.4 to 18 μ region using $^{15}N_2O$, $^{14}N^{15}NO$, $^{15}N^{14}NO$, HCOOH[13] CS_2[22], $^{15}NH_3$[23] and $^{12}C^{18}O$, $^{12}C^{16}O^{18}O$, $^{13}C^{16}O^{18}O$ [24] molecules. Hot band lasing at 16 μm and difference band at 14 μm was also obtained[12] using an HBr laser to pump 1:1 mixture of HBr and CO_2.

Laser emissions following two photon excitation have been reported from SF_6[25], NH_3[9-11], and CH_3F[26], most notable of which were the 10 emissions from NH_3 in the 6-35 μm region with energy extraction of a few milli joules following excitation of the 2a $v_2(5,4)$ level of the molecule. Other hot band lasers include OCS[27,28] (type A_2), CF_3I[29] and COF_2[2]; maser action also being obtained in OCS[74].

Recent developments have identified the significant role played by FIR interactions on MIR generation arising in molecules with a permanent dipole moment. Referring to fig. (3) there are four basic possibilities

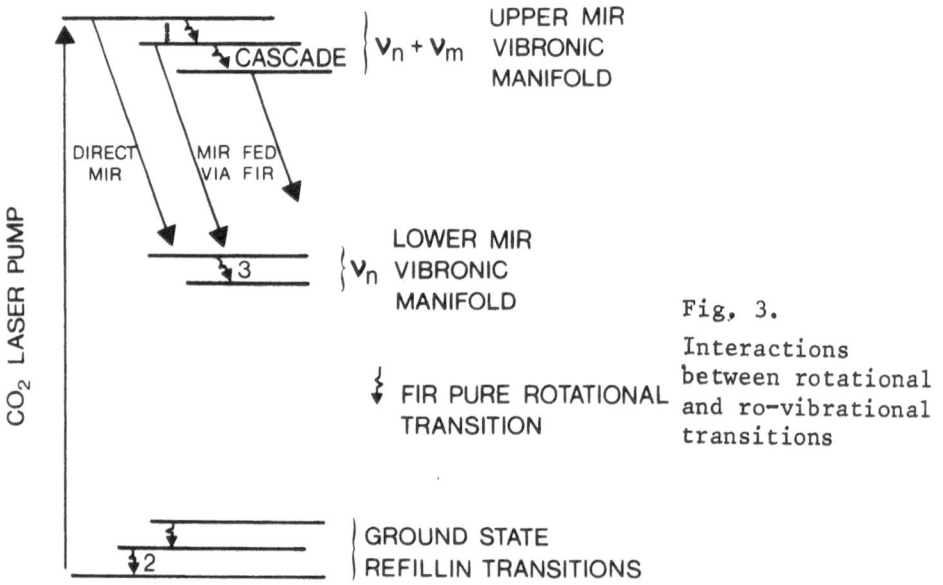

Fig. 3.

Interactions between rotational and ro-vibrational transitions

Firstly a pure rotational transition may compete directly with the desired MIR output line (1 in Fig. 1). Secondly refilling transitions may occur in the ground state. These will give access to additional molecules even under low pressure, collision free conditions (2 in Fig. 1). Thirdly FIR transitions may occur in the lower vibronic manifold of the MIR trnsitions; these will help to depopulate the lower MIR laser level, and enhance the MIR output (3 in fig. 1). Finally, FIR transition or transition cascade may be 'followed by' a MIR transition.   In the second, third and fourth cases these transitions may be either sequential or multi-photon events, depending on the system.

Transitions of the fourth type (MIR 'fed' by FIR) were inferred to occur in ammonia[57]; when pumped on the aR(6,0) transition outputs occurred on both the aP(8,0) direct pumped output and the aP(6,0) output, which was believed to be fed by an FIR two step cascade.   Introducing FIR loss into the MIR cavity suppressed the aP(6,0) output, but the FIR outputs were not directly observed.

Very strong FIR/MIR interaction has been observed in propyne (CH_3CCH)[58] which has a permanent dipole moment of .75 Debye and is a well known cw FIR laser when pumped on its γ_5 or γ_8 fundamentals.   Pumping of the $\gamma_9 + \gamma_{10}$ combination band also leads to 16 μm region emission, but the output energies and number of transitions observed in early experiments were low.

When operated in a metal cell with a tuned, stabilized pump source very strong FIR outputs were observed in the 400 μm - 1.4mm region when the 16 μm generating transitions were pumped.   The pulse shapes and wavelengths of these lines strongly suggests that they are of type 1 in fig. (3)   that is parasitic on the MIR outputs.

The FIR outputs can be partly suppressed by lining the OPML laser tube with convoluted steel tubing and using external mirrors some distance from the cell window to provide diffractive loss. This leads to a dramatic improvement in the MIR region outputs.   In propyne four outputs are completely suppressed by competition from superfluorescent FIR outputs.   For the other four MIR outputs a minimum increase of a factor of 10 is observed in the MIR output energy when the FIR transitions are suppressed as far as possible.

In principle it should be possible to selectively enhance ground state refilling transitions and lower MIR laser level transitions which increase the MIR output and suppress parasitic transitions.   In most cases this would be technically difficult, since for high J lines the FIR wavelengths are similar.

In one particular case this may however be easily achieved. For OCS pumped by CO_2 9P22 the parasitic upper state transitions[74] are all at 6mm and longer, whilst the ground state refilling transitions are at 4.1mm and shorter.   The use of a metal cell with a 5.5mm cut off wavelength (3.2mm diameter) should strongly enhance the MIR output.   In general in lines pumped on P or Q transitions the refilling transitions are of shorter wavelength than the parasitics,and other designs of 'high pass' cavities for the FIR may be possible.

The FIR output wavelengths can greatly assist assignment of the states in complex MIR lasers such as propyne. Observation of the FIR laser wavelengths in propyne has permitted partial assignment of all the previously unassigned MIR transitions in this molecule.[58]

These interactions can be expected to occur in all molecules with substantial dipole moments (e.g. CH_3CCH, OCS, NOCl, CF_3I) but to be absent in those with very low ($FClO_3$) or identically zero (CF_4) moments.

b) Lasers based on inversion in a difference band

In lasers belonging to group B a fundamental level is pumped and laser action occurs within a difference band to a lower level. Here the pump absorption cross section can be large, although, since difference band transitions are forbidden in the harmonic approximation, the gains of these systems can be small. Very high pressure operation is however possible for this type of laser due to the absence of significant self absorption and the long collisional relaxation times associated with fundamental molecular modes. Two molecules that have been made to lase efficiently on this scheme are CO_2 and N_2O. In both these cases very high pressure operation (30-40 atm) was achieved permitting generation of continuous tunable MIR output. For CO_2[30,31] the 4.3 μm line of HBr was used to pump the ν_3 band of CO_2 resulting in complete conversion of 4.23 μm photons to 10.6 μm photons. Due to the small penetration depth of the pump beam resonator lengths as short as 1 mm were used similar to that for the N_2O system[31,32] for which lasing occured via resonant collisional energy transfer to N_2O from CO_2 excited by an HBr laser. With this sytem continuous tuning of over 5 cm^{-1} near 10.5 μm was obtained with a resolution of 0.014 cm^{-1}.

The versatility of collisional energy transfer has also been demonstrated using v-v collisional transfer from CO, excited by the second harmonic of a CO_2 laser line, to obtain MIR emission in the difference band of 6 molecules[33]. The CO molecule is ideal for storage of vibrational energy because of its exceedingly slow vibration to translation transfer rate of 1.9×10^{-1} sec^{-1} $torr^{-1}$. It is also efficiently excited by the second harmonic of the CO_2 9P(24) line which falls within 0.003 cm^{-1} of the CO 0→1 P(14) transition. Of the lasers, OCS-CO was notable for grating tuned emission over 80 lines between 8.19 and 8.46 μm with maximum outputs of 1.3 mJ and efficiencies of ∿ 7% to be compared to 19% for direct pumping (2ω CO_2) of OCS.

c) Lasers based on fundamental bands

For this laser (Fig. 2 type C), the pump and lasing occur on different lines within the same fundamental band. For example resonant excitation of the R-branch line, can be followed by lasing on a P branch line. The cross section for both the pump and lasing transitions can be large in this case resulting in very efficient laser performance. Simple spectroscopic considerations show that this scheme is limited in application to light molecules for which the rotational level spacing is sufficiently large to permit gain in the presence of the thermal population of the terminal level, although even here the available transitions are found to be limited. As such inversion (as opposed to Raman laser action) can only be obtained at an absolute temperature T (Kelvin) if the separation of the initial and final states is > 0.6 kT. The most important of the molecules suitable to this scheme is NH_3[34] which remains the most powerful and efficient mid infrared laser, emitting in the 10.7 to 13.9 μm region[35,15]. Power conversion efficiencies of 40-80%[36] and energy extractions in excess of 1 J with efficiencies of > 20% have been obtained in some of the 50 emission lines[37], making this system the best of the OPML's to date. Many of the emissions have been obtained with the NH_3 buffered by large amounts of N_2, so ensuring rapid rotational thermalization in all the vibrational states and also depletion of the population of the terminal lasing states by resonant energy transfer to levels of N_2, resulting in population inversion for most of the vibrational-rotational transitions whose frequencies are lower than that of the pump[37,38]. Other emissions are attributed to lasing among the rotational levels of the ν_2 vibrationally excited mode involving far infrared emission and subsequent MIR generation[39]. Several emissions obtained in which both the pump and lasing frequencies are off-resonant, are attributed to Raman like processes[40]. Since the first report of the NH_3 laser[34] the system has received considerable attention due to its applications in tunable mid-infrared generation by nonlinear mixing techniques[42-44] and as pump for InSb Spin Flip Raman (SFR) laser[45] and in laser photo-chemistry[46,47].

d) C.W. OPML Systems

More recently attention has been directed to the development of c.w. systems with notable success using NH_3 gas. C.w. emission was first obtained at 12.08 μm with output powers of 180 mW[41] on near resonantly pumping the sR(5,0) transition of NH_3 with a 30 W CO_2 laser. Using similar open ring cavity arrangements[59,60] output powers have since been increased to 3.3 W. Lasing is clearly identified as a Raman process[41], the lasing frequency being offset by ∿ 190 MHz from line centre; identical to the pump offset frequency. Subsequently impressive output powers of ∿ 10 W with photon conversion efficiencies of > 40% have been obtained[61] for this emission by confining the NH_3 gas in a pyrex capillary tube which acts as a waveguide for both the 9 μm and 12 μm radiation. The system facilitates high pump intensities over long interaction

lengths (∿ 1 m) greatly enhancing the single pass Raman gain at
12 μm.   Using a similar system with the pump signal down shifted
in frequency by acousto optic modulators to pump the sR(5,0) transition on line centre;c.w. oscillation, resulting from vibrational
inversion (rather than Raman processes) has since been obtained[62]
on 20 different NH_3 lines between 10.7 and 13.3 μm.   Operation is
obtained under conditions of rotational thermalization ensured by
buffering the NH_3 with a high concentration of N_2 similar to that
adopted for pulsed systems[37].   Single line output powers as high
as 760 MW were obtained for input powers of ∿ 9W.

To date CF_4 is the only other gas for which cw emission in
the mid infrared has been reported[63] yielding ∿ 2mW on the P(31)
transition at 16 μm pumped on the R(29) line by a 3W CO_2 laser.

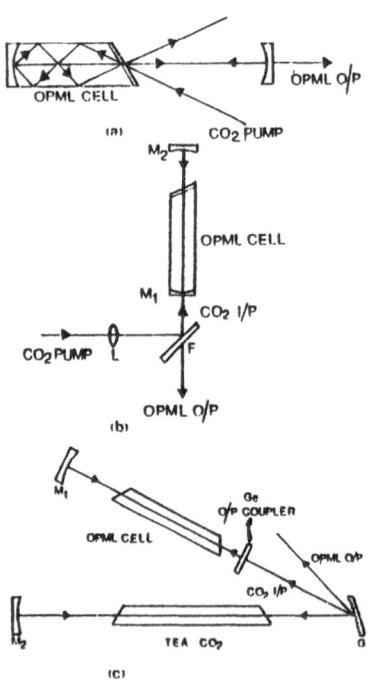

Fig. 4.   Cavity configuration
for mid infrared OPML's.

EXPERIMENTAL CONSIDERATIONS

a) Optical Cavities

The high gain of optically pumped ro-vibrational transitions
is often sufficient to sustain superflourescent emission, thereby
negating the need for optical cavities at least in exploratory
searches for new emissions.   Subsequent optimization of laser
action involving optical resonators poses the simultaneous requirement of injecting a high power pump laser beam into an
absorbing gas while maintaining a high Q cavity at the pumped laser
wavelength.   Early systems[34] utilizing off axis injection (fig 4a)
of the pump beam in the OPML cavity resulted in critical and inefficient operation, chiefly due to the nonuniformity of pumping
which can result in self absorption of stimulated emission by the
unexcited gas.   This problem is minimized by having the CO_2 pump
propogate collinear to the axis of the optically pumped cavity,
(fig. 4b) using for example specifically coated dielectric mirrors
and filters to transmit the CO_2 pump and form the OPML cavity[48],
although such schemes are prone to optical damage.   Subsequent
schemes have used cavity configurations with damage resistant
original gratings.   A particularly efficient and simple one[36] is
shown in Fig. 4c.   Based on the use of a common output coupler
for both the CO_2 laser and the OPML (the gains of which are
similar) the beams are maintained collinear and the output is
efficiently coupled directly off the TEA CO_2 intra-cavity grating.

In overcoming the limitations to pump energy inherent to the
conventional stable resonator configurations commonly used for
TEA CO_2 lasers alternative systems have been used and include
oscillator-amplifiers[13] and unstable resonator cavities[49].

b) Mode Control of Pump Laser

For the majority of combination band pumped (type A_2) lasers
output energies obtained by pumping 1-2 m long cells with a conventional TEA laser (typically 5J) are very low, and very severe
shot-to-shot jitter is experienced.   This low efficiency and jitter
results in part from the bandwidth miss match between the CO_2 TEA
laser (typically 500 MHz - 2 GHz) and the OPML pump line, which
is usually Doppler limited and $\sim$ 50 MHz FWHM.   If the CO_2 TEA
lasers axial mode structure is uncontrolled severe jitter results,
a good shot resulting when an axial mode frequency happens to
match the laser material absorption.   Dramatic improvement in the
performance of these systems was first demonstrated for the
615 cm^{-1} 9R12 pumped $^{12}CF_4$ laser[64] when the pump laser was line
narrowed and at CO_2 line centre, which is fortuitously in close
coincidence with the CF_4 absorption.   Other workers[65,66] have
demonstrated the necessity to provide frequency stabilization of
the pump laser to much less than the Doppler FWHM if reliable

output is to be obtained; stability of better than ± 5 MHz is desirable.

Of the various methods that have been used to simultaneously tune and narrow the CO_2 laser the most successful technique is to use an unstable resonator main CO_2 TEA laser which is injection locked for narrowing and tuning. The injection source may be a waveguide laser or a miniature TEA laser. Gascoyne et al 1983[67] have described an opto acoustic lock system for a miniature TEA laser with a single mode tuning range of ± 350 MHz. The miniature laser is run at twice the required output pulse rate and every other pulse is used by the locking circuit. The opto-acoustic cell is fitted with the gas to be pumped. When the absorbing transition is very weak or overlapped an alternative offset locking technique can be used which provides a known offset with respect to CO_2 line centre[68]. An extremely powerful technique is to measure the offset frequency on every shot by heterodyning the miniature TEA laser against a cw low pressure CO_2 laser, and using this information to feedback control the pulsed laser frequency[69].
All of these techniques provide better than ± 1 MHz long term stability. Whilst the waveguide laser has the advantage that no synchronisation is required, the miniature TEA laser provides much higher output powers which make injection locking extremely easy; and since it operates at atmospheric pressure there is no reduction in output power on tuning away from line centre.

The advantages of such a pump laser system are striking. Virtually all lasers of type A_2 with weak pump bands benefit enormously from its use; strong lines are typically enhanced in output by a factor 10 or more, and weak lines by much larger amounts, and numerous new pump transitions are observed. It is often found that each CO_2 line can span more than one pump transition, especially in the heavier OPML molecules, and CO_2 lines which did not produce OPML output frequently do so when injection locked and optically tuned.

c)   Gas Cells

The low absorption coefficient of type A OPML's, typically .001 cm^{-1} or less dictates the use of long cells if efficient operation is to be achieved. This has been demonstrated by several workers[70,71,72]. Most of these cells were cooled by liquid nitrogen boil off, and had only limited temperature uniformity and stability (up to 30 K non uniformity[70]) preventing the gas from being uniformly operated at optimum temperature, in most cases the lowest temperature possible as determined by the saturated vapour pressure curve at the operating pressure.

An improved design of OPML cryogenic cell has been described[51] which achieves temperature stability and uniformity of 1.5K over many hours. The design is readily scaled to large diameters capable of providing multi-pass cells with very long absorption lengths.

THEORETICAL CONSIDERATIONS

The gain for OPML may be due to either coherent Raman type processes or to pure population inversion effects or to both. For a complete description of the situation a quantum mechanical approach is required as has been used for FIR lasers[52,53]. However, in situations where the coherent effects are not important i.e. gain is purely due to population inversion, a rate equation approach is adequate to describe the basic gain features of these lasers[4,54,55]. This approach is justified to the extent that the pump sources are generally multimode lasers and the excited molecules are at pressures for which collisional energy transfer invariably plays a significant role.

The development of a theoretical basis for the operating characteristics of optically pumped mid-infrared gas lasers is still in an embryonic stage. Thus so far only two attempts to model a pulsed mid-infrared OPML have been reported, the molecules being OCS[28] and NH_3[56] for both of which adequate information about their spectroscopic and energy transfer aspects were available. Both the attempts have used the rate equation approach and result in good agreement between experimental and theoretical data.

Recently this approach has been successfully applied to characterise the performance of the c.w. multi line NH_3 system discussed earlier.

CONCLUSIONS

The considerable progress in optically pumped molecular lasers over the last eight years has proved their potential as an efficient source of high power pulsed MIR radiation. They are by far the most convenient laser source in this wavelength region providing simple and low cost systems especially attractive as wavelength converters for use with CO_2 TEA lasers. Recently improved frequency control of this pump source has already lead to further dramatic improvement in performance and spectral coverave of these systems. With the recent achievement of efficient high power c.w. operation of NH_3 future prospects for extending the spectral range of c.w systems using other molecules is very promising and will undoubtedly lead to a rapid progress in this area.

REFERENCES

1. T.Y. Chang, in Nonlinear Infrared Generation, Y.-R. Shen, ed. Springer-Verlag, Berlin (1977), pp215-272.
2. C.R. Jones, Laser Focus, 14, No.8, 68 (1978).
3. A.Z. Grasiuk, V.S. Letokhov, and V.V. Lobko, Prog.Quant. Electr. 6, 245, (1980).
4. R.G. Harrison, P.K. Gupta, in Infrared and Millimeter Waves, vol. 7, K.J. Button, ed. to be published by Academic Press.
5. J.J. Tiee and C. Wittig, Appl. Phys. Lett. 30, 420 (1977).
6. H.N. Rutt, Opt. Commun. 34, 434 (1980).
7. T.A. Fischer, J.J. Tiee and C. Wittig, Appl. Phys. Lett. 37, 592 (1980).
8. T.A. Fischer and C. Wittig, Appl. Phys. Lett. 39, 6 (1981).
9. R.R. Jacobs, D. Prosnitz, W.K. Bischel and C.K. Rhodes, Appl. Phys. Lett. 29, 710 (1976).
10. J. Eggleston, J. Dallarosa, W.K. Bischel, J. Bokor and C.K. Rhodes, J. Appl. Phys. 5½, 3867 (1979).
11. A.N. Bobrovskii, A.A. Vedenov, A.V. Kozhevnikov and D.N. Sobolenko, JETP Lett. 29, 537 (1979).
12. R.M. Osgood Jr., Appl. Phys. Lett. 32 564 (1978).
13. J.J. Tiee, T.A. Fischer and C. Wittig, Rev. Sci. Instrum. 50, 958 (1979).
14. V. Yu Baranov, S.A. Kazakov, V.S. Mezhevov, A.N. Napartovich M. Yu Orlov, V.D. Pismennyi, A.I. Starodubtsev and A.N. Startostin, (1980), Sov. J. Quantum Electron 10, 47.
15. Handbook of Laser Science and Technology, vol. II, M.J. Weber, ed., to be published by CRC Press, Boca Raton, EL.
16. V. Varanov, B.I. Vasilev, E.P. Velikhov, Yu. A. Gorokhov, A.Z. Grasyuk, A.P. Dyad'kin, S.A. Kazakov, V.S. Lotokhov, V.D. Pismenny and A.I. Starodubtsev, Sov. J. Quantum Electron 8, 544 (1978).
17. J.M. Green, J. Phys. D: Applied Physics 12, 489 (1979).
18. J.J. Tiee and C. Wittig, Opt. Commun. 27, 377 (1978).
19. J.A. Horsley, P. Rabinowitz, A. Stein, D.M. Cos, R.O. Brickman and A. Kaldor, IEEE J. Quantum Electron. QE-16, 412 (1980).
20. H.N. Rutt and J.M. Green, Opt. Commun. 26, 422 (1978).
21. B.K. Deka, P.E. Dyer and R.J. Winfield, Opt. Lett. 5, 194, (1980).
22. A.H. Bushnell, C.R. Jones, M.I. Buchwald and M. Gundersen, IEEE J. Quantum Electron. QE-15, 208 (1979).
23. C.R. Jones, M.I. Buchwald, M. Gundersen and A.H. Bushnel, Opt. Commun. 24, 27 (1978).
24. M.I. Buchwald, C.R. Jones, H.R. Fetterman and H.R. Schlossberg, Appl. Phys. Lett. 29, 300 (1976).
25. W.E. Barch, H.R. Fetterman and H.R. Schlossberg, Opt. Commun. 15, 358 (1975).
26. D. Prosnitz, R.R. Jacobs, W.K. Bischel and C.K. Rhodes, Appl. Phys. Lett. 32, 221 (1978).

27. H.R. Schlossberg and H.R. Fetterman, Appl. Phys. Lett. 26, 316 (1975).
28. E. Armandillo and J.M. Green, J. Phys. D: Appl. Phys. 11, 421 (1978).
29. J.J. Tiee and C. Wittig, J. Appl. Phys. 49, 61 (1978).
30. T.Y. Chang and O.R. Wood II, Appl. Phys. Lett. 23, 370, (1973).
31. T.Y. Chang and O.R. Wood II, IEEE J. Quantum Electron. QE-13, 907 (1977).
32. T.Y. Chang and O.R. Wood II, Appl. Phys. Lett. 24, 182 (1974).
33. H. Kildal and T.F. Deutch, Appl. Phys. Lett. 27, 500 (1975).
34. T.Y. Chang and J.D. McGee, Appl. Phys. Lett. 28, 256 (1976).
35. S.M. Fry, Opt. Commun. 19, 320 (1976).
36. P.K. Gupta, A.K. Kar, M.R. Taghizadeh and R.G. Harrison, Appl. Phys. Lett. 39, 32 (1981).
37. B.I. Vasilev, A.Z. Grasyuk, A.P. Dyadkin, A.N. Sukhnov and A.B. Yasterbkov, Sov. J. Quantum Electron 10, 64 (1980).
38. H. Tashiro, K. Suzuki, K. Toyoda and S. Namba, Appl. Phys. 21, 237 (1980).
39. T. Yoshida, N. Yamabayashi, K. Miyazaki and K. Fujisawa, Opt. Commun. 26, 410 (1978).
40. T.Y. Chang and J.D. McGee, Appl. Phys. Lett. 29, 725 (1976).
41. C. Rolland, B.K. Garside and J. Reid, Appl. Phys. Lett. 40, 655 (1982).
42. R.G. Harrison and F.A. Al-Watban, Opt. Commun. 20m 225 (1977).
43. R.G. Harrison, R.A. Wood and S.R. Butcher, Opt. Commun. 27, 157 (1978).
44. R.G. Harrison, P.K. Gupta, M.R. Taghizadeh and A.K. Kar, IEEE J. Quantum Electron QE-18, 1239, (1982).
45. C.K.N. Patel, T.Y. Chang and V.T. Nguyen, Appl. Phys. Lett. 28, 603 (1976).
46. R.V. Ambartsumian, Z.A. Grasiuk, A.P. Dyadkin, N.P. Furzihov, V.S. Letokhov and B.O. Vasil'ev, Appl. Phys. 15, 27 (1978).
47. J.J. Tiee and C. Wittig, Appl. Phys. Lett 32, 236 (1978).
48. B. Walker, G.W. Chantry and D.G. Moss, Opt. Commun. 23, 8, (1977).
49. B.K. Deka, P.E. Dyer and I.K. Perera, Opt. Commun. 32, 295 (1980a).
50. T. Stamatakis and J.M. Green Opt. Comm. 30, p413 (1979).
51. H.N. Rutt, (1981), UKAEA Culham Lab. Reprint CLM-P660.
52. T.A. DeTemple (1979) in Infrared and Millimeter Waves Vol.1 ed. Button K, (Academic Press, N. York, San Francisco, London) p129.
53. T.Y. Chang, IEEE J. Quantum Electron QE-13, 937 (1977).
54. A.L. Golger and V.S. Letokhov Sov. J. Quantum Electron 3, 15, (1973).
55. A.L. Golger and V.S. Letokhov, Sov. J. Quantum Electron 3, 428, 1974).
56. P.K. Gupta and R.G. Harrison, IEEE J. Quantum Electron QE-17, 2238 (1981).

57. L.Y. Nelson, M.I. Buchwald and C.R. Jones, Appl. Phys. Lett. 37. 765 (1980).

58. H.N. Rutt, D.N. Travis and K.C. Hawkins, Int. J. IR and MM Waves (submitted).

59. P. Wazen and J.M. Louritsz, Opt. Commun. 47, 137 (1983).

60. P. Wazen and J.M. Lourtisz, Appl. Phys. B32, 105, (1983).

61. C. Rolland, J. Reid and B.K. Garside, Appl. Phys. Lett. 44, 725 (1984).

62. C. Rolland, J. Reid and B.K. Garside, Appl. Phys. Lett. 44, 380 (1984).

63. J. Telle, IEEE J. Quantum Electron, QE-19, 1469 (1983).

64. A. Stein, P. Rabinowitz and A. Kaldor, Opt. Lett. 3, 97 (1978).

65. T.L. Stamatakis, D.A. Aldcroft and J.M. Green, J. Phys. B. 15, 3639 (1982).

66. M.A. Gunderson and T.A. Yocom, J. Quantum Electron, QE-18, 1237 (1982).

67. P.F. Gascoyne and H.N. Rutt, J. Phys. E. Sci. Instrum. 16, 31 (1983).

68. P.F. Gascoyne, K.C. Hawkins and H.N. Rutt, J. Phys. E.: Sci. Instrum. (submitted).

69. H.N. Rutt, J. Phys. E. Sci. Instrum. 17, – (1984).

70. J.J. Tiee, T.A. Fischer and C. Wittig, Rev. Sci. Instrum. 50, 958 (1979).

71. J. Telle, Opt. Lett. 7, 201 (1982).

72. H. Tashiro, H. Soyma, K. Toyoda and S. Samba, Appl. Phys. B, 34, 37 (1984).

73. H.N. Rutt, Infrared Physics 1984 (in press).

74. H.N. Rutt and K.C. Hawkins, Opt. Commun (1984) (in press).

OPTICALLY PUMPED FAR INFRARED LASERS

F. Strumia, N. Iolia, and A. Moretti

Dipartimento di Fisica
Università di Pisa and GNSM-CNR
Piazza Torricelli 2, 56100 Pisa, Italy

INTRODUCTION

The search for efficient and powerful sources of coherent radiation spanning the widest possible spectral range is a major task in the fields of Quantum electronics and Laser spectroscopy. Until the discovery in 1970 [1] of the optically pumped Far Infrared (FIR) lasers, the submillimeter range of the spectrum had been almost barren due to the lack of radiation sources. Before 1970, the FIR portion of the electromagnetic spectrum was only very sparsely covered by molecular glow discharge lasers (HCN, DCN, H_2O) and the list of available laser lines was rather poor[2] Since 1970 more than two thousand optically pumped FIR laser lines throughout the 0.03 to 2 mm region were discovered, some providing hundreds of Kilowatts of peak power in pulsed operation, others providing a few hundreds of milliwatts of continous power.
The emission lines are so closely spaced in the wavelength range between 1 mm and 0.03 mm that these sources of radiation can be considered to be step-tunable. Power fluctuation can be limited to a few percent and frequency stability can be better than 1 part in 10^9 . In fact the emitted lines are in general homogeneously broadened and the linewidths of the resonator modes are larger than the gain curve of the active medium. As a consequence, the frequency of the emitted radiation is determined by the molecular transitions and is only slightly perturbed by the cavity tuning via pulling effects. This makes it possible to use FIR sources in high resolution spectroscopy without the need of active frequency stabilization.

While for the pulsed regime other sources (FEL) are forecast, the optically pumped FIR laser will remain the only practical CW source for some years to come at least in the region 1 to 10 THz. A number of review papers were published in the past[3-9]. A critical review of the laser emissions from 30 of the most useful molecules has also been published[10].

FIR LASER SYSTEM

The FIR lasers need as a pump source a laser emitting photons with an energy well above the KT in order to excite the molecules in an empty vibrational state. The laser action takes place between the rotational states of this vibrational manifold as shown in fig. 1.

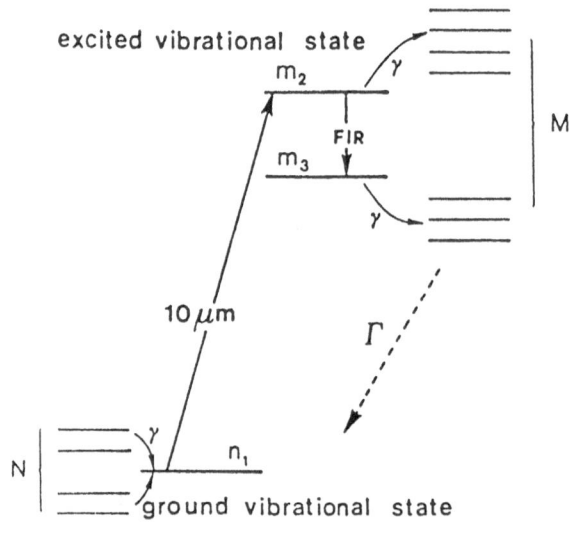

FIG. 1

The most used pump laser is the CO_2 , in some cases the isotopic CO_2 lasers, the CO_2 sequence lines laser and the N_2O laser were also used. A typical long pulse or continous-wave optically pumped FIR laser system is shown in fig. 2. The pump source must have an output power in the 3-60 W range. Only in the case of very efficient FIR laser lines the threshold pump power can be as low as a few hundred milliwatts. The pump laser is line selected by means of a diffraction grating while tuning resonance with the absorption line is obtained by a PZT tuner.

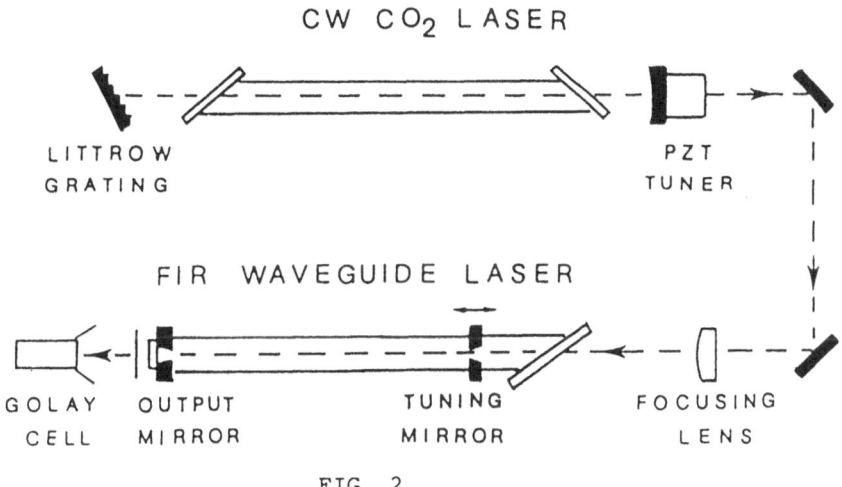

FIG. 2

In many cases a waveguide resonator is used for the FIR. A hollow dielectric cylindrical guide (a Pyrex tube) is a popular solution since the lowest attenuation mode is the EH_{11} , linearly polarized without a preferred direction[11] . The FIR electromagnetic field can be excited either parallel or orthogonal to the polarization of the pump radiation according to the ΔJ selection rule [3,4]. In the case of a cylindrical metallic waveguide the lowest attenuation mode is the TE_{01}. The electric field has a minimum on the axis and a circular polarization that misfits the pump radiation and causes a lower conversion efficiency. The rectangular hybrid waveguide consisting of two metallic and two dielectric sidewalls is also used since in this case a homogeneous electric field can be easily applied on the FIR laser active medium. By means of the Stark effect new FIR laser lines, a significant power enhancement and a moderate frequency tuning have been obtained for the most powerful CH_3OH laser lines [5].The waveguide resonators are terminated by flat mirrors that must be oriented orthogonal to the waveguide axis. Metal mirrors have a very good reflectivity at FIR wavelengths.

Also a confocal mirror cavity can be used. The fundamental mode is the TEM_{00} that combines very well with the pump radiation. The mirrors resonator is convenient for long wavelength FIR lasers $(\lambda > 0.5$ mm) since in this case it has a lower threshold pump power than the W.G. lasers. On the other hand W.G. lasers have a better conversion efficiency and a larger output power. In fact the vibrational relaxation in the FIR laser is dominated by the molecular diffusion toward the wall wich depends critically upon the resonator diameter.

The pump laser power is coupled in the FIR laser by means of a small hole ($d \simeq 1 - 3$ mm) at the center of the input mirror. A larger ($d = 4 - 15$ mm) hole in the output mirror is often used for extracting the FIR power. A variety of more sophisticated output devices have been used[7] , however a hole of the proper diameter drilled in a metal mirror can always approach the best extraction efficiency independently of the FIR wavelength.

One mirror must be translatable along the laser axis in order to tune the FIR cavity in resonance with the FIR laser line. In fact the laser line is much narrower than the cavity free spectral range. The recording of the FIR output power as a function of the cavity length is a simple and convenient method to measure the FIR laser wavelength with a maximum precision of about one part in 10^4 . Moreover in several molecules two, three or more FIR laser lines are obtained for each pump line. Again the separation of the different lines is obtained by changing the FIR cavity length.

RATE EQUATION MODEL

A rate equation model has proved to be quite useful in the description of the optically pumped FIR lasers[7,12,13] . The model is based on describing the kinetics as the sum of two separate laser interactions that share a common level and including all the collisional processes relevant for the relaxation of the perturbed states toward their equilibrium values. The rate equation model has proved to be in qualitative agreement with all the relevant FIR laser features with the obvious exception of effects like the AC Stark effect and the Raman gain that require a quantum mechanical treatment. A common lay-out in the description of the CW FIR laser takes into account that the pump transition is Doppler broadened while the FIR transition is homogeneously broadened (typical pressure broadening $\sim$ 20-30 MHz/Torr). Moreover it is demonstrated experimentally [15,16] that the velocity and M changing collisions are negligible compared to the $\Delta J, \Delta K, \Delta V$ changing collisions. As a consequence of the Doppler broadening in the CW FIR lasers, only a fraction of the molecules in the level n_1 (fig. 1) can be simultaneously excited and the velocity dependent excitation rate is

$$W(\nu) = \frac{I_P(\nu)}{f_i N} \, \alpha_{IR}^S(\nu) \, \exp\left[-4 \ln 2 \left(\frac{\nu - \nu_0}{\Delta \nu_D}\right)^2\right] \qquad 1)$$

where $\alpha_{IR}^S(\nu)$ is the saturated absorption coefficient for the pump transition, I_p the pump intensity, f_i the fractional rotational equilibrium population and N the ground state vibrational manifold population. Since the IR absorption per transit in the FIR cavity is small we can assume that the pump radiation is propagating in the cavity with about the same intensity in both directions. In this case, as a first order expansion in I_p/I_{IR}^S

$$\alpha_{IR}^S(\nu) = \alpha_{IR}^0 \left[1 + \frac{I_P}{2 I_{IR}^S}\left(1 + \frac{(\Delta\nu_H/2)^2}{(\nu - \nu_0)^2 + (\Delta\nu_H/2)^2}\right)\right]^{-1} \qquad 2)$$

where α_{IR}^0, I_{IR}^S, and $\Delta\nu_H$ are the unsaturated absorption, the saturation intensity and the homogeneous linewidth of the IR transition respectively. The relaxation between levels is a consequence of collisions. The rotational relaxation within a given vibrational state is very fast ($\gamma \sim 10^{10}$ sec^{-1}atm^{-1}) while the relaxation between the vibrational states is much slower ($\Gamma \sim 10^5 - 10^7$ sec^{-1}atm^{-1}). As a consequence also the wall collisions contribute significantly to the excited vibrational states relaxation, with a diffusion rate proportional to $p^{-1}d^{-2}$ (p: gas pressure, d: FIR resonator diameter). The spontaneous radiative emission being negligible we have

$$\Gamma = \Gamma_c \cdot p + \Gamma_D/p \simeq 10^3 - 10^4 \, s^{-1} \text{Torr} + 10^3 \, s^{-1} \text{Torr}^{-1}$$

In conclusion in the FIR laser there is a bottleneck with a fast thermalization within the vibrational states followed by a slower relaxation between the vibrational states. The fraction of molecules in the upper vibrational level is

$$M = M_r + M_e = \frac{f_i N W(\nu)}{\Gamma} + M_e \qquad 3)$$

where M_e is the population at thermal equilibrium. At the steady state

$$\left(\frac{dm_2}{dt}\right)_r = f_1 N W(\nu) = -\left(\frac{dm_2}{dt}\right)_c = \gamma (m_2 - f_2 M)$$

and

$$m_2 - m_3 = \frac{f_1 N W}{\gamma} + M\left(f_2 - \frac{g_2}{g_3} f_3\right) \qquad 4)$$

The unsaturated gain for the FIR line is then

$$\alpha^o_{FIR} = \hbar \nu_{FIR} \, \sigma_{FIR}\left[\frac{t_1 N W}{\gamma} + M\left(t_2 - \frac{g_2}{g_3} f_3\right)\right] \qquad 5)$$

where σ_{FIR} is the stimulated emission cross section and a homogeneous broadening has been assumed for the FIR transition. The FIR gain is positive only when

$$\frac{f_1 N W(\nu)}{\gamma} \geq M\left(\frac{g_2}{g_3} f_3 - f_2\right) \qquad 6)$$

which is a threshold condition for the FIR laser. When $M_p \gg M_e$ eq. 6 becomes

$$1 \geq \frac{\gamma}{\Gamma}\left(\frac{g_2}{g_3} f_3 - f_2\right) \simeq \frac{\gamma}{\Gamma} f_2 \frac{\hbar \nu_{FIR}}{KT} \qquad 7)$$

that in most cases cannot be satisfied without considering the diffusion relaxation. There is a cutoff pressure given by

$$p_c \leq \left[\frac{\Gamma_D}{\gamma'\left(\frac{g_2}{g_3} f_3 - f_2\right)}\right]^{1/2} \qquad 8)$$

where $\gamma = \gamma p$.
The typical operating pressure, around $p_c/2$, is in the range 0.3-0.03 Torr.
The output power from a homogeneously broadened laser is approximately given by

$$P_o \simeq V \frac{T}{\delta + T} \alpha^o I^s \qquad 9)$$

where V is the volume of the active medium, T the output mirror transmission and δ the round trip cavity losses. By using eq. 5

$$\frac{P_{FIR}}{V} = \frac{T I^s_{FIR}}{\delta + T} \cdot \sigma_{FIR} \hbar \nu_{FIR} f_1 N W\left[\frac{1}{\gamma} - \frac{1}{\Gamma}\left(\frac{g_2}{g_3} f_3 - f_2\right)\right] \qquad 10)$$

where fNW(ν) can be conveniently written as

$$f_1 N W = \alpha^s_{IR} I_{IR} = \frac{P_{IR}}{V} \frac{\alpha_{IR} L}{\delta_{IR} + \alpha_{IR} L} \frac{1}{\hbar \nu_{IR}} \qquad 11)$$

where L is the FIR cavity length and δ_{IR} the FIR cavity losses at IR frequency. Since

$$\sigma_{FIR} I^s_{FIR} = \gamma / \left(1 + \frac{g_2}{g_3}\right)$$

eq. 10) can be finally written as[16]

$$\frac{P_{FIR}}{V} = \frac{P_{IR}}{V} \frac{\gamma_{FIR}}{\nu_{IR}} \cdot \frac{\alpha_{IR}L}{\delta_{IR}+d_{IR}L} \cdot \frac{T}{\delta_{FIR}+T} \left[1 - \frac{\gamma}{\Gamma}\left(\frac{q_2}{g_3}t_3 - t_2\right)\right]\left(1 + \frac{q_2}{g_3}\right)^{-1} \quad 12)$$

The above equation represents a guideline for evaluating the maximum expected conversion efficiency of an optically pumped FIR laser. The meaning of the various terms in eq. 12 is straightforward. The frequency ratio expresses the quantum conversion efficiency multiplied by a factor of order 1/2 since $g_2 \ne g_3$ and the rotational lifetime has been assumed equal for all the levels. P_{IR} is the pump power injected in the FIR cavity. The next two terms express the cavity efficiency at IR and FIR wavelengths respectively. The final bracketed term includes the molecular dynamics as discussed in eq. 5-8.

I^s_{FIR} and α^0_{FIR} do not appear in eq. 12 as a consequence of oversimplification of eq. 9, leading to the conclusion that all the lines of a given molecule would have the same efficiency. In fact it is possible to assume $2\alpha^0_{FIR}L \gg \delta_{FIR}+T$ only in the case of high gain FIR lines. A less approximated eq. 9 must be used when the gain is low [11]. It is worth noting that even for the strong IR lines α^0_{IR} is of the order of unity in m^{-1} $Torr^{-1}$.

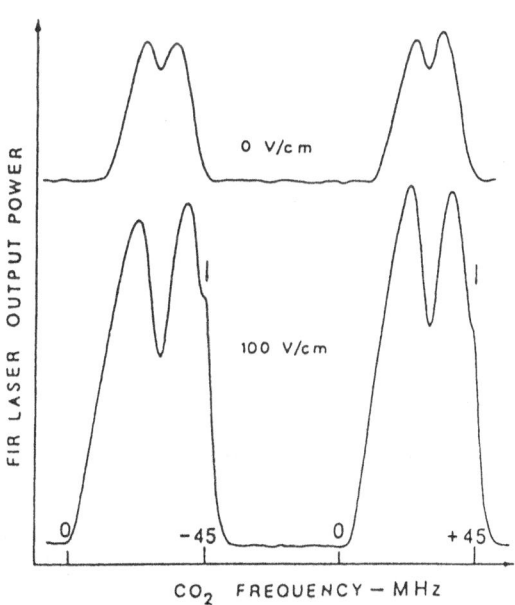

FIG. 3- Transferred Lamb dip observed in the 0.119 mm FIR laser line as a function of the CO_2 frequency tuning (after ref.24)

223

As a consequence it is important for an efficient conversion to have a FIR cavity with small IR absorption. The rate equation model can explain other observed effects. As an example from eq. 2 a saturation pump dip is expected when the pump laser frequency is tuned in resonance with IR absorbing line. As a consequence the FIR laser output must have a corresponding power decrease as is clearly shown in fig. 3. This effect is known as the transferred Lamb dip and is particularly useful for a precise measurement of the pump offset[5,17].

SELECTION CRITERIA FOR THE FIR LASER LINES

Up to the present time FIR laser lines have been obtained from about 100 molecules. However the observed frequencies and the efficiency are quite different and some further criteria may be useful for the selection of the molecules.

Frequency

To a first order the rotational energy of a symmetric top molecule is

$$W_R = B J (J+1) + (A-B) K^2 \qquad 13)$$

The selection rules for rotational lines are $\Delta J = 1$, $\Delta K = 0$ and the emitted frequency is given by $\nu = 2BJ$. Since the B constant is inversely proportional to the moment of inertia the FIR laser lines are confined in the region 1 - 0.4 mm with the only exception of some special molecules like NH_3. In asymmetric as well as slightly asymmetric top molecules the permanent electric dipole moment may have a non vanishing component along all the principal axes of inertia and transitions with $\Delta K = \mp 1$ (or equivalent for K_{+1} and K_{-1}) are allowed adding a new energy term to the transitions. In the case of lightweight nearly symmetric top molecules like CH_3OH and CH_2F_2 the FIR laser lines are shifted mainly in the 0.4 0.1 mm region (fig. 4). A further effect like inversion or internal rotation may add a new energy term to the levels. The most important case is that of CH_3OH where the internal rotation of the OH group with respect to the CH_3 methyl group needs the introduction of two new quantum numbers τ, and n. As a consequence the number of levels populated at room temperature is increased by about a factor six and the selection rules are relaxed to $\Delta J = \pm 1,0$, $\Delta K = \pm 1,0$, Δn any. In particular

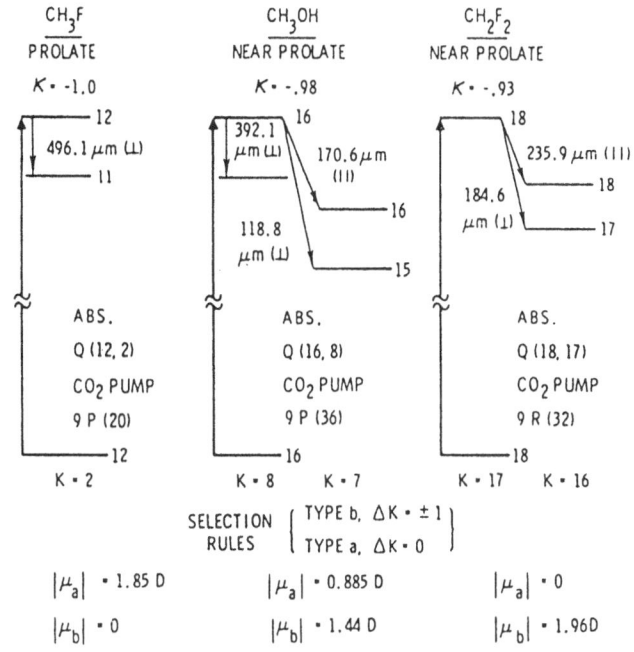

FIG. 4 - (After E.J. Danielewiecz in ref. 10)

the levels with n = 0 and n = 1 have an energy separation of about 150 cm^{-1} and the corresponding FIR laser lines a wavelength in the 0.1 - 0.03 mm region. For this reason the CH_3OH and its isotopically substituted forms are the most important molecules to obtain FIR laser lines with a wavelength in the 0.3 - 0.03 mm region.

Efficiency

Equation 12) is the starting point for selecting molecules candidate for high conversion efficiency FIR laser lines. In particular α^0_{IR} must be large, a condition satisfied when the electric dipole moment of the vibrational transition $|\mu_{IR}|$ is large and the rotational partition function Z is small. The ratio γ/Γ must be on the contrary as small as possible. The permanent electric dipole moment $|\mu|$ must be also large in order to have a large gain at the FIR transition. In table I are shown the values of the significant parameters for some of the most popular FIR lasing molecules. S is the band intensity in cm^{-2}atm^{-1}, Z the rotational partition function, X=S/Z, and Y=XZ (the data of table

TABLE I

MOLECULES	S	Z	X (x10^3)	Y (x10^6)
CH_3I	36	9294	4	3
CH_3F	440	2730	161	4.2
$C_2H_2F_2$	200	34772	6	30
CH_2F_2	1196	12404	96	159
CH_3OH	832	14280	58	1280
D_2O	400	225	2000	90000
NH_3	900	210	3000	170000

I are from ref. 7). The numbers in the last column are the guideline for the molecules selection. In fact the important molecule CH_2F_2 was successfully selected in this way[7,18] However it must be noted that a large number of coincidences with the CO_2 pump lines is possible only when Z is large. For this reason CH_2F_2 and CH_3OH are very rich in FIR laser lines while the more promising $^{14}NH_3$ have no coincidences with the CO_2 laser. Moreover not all the strong absorbing lines of a good molecule may produce strong FIR laser lines since the effective dipole moment $|\mu_{FIR}|$ is a function of the quantum numbers J and K and of the selection rules. As an example in the case of a symmetric top molecule the intensity of a IR line with $\Delta J=0$ (Q branch) is proportional to K^2 while the intensity of the associated FIR laser line is proportional to J^2-K^2. In table II are listed the most efficient FIR laser lines. The quoted powers have been obtained in different experimental conditions and are only indicative of the expected output power in optimized conditions and with an average pump power. The experimental evidence shows that the output power scales linearly with the pump power and the FIR cavity length. The conversion efficiency of the lines listed in Table II is of the order of 10~25%. This figure must be multiplied by the $\nu_{FIR}/2\nu_{IR}$ factor which reduces the absolute power proportionally to the FIR wavelength.

TABLE II

MOLECULE	CO_2 LINE	FIR (μm)	POL	POWER (mW)
$^{15}NH_3$	1OR 18 ($^{13}CO_2$)	152.74	$\perp$	180
CH_3I	1OP 18	477.15	//	40
HCOOH	9R 18	393.63	//	50
CH_2F_2	9R 32	184.30	$\perp$	150
	9R 20	117.73	$\perp$	70
CH_3OH	9P 36	118.83	$\perp$	400 (750[a])
	9R 10	96.52	//	300
	9P 34	70.51		100
	9P 32	42.16	//	50
	9P 16	570.57	//	38
CD_3OH	1OR 18	41.35	$\perp$	60
$^{13}CD_3OH$	1OP 8	127.02	$\perp$	b

a- High pumping power; see ref. 20
b- Stronger than the CH_3OH 119 μm; see.ref. 29

IMPROVING THE EFFICIENCY OF THE LASERS

The output power of the optically pumped FIR laser can be improved in several ways.

1- Cavity length, pump power

By increasing the FIR cavity length the quantity of lasing molecules incresases proportionally, thus increasing also the output power as long as the single pass attenuation of pump power remains small. Experimentally a power increase proportional to the cavity length has been observed up to 3-4 m in the case of CH3OH 119 and 71 μm laser lines. In this way it is possible to bypass the vibrational relaxation bottleneck discussed in the previous section. The limit in the useful cavity length depends also on the available pump intensity and on the pump saturation effect.

2- CO_2 tunability

The gas absorption depends on the frequency detuning of the CO_2 laser from the absorbing line center. FIR line emission can be usually observed only for detunings smaller than the Doppler linewidth as shown in fig. 3. Since a perfect coincidence with the CO_2 line is unlikely, a CO_2 laser with a TEM_{00} mode frequency tunable over all the free spectral range is more convenient than a higher power but multimode laser that cannot be tuned as a consequence of mode hopping. In many cases FIR laser lines classified as weak were found to be much stronger when pumped by a short cavity CO_2 laser with less power but larger tunability.

3- FIR temperature

From the rate equation model the conversion efficiency is proportional to the number f_1N of molecules in the ground state. The fraction f_1 is a function of the gas temperature and can be maximized either by heating or cooling the FIR laser. As an example both the 496 μm line of CH_3F and the 119 μm line of CH3OH have an optimum temperature around 0°C with a power increase of about 30% with respect to the room temperature 19,20,8.

4- Buffer gases

The VV and VT relaxation rates can be increased by adding a buffer gas. A lightweight and transparent buffer gas like H_2, and He is useful in cooling the gas that is heated by the absorbed pump power and by increasing the V-T relaxation rate. In fact a power increase of about 30-50% was found in the case of the 119 μm CH3OH laser [20] . On the contrary no beneficial effect was observed in the case of the 496 μm CH3F laser [21,9] A power increase of about 80 % was observed for this laser when SF_6 was used as a buffer gas [22] (fig.5). In this case there is a very fast near resonant VV transfer between the two molecules and the SF_6 has

228

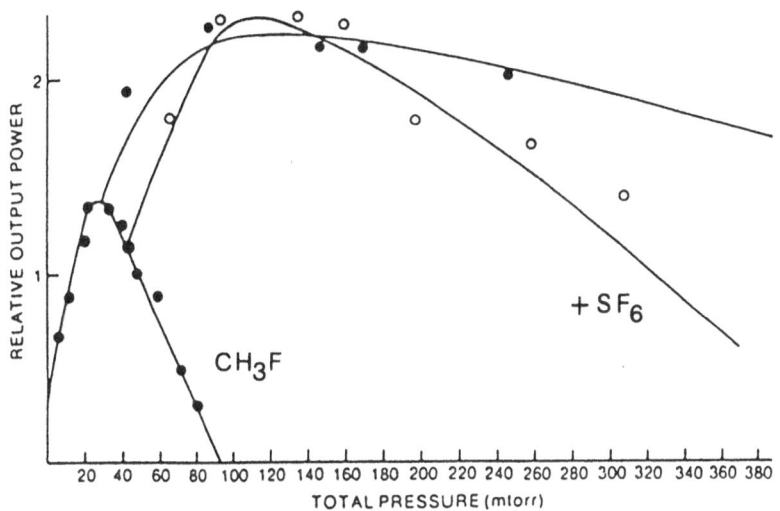

FIG. 5 - Relative power emitted by 0.496 mm CH₃F FIR laser as a
function of the total gas pressure with SF_6 as a buffer
gas (after ref. 22)

also a fast VT deactivation [9]. The SF_6 gas has proved to produce
the greatest increase in output among all the buffer gases
reported.Unfortunately it is also a strong absorber and must be
used only with a fraction of CO_2 laser lines.

5- Two photon pumping

In some cases the ground state is connected to the other
states of the ground vibrational manifold by microwave or
radiofrequency transitions. The pumping rate can then be increased
by transferring molecules from these levels by means of a resonant
MW or RF pump field (fig. 6). An output power increase was also
observed when the RF field was used to depopulate the lower FIR
laser level (fig. 6). The two effects are cumulative and can be
used to obtain a power increase of the order of 15-30% [23].

6- Non linear Hanle effect (NLHE)

The molecules candidates for FIR laser sources must be polar
molecules with a large permanent electric dipole moment. For many
of them, the Stark effect splitting of the M sublevels is larger
than the homogeneous linewidth even when the applied electric

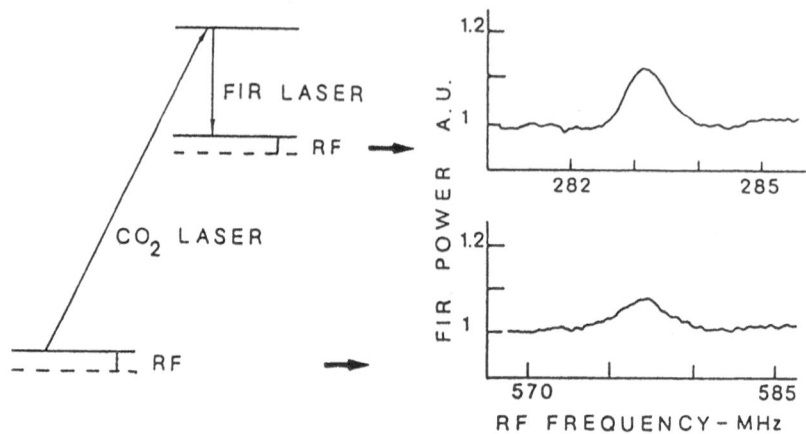

FIG. 6 - Output power increase of the CH₃OH 251 μm FIR laser with resonant RF pumping (after ref.23)

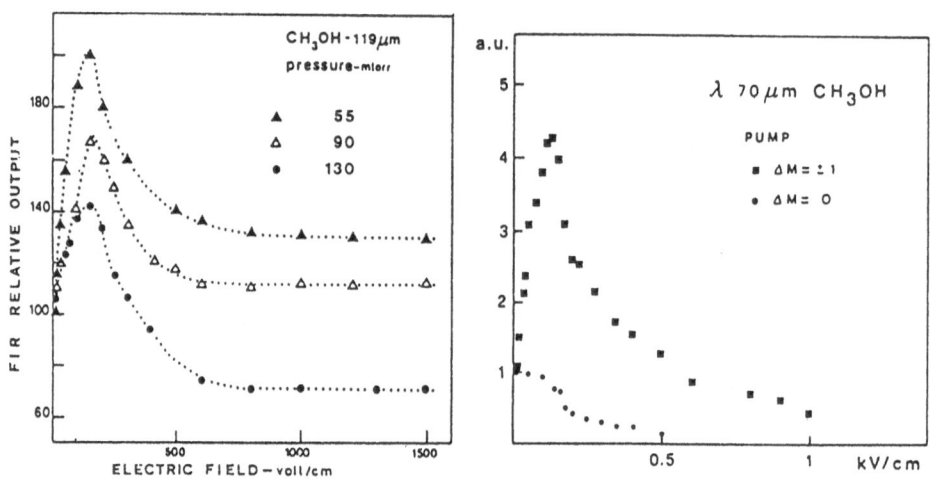

FIG. 7 - NLHE power enhancement effect as a function of the applied electric field (after ref. 24)

field intensity is of the order of a few hundred V/cm. In this case a large power enhancement can be obtained for many FIR laser lines when the polarization of the CO_2 pump radiation is orthogonal to the Stark field. Such power increase is explained in terms of the non linear Hanle effect (NLHE)[6,24,25] that predicts a saturated absorption increase for any transition when the degeneracy between $\Delta M=+1$ and $\Delta M=-1$ transitions is removed with respect to the homogeneous linewidth by an external field. In the case of the FIR laser the increase in the pumping rate may be of the order of 30-40 % and the increase in the FIR output power is as large as 100 % to 500% as shown in fig. 7. The NLHE, being a coherent effect between the M sublevels, is independent of the frequency tuning of the CO_2 laser around the IR absorbing line. As a consequence the power enhancement can be observed independently of the detuning and the lasing frequency interval is increased as is clearly shown in fig. 3. The NLHE enhancement has been observed in the case of many laser lines including all the methyl alcohol lines reported in table II.

It is worth to note that the effects described act independently of the FIR laser and as a consequence can be used simultaneously in order to obtain a substantial power increase. As an example methods 3 and 4 nearly doubled the output power of a 2.7 meters long 119 μm CH_3OH FIR laser[20] .

FREQUENCY TUNING

The FIR lasers are nearly fixed frequency laser sources. However the frequency tuning of some lines has been increased by about two orders of magnitude by using the linear Stark effect [6]. A fast frequency modulation (1 MHz) was also obtained[6] . For many powerful methyl alcohol FIR laser lines the Stark effect appears as a splitting in only two laser lines at respectively lower and

TABLE III [a]

Line (μm)	CO_2 pump	Pump offset (MHz)	Tuning (MHz/kV cm^{-1})	Stark lines
37.9	9-P(32)	−16	59	2
70.5	9-P(34)	+36	44.2	2
96.5	9-R(10)	0	12	2
99.3	9-P(36)	−89	29	2
110.8	9-P(36)	−80	24.6	2
118.8	9-P(36)	+29	26.4	2
133.1	9-P(24)	+6	101	2
193.1	9-P(38)	+13	34	2
205.6	9-P(34)	+120	35	4
208.3	9-P(34)	>130	135	2
261.5	9-P(12)	+85	18.3	2
292.2	9-P(38)	+13	20.5	2

a) After ref. 6

higher frequency than that of the unperturbed line, as a consequence of the molecular selection rules (transitions with $\Delta J = \Delta K = 1$ and $K \simeq J/2$). In table III are reported the observed Stark tunings.

NEW EFFICIENT FIR LASER LINES

New FIR lasing molecules are continuously reported, however efficient molecules satisfying the selection criteria previously discussed are very few. New efficient laser lines are more likely to be found by using isotopic substitutes of already well known molecules like CH_3OH, NH_3, CH_2F_2. A similar approach is to use the isotopic CO_2 laser as a pump source. Recently powerful waveguide CO_2 lasers have been developed with an extended tunability with respect to the conventional one. In this way new coincidences with strong absorbing lines of "efficient molecules" have become available and a new set of strong FIR laser lines has been discovered in CH_3OH, $^{13}CH_3OH$, CH_2F_2 and many others will be available in the future[26,27,28].

REFERENCES

1. T.Y. Chang, T.J. Bridges, Opt. Commun. 1, 423 (1970)
2. P.D. Coleman, IEEE J. Quant.Electron. QE 9, 130 (1973)
   G.W. Chantry, G. Duxbury, in " Methods Exp. Phys.", vol. 3A, L. Marton ed., Academic Press, 1974 pagg. 302-394.
3. T.Y. Chang, " Optical Pumping in gases" in Y.R. Shen " Nonlinear Infrared Genration" Springer Verlag, Berlin, Heidelberg, New York, 1977, pagg. 215-272.
4. T.A. De Temple,"Pulsed optically pumped far infrared lasers", in "Infrared and millimeter waves" vol.1, K.J. Button ed., Academic Press, New York, pagg.129-184 (1979)
5. J.O. Henningsen,"Molecular Spectroscopy by far infrared laser emission",ibid., vol. 5, pagg. 29-129. (1982)
6. F. Strumia, M. Inguscio," Stark Spectroscopy and frequency tuning in optically pumped far-infrared lasers" ibid.,vol.5, pagg. 130-214 (1982)
7. T.A. De Temple, E.J. Danielewicz, " Continuous wave optically pumped lasers", ibid., vol 7, pagg. 1-41 (1983)
8. K. Walzer, Optimization of optically pumped far-infrared lasers,ibid. vol.7, pagg. 119-163 (1983)
9. J.P. Picharruthu, Submillimeter lasers with electrical, chemical and incoherent optical excitation,ibid. vol.7, pagg. 165-244 (1983)

10. K.J. Button, M. Inguscio, F. Strumia eds., Optically pumped far-infrared lasers, "Reviews of infrared and millimeter waves" , vol.2, Plenum Press, New York (1984)

11. F. Strumia,N. Ioli, High power, tunable waveguide CO_2 lasers, this book.

12. J.R. Tucker, IEEE Trans. Microw. Theory Tech., MTT-22, 1117 (1974)

13. D.T. Hodges, J.R. Tucker, Appl.Phys.Lett. 27, 667 (1975)

14. D.T. Hodges, Infr. Phys. 18, 375 (1978)

15. T. Oka in "Advances in Atomic and Molecular Physics", D. Bates, B. Bederson eds., vol. 9; Academic Press, New York (1973)

16. M. Inguscio, F. Strumia, K.M. Evenson, D.A. Jennings, D.A. Scalabrin; S.R. Stein, Opt. Lett. 4, 9 (1979)

17. M. Inguscio, A. Moretti, F. Strumia, Opt. Commun. 30, 355 (1979)

18. E.J. Danielewicz, C.O. Weiss, IEEE J. Quant. Electron. QE14, 705 (1978)

19. K.Walzer, M. Tacke, IEEE J. Quant. Electron. QE16, 255 (1980)

20. D.K. Mansfield, L.C. Johnson, R. Chouinard, Conf. Digest 8th Int. Conf. on Infr. Millim. Waves, Miami 1983, paper W5.4 IEEE cat. N° 83 CH 1917 - 4.

21. T.Y. Chang, C. Lin, J· Opt. Soc. Am. 66, 362 (1976)

22. N.M. Lavandy, G.A. Koepf, Opt. Lett. 5, 336 (1980)

23. N. Ioli, A. Moretti, G. Moruzzi, P. Roselli, F. Strumia, J. Mol. Spectr. 105, 284 (1984)

24. M. Inguscio, A. Moretti, F. Strumia, IEEE J. Quant. Electr. QE 16, 955 (1980)

25. F. Strumia, J. de Physique-Colloque C7, 117 (1983)

26. M. Inguscio, N. Ioli, A. Moretti, G. Moruzzi, F. Strumia, Opt. Commun. 37, 211 (1981)
    M. Inguscio, F. Strumia, J.O. Henningsen, in ref. 10

27. M. Inguscio, N. Ioli, A. Moretti, F. Strumia, "New FIR laser emissions in CH_3OH" Proc. Third CIRP Conf. Zurich 1984,p.448
    N. Ioli, A. Moretti, G. Moruzzi, F. Strumia, F. D'Amato, "New large offset FIR laser lines in $^{13}CH_3OH$, $^{13}CD_3OH$ and CH_2F_2" Proc. Ninth Int. Conf. Infrared and Millim. Waves Osaka 1984

28. G. Merkle, J. Heppner, Opt. Commun. in press (CH_3OH) and Opt. Lett. in press (CH_2F_2)

29. M. Inguscio, K.M. Evenson, F.R. Petersen, F. Strumia, E. Vasconcellos, Int . J. Infr. and Millim. Waves, in press.

FREE-ELECTRON LASER EXPERIMENTS

Charles A. Brau

University of California
Los Alamos National Laboratory
Los Alamos, NM 87545

ABSTRACT

Using an electron beam in a magnetic field as the gain medium, free-electron lasers offer a powerful new source of tunable, coherent radiation. Results are just becoming available from several new experiments, but they are confirming the theoretical promise of these devices. Backed up by a strong base of accelerator technology, free-electron lasers are expected to become important in applications requiring wavelengths from the far infrared part of the spectrum to the vacuum ultraviolet.

PRINCIPLES

Free-electron lasers represent a totally new concept for producing coherent radiation, and offer a variety of advantages over conventional lasers. In place of a solid, liquid or gas as the gain medium, free-electron lasers use a high-energy electron beam in a magnetic field. First invented by Robert Phillips in 1960 as a microwave tube called the ubitron, the free-electron laser was independently rediscovered as an optical device by John Madey in 1970 as an outgrowth of research in synchrotron-radiation sources.

Reduced to its essentials, a free-electron laser consists of an accelerator to produce the electron beam, a "wiggler" magnet to force the electrons to oscillate and radiate, and an optical system to form the laser beam. These are shown schematically in figure 1. As shown there, the wiggler magnet consists of a series of alternating magnetic poles which form a magnetic field directed up

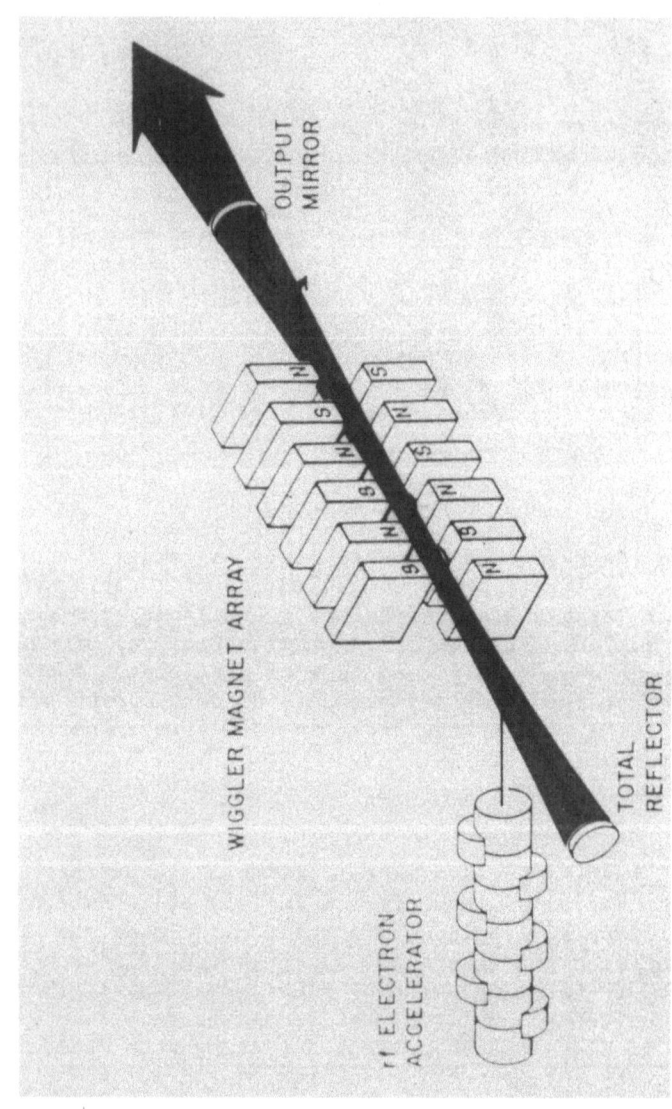

Fig. 1. In a free-electron laser, electrons wiggling in the magnetic field of the wiggler emit light which amplifies the laser beam entering the device from the left. If mirrors are used as shown, the light reflects back and forth between them, getting stronger on each pass through the wiggler. The light is then extracted as a laser beam through the output coupling mirror. In this configuration, the laser is referred to as an oscillator.

and down along the length of the wiggler. As the electrons pass
through this magnetic field, they are deflected alternately left
and right. Because of this wiggle motion, the electrons emit
radiation at the wiggle frequency. Due to relativistic effects,
the radiation is strongly forward directed, and appears at a
frequency which is doppler shifted to much shorter wavelengths. In
fact, to a good approximation the wavelength λ_L of the light is
related to the wiggler period λ_W by the expression

$$\lambda_L = \lambda_W / \gamma^2 \quad , \tag{1}$$

where γ is the energy of the electrons in units of their rest
energy, which is 0.511 million electron volts. Thus, for
100-million-volt electrons ($\gamma=200$), travelling through a wiggler
with a period of 2 centimeters, the wavelength is about 0.5
micrometers. This corresponds to green light, in the middle of the
visible part of the spectrum. Since 100-million-volt electrons are
conveniently obtained from conventional radio-frequency (rf) linear
accelerators, free-electron lasers promise to be a source of
radiation at important wavelengths. In fact, by using different
types of accelerators to produce electrons of different energies,
it is possible to obtain wavelengths from the microwave region to
the ultraviolet, as shown in figure 2. Free-electron lasers have
already been operated at wavelengths from the microwave region to
the visible portion of the spectrum, and experiments are underway
to extend this to even shorter wavelengths. Of course, the
electron radiation described above is just incoherent emission,
similar to the spontaneous radiation from conventional lasers.
However, when a coherent optical field from a laser (even the free-
electron laser itself, as shown in figure 1) is superimposed on the
electrons, the electric field of the optical beam acts on the
electrons. At resonance, that is, when the laser satisfies
equation (1), the interaction becomes strong and the electrons are
forced to oscillate in phase with the optical field. The electron
emission then adds coherently to the incident beam and amplifies it
as in a conventional laser. When mirrors are used, as in figure 1,
the light reflects back and forth between the mirrors, gaining
strength on each pass through the wiggler. The laser beam is
extracted through the output mirror, as in conventional laser
oscillators. Thus, in most respects, free-electron lasers behave
like other types of lasers, and share their properties.

STATUS

The first free-electron laser was built and operated by Madey
and his co-workers at Stanford University in 1976. Using
23-million-volt electrons from an existing rf accelerator, they
were able to amplify 10-micrometer (infrared) radiation from a CO_2
laser, and demonstrate stimulated emission (laser action). Shortly
afterward, they increased the electron energy to 43 million volts

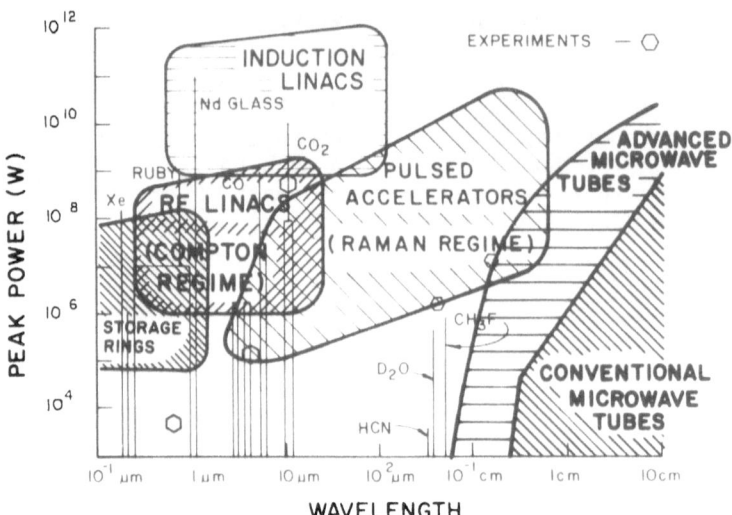

Fig. 2. To change the wavelength, or color of the beam from a free-electron laser, it is necessary to change the electron velocity, or energy; short wavelengths require higher energy. Different types of accelerators are used for low- and high-energy electrons. To operate in the visible portion of the spectrum, storage rings, rf linacs, and possibly induction linacs can be used.

and added mirrors as shown in figure 1, and were able to make the free-electron laser oscillate at 3.3 micrometers without the support of a conventional laser. The laser produced about 5 watts of power averaged over the millisecond pulses; the efficiency was about 0.2 percent for converting electron-beam power to laser power. Later experiments demonstrated that the optical beam quality was nearly perfect (diffraction limited). Although the laser power was small, considering the size and cost of the accelerator, the potential was obvious, and the results stimulated considerable interest.

The principal shortcoming of these experiments was their low efficiency. This was predicted theoretically, and was due to the fact that as the electrons lost energy to the optical beam, they slowed down and no longer satisfied the resonance condition, equation (1). To compensate for this, Phillip Morton of SLAC proposed that the change of electron energy, γ in equation (1), could be compensated by a corresponding change in λ_w. In fact, he showed that the process should be stable, with the electrons losing just enough energy to satisfy the resonance condition as the wiggler period is "tapered" to a smaller value. To prove this, an experiment was conducted at Los Alamos in 1981. An existing accelerator was modified to provide a beam of 21-million-volt electrons, and a wiggler was constructed to be resonant with 10-micrometer (infrared) radiation. The experiment was configured in the form of an amplifier: instead of mirrors, a CO_2 laser was used to provide the optical input. Photographs of the accelerator and the wiggler are shown in figures 3 and 4, respectively. Both the energy removed from the electrons and that given to the laser beam were measured, although the optical measurements were much more difficult. Both measurements gave the same result, and confirmed the theoretical predictions. Some of the results are summarized in figure 5, where we see that an efficiency approaching four percent was achieved. These results represent the best confirmation yet obtained of the theory of free-electron lasers, and support the largely theoretical promise of these devices. In addition to the 10-micrometer radiation, harmonic radiation was observed at 5, 2.5, and 1.25 micrometers. This radiation derives from nonlinearities of the interaction between the electrons and the incident optical beam, and has the coherence properties of the incident beam. Although it was very weak, a million times less powerful than the 10-micrometer radiation, such coherent harmonic radiation may someday be useful for very short wavelengths, not achievable by any other techniques.

These results stimulated considerable interest because they pointed the way toward useful, high-efficiency lasers. However, for most applications, it will be necessary to operate the free-electron laser as an oscillator, as shown in figure 1, rather than as an amplifier. For this reason, the experimental results obtained by a number of groups in 1983 are of particular interest. The first results were obtained by a team of French scientists using the electron storage ring at LURE, near Paris. The ring was originally designed as a synchrotron-radiation source, and operates at a rather low current. Measurements in 1981 showed a free-electron laser amplification of only four parts in ten thousand per pass. Thus, it was a considerable triumph when a specially designed wiggler, called an "optical klystron", achieved oscillation in the summer of 1983. With the storage ring operating at an energy of about 150 million volts, the free-electron laser produced about 100 microwatts of output power at a wavelength of 0.6 micrometers, in the red part of the spectrum. Although Soviet scientists at Novosibirsk had previously observed gain in the same wavelength region, using a similar wiggler (which they invented), the results obtained at LURE represented the first real free-electron laser operating at a visible wavelength, and the first operating with the electron beam from a storage ring. While the laser power was small, it lasted for more than an hour, limited only by the storage time of the ring.

Fig. 3. This photograph shows the 21-million-volt rf linac used in the Los Alamos free-electron laser experiments. The accelerator itself is the 10-foot long copper structure shown in the center of the photograph. To accelerate electrons to higher energies, proportionately longer accelerator structures are used.

The next results were obtained by a team of scientists from TRW, working with the Stanford accelerator. Using a tapered-wiggler design incorporating ideas from the optical klystron, they obtained laser oscillation at a wavelength of 1.6 micrometers, with a maximum efficiency of 1.2 percent. These results show clearly that tapered wigglers can be used to enhance the efficiency of free-electron laser oscillators as well as amplifiers. The output power averaged over the millisecond pulses was as high as 10 watts. Unexpectedly strong third-harmonic radiation was observed at 0.5 micrometers (green).

Most recently, the Los Alamos amplifier experiment, described above, has been reconfigured as shown in figure 1, and operated as a free-electron laser oscillator, in the infrared. Because of the high current (100 amperes peak) produced by the Los Alamos accelerator, the laser produced 3000 watts average power during the 100-microsecond pulses; even higher powers are expected in the near future. By varying the electron energy, it was possible to tune the wavelength from nine to eleven micrometers. The tuning range

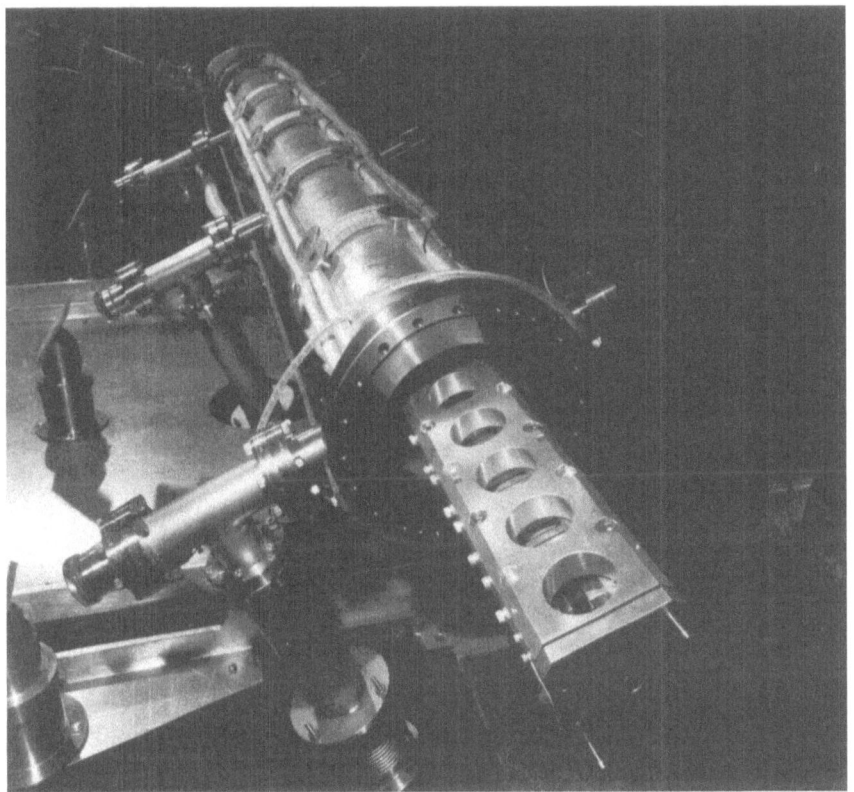

Fig. 4. This photograph shows the wiggler magnet used in the Los Alamos free-electron laser experiments. The wiggler is one meter long, and contains about 340 small permanent magnets which form the magnetic field.

was restricted only by the reflective coatings of the mirrors. Measurements showed that the optical quality of the output beam was essentially perfect, even when the mirrors were deliberately misaligned.

Although most of the recent emphasis has been on free-electron laser oscillators, work at Livermore has focussed on high-gain amplifiers. Using a linear induction accelerator, rather than an rf accelerator, much higher peak current (as high as 10,000 amperes) can be obtained. This makes it possible to obtain sufficient laser amplification to make a useful free-electron laser amplifier. Besides the wavelength restrictions imposed by the need for a conventional input laser, this approach is limited by the fact that high-current electron beams are difficult to focus, making it difficult to operate the free-electron laser at short wavelengths. Nevertheless, induction linacs have advantages for

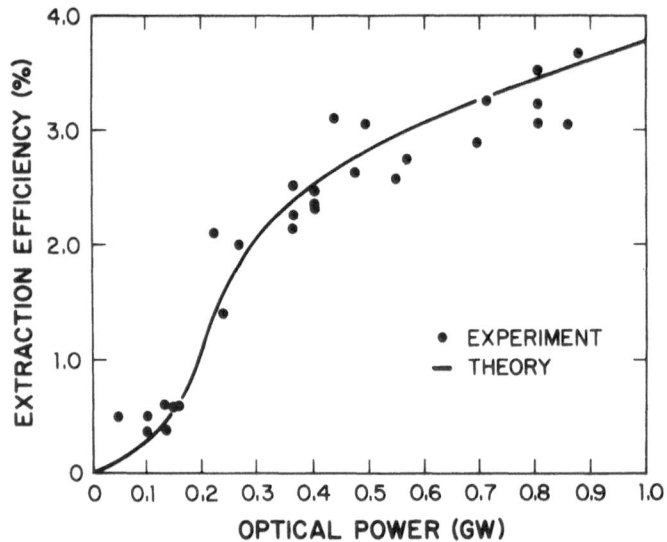

Fig. 5. This figure shows some of the results of the Los Alamos free-electron laser experiment. As the power of the input laser beam is increased, more of the electron energy is converted to light, increasing the efficiency. The results agree quite well with the theoretical predictions, and show an efficiency for extracting light from the electrons which approaches four percent.

applications which require high peak power. Recently, an efficiency of five percent was demonstrated at a wavelength of 9 millimeters. Although this is a rather long wavelength, the physics and technology should scale to much shorter wavelengths, as indicated in figure 2. The peak power in these experiments was about 100 million watts, in pulses lasting about 10 billionths of a second.

In the near future, several more experiments are expected to come on line. At the University of California, Santa Barbara, an interesting experiment using a five-million-volt Van de Graaff accelerator is about to begin operation. It is designed to operate

at wavelengths in the far infrared, near 400 micrometers. The
experiment is particularly interesting because it is designed to
operate with improved efficiency by recovering the energy remaining
in the electrons after they have passed through the free-electron
laser. At the National Synchrotron Light Source at Brookhaven, a
free-electron laser is being developed for the ultraviolet part of
the spectrum, around 0.3 micrometers, using the uv storage ring.
Meanwhile, at AT&T Bell Laboratories a compact type of accelerator
called a microtron is being used to develop a free-electron laser
for the far infrared part of the spectrum from about 100 to 400
micrometers. All these experiments illustrate the interest in
free-electron lasers for wavelength regions not easily accessible
to conventional lasers. Finally, a free-electron laser oscillator
for the mid-infrared region, from perhaps 5 to 50 micrometers, is
being developed at the Kelvin Laboratory in Glasgow. This
experiment which is intended mostly for research in free-electron
lasers themselves, is undergoing preliminary tests.

In the future, the emphasis will be on free-electron lasers for
very short wavelengths, reaching down into the extreme ultraviolet
and soft X-ray regions. The problems in these regions are
formidible: laser gain decreases at short wavelengths, mirrors are
poor, and it is difficult to collimate electron beams well enough
to operate at such short wavelengths. But the difficulty of
obtaining coherent radiation at these wavelengths by any other
techniques makes free-electron lasers interesting despite the
difficulties, and it may even be possible to make use of short-
wavelength harmonics, as discussed above.

ISSUES

The further development of free-electron lasers must address a
variety of problems involving the basic operation of the free-
electron lasers themselves, as well as the technology of optics and
accelerators. For free-electron lasers operating from linear
accelerators, the most important fundamental issue is that of
so-called "synchrotron" instabilities. As the electrons proceed
down the length of the wiggler, they tend to oscillate about the
resonant energy, which satisfies equation (1). These motions,
which are called synchrotron oscillations, predicted to cause
radiation at wavelengths shifted from the laser wavelength by the
frequency of the synchrotron oscillations. This radiation may rob
power from the laser radiation, and degrade the performance of the
device. So far, attempts to see this effect have not been successful, but if it appears in future experiments it should be
possible to suppress it by various techniques. For free-electron
lasers operating in storage rings, the principle problem is the
tendency of the free-electron laser to disrupt the electron beam,
stretch out the electron-beam pulses, and make the beam unsuitable
for the free-electron laser on the next pass. Although the natural

properties of the ring tend to damp out the effects of the free-electron laser, the interaction is not well understood at the present time. In fact, under some conditions the storage ring works better than expected due to anomalous shortening of the electron pulses in the storage ring.

Optics technology must also see some development for free-electron laser applications. In particular, optical damage to the mirror coatings by the high-intensity laser beam is expected to be a problem. The problem is complicated by degradation of the coatings by radiation other than the laser beam itself. This radiation includes harmonics of the laser radiation, as discussed above, as well as X-rays, gamma rays, neutrons, and stray electrons from the accelerator. Such degradation has already proved to be a problem in the storage-ring free-electron laser experiments at LURE.

Accelerator technology must also be advanced for free-electron laser applications. Both the peak and average current capability of rf accelerators and storage rings must be increased. As the current increases in rf accelerators, two problems arise. First, the ability of the electron beam to be collimated declines as the peak current increases. Second, instabilities caused by the interaction between the electrons and the accelerator structure develop at high average current. Interesting programs are underway at Los Alamos to address both of these problems, using photocathode guns to provide high peak current and advanced accelerator structures to avoid instabilities. The most stressing problem for induction linacs is collimation of the high-current beams which these machines produce, but progress is being made at Lawrence Livermore Laboratory. Storage rings available for free-electron laser research in the past have all suffered from insufficient peak current to provide adequate optical gain. When it is fully operational, the uv ring at Brookhaven should have significantly higher current than older rings, and at Stanford, Madey and his co-workers are about to begin construction of a new ring designed specifically for free-electron laser applications. This will provide sufficient current for operation at very short wavelengths, down into the far ultraviolet.

APPLICATIONS

Due to their unique technology, free-electron lasers offer several advantages relative to conventional lasers. These include wavelength tunability, high power, good optical-beam quality, and high efficiency. Wavelength tunability follows, of course, from the relationship (equation (1)) between the wavelength and the electron energy. Since the electron energy can typically be varied over a factor of two in a given accelerator, it is possible to vary the wavelength over an even wider range. Generally it will be necessary to change the mirrors, which have a narrow range over which they are reflective, but this is not difficult. Free-electron lasers have two advantages for achieving high power. In the first place, conventional lasers are limited by the difficulty of removing the waste heat. The most powerful lasers, gas lasers, depend on high-speed (even supersonic) flow to remove the hot laser medium. In free-electron lasers the waste heat resides in the electron beam, which is moving at nearly the speed of light and exits the optical cavity in a few billionths of a second. Moreover, high-power accelerators already exist. SLAC, at Stanford, accelerates an average electron-beam power of hundreds of kilowatts, twenty four hours each day. Good optical-beam quality is another advantage of free-electron lasers. This follows from the nature of the optical resonators used in free-electron lasers: they are long, slender devices, and do not support undesirable optical modes. In addition, the optical-beam distortion due to the electron beam is very small. In the Los Alamos experiments, the electron beam is equivalent to a lens with a focal length of more than a kilometer. Finally, free-electron lasers may someday offer high efficiency. When used to accelerate high-current electron beams, accelerators can be made as much as 50 percent efficient. Combined with the demonstrated four percent conversion to laser radiation, this corresponds to two percent overall efficiency. In fact, it should be possible to improve on this by recovering a substantial fraction of the energy remaining in the electron beam after it passes through the wiggler. Experiments of this type are already being conducted at Santa Barbara, as discussed above, using a Van de Graaff accelerator. At Los Alamos it is proposed to use an rf linac to both accelerate the electrons and then to convert their energy back into rf power after they have passed through the free-electron laser. Experiments have demonstrated that this process is indeed possible, and application of this idea to free-electron lasers is planned for the near future.

These advantages suggest a number of applications. The easiest advantage to achieve, and the one which will be the first to be exploited, is the wavelength tunability. Long wavelengths, in the far infrared from about 100 to 400 micrometers, are of interest for solid-state materials research. For this reason, AT&T Bell Laboratories is building a free-electron laser for this wavelength region. Shorter wavelengths, particularly in the ultraviolet, are of interest for chemistry, and high power may someday make photo-chemical processes of interest for industrial applications. For infrared photochemistry, where molecular spectra are strongly structured, the tunability of free-electron lasers is particularly important. Likewise, the ability of free-electron lasers to tune into spectral regions inaccessible to conventional lasers makes them of interest for medical applications. This is one of the principle motivations for the short-wavelength storage-ring free-electron laser program which is now beginning at Stanford.

Finally, the advantages which free-electron lasers offer for the production of high-power laser beams of good optical quality has made them interesting for strategic defense, also known as "star wars". While such applications are only distant possibilities at the present time, and face a variety of problems of which the laser may be the simplest part, the possibility of enormously powerful laser beams still excites the imagination, and work in that direction is providing spin-offs for a variety of less ambitious applications.

PRELIMINARY MEASUREMENTS OF SPONTANEOUS RADIATION OUTPUT FROM THE

UK FREE ELECTRON LASER

J.S. MacKay and D.M. Tratt

Physics Department
Heriot-Watt University
Riccarton, Edinburgh, U.K.

INTRODUCTION

The UKFEL is a major collaborative project between (principally) Heriot-Watt University, Glasgow University and Daresbury Laboratory. One of the key features of the project is that the electron linear accelerator at the Glasgow Kelvin Laboratory is virtually fully dedicated to the project.

The aim of the project is, in the first instance, to mount a comprehensive series of experiments to characterise the performance of the laser in both amplifier and oscillator configurations. Relatively high gain is expected over a wide range of output wavelengths (2-20 µm) for various combinations of linac energies and wiggler fields[1]. Harmonic generation provides the exciting possibility of extension to even shorter wavelengths in the visible and UV[2].

Initial investigations have been carried out into the spatial and spectral distributions of the spontaneous radiation generated by the interaction of a 100 MeV electron beam with a two-section 2.47m wiggler magnet assembly.

BASIC OPERATING PARAMETERS

As can be seen from equation (1) the FEL wavelength is dependent on the inverse square of the linac energy:

$$\lambda_{FEL} = \frac{\lambda_w}{2\gamma^2} \left[1 + \frac{K^2}{2} \right]$$

(1)

where λ_w is the wiggler period (65mm), γ is the electron energy expressed in units of its rest mass and K is the dimensionless wiggler parameter.

Thus it was seen as crucial that in order to demonstrate and investigate maximum wavelength tunability of the laser, it was necessary to choose the electron beam energy in the centre of the linac uniform current range ($\sim$50 MeV) at the nominal output wavelength. This wavelength was chosen to be 10.6 µm in view of the associated well-developed laser technology (fast detectors, optical components, etc.).

With the emphasis on __gain__ rather than efficiency, and as a compromise between increasing gain and restrictions on both space and cost, a uniform wiggler of length 5m has been built. It was also decided that the wiggler should be planar rather than helical despite the significant reduction in gain. Advantages of the planar option included design flexibility and ease of access to, for example, the vacuum components. Moreover, the choice of a helical wiggler restricts FEL oscillation to the fundamental harmonic, since only this harmonic appears on axis with the helical geometry. Rare earth cobalt permanent magnets, rather than superconducting magnets, were chosen from both technical and economic viewpoints. With the minimum pole separation set at 22mm, optimum gain was predicted for a magnet period of 65mm. For these chosen central operating parameters, the FEL gain should be >30%/pass at 10.6µm. Optimistically assuming full coupling between radiation and electron beams at the operating range extremities (2 µm, 20 µm), there should only be a twofold gain reduction at these wavelengths.

The cavity length is governed by the requirement for synchronism between the electron and optical pulses __i.e.__ it must comprise an integral no. of 0.315m units ($\equiv$ RF wavelength/12); we have chosen a cavity length of 7.878m. This will allow 25 optical pulses to circulate within the cavity, each with a round trip time $\sim$50ns resulting in average powers of some tens of watts for saturation over the last few µs of each electron macropulse. Megawatt peak powers should be obtainable despite the low extraction efficiency (0.6% max) of such an untapered wiggler magnet.

The output coupler is remotely translatable in 1 µm steps over the electron microbunch length (2 mm, 20 mm with ECS operation[3]) to test the effects of mismatch. Mirror curvatures of 4.5 m have been chosen to conform with the radiation phase-front curvature, resulting in a near-concentric symmetric cavity.

LINAC/ ELECTRON BEAM

The Glasgow electron accelerator is an RF linac of conventional design operating at 2.856 GHz with an energy range of 30-160 MeV. At present the linac readily achieves a pulsed current of 250 mA in a 4 μs macropulse at repetition rates up to 100 Hz. The micropulse structure within the macropulse consists of a train of 5 ps pulses separated by 350 ps.

At the principal operating energy of 56 MeV, corresponding to λ_{FEL} = 10.6 μm, the primary energy spectrum has a width ~1%. An energy compression system (ECS) installed on the linac should reduce this width to 0.2%[3], thereby improving the FEL linewidth at a given setting. The ECS however stretches the micropulse by a factor of ten, to 50 ps, also reducing the peak current tenfold to 1.5 A.

At present a new triode gun is being fitted to allow subharmonic bunch mode operation. The chosen bunching factor of 6 should restore the peak current to 10 A. It is further anticipated that the macropulse duration, which determines the available gain in the optical cavity, will also in time be increased to 10 μs by modifying the linac pulse forming networks.

Recent experiments suggest a beam emittance of 0.5π mm.mrad[4], in line with projected requirements for FEL operation.

WIGGLER

A plane magnet (constant peak field) of length 4.94 m and period 65 mm has been constructed using $SmCo_5$ material and is now installed at Glasgow. The magnet is split equally into 4 independently operable sections each with 19 periods and a pole separation remotely variable over the range 22-200 mm, giving a maximum K value of 2.68 at the minimum gap. At 200 mm separation, the peak field is less than 0.1 mT and the magnet can be considered "switched off". Both magnet period and length were chosen following extensive computations to optimise the FEL gain[1].

The wiggler has been orientated to allow a vertical wiggling plane in order to simplify the design requirements of diagnostic and support structure assemblies.

The relatively low energy of the electron beam in the <50 MeV regime makes it particularly sensitive to magnetisation errors, and so a procedure for arranging the blocks in a specific order to minimise axial field errors was devised and implemented in the construction[5]. Horizontal and vertical correction coils fitted to each section will be used to correct for small cumulative axial discrepancies.

EXPERIMENTAL ARRANGEMENT

The experimental layout is shown in fig. 1. A single BBC-B microcomputer performs remote control activities and also receives incoming data from a separate diagnostics microcomputer sited in the experimental area. Communication between the BBC-B and the REMDACS digital system is through a twisted pair cable; the system then distributes the control information in pulse-form to the individual pieces of equipment. The data-logging microcomputer synchronises the linac pulses with the FEL pulses by way of a pulse-gating system and transmits the data via a fibre-optic link to the BBC-B control unit. Archiving of data can be executed via a direct serial link to the VAX minicomputer facility at the Kelvin Laboratory.

SPONTANEOUS RADIATION MEASUREMENTS

Preliminary measurements have been made of both the wavelength and linewidth of the spontaneous emission from the laser.

With a very low current (1 μA time-averaged) electron beam of 100 MeV and using only one of the four wiggler sections, both on-axis odd harmonics and off-axis even harmonics were observed with a sensitive color TV camera. As the K-value of the wiggler was varied from 1 to 1.8, increasing the fundamental wavelength from 1.3 μm to 2.0 μm, the visible higher order harmonics were seen to tune across the visible spectrum (see figs. 2 and 3). Both spatial and spectral distributions of the odd and even harmonics agree well with distributions predicted by the expression[6]:

$$\lambda_i = \frac{\lambda_w}{2\gamma^2} \frac{1}{i} \left[1 + \frac{K^2}{2} + \gamma^2\theta^2 \right] \tag{2}$$

in which i is the harmonic order and θ the detected emission half-angle with respect to the electron beam/ wiggler axis.

With two wiggler sections in place and using a photovoltaic detector to observe the output from a remotely-controlled scanning prism spectrometer, we successfully tuned the 3rd harmonic across the visible wavelength region. In the absence of angular broadening we expect the natural homogeneous linewidth to be given by the relation:

$$\Delta\lambda/\lambda_i = 1/iN \tag{3}$$

For our configuration $\Delta\lambda \sim 5$ nm.

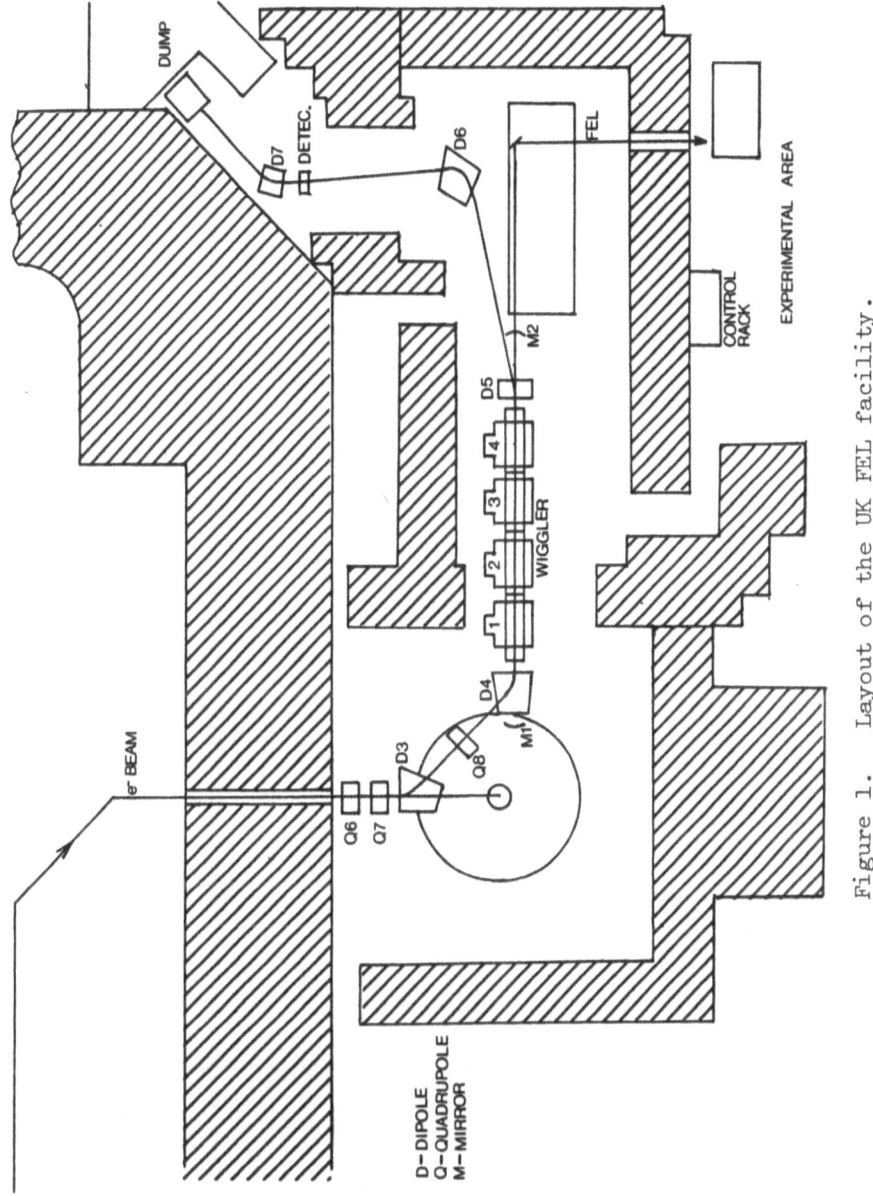

DUMP

D7

DETEC.

D6

FEL

M2

D5

4

3

2

1

WIGGLER

D4

M1

Q8

D3

Q6

Q7

BEAM

EXPERIMENTAL AREA

CONTROL RACK

D – DIPOLE
Q – QUADRUPOLE
M – MIRROR

Figure 1.   Layout of the UK FEL facility.

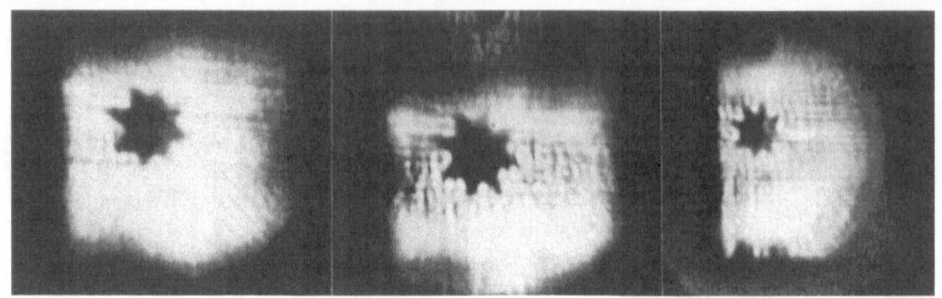

Figure 2. Spatial distribution of the visible spontaneous
output pattern for K=0.99 (left), 1.34 (centre)
and 1.64 (right). The horizontally-oriented divisions
within each pattern are a manifestation of the
multiple-harmonic nature of the radiation.

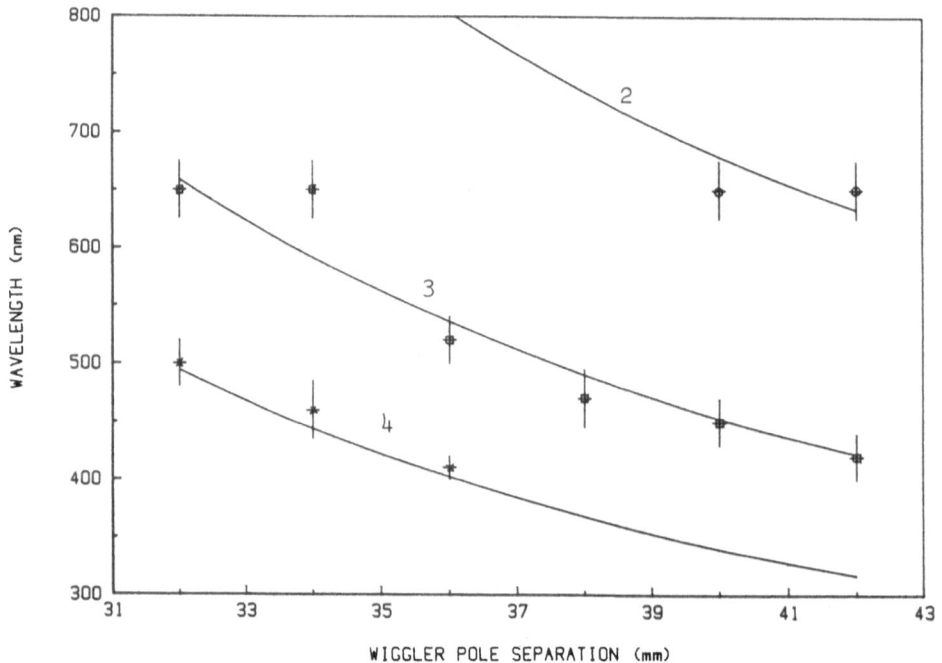

Figure 3. Theoretical dependence of the on-axis spontaneous
output wavelength on the wiggler pole separation
for the 2nd (o), 3rd (□) and 4th (*) harmonics. Also
shown are the corresponding data points derived from
fig. 2.

Using a variable limiting aperture positioned just before the main deflection mirror inside the cave (fig. 1), it was possible to measure the radiation linewidth for a number of aperture settings and in this way control the acceptance angle θ of the detection system.

The results from two such runs are reproduced in fig. 4, from which it is evident that the emission linewidth from our system does indeed increase with θ, as expected from the undulator radiation equations[6].

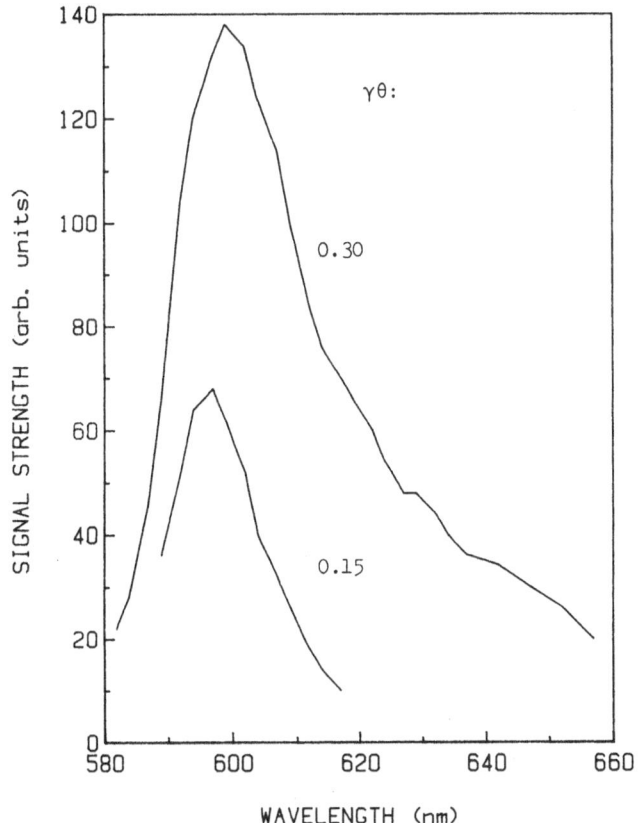

Figure 4.   Dependence of spontaneous third harmonic linewidth on collection half-angle θ, showing qualitative agreement with theoretical predictions (cf. ref. 6).

PROPOSED EXPERIMENTAL PROGRAM

In the light of the encouraging results presented here, the original UKFEL proposal[7] has been extended to include a feasibility study into possible oscillation in the visible wavelength region utilising higher order (3rd or 5th) harmonic generation at high values of K.

ACKNOWLEDGEMENTS

The results presented here were obtained in collaboration with Profs. C.R.Pidgeon and S.D.Smith (Heriot-Watt University- Project Directors), M.G.Kelliher, W.A.Gillespie and J.M.Reid (Glasgow Kelvin Laboratory- Linac), M.W.Poole and R.P.Walker (Daresbury Laboratory - Wiggler) and in particular, on the experimental effort, M.F.Kimmitt (Essex University, UK) and D.A.Jaroszynski (Heriot-Watt University). Theoretical modelling of the UK FEL is being carried out by a collaboration between W.J.Firth (Heriot-Watt University), J.N.Elgin and C.Penmann (Imperial College, UK).

REFERENCES

1. M.W.Poole, G.Saxon, R.P.Walker, S.D.Smith, C.R.Pidgeon, W.J.Firth, J.M.Reid, M.G.Kelliher and W.A.Gillespie, IEEE Trans. Nucl. Sci. NS-30 (1983) 3091.
2. G.Dattoli, T.Letardi, J.M.J.Madey and A.Renieri, IEEE J. Quant. Electron., QE-20 (1984) 637.
3. W.A.Gillespie and M.G.Kelliher, Nucl. Instr. and Meth., 184 (1981) 285.
4. W.A.Gillespie, private communication.
5. M.W.Poole, R.J.Bennett and R.P.Walker, J. Phys. Colloq., 45, no. C-1 (1984) 325.
6. S.Krinsky, "Study of a Coherent Wiggler for the 700 MeV VUV-Ring of the National Synchrotron Light Source", Informal Report, NSLS, Brookhaven, (February 1979).
7. S.D.Smith, C.R.Pidgeon, M.W.Poole, and K.Hohla, Phys. Quant. Electron., 8 (1982) 275.

THE FREE-ELECTRON LASER

Alberto Renieri

Frascati Center, Comitato Nazionale per la Ricerca e per lo
Sviluppo dell'Energia Nucleare e delle Energie Alternative
Rome, Italy

Abstract

The generation of coherent synchrotron radiation using beams of
relativistic electrons passed through magnetic undulators opens up
tremendous scope for photochemistry in general and isotopic separation by laser in particular. Applications for plasma heating in
Tokamak-type installations to trigger off the controlled nuclear
fusion process are also conceivable.

INTRODUCTION

Early in 1977 John Madey and his team at Stanford University
in California succeeded in generating coherent infrared light with
a wavelength (λ) of 3.417 μm, a line width ($\Delta\lambda$) of 8 nm, and an
average output (P) of 0.36 W. This new laser could now join the
vast family of coherent radiation sources which - ever since a beam
of red light with a wavelength of 694.3 nm was produced by the first
ruby laser in 1960 - had been gradually growing by the addition of
new devices operating within the band ranging from the far infrared
to the ultraviolet. Viewed in this perspective, the parameters of
the laser developed at Stanford were nothing exceptional. Nevertheless, the new source immediately aroused very considerable
interest, because the active medium instead of being made up of
atoms or molecules was nothing other than a beam of high-energy free
electrons.

Reprinted with permission from Endeavour, New Series, Vol.8, No.1
(1984). Copyright © 1984 Pergamon Press Ltd.

255

The Emission of Electromagnetic Radiation by Free Charges

A laser based on light emission by free charges might be considered in principle as a contradiction of the laws of physics. We know that a free charge in a vacuum can neither emit nor absorb radiation, because the laws of energy and momentum conversion cannot be satisfied at one and the same time. However, this charge (and in what follows we virtually always understand charge as electron) is moving in an external field, the field itself absorbing some of the momentum and consequently rendering emission possible. Such a field could be the magnetic bending field in a high-energy accelerator (in which case the emission is of synchrotron radiation) or a Coulomb field in a nucleus (Bremsstrahlung) or a field generated by the image charges on a metal lattice (Smith-Purcell effect). Finally, electromagnetic radiation can be produced by causing a beam of electrons to move in a medium (transparent for this radiation) at a velocity greater than that of light in the medium so that the beam can generate the momentum required to satisfy the conservation laws (Čerekov effect).

It could be argued that under these circumstances the electron is no longer free, but this is really a question of definition. We shall apply the term 'free' from now on to any electron not bound to a nucleus in a stationary state. The remainder of this article will concentrate on radiation in a magnetic field, because the operation of the free electron laser (FEL) developed at Stanford is based on this principle. It is interesting to note, however, that lasers have recently been produced which operate with free electrons in the millimeter wavelength range which are based on the Smith-Purcell effect (Orotron). Coherent absorption by the inverse Čerekov effect has also been observed during experiments as a by-product of stimulated Čerekov radiation.

Synchrotron Radiation

The generation of radiation by charged particles moving in a magnetic field was first achieved in the 1940s when the first high-energy electron accelerators (such as the betatron and synchrotron) were built. When Ivanenko and Pomeranchuk were investigating in 1944 the limits of energy obtainable with a betatron they came to the conclusion that the output from a charge moving in a magnetic field increases rapidly with energy and decreases with the particle's mass. It was clear, however, that appreciable synchrotron radiation could be obtained only from ultra-relativistic electrons and positrons ($E \gg m_0 c^2$, where E and m, are the energy and mass of the particle and c is the velocity of light in a vacuum).

An interesting fact is that recently weak proton radiation has been observed in the visible range (energy level approximately 300 GeV) in the SPS protosynchrotron at CERN in Geneva.

A typical feature of synchrotron radiation is its considerable bandwidth. To illustrate this more clearly Figure 1 shows the spectral distribution of synchrotron light as a function of ω/ω_c where ω is the frequency of the radiation emitted and δ_c the 'critical frequency' dependent on the curvature radius of the trajectory and the energy of the electron.

As shown in Figure 1, radiation stretches continuously from the low frequencies to those approaching the critical frequency ω_c. In the case of ultrarelativistic electrons this frequency is extremely high. To give an example, between 100 MeV and 1 GeV the radiation emitted ranges from the visible frequency ($\omega_c \sim 3 \times 10^{15}$ Hz to hard X-rays ($\omega_c \sim 3 \times 10^{18}$ Hz) for radii of curvature of approximately 1 meter.

This considerable bandwidth is due to the dynamics of synchrotron radiation. To give a clearer illustration of this point, which is particularly important to us as it is closely linked to problems associated with the free-electron laser, Figure 2 shows the emission process for an electron moving in a circular orbit. The radiation emitted due to the acceleration induced by the magnetic field is concentrated forwards in the direction of movement in a small cone

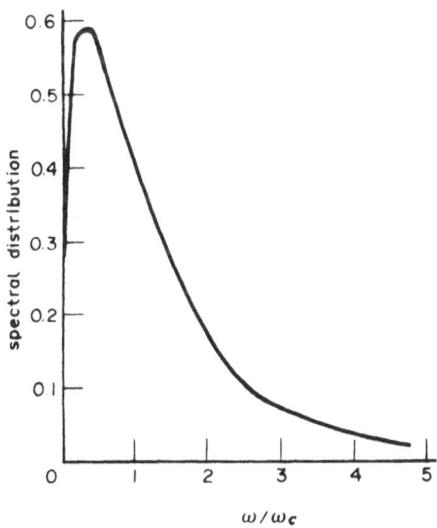

Fig. 1. Spectral distribution of electromagnetic radiation emitted by a high-energy charged particle moving in a uniform magnetic field (synchrotron radiation). The abscissa shows the ratio between frequency emitted (ω) and critical frequency (ω_c).

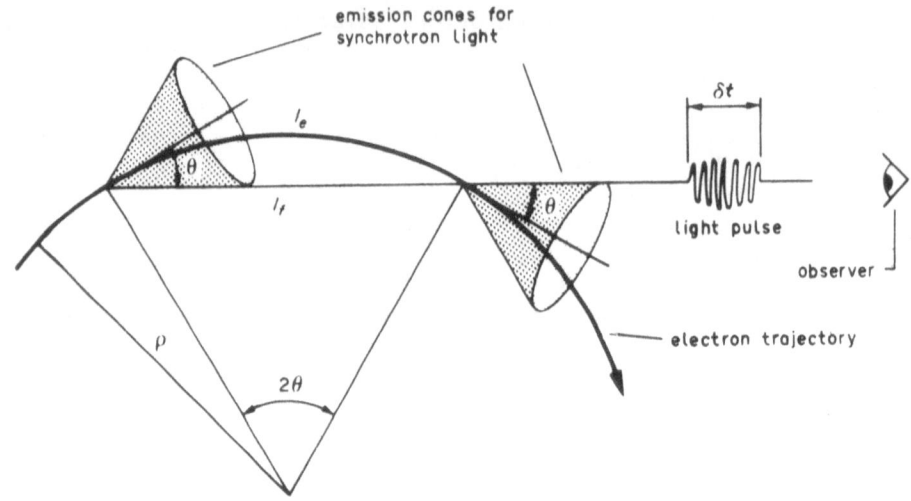

Fig. 2.
Process of synchrotron radiation in a uniform magnetic
field. Light is emitted forward within an aperture cone
given by Equation (1). An observer at the height of the
orbit plane sees the radiation emitted by the particle only
within arc l_e subtended by an angle equal to twice the
value of the emission angle. Duration of the light pulse
(δt) is thus given by the difference in passage times of the
particle on arc l_e and the photons on the chord l_f subtended
by l_e.

with the following aperture:

$$\theta \sim \frac{m_0 c^2}{E} . \tag{1}$$

For ultrarelativistic electrons this aperture is very small.
Consequently an observer standing some distance away on the orbit
plane will see only the light emitted by a small portion I_e of the
trajectory (see Figure 2) subtended by an angle twice that of the
emission cone aperture represented by Equation (1). The duration δl
of the light pulse seem by the observer is determined as will be
readily understood, by the difference between the passage times of
the electron on arc I_e and the photons on chord I_f subtended by I_e
(see also Figure 2) and is also extremely short. The substantial
bandwidth of the synchrotron radiation is due to the short duration
of the light pulse. From Fourier analysis of a wave train of finite
length it can be seen that the radiation spectrum emitted extends to

frequencies as high as the inverse of δt:

$$\omega \sim \frac{1}{\delta t} \ .$$

This frequency is the critical frequency (ω_c) referred to earlier (Figure 1).

Synchrotron Radiation from a Magnetic Undulator

Synchrotron light is now widely used in many areas of research (basic and applied physics, chemistry, biology, etc). Its considerable bandwidth extends its scope of application by providing a light beam with a wide range of frequencies (from infrared to X-rays) but at the same time has the effect of limiting output per unit of available frequency. In order to improve this parameter Motz proposed in 1951 the irradiation of an electric beam in a static periodic magnetic field (magnetic undulator). A field of this type can be produced, for example, by a sequence of alternate magnetic dipoles, as shown in Figure 3. Under the influence of this field the electron does not follow a circular trajectory but oscillates along the axis of the magnet. If the intensity and period of the magnetic field are selected in such a way that the angle of deviation of the trajectory from the axis is always less than the angle of cone aperture within which the radiation is emitted, an observer on the axis of the undulator (Figure 3) would see the light emitted along the entire length of the trajectory and not merely from a small fraction of it. This experimental arrangement produces what is commonly known as 'undulator operation'.

The light pulse duration is much longer than that of the example considered in the previous paragraph (circular trajectory). Consequently the emission band is considerably narrower and its relative width is inversely proportional to the number N of undulator periods.

$$\left[\frac{\Delta\omega}{\omega} \right]_0 = \frac{1}{2N} \ . \qquad (2)$$

The significance of Equation (2) is obvious. The longer the undulator (i.e. the larger N), the longer will be the duration of the light pulse and the narrower the bandwidth. There is a lower limit to the bandwidth due to the quality of the electron beam. In fact an inhomogeneous widening of the spectrum occurs which is a function of emittance (essentially given by the product of transverse dimension multiplied by angular divergence) and the dispersion in energy of the particle).

This widening is similar to the Doppler widening which atoms or molecules undergo during emission. The theoretical features of the radiation emitted in the undulator were studied by Alferov, Bashmakov, and Bessonov in 1973.

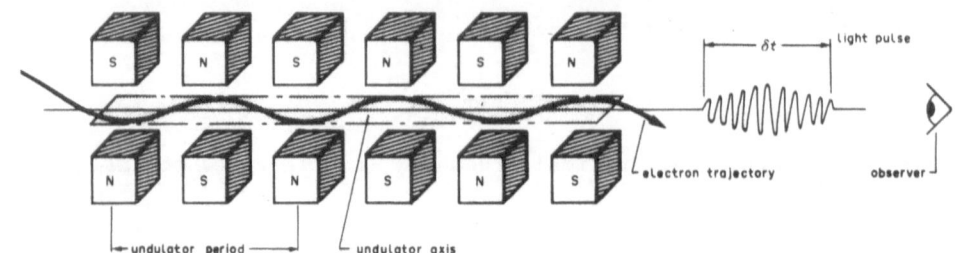

Fig. 3.   Diagram of the magnetic undulator, in which the field is
generated by a series of alternate magnetic dipoles. S and
N denote the south and north alternation of the magnetic
dipoles. The trajectory of a charged particle oscillates
about the axis of the undulator. By controlling period and
magnetic field intensity of the angle of deviation of this
trajectory from the magnetic axis is less than the width of
the emission cone. Consequently an observer on the axis of
the undulator would see the radiation emitted along the
entire length of the magnet. The duration δt of the light
pulse is thus considerably greater than that associated with
synchrotron radiation in a homogeneous field (for the same
magnetic field and electron energy).

Figure 4 (Graph a) shows the light spectrum radiated along the
axis of the magnet, in a homogeneous line condition. The spectrum is
symmetrical about the central frequency ω_U, the dependence of which
on the energy E of the particle and on the magnet's parameters can be
obtained from purely physical considerations. Application of the
Lorentz equation shows that in the electron's reference system the
static field of a λ_q period undulator (Figure 3) is very similar to
an electromagnetic wave with the following frequency

$$\omega^1 = \frac{2\pi c}{\lambda_q} \quad \frac{E}{m_0 c^2} \tag{3}$$

which is propogated in the opposite direction to that of the electron. The merging of the undulator field with an equivalent electromagnetic wave (Weizacker and Williams virtual photon method) improves
with the energy level of the electrons. This wave is scattered by
the Compton effect and the photons scattered will have the same
frequency as the incident photons. The back-scattered light in the
laboratory frame undergoes a Doppler shift towards the higher frequencies. The frequency of this radiation, which is the undulator
synchrotron radiation, can thus be obtained by the following equation
(for ultrarealistic electrons)

$$\omega_U \sim 2 \left[\frac{E}{m_0 c^2} \right] \omega^1 \sim \frac{4\pi c}{\lambda_q} \left[\frac{E}{m_0 c^2} \right]^2 . \tag{4}$$

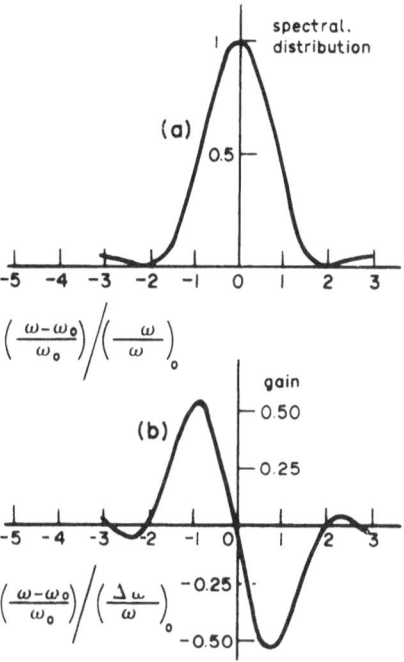

Fig. 4.   Graph a shows the spectral distribution of the electro-
          magnetic radiation emitted at angle zero along the axis of
          an undulator.   The abscissa shows in units of width of the
          homogeneous frequency band (cf. Equation (2)), the differ-
          ence between the frequency emitted ω, and the central fre-
          quency ω_0, as produced by Equation (5).   Graph b shows the
          free electron laser gain as a function of frequency.   The
          abscissa shows the same quantity as shown in Graph a.   The
          gain in proportional to the derivative of spectral distri-
          bution (Graph a).

The bandwidth (Equation (2)) can be obtained by taking into account
that it must be equal to that of the incident radiation which is
composed of a wave train of N periods.

By continuing the analogy with the Compton scattering, the
polarization of the radiation generated can also be established and
it will be the same as that of the incident wave.   For example, in
the case of the undulator shown in Figure 3 linear polarization
occurs at right angles to the magnetic field.

For a more accurate determination of central frequency ω_0 allowance has to be made for the energy associated with the transverse movement of oscillation. This involves a small correction to Equation (4), which now becomes

$$\omega_0 = \frac{4\pi c}{\lambda_q} \left[\frac{E}{m_0 c^2} \right]^2 \frac{1}{1+K^2},$$ (5)

where K, known as the undulator parameter, is linked to the average quadratic magnetic field B and to period λ_q by the following Equation:

$$K = \frac{B\lambda_q}{2\pi m_0 c^2}$$ (6)

Parameter K is extremely important for the definition of the properties of the radiation emitted. In particular, the maximum angle of deviation of the electron's trajectory from the magnetic axis is given by

$$\theta_{max} = K \left[\frac{m_0 c^2}{E} \right]$$ (7)

for which reason the operation of the undulator ($\theta_{max} \lesssim \theta$, cf. Equation (1)) is obtained by $K \lesssim 1$.

As can be seen from Equation (5) the frequency of the radiation produced can be varied by altering the energy of the electrons, the period, or the magnetic field of the undulator. With $\lambda_q \sim 10$ cm and $K \sim 1$, for example, the light emitted ranges from infrared to ultraviolet when the energy of the electrons is between 10^2 and 10^3 MeV. Under these conditions by using a 5 m long undulator (N=50) the relative width of the homogeneous band Equation (2) is approximately 1 per cent.

The first undulator, which was based on permanent magnets, was built by Motz, Thon, and Whitehurst in 1953. Its main features were as follows: λ_q = 4 cm, l = 50, cm B = 3.9-5.6 kGs, where L is the length of the undulator.

Using the electron beam produced by the Stanford linear accelerator, first operated a short while before, they generated, in accordance with the theory, radiation in the range between green (λ = 550 nm) and blue (λ=340 nm) for electron energy levels of between 95 and 120 MeV.

Using the same equipment but with a lower-energy electron beam (3-5 MeV) this research team succeeded in generating coherent radiation in the millimeter wavelength range from a number of electrons bunched in packets, the size of a packet being comparable to the

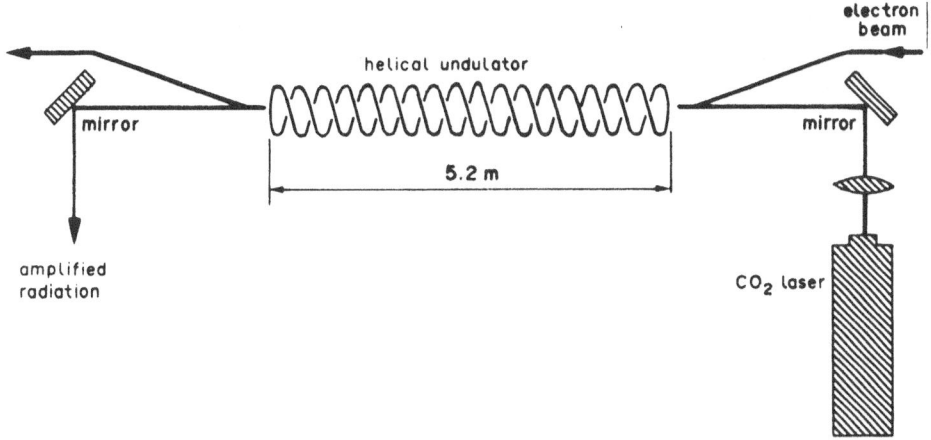

Fig. 5. Diagram of the Standford FEL amplifier. The electron beam
produced by the superconducting linear accelerator is passed
along the axis of a superconducting helical undulator. The
radiation for amplification is generated by a TEA CO_2 laser.

wavelength emitted. Under these experimental conditions the electrons actually emit in phase together and consequently the intensity
of the radiation, instead of being proportional to the total number
of electrons N, is proportional to N^2. The output observed was
approximately 1 watt, or some 6 orders of magnitude greater than that
obtainable with a non-modulated beam. The authors also indicated
that this method was unsuitable for shorter wavelengths and there was
no equipment available capable of bunching electrons into packets of
less than 1 millimeter. Later, we shall see that the FEL offers
precisely this possibility.

In the past few years a considerable number of magnetic undulators has been fitted to electron storage rings in order to produce
sufficiently monochromatic tunable synchrotron radiation. In the
undulator (electromagnetic) fitted on the Adone storage ring at the
laboratories of the Italian National Nuclear Physics Institute (INFN)
in Frascati the magnet has three periods (λ_q = 65.4 cm), corresponding to a homogeneous bandwidth of approximately 17 per cent cf
Equation (2). The system can operate with a very strong magnetic
field on the axis (up to 18.5 kGs), and in this configuration it is
used as an X-ray source. For weaker magnetic fields (less than
200 Gs) the magnet operates as an undulator (K $\lesssim$1). With electrons
with energy levels of 500-700 MeV, visible spectrum radiation is
produced.

Stimulated Compton Scattering

In 1968 Pantell, Soncini, and Puthoff put forward the idea of using stimulated Compton scattering to generate coherent radiation which could be easily tunable on a wide spectrum of frequencies by varying the energy of the electrons. A simplified description of this procedure was given by Schrödinger as early as 1927, while in 1933 Kapitza and Dirac proposed an experiment to observe stimulated Compton scattering from non-relativistic electrons.

On the basis of the proposal put forward by Pantell and his collaborators, Sukhatme and Wolff calculated in 1973 the gain for this type of laser and came to the conclusion that with the electron beams and electromagnetic radiation sources available at that time, the gain for stimulated Compton scattering was too low compared with typical loses for resonant cavities operating in the infrared and visible ranges for a laser oscillator to be worth building.

In the meantime, however, Madey had established in 1971 that using a static magnetic undulator instead of a real electromagnetic radiation source (a klystron, for example) it was possible to obtain sufficient gain to produce a laser effect. This is explained by the fact that the power density of the radiation corresponding to the undulator can be extremely high compared with that of a real wave, because it is possible to generate experimentally very strong static magnetic fields. The undulator built by Motz and collaborators, for example, was the equivalent of a radiation with a wavelength of $\lambda = 8$ cm and an output density of between 1 and 2 GW/cm^2, some orders of magnitude greater than anything which could be obtained with a conventional source for that wavelength. The dependence of gain on the operating frequency of a laser based on stimulated Compton scattering (or alternatively on stimulated synchrotron radiation in an undulator) is very peculiar (Figure 4) (Graph b)). Instead of following the spectral distribution of the spontaneous radiation (Graph a in Figure 4) as in conventional lasers, gain is proportional to its derivative. Such behavior is due to the nature of the interaction which is based on a scattering process and not a process of radiation from bound states.

As Graph b in Figure 4 shows, the gain for this particular type of FEL is zero when the frequency is the same as central frequency ω_0 when emitted spontaneously. On the other hand, positive gain is obtained whenever the wave to be amplified is of a frequency less than ω_0, and there is negative gain if the reverse is the case. Maximum amplification occurs in the area where the spontaneous emission spectrum curve rises most sharply, while the positive part of gain curve is equal in width to that of the homogeneous band. It is easy to estimate, and subsequently confirm by more accurate calculations, the maximum power that can be transferred from the electron beam to the laser beam. If allowance is made for the fact that the radiated frequency is dependent on the energy levels of the electrons cf (Equation (5)), the maximum energy obtainable from the electrons themselves is that which induces a shift of the operating frequency to a level of the gain graph. Consequently the maximum efficacy obtainable, η max, defined as the ratio between energy transferred to the laser beam and the initial energy level of the electrons, is given by the relative width of the gain graph:

$$\eta \text{ max } \backsim \frac{\Delta\omega}{\omega} = \frac{1}{2N} \qquad\qquad (8)$$

First Observation of Stimulated Synchrotron Emission

The shape of the gain curve (or stimulated synchrotron emission in a magnetic undulator (Graph b, Figure 4) was confirmed experimentally by the first test carried out at Stanford by Elias, Fairbank, Madey, Schwettman, and Smith in 1976. The experimental equipment is shown in diagrammatic form in Figure 5. A superconducting magnet in the form of a double helix would round a copper tube was used as the undulator. Figure 6 shows a photograph of a number of the magnetic periods. In the two helical coils current flows in opposite directions and the magnetic field generated is perpendicular to the axis of the undulator and rotates at a period equivalent to the period of the helix.

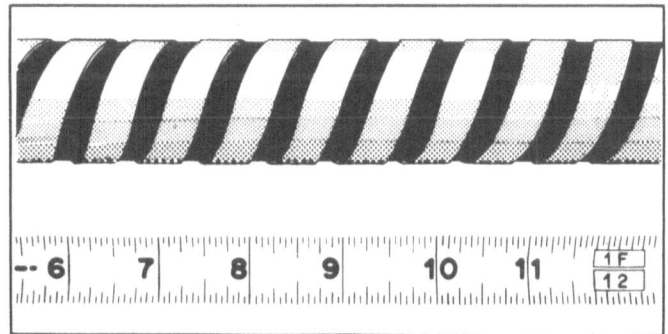

Fig. 6. Some periods of the superconducting double helical undulator built at Stanford. In the two helixes the current passes in opposite directions, and the magnetic field generated is thus perpendicular to the axis of the undulator and rotates with a period corresponding to the period of the helix.

According to Weizäcker and Williams' approximation, this type of field corresponds to an electromagnetic wave with circular, right-hand polarization. The main features of the undulator are shown in Table 1.

A beam of electrons produced by the Stanford superconducting linear accelerator with an energy of E = 24 MeV and a peak current I_p = 70 mA and a beam of infrared light with a wave length of λ^p= 10.6 μm and an output density of 1.4 x 10^5W/cm^2, generated by a carbon dioxide laser (TEA CO$_2$ laser (TEA standing for transversally excited at atmosphere pressure)) are passed along the axis of the magnet.

Figure 7 shows the spontaneous radiation output (Graph a) and the gain (Graph b) as a function of the energy of the electron beam at the CO$_2$ laser wavelength (∿ 10.6 μm). The experimental graphs

Table 1. The Superconducting Helical Undulator
of Stanford University

Period (cm) 3.2
Length (m) 5.2
Width of homogeneous band (%) 0.3
Magnetic field on the axis (kGs) 2.4

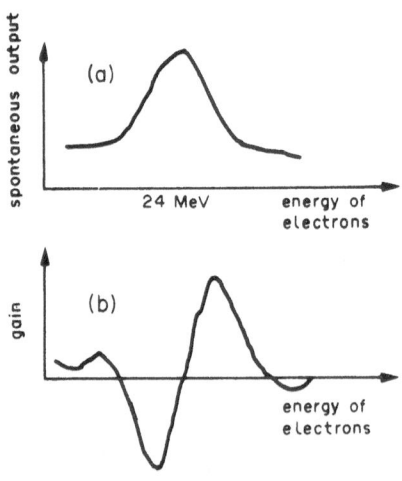

Fig. 7. Radiation output by spontaneous emission (Graph a) and gain (b) measured (with the equipment built at Stanford) at the wavelength of the TEA CO$_2$ laser (λ ∿10.6 μm) as a function of the energy of the electron beam.

confirm the theoretical estimates (Figure 4). It is worth noting that the preferred method was to vary the energy of the electrons, as this operation, which corresponds to varying the frequency of the laser (cf. Equation (5)) is the easier operation under experimental conditions.

The maximum measured gain was approximately 7 per cent passage, which is sufficiently close to the theoretical estimate. This gain corresponds to a stimulated output emission of approximately 4×10^3W, compared with the spontaneous radiation (that is, radiated without an incident laser beam) which was about 4×10^{-6}W. The increase in output was thus of some nine orders of magnitude. Lastly it was established that only the radiation with the same polarization as the undulator is amplified, zero gain was measured for radiation with circular left-hand polarization.

The full evaluation of the experimental data involved a considerable amount of work. In particular, it became clear that the stimulated Compton scattering can be considered as two distinct phases. In the first instance, the action of the laser beam and undulator beam modulates the energy of the electron beam and this energy modulation is then transformed into density modulation at the same wavelength as that of the laser; in other words, there is coherent radiation which intensifies the existing laser field.

As this point a comparison with the millimeter wave source built by Motz and collaborators referred to earlier may be of interest. In that experiment the density of the electron beam was modulated as it entered the magnet. In the FEL, on the other hand, the entire process (modulation and emission) automatically occurs in the undulator, which allows laser operation at wavelengths ranging from submillimeter to ultraviolet.

First Operation of an FEL as an Oscillator

In 1977, one year after the first amplification experiments, Deacon, Elias, Madey, Ramian, Schwettman, and Smith, built the first FEL oscillator, referred to at the beginning of this article.

The experimental device, which is depicted in Figure 8, uses the helical undulator successfully tried in the amplification experiments (Figure 6 and Table 1). The electron beam is still the same one produced by the Stanford superconducting accelerator, but this time the energy level is higher (E~43 MeV). Consequently, the operating wavelength is shorter ($\lambda \sim 3.4$ μm). Radiation is initially produced by spontaneous synchrotron radiation and then, reflected by the two mirrors located at the ends of the undulator, it is amplified in the course of subsequent passes with the electron beam inside the magnet. The main characteristics of the electron and laser beams, updated by latest (1981) experimental data, are shown in Table 2.

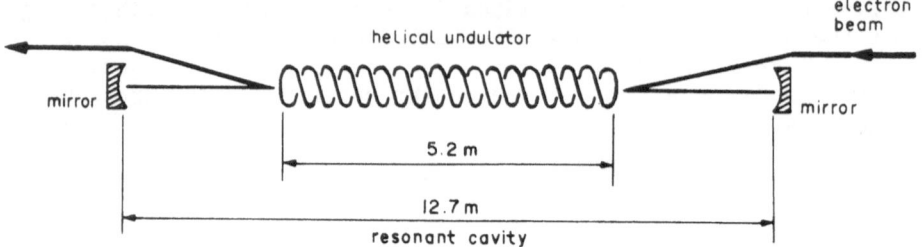

Fig. 8. Operation diagram of the Stanford FEL oscillator. The radiation is produced by spontaneous synchrontron irradiation and then reflected by the two mirrors located at the ends of the undulator. It is amplified during subsequent passes with the electron beam inside the magnet.

Table 2. Main Features of the Stanford FEL

Electron Beam (Superconducting Linear Accelerator)	
Energy (MeV)	43
Energy spread (%)	5×10^{-2}
Emittance (mm x mrad)	4×10^{-2}
Peak current (A)	1.3
Average current (μA)	60
Laser Beam	
Wavelength (μm)	3.3
Line width (nm)	2.5 - 6.6
Average output (W)	5
Peak output (kW)	130

The effect of the laser is most apparent if the outputs above-threshold (gain greater than losses) and below-threshold (gain less than losses) are compared. The ratio between these two levels is extremely high ($\sim 10^8$). Another important parameter is the narrowing of the emission line from $\delta\lambda \sim 36$ nm (below threshold) to $\Delta\lambda \sim 8$ nm (above threshold) as shown in the spectrum in Figure 9. The width of the spontaneous emission line (lower graph in Figure 9) is due mainly to the homogeneous widening resulting from the excellent characteristics of the electron beam (slight spread of energy and emittance (Table 3).

The width of the laser line is determined by the variation with time of the electron beam. This is composed of a continuous train of electron packets with a duration of $\tau \sim 3.2$ ps at approximately 84.7 ns

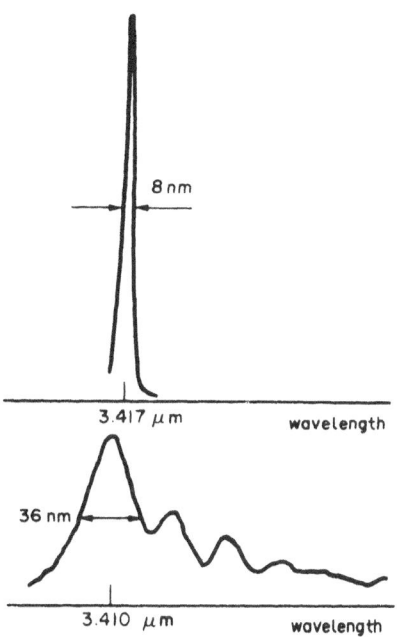

Fig. 9. Spontaneous emission line (lower graph) and laser line
         (upper graph) for the Stanford FEL (cf. Figure 8 and
         Table 2).

Table 3. Main Features of the Orsay FEL Amplifier Superconducting
          Undulator

Period (cm)	Length (cm)	Homogeneous band width (%)	Maximum magnetic field on axis (kGs)
4	94	2	4

Electron Beam (ACO)		Laser Beam (argon)	
Energy (MeV)	Average current per electron packet (mA)	Wavelength (nm)	Output density (W/cm^2)
150	10–20	488.0 514.5	1.6 x 10^2

intervals. This interval was chosen in order to be approximately the
same as the time taken by the light to travel back and forth in the
optical cavity.

The radiation moving back and forth in the cavity and the subsequent packets of electrons are thus synchromized. A sort of mode-locking device is thus obtained. The result is that the laser radiation is no longer continuously distributed but grouped into light pulses, the duration of which is approximately the same as that of the electron packet. The width of the laser line under these conditions is inversely proportional to the duration of the micropulse

$$\Delta\omega \sim \frac{2\pi}{\tau} \sim 1500 \text{ GHz} \tag{9}$$

which in terms of wavelength corresponds to $\Delta\lambda \sim 9$ nm and confirms the experimental results (Figure 9).

Table 2 shows the energy transfer efficacy from the electron beam to the laser beam, as represented by the following equation:

$$\eta = \frac{P_L(W)}{1(A) \times E(eV)} \sim 0.2 \text{ per cent} . \tag{10}$$

When 1 and E are the average current and energy of the electron beam and P_L is the average output of the laser. These results are in complete harmony with the theoretical limit given by the homogeneous bandwidth (cf. Equation (8)), which for the Stanford FEL is approximately 0.3 per cent (Table 1).

At the beginning we spoke of evidence of the production of laser light, intending this rather general term to mean coherent light. A theoretical study of the process by Bonifacio, Dattoli, Renieri and Romanelli, in 1980, confirmed that FEL radiation shows coherence properties similar to those shown by the light from conventional lasers.

The FEL in Storage Rings

Amplification experiments have recently been carried out at Orsay in France, at Frascati in Italy and an Novosibirsk in the USSR. In all these cases the equipment used an electric beam circulating in a storage ring. A diagram of the FEL storage ring configuration is given in Figure 10. After they have given off energy to the laser beam, the electrons are re-accelerated by the machine's radio-frequency system and subsequently re-injected into the undulator to interact again with the radiation beam. Ultimately, there is a continuous transfer of energy from the storage ring accelerator system to the laser beam via the electron beam. Contrary to appearances this process is not more efficient than the single-pass FEL system (Stanford model), where the electron beam is trapped in a beam damp after it has given off a small portion of its energy to the laser beam. The reason for this is that the successive interactions

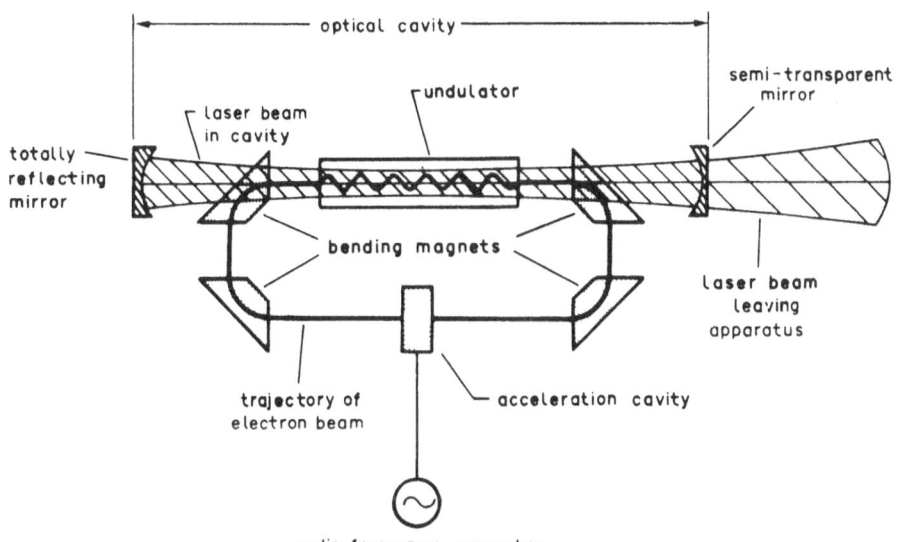

Fig. 10. Operating diagram for a FEL in a storage ring. After they have given off energy to the laser beam in the undulator, the electrons are reaccelerated by the machine's radio-frequency system and made to ineract again with the radiation beam. The particles thus circulate continuously in the ring and transfer energy from the radiofrequency system to the laser beam.

between the electron beam and the laser beam continuously increase the electron energy spread (that is, they 'heat up' the electron beam). This process causes the spontaneous emission line to be widened inhomogeneously and lowers the system's gain to the point where it becomes less than losses from the resonant cavity, thus terminating the laser process.

The only possibility of limiting this effect involves synchro-tron irradiation in the bending magnets of a storage ring. The greater this radiation, the more efficient the cooling of the electron beam becomes since the output obtainable from an FEL operating in a storage ring is linked to the output radiated (P_s) by synchrotron emission. It can be demonstrated (Dattoli and Renieri (1980)) that the maximum laser output is described by the following equation:

$$P_{max} \sim \left[\frac{\Delta\omega}{\omega} \right]_0 P_s . \tag{11}$$

In this case $(\Delta\omega/\omega)_0$ is the homogeneous bandwidth. Maximum efficacy is thus described by:

$$\eta_{max} \sim \frac{P_{max}}{PN} \sim \left[\frac{\Delta\omega}{\omega} \right] = \frac{1}{2N} \tag{12}$$

where N is the number of periods in the undulator. This equation is consonant with that already obtained for the single-pass FEL (8).

Let us now describe in more detail some of the experiments mentioned earlier. The characteristics of the Orsay amplifier are given in Table 3. The magnet is in the form of superconducting helixes which create a periodic field parallel to the plane of the apparatus: the polarization of the wave generated is thus linear. A maximum gain per pass through the magnet of 4×10^{-4} was registered (at a wavelength of = 488 mm). This extremely low value corresponds to the limit for losses from a typical resonant cavity operating at that wavelength.

The slight gain on the FEL at Orsay is due, among other things to the low number of magnetic periods (N=23). One of the greatest hindrances to the operation of an FEL in a storage ring is in fact the absence of sufficiently long straight sections. To overcome this difficulty, Vinokurov and Skrinsky proposed a variant to the traditional free electron laser in 1977. This device, which the two authors called an optical klystron, is shown in Figure 11. It is composed of two undulators each with an equal number of periods, separated by a triplet of bending magnets of inverted polarity so that total deflection is zero. The FEL process can thus be divided into its three separate stages. In the first undulator the electron beam is energy modulation is transformed in the magnetic triplet into

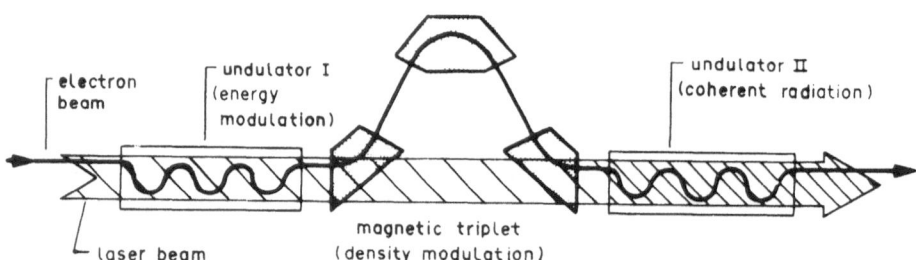

Fig. 11. Operating diagram for the optical klystron. In the first undulator the energy modulation of the electron beam is obtained and this is then transformed in the magnetic triplet into powerful density modulation. In the second undulator the coherent radiation process finally occurs.

density modulation. Finally, in the last undulator the process of coherent radiation occurs. This configuration is much more efficient that the traditional one, in that for the same length the density modulation is much more intensive due to the dispensive characteristics of the magnetic triplet (electrons with different energy levels describe orbits of different lengths). The Novosibirsk FEL operates on the basis of this principle. The optical klystron was fitted to the VEPP-3 storage ring and the first spontaneous emission and gain measurements were carried out using an HeNe laser which confirmed the theoretical estimates. Finally an optical klystron FEL oscillator is now in operation at Orsay.

The FEL (Raman Operation)

At the same time as the work on FEL apparatus using high-energy electron beams (E>10 MeV) and low current (U $\sim$1-10^2 A) was in progress laser sources operating in the millimeter wavelength range, which use low-energy electron beams (E $\sim$1-2 MeV) and high current (I $\sim$10-20 kA) produced by accelerators such as the Mars generators and linear indication accelerators, were also undergoing development. In these machines there is considerable interaction between the charged particles. In particular, the electromagnetic radiation and the undulator field can excite collective oscillation modes in the electron beam (plasma waves). The frequency of the radiation generated can no longer be calculated simply on the basis of Equation (5) which essentially reflects the Doppler shift of the simulated undulator wave. Instead, it is shifted towards the low frequencies owing to energy absorption on the part of the plasma oscillation modes.

These radiation sources have been termed Raman FELs in view of the similarity to the Ramnan light diffusion process by molecules. The FELs operating at low density and high energy (the Stanford model) are accordingly sometimes called Compton FELs.

The first Raman FELs were built at the Naval Research Laboratory (NRL) in Washington, Columbia University in New York; the laboratories of the TRW in California; and the Institute of Nuclear Physics, Tomsk, USSR. A diagram of the Raman FEL built by the NRL is shown in Figure 12 as an example of a spontaneous radiation amplifier.

The large structure to the left is the acceleration for intensive electron beams (VEBA, E = 1÷2 MeV, I = 1-100 kA, pulse duration $\sim$50 ns). The electron beam is passed into a double-helix pulsed undulator. This type of undulator has a 3 cm period, a length of 1 m, and a maximum field of 4 kGs. This apparatus has been used to generate millimeter and submillimeter radiation with an output of $\sim$1÷10 MW.

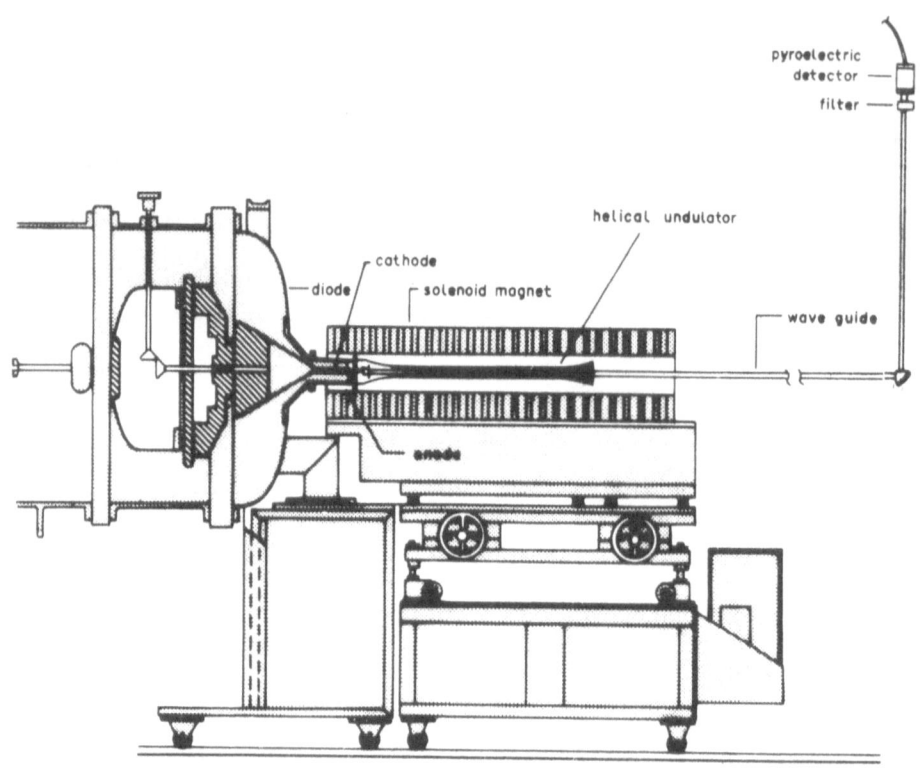

Fig. 12.   Diagram of the single-pass spontaneous radiation amplifier
built at the Naval Research Laboratory, Washington, DC.
The electron beam produced by the VEBA accelerator (device
on left) is passed into the helical undulator shown at the
center of the photograph. A solenoid magnet is used to
focus the electrons inside the undulator. On the right is
the detector for the radiation produced.

The FEL Special Feature and Future Prospects

    To understand fully why the FEL aroused so much interest when it
first appeared, one must take a close look at its main features; it
can be modulated, it has a good linewidth, it is efficient, and it
can be boosted to high average output levels. Let us also review the
enormous range of activities at international level aimed at producing new FEL sources specifically designed to optimize particular
operating parameters for specific purposes.

Tunability. The free electron laser's most striking characteristic is without doubt its ability easily to accommodate changes in
it operating frequency. Whereas in the case of lasers based on
emission stimulated by atoms or molecules this parameter is determined by the difference in energy between the states affected by the
transition, the FEL's frequency can be varied by changing the energy
of the electron beam, the period and the magnetic filed of the undulator (cf. Equation (5)). As a result of this the FEL can potentially
cover continuously the entire spectrum from millimeter waves to
ultra-violet waves. The importance of this type of laser is most
obvious where tunable sources are not available, for example in the
far infrared and in the ultra-violet range. It is less important in
the visible range, where the dye lasers can cover the entire spectrum
with excellent results.

The problem of extending the operating range of the FEL to
include very short wavelengths ($\lambda \sim 100$ nm) is one which deserves
special attention. In the very short wavelength range the reflectivity of the mirrors is greatly reduced (usually by something
approaching 50 per cent). Consequently, to build an FEL oscillator
operating in that range of the spectrum, very high gain will be
needed, and this can be obtained only with very high peak currents
($I \sim 10^2$-10^3 A). Such high currents can be readily obtained in the
high-energy storage rings and these machines would thus appear to be
the best sources of electrons for FELs operating in the far ultraviolet.

There are currently many projects underway aimed at building
free electron lasers on storage rings. Earlier, we described the
experiments conducted at Orsay and Novosibrisk. At the present time
experiments on FEL amplification and oscillation are being conducted
at the Brookhaven laboratories in the United States, where a synchrotron light storage ring will be used to produce laser radiation at
wavelengths of 250-450 nm by means of a permanent magnet undulator
(period = 6.5 cm, number of periods = 38) and also at INFN at
Frascati. In this experiment the electron beam from the Adone ring
is used. The undulator is in the form of an electromagnet, the main
feature of which are shown in Table 4, together with the main feature
of the electron beam. Initially, experiments on radiation amplification produced by an argon laser with a wavelength of 514.5 um has
been carried out.

The gain per pass is approximately 10^{-4}. At a later stage laser
oscillator will be built both in the traditional way and in the
high-gain configuration of the optical klystron type.

Line width. The variation with time of the laser beam is
closely related to that of the electron beam. Consequently, as
was demonstrated in connection with the description of the Stanford
experiments, if the electrons are bunched in packets the photon beam

Table 4. Principal Parameters for the FEL Source being Built at the
Italian National Nuclear Physics Institute at Frascati

| | Undulator (electromagnetic) | | | Electron Beam (Adone) | |
Period (cm)	Length (m)	Homogeneous Band width (%)	Max. magnetic field on axis (kG)	Energy (MeV)	Average Current per electron packet (mA)
11.6	2.32	2.5	4.459	610	100

will also be in the form of a pulse train having a duration approximately the same as that of the electrons. This situation occurs when the acceleration is provided by systems using resonant cavities at radio frequencies, which is the case of linear accelerators, microtrons and storage rings. Under these circumstances, the duration of the light micropulse is very short (normally between 10^{-12} and 10^{-9}s). This leads to a relatively large linewidth, as shown by Equation (9).

A recent proposal (currently being developed at the University of California in Santa Barbara) was for an FEL source in the far infrared ($\lambda \sim 100-400$ µm) using a continuous electron beam with an energy of 3 MeV and current density of 16 A cm^2 produced by an electrostatic generator. Under these conditions the linewidth of the laser is much less than that of sources which use radio-frequency accelerators. An interesting aspect of the experiment is the recovery of the electron beam after the FEL interaction in the undulator. This aspect of free electron laser technology, which determines the overall efficacy of the system, will be considered in more detail in the next section.

Efficacy. As was seen earlier, only a small fraction of the electron beam's energy is transferred to the laser beam. A number of ideas have recently been put forward with a view to improving the efficacy of the FEL source. They fall into two distinct groups. The first group comprises the system which recover the residual energy from the electron beam, while the second group concentrates on improving the efficiency of the energy transfer between electrons and laser beams.

The FEL project under development at the University of California belongs in the first category. In this apparatus, the electron beam is decelerated after passing through the undulator and its energy and charge are for the most part recovered.

In the second group we find the designs which use special, variable-parameter undulators. In these systems the period and the magnetic field are changed along the undulator so as to cancel the frequency shift due to energy losses from the electron beam. Theoretical estimates predict an energy transfer efficacy between electrons and photons of some 50 per cent. Apparatus for the experimental verification of the validity and applicability of this model has been developed successfully at the Los Alamos Laboratories of Mathematical Science Northwest and TRW in the United States.

Boosting to high output levels. The FEL oscillator at Stanford produced a laser beam with an average output of around 5 W (Table 2). It should be remembered, however, that this result was obtained with a very low average current ($\sim$60 µA). Using a conventional linear accelerator as an electron source it is possible, however, to produce average currents in the 10 to 100 MeV range that are ten times greater than that obtained at Stanford. The use of electron beams with these characteristics allows laser beams to be generated which have an average output of some several hundred watts.

As an example, Table 5 shows the major parameters for the FEL oscillator being built at the ENEA's (formerly CNEN) energy research center at Frascati, which uses a microtron as an electron source. The long-term objective of this project is to build a source which is tunable in the near infrared ($\lambda = 10$-30 µm) with an average output of between 50 and 100 W.

Some Possible Uses for the Free Electron Laser

It is possibly still somewhat premature to discuss in detail the possible applications of the FEL. What is clear is that its major feature – that it can be tuned – renders it an ideal source for photochemistry in general, and isotopic separation by laser in particular.

In addition to this extensive field of application we could also mention two possible uses particularly relevant at a time like the present in which the search for new energy sources appears to be one of the most urgent problems facing our industrial society.

One application concerns the possibility of building high-efficacy amplifiers with variable-parameter undulators. The high laser energies obtainable in this way ($\sim$100-1000 kJ) concentrated in very short pulses ($\sim$100 ns) make these devices suitable for use as energy production systems based on inertial nuclear fusion by laser.

The second application concerns the potential production (using the Raman FEL sources) of tunable radiation on the 100-1000 µm band with high peak output (100 1000 MW) in pulses lasting $\geqslant$ 1 µs and good efficiency ($\sim$10 per cent). With such apparatus it will be possible to heat plasmas (using cyclotron resonance or its harmonics) in magnetic confinement machines (of the high-field Tokamak type) for triggering the process of controlled nuclear fusion for energy production.

Table 5. Main Parameters of the FEL Sources under Development at the Energy Research Center in Frascati

| Undulator (SmCo permanent magnets) | | | | Electron beam (microtron) | | | | |
Period (cm)	Length (cm)	Homogeneous band width (%)	Maximum magnetic field on axis (kGs)	Energy (MeV)	Average current (mA)	Peak current (A)	Current pulse duration (µs)	Repetition frequency (Hz)
5	2.25	1.1	3	20	350	6.5	12	150
				30	250	4.5		

REFERENCES

Deacon, D. A. G., Elias, L. R., Madey, J. M. J., Ramian, G. J., Schwettman, H. A., and Smith, T. I., 1977, Phys.Rev.Lett., 38:892.

Elias, L. R., Fairbank, W. M., Madey, T. M. J., Schwettman, H. A., and Smith, T. I., 1976, Phys.Rev.Lett., 36:717.

Jacobs, S. F., Moore, G. T., Piloff, H. S., Sargent, M., Scully, M. O., and Spitzer, R., eds., "Free-Electron Generators of Coherent Radiation," Addison-Wesley, Reading, Mass., Vol.7 (1980), Vols.8,9 (1982).

Kapitza, P. L., and Dirac, P. A. M., 1933, Proc.Phil.Soc.(Cambridge), 29:297.

Madey, J. M. J., 1971, J.Appl.Phys., 42:1906.

Motz, H., Thon, W., and Whitehurst, R. N., 1953, J.Appl.Phys., 24:826.

Pantell, R. H., Soncini, G., and Puthoff, E., 1968, IEEE J.Quantum Electr., 11:905.

Pellegrini, C., ed., 1981, "Developments in High-power Lasers and their Applications," North-Holland, Amsterdam.

Szöke, A., ed., 1981, IEEE J.Quantum Electr., 8.

SYNCHROTRON RADIATION SOURCES AND USES IN THE UV-VIS

Mario Piacentini

Istituto di Struttura della Materia del CNR
Via E. Fermi 38
00044 Frascati, Italy

INTRODUCTION

We call synchrotron radiation the radiation emitted by electrons moving along circular paths at relativistic speeds. The fact that electrons in circular trajectories radiate is well known since approximately one century. For example, this effect caused serious difficulties in the early, classical atomic theories. According to the Rutherford model, atoms could not be stable objects, for the electrons orbiting around the nuclei were radiating and, thus, losing their energy. Such a difficulty was overcome by N. Bohr, who postulated that the electrons could move on stable orbits, without radiating, provided the electron angular momentum were quantized in multiples of the constant $h/2\pi$.

Practically, synchrotron radiation was discovered in 1947, when F. Haber, H. C. Pollok and R. V. Langmuir ventured a quick glimpse to the radiation coming out from the first electron-synchrotron, the 70 MeV machine at the General Electric Laboratories.[1] Since then, the properties of the radiation emitted by relativistic particles accelerated centripetally by a magnetic field were studied in great detail. It is interesting to note here that the peculiar, but well known emission of light from an astronomical object, the Crab Nebula, has been finally identified and attributed to the emission of synchrotron radiation by the electrons orbiting in the galactic magnetic fields.

Table 1. Principal storage rings used as synchrotron radiation sources; most of them are dedicated as light sources. E, particle energy (GeV); R, magnetic radius (m); I, maximum current (mA); P, total radiated power (kW); $\gamma = E/m_o c^2$; λ_c, characteristic wavelength (Å); $N(\lambda_c)$, photon flux at λ_c (phot/sec/mrad/1% bandpass $\times 10^{12}$); $N(\lambda)$, photon flux at 100 nm (phot/sec/mrad/1% bandpass $\times 10^{12}$).

Name	Location	E	R	I	P	γ	λ_c	$N(\lambda_c)$	$N(\lambda)$
TANTALUS	Stoughton	0.24	0.64	200	0.1	480	260	7.6	16
SURF II	Washington	0.24	0.83	30	0.01	480	335	1.15	2.64
INS-SOR	Tokyo	0.4	1.1	250	0.5	800	96	16	24
ACO	Orsay	0.55	1.11	100	0.73	1100	37	8.8	9.7
VEPP-2M	Novosibirsk	0.67	1.22	100	1.46	1340	22.4	10.7	10.3
NSLS VUV	Brookhaven	0.7	1.9	500	5.6	1400	31	56	58
BESSY	Berlin	0.8	1.83	500	9.9	1600	20	64	57
ALADDIN	Stoughton	1.0	2.8	500	15.8	2000	15.7	80	66
ADONE	Frascati	1.5	5.0	100	9.0	3000	8.3	24	16
DCI	Orsay	1.8	3.82	250	61	3600	3.66	72	36.5
SRS	Daresbury	2.0	5.55	500	128	4000	3.88	160	83
NSLS X-RAY	Brookhaven	2.5	8.17	500	210	5000	2.92	200	94
PHOTON FACTORY	Tsukuba	2.5	8.33	500	207	5000	3.70	200	95
SPEAR	Stanford	4.0	12.7	100	178	8000	1.10	64	22
ESRF		5.0	20.0	100	280	10000	1.0	80	25
DORIS	Hamburg	5.7	12.1	100	770	11400	0.366	91	22

The most interesting features of synchrotron radiation are: intense, continuous spectrum extending over a very broad range from infrared to X-rays; high directionality; high degree of linear polarization; fast time structure. All these properties can be calculated exactly and depend only on the electron accelerator parameters, so that synchrotron radiation is an ideal calibration source.

Synchrotron radiation was soon recognized as a unique light source for studies in atomic and solid state physics. Near the electron accelerators constructed for high energy physics, synchrotron radiation facilities began to appear. Nowadays, a large number of storage rings are employed as synchrotron radiation sources all over the world (Tab. 1). Several of them have been constructed or are in project for this purpose only. The applications of synchrotron radiation are numerous and limited only by the imagination of the users. Even if the present demand is shifting from the low-energy spectral region to the hard X-ray region, the importance of synchrotron radiation as a vacuum ultraviolet (VUV) source is well established and it has not been fully exploited yet.

SYNCHROTRON RADIATION PROPERTIES

Total radiated power

The total power radiated by a single, relativistic particle with charge e, rest mass m_o, and total energy E, moving with velocity $\beta=v/c$ in a magnetic field of induction B along a circular orbit of radius R, is given by:[2]

$$P = \frac{2}{3} \frac{e^2 c}{R^2} \beta^4 \left(\frac{E}{m_o c^2} \right)^4 .$$ (1)

The total radiated power is inversely proportional to the fourth power of the particle rest mass. For this reason, the emission of synchrotron radiation is important only for electrons and not for protons with the same total energy. From now on we shall consider only the emission from electrons accelerated in a storage ring. Eq. (1) can be written in the more practical form:

$$P \text{ (kW)} = 88.5 \text{ } E^4 I/R = 2.65 \text{ } E^3 IB ,$$ (2)

where I is the electron current in the accelerator. The quantities appearing in Eq. (2) and in the following equations are expressed in the units: E (GeV); R (meters); I (amperes); B (tesla). P increases as the fourth power of the electrons energy. For an average machine, such as Adone, P is 9 kW at the maximum operation conditions, i.e. E=1.5 GeV and I=100 mA.

Angular and spectral distribution

In a reference frame instantaneously fixed with the electrons, the emitted radiation follows the dipole pattern. The intensity distribution has the typical shape of a doughnut, shown on the left of Fig. 1, with zero emission along the acceleration direction and with maximum intensity in the velocity direction. In the laboratory frame all emission angles are transformed according to the relation:

$$\text{tg } \theta' = \frac{\sin \theta}{\gamma(1+\cos \theta)} ,$$ (3)

with $\gamma=E/m_o c^2=2x10^3$ E(GeV). The direction of zero emission, $\theta=90°$, is pushed forward to form an angle $\theta'=1/\gamma$ with respect to the direction of motion. The emission pattern takes the sigar shape shown on the right of Fig. 1: at each instant the electron is emitting along the orbit tangent within a very narrow cone of opening $2/\gamma$. For Adone, operating at 1.5 GeV, the cone is 0.6 mrad wide. The observer receives the radiation produced by an electron as it moves along the arc of angle $\theta_c = 2/\gamma$. The duration of the radiation pulse is just the difference between the time required by the electron to traverse this arc and that required to the light to travel along the chord of the arc:

$$\tau \sim \frac{2R\theta_c}{v} - \frac{2R\sin\theta_c}{c} \simeq \frac{R}{c\gamma^3} . \tag{4}$$

A light pulse of this duration has frequency components up to about $\omega_c \sim 1/\tau = c\gamma^3/R$. For Adone $\omega_c = 1.62 \times 10^{18}$ sec^{-1}, corresponding to a wavelength of 11.6 Å, in the soft X-ray region.

A detailed calculation of the basic equations of synchrotron radiation gives the following expression for the number of photons per sec emitted into unit solid angle $d\Omega$, along a direction forming an angle θ with respect to the orbit plane, and constant percentage bandwidth $\Delta\lambda/\lambda$:

$$\frac{dN(\theta,\lambda)}{dt\, d\Omega\, (\Delta\lambda/\lambda)} = \frac{3e^2\gamma^2}{2\pi\, hR} \left(\frac{\lambda_c}{\lambda}\right)^2 (1+\gamma^2\theta^2)^2 \cdot$$

$$\cdot \left[K_{2/3}^2(y) + \frac{\gamma^2\theta^2}{1+\gamma^2\theta^2} K_{1/3}^2(y) \right] . \tag{5}$$

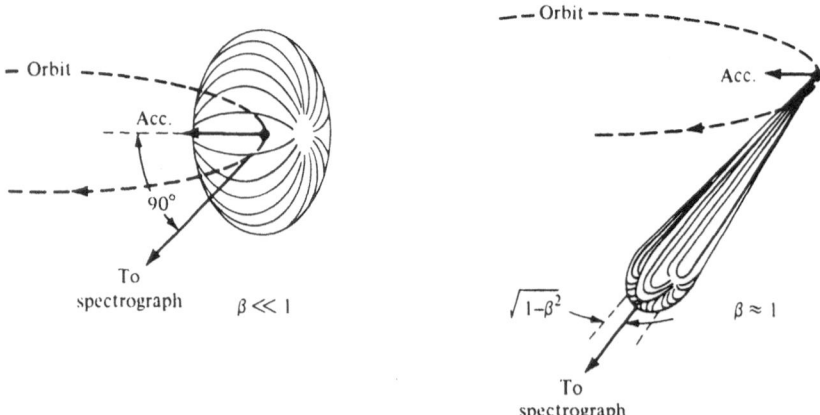

Fig. 1. The dipole pattern of the emitted intensity from a slow electron on a circular orbit (left) is distorted into a narrow cone in the instantaneous direction of motion for an ultrarelativistic electron ($\beta = v/c \simeq 1$) (right) (from Ref. 3).

Table 2. Table of various modified Bessel functions and integrals.

y	$K_{2/3}$	$K_{1/3}$	G
0.001	107.46	16.715	0.2136
0.01	23.093	7.486	0.4450
0.03	11.017	4.932	0.6135
0.05	7.762	3.991	0.7015
0.1	4.753	2.900	0.8182
0.2	2.802	1.979	0.9034
0.3	1.987	1.509	0.9177
0.5	1.206	0.989	0.8710
1.0	0.4945	0.4384	0.6514
1.5	0.2402	0.2202	0.4506
2.0	0.1243	0.1165	0.3016
4.0	0.01173	0.0113	0.05284
10	1.816 E-5	1.787 E-5	1.922 E-4

$K_{2/3}$ and $K_{1/3}$ are modified Bessel functions, some values of which are given in Tab. 2. The argument y of the two Bessel functions is expressed in terms of the reduced wavelength $\Lambda=\lambda/\lambda_c$ and the reduced angle of emission $\vartheta=\gamma\theta$:

$$y = \frac{1}{2\Lambda} (1+\vartheta^2)^{3/2} . \tag{6}$$

λ_c is the critical wavelength of the emission spectrum, given by:

$$\lambda_c \ (\mathring{A}) = \frac{4\pi R}{3\gamma^3} = 5.59 \ R/E^3 = 186/(BE^2) . \tag{7}$$

From Tab. 2 we see that both $K_{2/3}$ and $K_{1/3}$ become very small with respect to their maximum values for $y\sim1$. This means that, at $\lambda=\lambda_c$, the emission occurs in a cone of half-opening $1/\gamma$, or that $\lambda_c/2$ corresponds to the cutoff wavelength for $\theta=0$. We find in a quantitative form the same results as those derived above in a qualitative manner. The cone aperture increases with wavelength, as it appears by setting y=1 in Eq. (6).

Eq.(5) can be written in practical form as

$$N(\theta,\lambda) = 1.95\times10^{14} IE^2 \Phi(y) \ \frac{photons}{sec \ mrad^2 \ 1\% \ bandwidth} , \tag{8}$$

where the function

$$\Phi(y) = \frac{2.71\, y^2}{1 + \vartheta^2}\ [K_{2/3}^2(y) + \frac{\vartheta^2}{1 + \vartheta^2}\, K_{1/3}^2(y)\,] \qquad (9)$$

is a universal function, plotted in Fig. 2.

Total spectral distribution

The total power emitted over all vertical angles, i.e. collected within an angle $>> 2/\gamma$, can be obtained by integrating Eq. (5) over all θ to obtain

$$N(\lambda) = 2.23 \times 10^{14}\ I E\, G(y)\ \frac{\text{photons}}{\text{sec mrad(1\% bandwidth)}}\ , \qquad (10)$$

where also

$$G(y) = y \int_y^\infty K_{5/3}(x)\ dx \qquad (y = \frac{\lambda_c}{\lambda}) \qquad (11)$$

is a universal function that does not depend on the characteristics of the machine. $K_{5/3}$ is a modified Bessel function. The function $G(y)$ is plotted in Fig. 2 and tabulated in Tab. 2. The intensity at $\lambda = \lambda_c$ is given by:

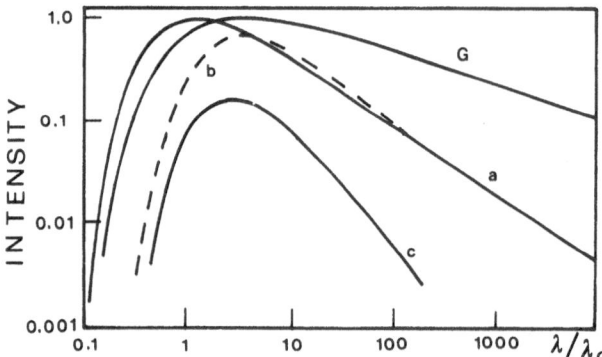

Fig. 2. Normalised spectral distribution of the universal functions $G(y)$ and $\Phi(y)$. The $\Phi(y)$ function is shown for $\theta=0$ (curve a) and for $\theta=1/\gamma$. The latter is separated into the two polarization components parallel and perpendicular to the electron orbit plane (curves b and c, respectively).

$$N(\lambda_c) = 1.6 \times 10^{14} \, IE \quad \frac{\text{photons}}{\text{sec mrad (1\% bandwidth)}} \cdot \qquad (12)$$

The maximum of the spectral distribution occurs at $\lambda \approx 4\lambda_c$. For $\lambda \ll \lambda_c$ $G(y)$ decays exponentially; useful photon fluxes are available approximately up to $\lambda \sim \lambda_c/8$. In the long wavelength region the spectral distribution decays slowly as

$$N(\lambda) = 9.35 \times 10^{13} \, I \left(\frac{R}{\lambda} \right)^{1/3} \quad \frac{\text{photons}}{\text{sec mrad (1\% bandwidth)}} \qquad (13)$$

and depends mostly on the electron current. The principal parameters related with the emission of synchrotron radiation are listed in Tab. 1 for a number of storage rings.

Polarization properties

The two terms in the square brackets of Eq. (5) correspond to the intensity of the spectral distribution for the polarization components parallel and perpendicular to the electron orbit plane, respectively. The factor before the second term assures that the radiation is 100% linearly polarized in the orbit plane. From Tab.2 we see that the perpendicular component is much smaller than the parallel one, except far from the orbit plane. Thus, the polarization is elliptical, with opposed ellipticity above and

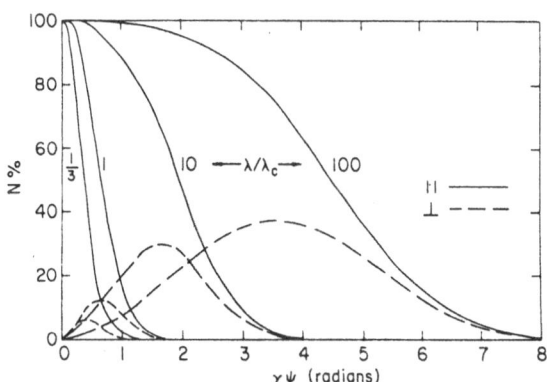

Fig. 3. Dependence on the reduced vertical angle $\vartheta = \gamma\theta$ of the intensities of the parallel (solid line) and perpendicular (dashed line) polarization components of the photon flux for several values of $\Lambda = \lambda/\lambda_c$. Each curve has been normalised to the intensity in the orbit plane ($\theta = 0$). The use of the abscissa ϑ makes these curves universal (from Ref. 2).

below the orbit plane. In Fig. 3 we show the angular dependence of the two polarization components for several wavelengths. The spectral distributions of the parallel and perpendicular components of polarization are plotted in Fig. 2 for $\theta = 1/\gamma$.

Time structure and further comments

The electrons circulate in storage rings grouped in almost monoenergetic bunches. These bunches are in synchronism with an applied radio-frequency field which replaces the energy lost by the emission of synchrotron radiation. A flash of radiation arrives to the observer whenever an electron bunch moves along the portion of observed orbit. The pulse duration is determined by the length of the bunches. For Adone it is approximately 2 nsec; in the new storage rings it is as short as 0.1 nsec. The repetition rate of the pulses depends on the number of bunches simultaneously stored. In the single bunch mode it corresponds to the fundamental frequency of the machine (2.85 MHz for Adone), otherwise it is a harmonic of it. An important feature is the very stable intensity from bunch to bunch.

In their motion, the electrons perform small oscillations around the ideal orbit, so that the electron beam has a finite transversal dimension and an intrinsic angular divergence. For many purposes, for istance when the radiation source must be focused onto the narrow entrance slit of a monochromator, the **source** brightness, defined as the flux emitted by the source per unit area and unit of solid angle, becomes an important parameter. For synchrotron radiation, the source brightness is related to the size and angular spread of the electron beam, and the angular divergence of synchrotron radiation. Most machines, in particular those designed as dedicated storage rings, have been optimized for the best brightness. For example, the ESRF storage ring at 100 nm has a photon flux 50% higher than Adone, but it becomes 20 times higher in terms of brightness.

Wigglers and undulators

The properties of the synchrotron radiation emitted from a particular machine depend on the machine parameters, in particular on the maximum energy attainable. For designing new, dedicated storage rings, this is a problem, since high energy machines, that supply a spectrum extending far in the hard X-ray region, are very expensive. It is more convenient to build 1-2 GeV machines and to insert in the straight sections special devices that allow either to shift the spectral distribution to higher energies, or to concentrate the emission within a narrow spectral range.

Wigglers are devices where the electrons are forced to perform an S-like path around the ordinary straight orbit by a magnetic pole sandwiched between two half poles of opposed field, in order to obtain a vanishing field integral, see Fig. 4a. The magnetic field can be done much stronger than that in the ordinary bending magnets, mainly by narrowing the gap between the magnet poles. The radius of curvature of the orbit is reduced, resulting in a shorter λ_c, as from Eq. (7). The spectral distribution of the radiation emitted by a wiggler is still given by Eqs. (5) and (10), but with a higher content of hard X-rays with respect to the bending magnets. Usually wigglers are constructed with two-three poles in series in order to multiply the photon flux from a single oscillation by the number of poles. In Fig. 5 we compare the spectral distributions of the radiation emitted from a bending magnet and from the wiggler mounted in Adone.

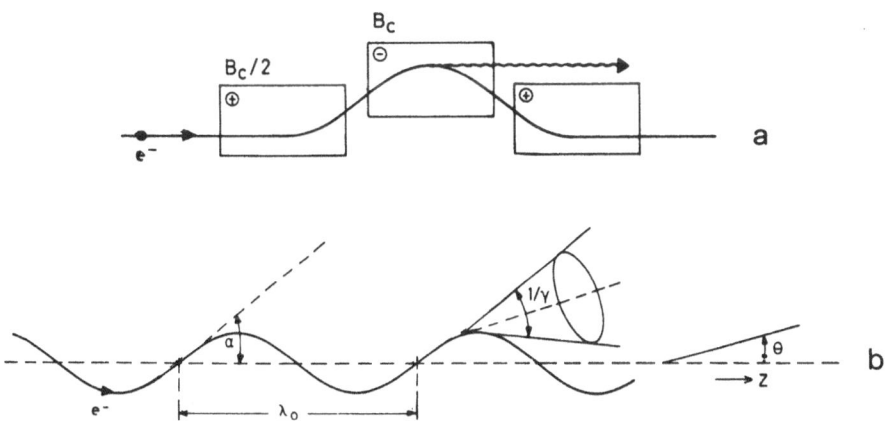

Fig. 4. a) A three pole wiggler with strong central field.
b) Description of a planar undulator. A perpendicular, periodic magnetic field of period λ_0 forces the electrons to oscillate on the plane of this sheet with the same period. Instantaneously, the light is emitted in a $1/\gamma$ cone centered on the tangent of the trajectory. θ is the angle of observation. α is the maximum deflection angle of the trajectory with respect to the z-axis (from Ref. 4).

If the number of magnetic poles becomes large, 20 to 100, then one speaks of undulators. Looking from far away, the electrons are seen to move back and forth around the equilibrium orbit, as if they were an electric dipole. It is intuitive to recognize that the emission is thus concentrated within the fundamental harmonic, or a few multiples of it. In Fig. 4b we show the electron path and we define the most important parameters of the undulator. There is interference of the radiation emitted by a single electron while travelling along equivalent regions of the poles. The difference between the time taken by an electron to travel over a period of the undulator and that taken by the emitted photons must be a multiple of the light period. For forward emission this is:

$$\tau = \frac{k\lambda}{c} = \frac{\lambda_o}{c} - \frac{\lambda_o}{v} \quad \Rightarrow \quad k\lambda \approx \frac{\lambda_o}{2\gamma^2} . \tag{14}$$

λ_o is the undulator period and k identifies the harmonics. The above relation is an approximation valid for a small deflection angle α. A detailed calculation shows that, at a given angle of observation θ the emission is quasi-monochromatic, with wavelength:

$$\lambda_k = \frac{\lambda_o}{2k\gamma^2} (1 + \frac{1}{2}K^2 + \gamma^2\theta^2) . \tag{15}$$

Only odd values of k are allowed for $\theta=0$.

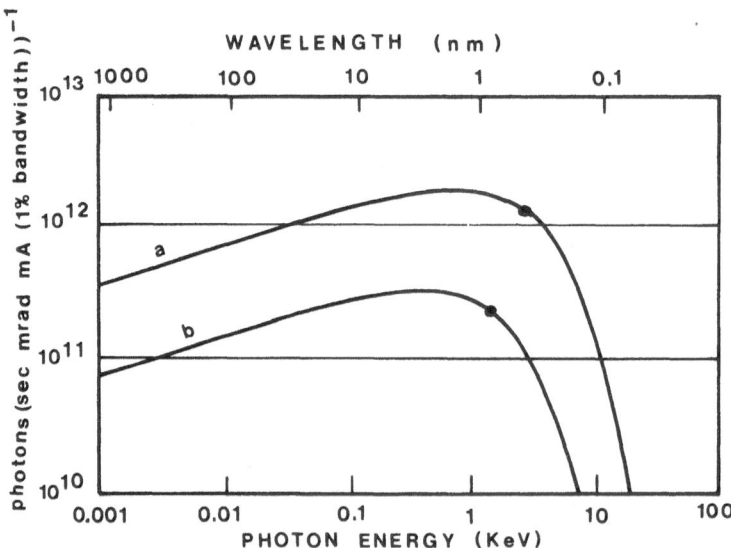

Fig. 5. Spectral distributions of the radiation emitted from a bending magnet (curve b) and from the wiggler magnet (curve a) in Adone, calculated for 1.5 GeV electrons (from Ref.5).

$$K = \alpha\gamma = \frac{e}{m_o c} \frac{\lambda_o B_o}{2\pi} = 0.0934\ B_o\lambda_o \quad (\lambda_o \text{ in mm}) \tag{16}$$

is the deflection parameter and compares the maximum orbit deflection with the emission angle. The wavelength λ_k can be changed either by changing the magnetic field, i.e. varying K in Eq. (15), or by changing the electrons energy. Since the emitted wavelength depends also on the observation angle θ, the experimental spectrum is a function of the observation area as well, as schematically illustrated in Fig. 6. In Fig. 6a the emission from an undulator with a small K and seen through a small observation area is depicted: only the fundamental frequency is observed. If the observation area is increased, the integration of Eq. (15) over all angles yields the broad band of Fig. 6b. In the case of a strong magnetic field (K>>1), but a small observation area, also the higher harmonics begin to appear, as in Fig. 6c. Finally, when the radiation is collected over a large area, the observed spectrum approaches that of a uniform magnet with field B_o, multiplied by the number of poles, Fig. 6d. The spectral width of the emission at $\theta=0$ is the natural one associated with an interference phenomenon, $\Delta\lambda/\lambda \sim 1.8/k\mathcal{N}$, where $\mathcal{N}$ is the number of poles of the undulator. We have already seen that the collection geometry may give an additional wavelength spread. Also the finite

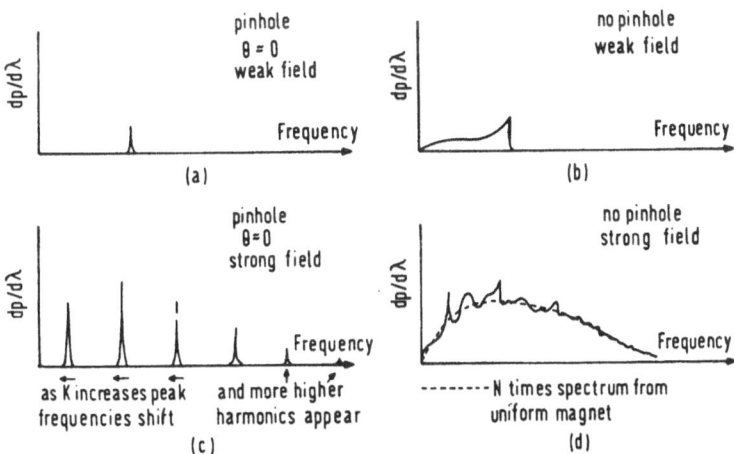

Fig. 6. Expected spectral distribution of the light emitted by an undulator in different modes of operation. Low field case (K<<1): observation through a pinhole (a) or through a large aperture (b). High field case: observation through a pinhole (c) or through a large aperture (d) (from Ref. 6).

Table 3. Table of some values of the function $F_k(K)$.

K	F_1	F_3	F_5	F_7	F_9	F_{11}
0.2	0.038	0	0	0	0	0
0.4	0.13	0.004	0	0	0	0
0.6	0.24	0.027	0.002	0	0	0
0.8	0.32	0.087	0.015	0.002	0	0
1.0	0.37	0.18	0.055	0.015	0.004	0
1.3	0.38	0.28	0.17	0.081	0.036	0.015
1.7	0.33	0.42	0.34	0.25	0.16	0.11
2.0	0.29	0.42	0.41	0.35	0.29	0.22
2.5	0.22	0.37	0.43	0.43	0.41	0.41
3.0	0.17	0.31	0.38	0.42	0.44	0.44

dimensions and divergence of the electron bunch contribute to the total line width. In order to minimize the latter contribution, it is important that both the beam dimensions and divergence are small inside the undulator, which makes undulators very bright sources. The intensity emitted in the forward direction within a very narrow cone is given by:

$$\frac{dN}{dt\, d\Omega\, (\Delta\lambda/\lambda)} = 4.56 \times 10^6 \mathcal{N}^2 \gamma^2\, I\, F_k(K) \tag{17}$$

photons/(sec $(0.1\ \text{mard})^2$ 1% bandwidth), where $F_k(K)=0$ for k even, while

$$F_k(K) = \frac{K^2 k^2}{(1+0.5K^2)^2} \left[J_{(k+1)/2}\left(\frac{kK^2}{4+2K^2}\right) - J_{(k-1)/2}\left(\frac{kK^2}{4+2K^2}\right) \right]^2 \tag{18}$$

for k odd. $J_{(k\pm1)/2}$ are Bessel functions. A few values of $F_k(K)$ are reported in Tab. 3. For increasing K, more harmonics appear and the maximum intensity shifts to higher k values. In Adone an undulator with $\lambda_0=11.6$ cm and K=3.51 has been constructed.[7] The first harmonic occurs at 1038 Å for 1 GeV electrons, and the emitted intensity is 9×10^{13} photons/(sec $(0.1\ \text{mrad})^2$ 1% bandwidth). This value is almost three orders of magnitude higher than the emission from the bending magnets.

In the past twenty years, the impact of synchrotron radiation on research in physics, chemistry, biology, and in applied sciences has become pervasive, both in the VUV and X-ray regions. In Tab. 4 a variety of possible applications of synchrotron radiation are presented with particular emphasys for the VUV region. When a beam of photons reaches a system (atoms, molecules or a solid), some photons are absorbed by the system, that is brought to an excited state. Afterwards the system reacts and releases the absorbed energy by emitting different types of particles, as illustrated in Fig. 7. Absorption spectroscopy gives informations mostly on the excitation mechanisms of the system. Different types of measurements can be performed on the emitted particles, that give indications on the decay mechanisms. Absorption spectroscopy received a lot of attention in the past years: the importance of synchrotron radiation in this field is well established. In the next sections I shall discuss briefly other fields of recent development, where the unique properties of synchrotron radiation are essential.

Table 4. Research fields in which synchrotron radiation is used and corresponding techniques.

	Atomic physics	Molecular physics	Electronic properties of solids	Surfaces and adsorbates	Defects in solids	Chemistry	Photochemistry	Molecular biology	Geology
Absorption/reflection spectroscopy	X	X	X	X	X		X	X	X
Emission spectroscopy (energy and time resolved)	X	X	X	X	X		X	X	X
Photoelectron spectroscopy (UPS-XPS-ARUPS)	X	X	X	X		X	X	X	X
Photoion spectroscopy	X	X		X		X	X		X
Photoconductivity			X	X		X	X		
Inelastic scattering of photons	X	X	X			X			
X-ray absorption fine structure			X		X	X	X	X	X
Other X-ray techniques	X	X	X	X	X	X	X	X	X

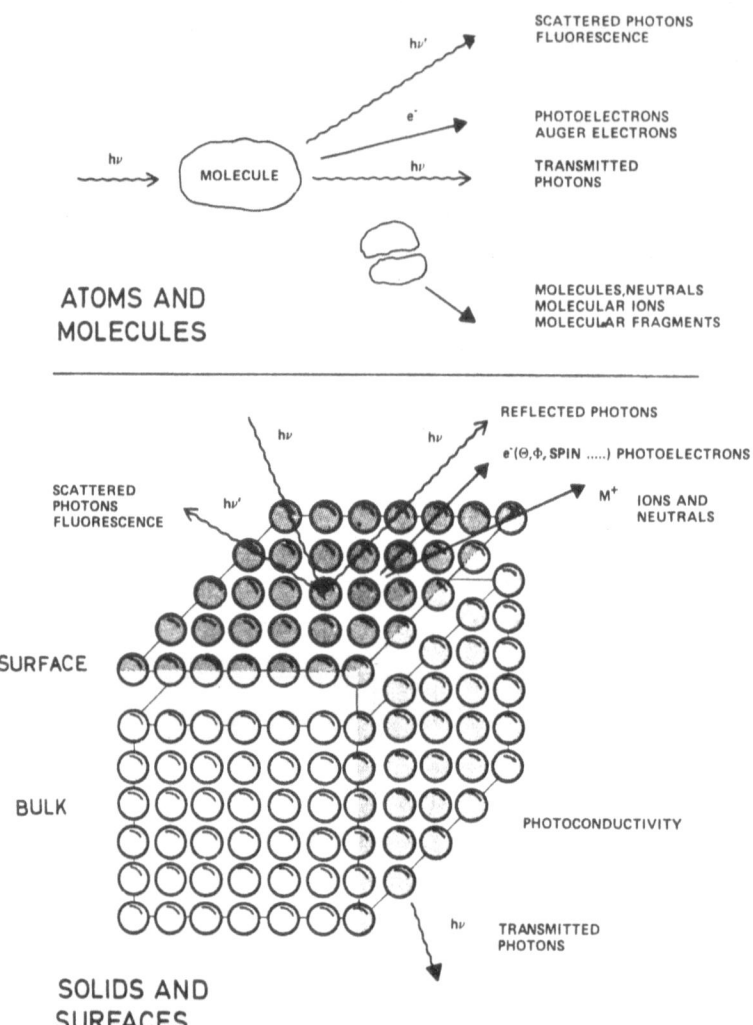

Figure 7. Upper panel: the interaction of photons with molecules is depicted schematically indicating the many possible spectroscopies. Lower panel: photo-excitation spectroscopies applied to solids, surfaces and adsorbates (from Ref. 1).

PHOTOEMISSION

In photoemission experiments, monochromatic light hits the sample surface and excites electrons from the valence bands or from the core states. A fraction of the excited electrons leave the sample and are collected by a detector. The number of electrons collected per second, i.e. the intensity of the measured signal, is a function of several quantities:

$$dN/dt = I[\hbar\omega, \underline{e}, (\theta, \phi)_{phot}, E, \sigma, (\theta, \phi)_{el}, \ldots] \, . \qquad (19)$$

$\hbar\omega$ is the energy of the photons polarized along $\underline{e}$ and incident with polar angles $(\theta, \phi)_{phot}$. E is the kinetic energy of the excited electrons with spin σ, leaving the sample along a direction identified by the polar angles $(\theta, \phi)_{el}$. Thus, photoemission is a very powerful technique capable of yielding several types of information.

The three-step model for photoemission is simple and intuitive. Even if it is an approximation, it has been used often for interpreting successfully the experimental data. Within this

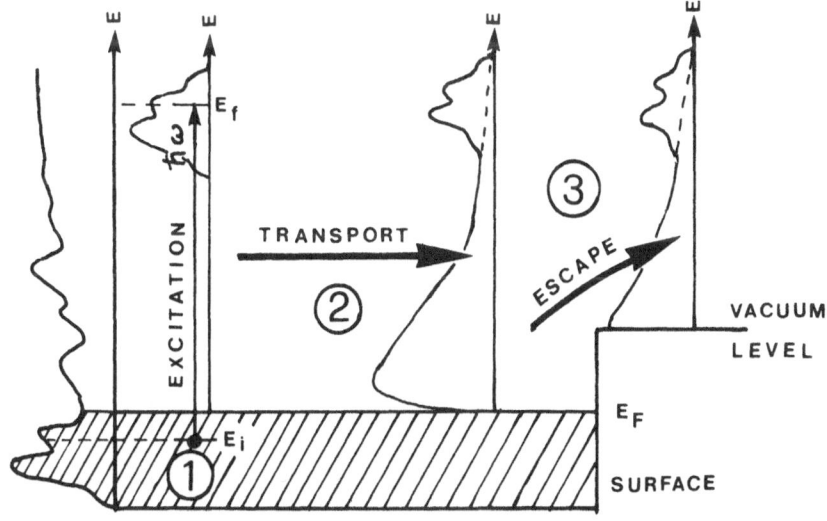

Fig. 8. Schematic of the three-step model for photoemission. A hypothetical density of states is drawn on the left side the other diagrams indicate how it changes at each step. The shaded area corresponds to the occupied states.

295

model, the photemission process is split into three successive events depicted schematically in Fig. 8. The first event is the optical excitation of an electron from an initial state of energy E_i to a final state of energy E_f, with the condition

$$\hbar\omega = E_f - E_i \tag{20}$$

and with probability approximately given by

$$W(\hbar\omega, E, \ldots) = |M_{if}|^2 \, \rho(E_f) \, \rho(E_i) \, . \tag{21}$$

$\rho(E_i)$ and $\rho(E_f)$ represent the densities of initial and final states, respectively, and M_{if} is the dipole transition matrix element. Within the approximation of a smooth final density of states and constant transition matrix element, the energy spectrum of the primarily excited electrons provides an image of the density of occupied states, as indicated on the left of Fig. 8. The optical excitation may occur as deep as 100 Å inside the solid, varying with material and photon energy. So, the second event is the transport of the electrons to the sample surface, during which the electrons may suffer from inelastic as well as elastic multi-scattering processes. The average distance travelled by an electron before it scatters is given by the electron mean free path $S(E)$. The dependence of $S(E)$ on the electron kinetic energy is reported in Fig. 9. The intensity of the primary electrons decreases and a tail of secondary electrons of low kinetic energy

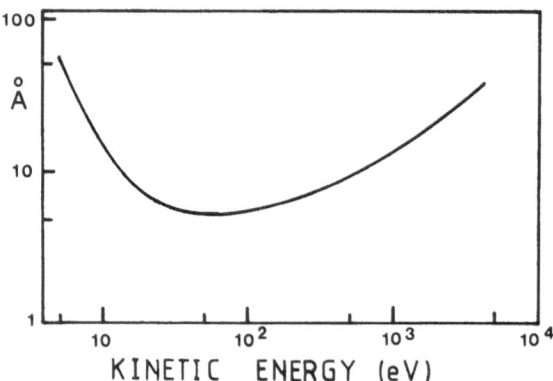

Fig. 9. Average electron inelastic mean free path as a function of the electron kinetic energy.

develops, as indicated at the center of Fig. 8. The third event is the escape of the electrons across the surface potential barrier into vacuum. In this process only the component of the electron momentum parallel to the surface is conserved.

In the most conventional photoemission experiments $\hbar\omega$ is kept fixed and the distribution in energy of the electrons photoemitted over a large solid angle is measured. The resulting curves are called energy distribution curves (EDC) and resemble the curve sketched on the right of Fig. 8. As an example, in Fig. 10 we present energy distribution curves for electrons emitted from the valence bands of gold, measured at several photon energies between 15 and 180 eV.[8,9] As customary, the EDC's are plotted as a function of the electrons binding energy referred to the Fermi energy, E_F, instead of the kinetic energy of the photoelectrons. In this way, structures associated with the same initial state appear lined up.

As shown in Fig. 9, S(E) has a broad minimum of about 5 Å for electron kinetic energies between 50 and 200 eV. The photoelectrons with such energies are excited mainly in the crystal surface or in

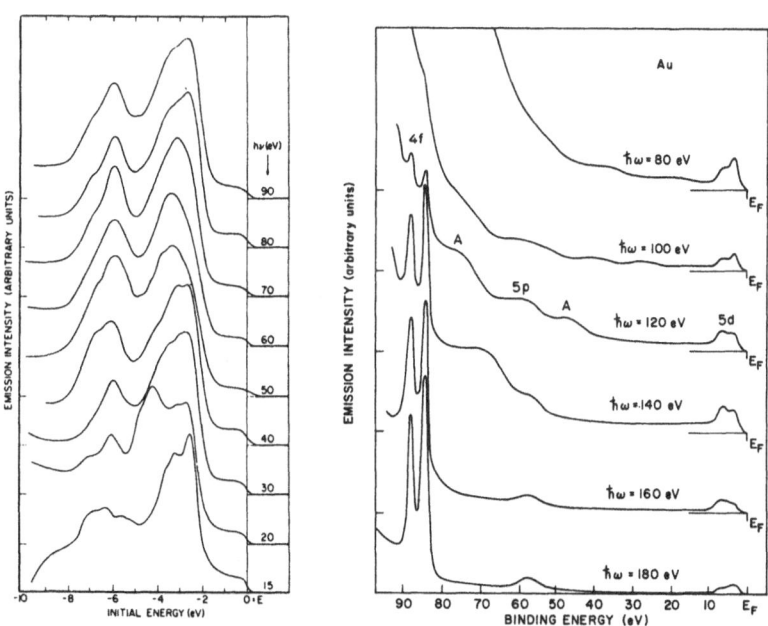

Fig. 10. EDC's of Au measured at several photon energies. (Left panel from Ref. 8. Right panel from Ref. 9.)

the first atomic layer behind the surface. We see immediately that EDC's measured with 100-200 eV photons are sensitive to the electronic surface states, and this opens to investigation the fields of surface physics and related phenomena, such as chemisorption, adsorption, catalysis, interfaces, etc. An example of the surface sensitivity is shown in Fig. 11, where the oxidation of the GaAs (110) surface is followed as the exposure to oxygen is increased.[10] It is noticeable the growing of the oxygen 2p bands and that of the chemically shifted Ga 3d and As 3d core levels. Two different oxidation processes have been discovered. When the GaAs (110) surface is exposed to molecular oxygen, only the As atoms, that stick outside the (110) surface, are bound to oxygen, as it can be inferred by the formation of only the 2.9 eV chemically shifted As 3d line (Fig. 11, left panel). In the case of activated oxygen, at first the 2.9 eV shifted As 3d peak appears. Then, increasing oxygen exposure gives a second As 3d peak together with a Ga 3d peak shifted by 1.0 eV. The latter two peaks become dominant at high levels of exposure (Fig. 11, right panel).

As already mentioned, EDC's should resemble very closely the densities of occupied states and should not vary with the energy of the exciting photons. A close examination of the curves of Fig. 10 shows us that this is not the case. For instance, the shape of the

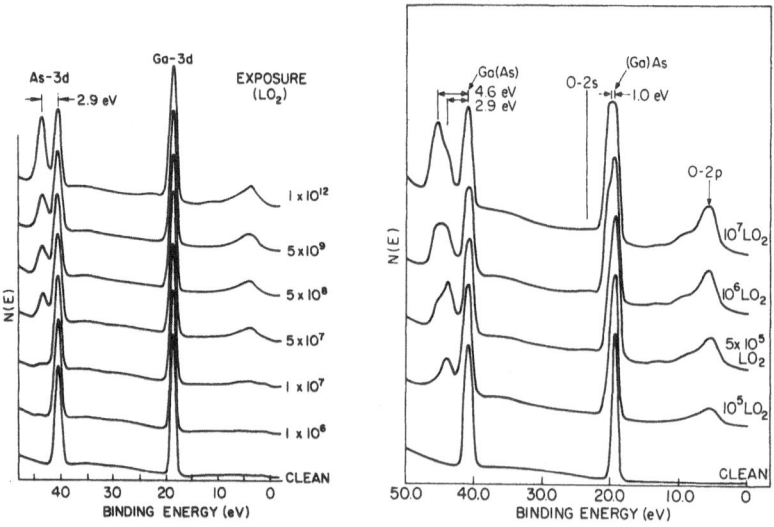

Fig. 11. EDC's of the GaAs (110) surface exposed to molecular oxygen (left panel) and to activated oxygen (right panel) (from Ref. 10).

first valence band peak at about 3 eV below E_F changes dramatically for $\hbar\omega=15$-20 eV; the intensities of the two peaks associated with the Au 5d bands are reversed between 120 and 180 eV; the Au 4f peaks become stronger and stronger with increasing photon energy, at least up to 180 eV. In the case of low photon energies, the intensity variation of the same valence band feature is due mostly to structures in the final density of states. At higher photon energies, $\rho(E_f)$ has almost levelled off. For a fixed, initial state the variation of $W(\hbar\omega,E)$ is caused mostly by the energy dependence of the transition matrix element as from Eq. (21). A plot of the intensity of a core peak versus $\hbar\omega$ gives the photoionization cross section of that particular core state. The behaviour of the Au 4f peaks is a typical example of a delayed threshold:[11,12] because of the centrifugal barrier in the potential for the g states, the oscillator strength for f→g transitions is very weak at threshold and it has a strong, broad maximum ∿100 eV above threshold.

In order to follow the intensity variation of the different structures as a function of $\hbar\omega$, it is necessary to take a large number of finely spaced EDC's. If in the experiment the initial state (either a core level or a valence structure) could be fixed, either M_{if} or $\rho(E_f)$ could be measured directly. As a matter of fact this is accomplished with a technique known as constant initial state photoemission spectroscopy (CIS), which consists in scanning synchronously the photon energy and the kinetic energy of the photoelectrons. From Eq. (20) we see that E_i is kept constant. A technique similar to CIS is the constant final state spectroscopy (CFS), also named partial yield spectroscopy, that consists in keeping the electron kinetic energy fixed and changing the photon energy. In this way also the initial state energy changes, according to Eq. (20), and in principle a better picture of the occupied density of states should be obtained, since the density of final states is not changed. In practice, the partial yield technique is applied by taking the electron energy analyzer window on the secondaries, the number of which is approximately proportional to the absorption coefficient of the sample. Thus, the partial yield technique is an alternative method for measuring the absorption coefficient of materials, and it is particularly useful for samples that cannot be made thin enough to perform transmission measurements, or for measuring the surface absorption coefficient.

Generally, the photoionization cross sections are smooth functions, with broad features, of $\hbar\omega$. There are cases where they undergo quick variations around particular values of the incident photons energy. This occurs, for example, in the 3d transition metal compounds when $\hbar\omega$ is scanned across the transition metal 3p absorption threshold (resonant photoemission).[13,14] Resonant photoemission needs still further investigation for a thorough understanding. Here we try to describe it through an intuitive, oversimplified model. Let us consider only the transition metal 3d states as localized and not hybridized with the ligand valence states. Resonant photemission results from the competition of two energetically degenerate channels, as sketched in Fig. 12. The first channel is the continuum of the electrons directly photoemitted from the localized filled 3d states. The second channel is quasi-discrete and originates from the excitation of a core 3p electron into an empty 3d final state, with a strong 3d

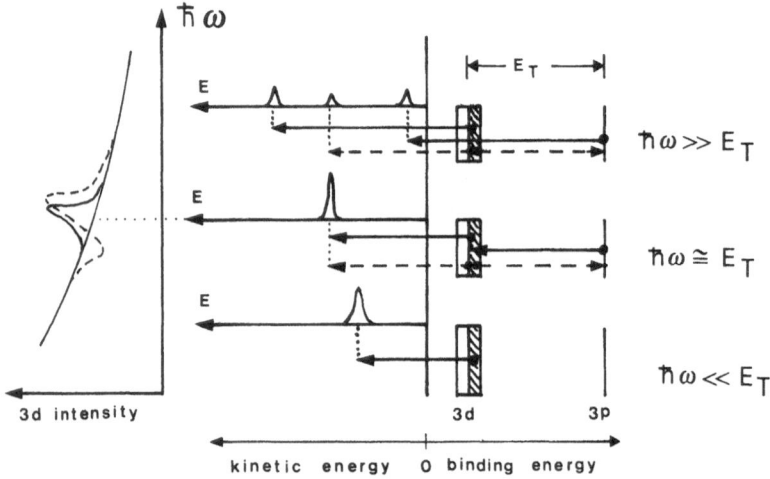

Fig. 12. Schematic of resonant photoemission in the case of a 3d transition metal ion. The solid arrows indicate photoexcitation. The broken arrows indicate the 3p hole decay, involving two 3d electrons. The expected EDC's are shown in the center. On the left, the expected intensity of the 3d line versus $\hbar\omega$ is sketched. At $\hbar\omega\approx E_T$ the peak due to the direct recombination (solid line) adds to the smooth 3d spectrum in the absence of interaction, but it evolves into the Fano line shape (dashed line) for the interaction with the direct 3d photoemission.

electron-core hole interaction. In this case the core hole can be filled by the direct recombination of the excited state accompanied by the emission of a 3d electron, which escapes with the same energy as the directly excited one. The interference of these two channels generates a typical Fano line shape. At photon energies above threshold, also the core 3p electron can escape directly from the sample, without forming bound final states. The core hole decays either radiatively or by Auger emission, and the photo-emission spectrum shows the respective three lines.

Let me discuss with some detail an application of the above considerations to the study of a group of materials of strong technological interest. The transition metal tiophosphates MPS_3 - M indicates a divalent transition metal ion, for example Mn^{++}, Fe^{++}, Ni^{++}, Zn^{++} - form a large family of layered materials of promising applications as cathodes in batteries based on Li^+ transport.[15,16] The operation of these batteries is sketched in Fig. 13.[17] A voltage develops between the Li anode and the cathode, made of a layered compound. During the discharge, Li^+ ions drift from the anode through the electrolyte and enter within the Van der Waals interlayer gaps. The electrons move in the external circuit through the load. In the case of the MPS_3 compounds up to three Li^+ ions per formula unit can be inserted within the interlayer gaps by electrochemical intercalation, and this is almost three times more than in the often studied TiS_2.[18] However, the intercalation capability of the MPS_3 compounds changes with the transition metal

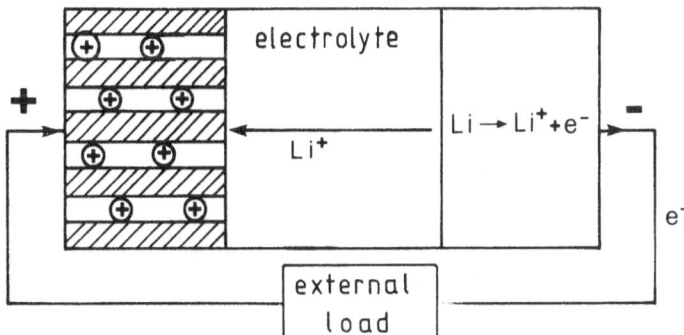

Fig. 13. Scheme showing an electrochemical battery using a Li anode (right) and a layered compound as cathode (left). The hatched regions on the left indicate the crystal layers, between which Li^+ ions are intercalated.

ion and seems to be related strongly with its ionicity and with the number of empty d states. In order to understand the origin of these differences and the changes induced by intercalation, a detailed study of the electronic states of the MPS_3 compounds is essential.

The MPS_3 compounds are partially ionic,[19] with formula M_2^{++} $(P_2S_6)^{4-}$. Their crystal structure is shown in Fig. 14, where the strong covalent bonds between the phosphorus and the sulphur atoms are indicated. The valence bands can be thought of to deriving from combinations of S 3s, $3p_xp_y$ and P 3s, $3p_z$ states of the $(P_2S_6)^{4-}$ cluster. The transition metal 3d states, split by the octahedral field, are located at the top of the valence bands or in the forbidden gap.[20,21] The valence bands of $NiPS_3$, $FePS_3$, and $ZnPS_3$ were investigated by means of X-ray photoemission spectroscopy.[21,22] Features common to all three compounds were assigned to states of the $(P_2S_6)^{4-}$ groups, and the remaining structures to the transition metal 3d levels. However, in interpreting the valence band spectra of these compounds, two difficulties arise. A photoelectron escaping from a partially filled, localized d state generates a multiplet structure corresponding to the allowed transitions from the ground state ($3d^n$ configuration) to states of

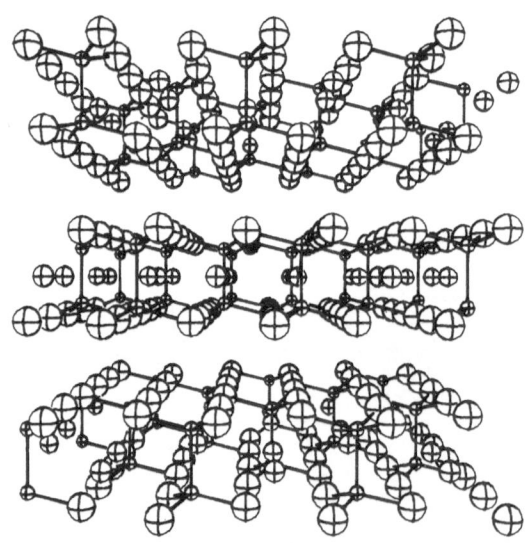

Fig. 14. Perspective view of the crystal structure of $NiPS_3$. The circles represent P, Ni, and S in order of increasing size (from Ref. 21).

the final $3d^{n+1}$ configuration. In addition, satellite structures associated with the excitation of the d electrons may overlap the $(P_2S_6)^{4-}$ valence features, as for other transition metal compounds.[23,24] In order to overcome these difficulties, resonant photoemission at photon energies around the transition metal 3p threshold was used.[25] The conduction band density of states was studied by measuring the absorption spectra of several core levels with the partial yield technique.[26]

Several EDC's measured far from resonance and in the resonance regime are shown in Fig.15 for $MnPS_3$, $FePS_3$ and $NiPS_3$. A remarkable variation of the intensity of some peaks can be clearly seen as the energy of the exciting photons is scanned across the Mn, Fe, and Ni absorption thresholds at ∿47 eV, ∿54 eV and ∿65 eV, respectively. In all three compounds, the structure labelled C in Fig. 15 becomes

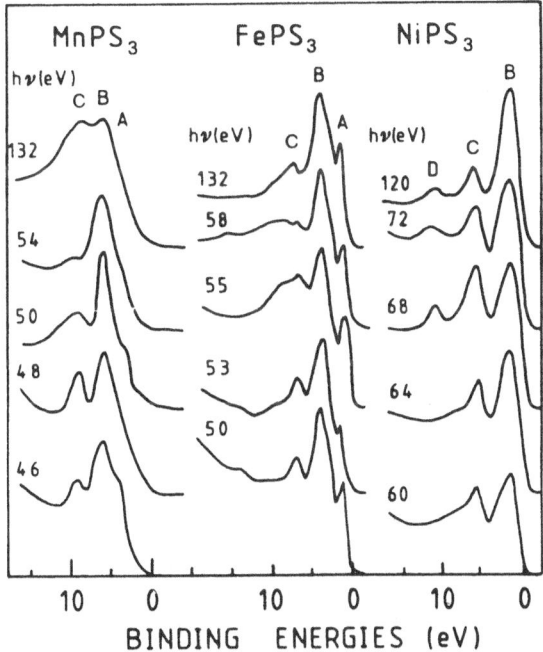

Fig. 15. Resonant photoemission for $MnPS_3$ (left), $FePS_3$ (center) and $NiPS_3$ (right) measured at photon energies around the Mn, Fe and Ni 3p thresholds (∿47 eV, ∿54 eV and ∿65 eV, respectively) (from Ref. 27).

almost comparable with the sharpest peak B at certain photon energies (∿68 eV for NiPS₃) and decreases again far from threshold. In MnPS₃ and FePS₃ the first structure A becomes very weak at $\hbar\omega$∿49 eV and 53 eV, respectively. In NiPS₃, the peak D develops at about $\hbar\omega$∿65 eV, grows fast and then decreases again. These variations are better followed in Fig. 16, where the intensities of some features of the FePS₃ and NiPS₃ spectra are reported as a function of the exciting photon energy, and are compared with the transition metal 3p absorption spectra.[26] The B and C structures of FePS₃ and the C and D structures of NiPS₃ follow the typical Fano antiresonance-resonance line shape, while those due to the structures A and B in the two compounds have a smooth behavior. The former have been identified univocally as satellites or d-features, the latter as true valence states of the $(P_2S_6)^{4-}$ cluster. After assigning the photoemission structures of the pure materials, in the next step of this research we shall investigate their evolution for different amounts of intercalated Li^+ ions.

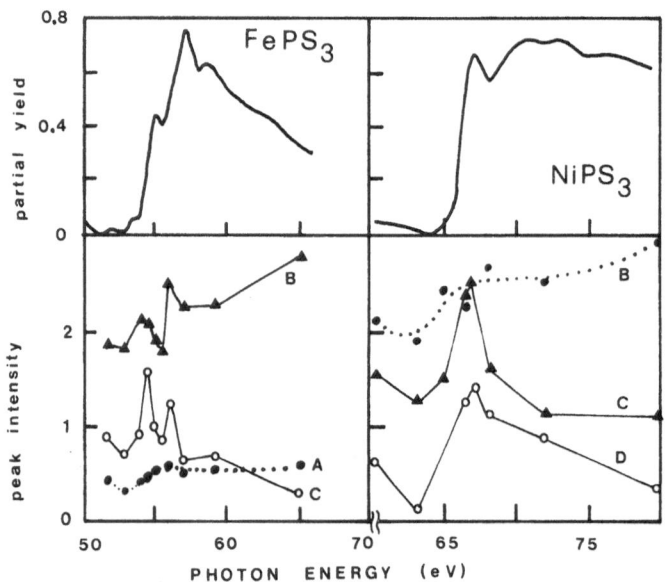

Fig. 16. Upper panels: absorption spectra of Fe and Ni 3p electrons in FePS₃ and NiPS₃ measured with the partial yield technique. Lower panels: variation of the intensities of some structures (labelled according to Fig. 15) of the EDC's as a function of the photon energy (from Ref.25).

Excited states of atoms, molecules or solids decay in several, different manners: radiative decay is one of the most important, since it gives informations on how the system relaxes, how the energy is transfered inside the system and the time evolution of these phenomena. A luminescence spectrum depends on the wavelength of the exciting radiation, λ_X, the wavelength of the emitted radiation, λ_M, and the time delay between excitation and emission:

$$\mathscr{I}_L = \mathscr{I}(\lambda_X, \lambda_M, t) \ . \tag{22}$$

Clearly, also the polarization and the directions of both the incident and the emitted radiation with respect to each other and to the crystallographic axes are important parameters, but we shall neglect them for the sake of simplicity. The simplest way of measuring luminescence is to average the intensity $\mathscr{I}_L$ for periods of time long compared with the lifetime of the decaying states of the sample, and either to scan λ_M keeping λ_X fixed (emission spectra), or to scan λ_X keeping λ_M fixed (excitation spectra). With the former technique the decay channels for each excitation wavelength are found; with the latter the excitation probability of the different decay machanisms are obtained. Time resolved fluorescence spectroscopy (λ_X and λ_M constant, t scanned) is the third, most valuable source of information, for it gives also the lifetimes of the different processes.

Until recently, luminescence spectroscopy has been limited to the visible or near ultraviolet regions for both excitation and

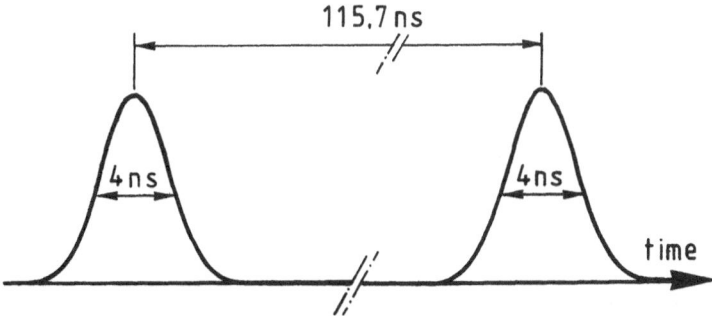

Fig. 17. The Adone synchrotron radiation pulses as measured at the PULS luminescence station L1 (courtesy by F.Fermi).

emission spectra. The use of synchrotron radiation has opened also the VUV region. In addition, synchrotron radiation is a naturally pulsed source, ideal for time resolved spectroscopy. The pulse duration depends mostly on the storage ring radio frequency power supply and on the circulating current. In the case of Adone it is approximately 1÷4 nsec long (see Fig. 17), but in other machines it may be as short as 100 psec. These values are not spectacularly short with respect to laser sources, but the repetition rate is very high ($\sim10^7$ pulses/sec) and stable. In addition, both the pulse intensity and shape (almost Gaussian) are totally reproducible from pulse to pulse during the time required for an experiment (typically 15-30 minutes), since each pulse is generated by the radiation emitted by the same circulating electron bunch while illuminating the observer[*]. With such a time pattern, lifetime measurements as short as 1/100 of the pulse duration can be done with the modulation and phase-shift technique described below.

In Fig. 18 the layout of the VUV beam line in the PULS laboratory is presented. The radiation from Adone is focused onto the entrance slit of the 1 m normal incidence monochromator HW (wavelength range 30-300 nm) by means of a grazing incidence cylindrycal mirror. The monochromatized radiation is refocused either in the reflectometer R or in the luminescence station L1. The reflectometer is used for reflectivity and/or transmission measurements in the VUV region. In the luminescence station the fluorescence light is collected through a quartz window at 90° with respect to the exciting radiation, focused on the analyzing monochromator (200-800 nm) and detected by a photomultiplier. The signal processing is performed with a conventional single photon counting technique. For time resolved experiments, a second detector is employed, that collects the radiation diffused inside the HW monochromator and supplies the Adone pulses for starting (stopping) a time-to-amplitude converter. The stop (start) signal is given by the fluorescence single photon pulse. The output of the time-to-amplitude converter is analyzed with a multichannel analyzer for obtaining a time dependent spectrum F(t). This results by the convolution between the experimental apparatus response function g(t) and the actual spectrum $\mathscr{I}_L(t)$:

---

[*] This is strictly true when the storage ring is operated in the single bunch mode. In the multi-bunch mode, the different bunches may have slightly different intensities.

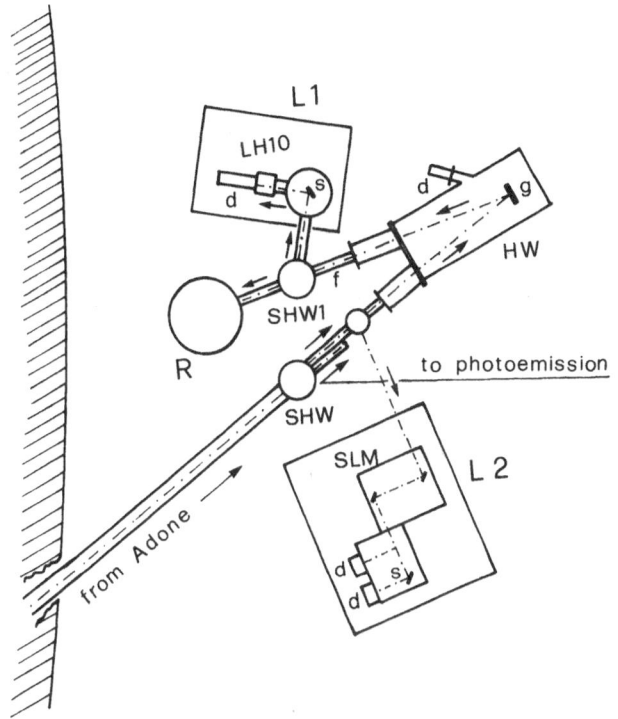

Fig.18. Layout of the VUV beam line of the PULS facility. SHW and SHW1: focusing mirrors. HW: 1 m normal incidence mono-chromator (30-300 nm). R: reflectivity station. L1: luminescence station with VUV exciting radiation and single photon counting detection. LH10: luminescence analyzing monochromator (200-800 nm). L2: luminescence station with visible exciting radiation and phase-shift method for life time measurements. SLM: visible monochromator (200-1000 nm). d: detectors. f: intensity monitors and filters. g: grating. s: sample position.

$$F(t) = g(t) * \mathscr{I}_L(t) . \tag{23}$$

$g(t)$ includes the exciting pulse shape as well as the response of the detector. For very short lifetimes, $\mathscr{I}_L(t)$ may be obtained by deconvoluting Eq. (23). With the Adone time pattern described above, we expect to measure decay times from approximately 0.5 nsec to some hundreds of nsec.

L2 in Fig 18 is a second station for lifetime measurements, using the visible and near-ultraviolet portions of the synchrotron radiation spectrum.[28] The radiation is collected through a saffire window, focused onto the entrance slit of the SLM monochromator (wavelength range 200-1000 nm), then it is split into a reference channel and a sample channel. The fluorescence light is collected at 90° with respect to the incident one and different spectral bands are selected with optical filters. The fluorescence lifetimes are obtained with the modulation and phase-shift method.[29] This method is based on the fact that, when a sample with a single exponential decay is excited with light of wavelength λ_X, the intensity of which is modulated sinusoidally at the frequency ν_0,

$$I_X(\lambda_X) = I(1+\sin\omega_0 t) \quad ; \quad \omega_0 = 2\pi\nu_0 , \tag{24}$$

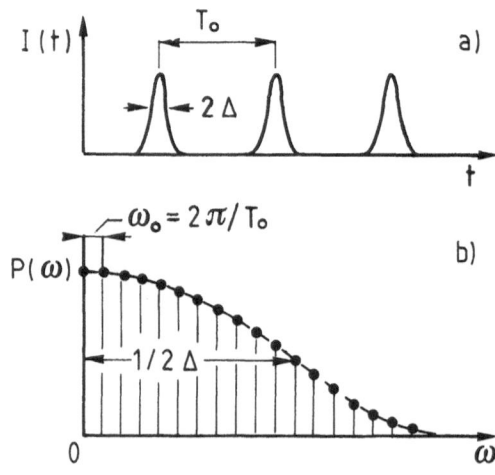

Fig. 19. a) The synchrotron radiation pulsed time structure. T_0: pulse period; 2Δ: pulse full width at half maximum. b) Harmonic content of synchrotron radiation (from Ref. 30).

also the fluorescence output is modulated at the same frequency

$$I_M(\lambda_M) = I'[1+M \sin(\omega_0 t+\phi)] , \qquad (25)$$

but with a reduced modulation amplitude M and a with a phase shift ϕ with respect to the incident light, both related to the lifetime τ by the expressions

$$tg \phi = \omega_0 \tau , \qquad (26)$$

$$M = [1 + (\omega_0 \tau)^2]^{-1} . \qquad (27)$$

The use of a single frequency is limited by the requirement of very high modulation frequencies for measuring very short lifetimes, and also to single exponential decays. Synchrotron radiation overcomes both difficulties.[30] In fact, as it consists of a series of equally spaced pulses (repetition rate $\nu_0 = 1/T_0$), with constant Gaussian shape of equal amplitude A and equal width 2Δ, it corresponds to a source sinusoidally modulated over a broad range of harmonics of the fundamental frequency, as shown in Fig. 19. The amplitude of the higher harmonics decreases following a Gaussian envelope of half-width $1/2\Delta$. For the Adone pulses shown in Fig. 17, the fundamental harmonic is 8.5 MHz, and the half width of the envelope Gaussian is 500 MHz. The zero frequency line gives the average power of the synchrotron radiation. If Δ is small, also high harmonics contain a noticeable intensity. Clearly, also the fluorescence spectrum contains all these harmonics. By measuring M and ϕ over a large number of frequencies, high resolution in the lifetime measurements can be achieved, and information on the multiexponential decay is available. For example, in test runs[28] we measured a decay time of 113±5 psec for bis-TNS in water ($\lambda_X=370$ nm; $\lambda_M>420$ nm). In the case of a tryptophan solution (pH = 6.9; $\lambda_X=280$ nm; $\lambda_M=313$ nm), two decay lifetimes of 3.12±0.28 nsec and 0.74±0.88 nsec were obtained with an intensity ratio of 0.9±0.1.

We have started a program of luminescence spectroscopy at Frascati, in order to investigate the excitation and the emission spectra of pure and doped alkali halides at several temperatures. KI:Tl is a typical phosphorus, that we shall discuss briefly. Below the KI fundamental absorption threshold at 5.6 eV leading to the first exciton at 5.83 eV,[31] there are several absorption bands due to the Tl^+ ion, shown in Fig. 20, the most prominent of which are

the A and the C bands at 4.40 eV and 5.30 eV, respectively, at
liquid nitrogen temperature (LNT).[32] The excitation of either band
gives a strong emission in the well known Stokes shifted A_x band,
centered at about 3 eV.[33,34] When a sample of KI:Tl is excited with
VUV radiation above the fundamental threshold, at room temperature
(RT) there is still strong emission in the A_x band.[35] This is
explained in the following way. The primary ionization produces an
electron-hole pair. The hole is easily trapped at a Tl^+ ion,
forming a Tl^{++} ion. The much more mobile electron wonderes around
the crystal, until it is trapped into one of the excited states of
Tl^{++} to form an excited thallium ion $(Tl^+)^*$. The system now behaves
exactly as if the Tl^+ ion has been excited directly. The emission
yield should present sudden increases at photon energies multiple
of the fundamental gap, so that the excited electron has enough
energy to generate other electron-hole pairs by inelastic
scattering. At low temperatures the transfer of energy from the
electron-hole pair to the Tl^+ center competes with the direct
recombination process, characteristic of the pure crystal. In the
latter case the hole is self trapped in the lattice, forming a V_k
center, and the electron-hole pair recombination gives rise to the
intrinsic luminescence at 3.34 eV and 4.15 eV in KI.[37]

The excitation spectra of pure KI and KI:Tl between 14 and 30
eV are presented in Fig. 21. Fig. 21a refers to the RT measurements and only the Tl^+ doped crystals gave a signal. The spectrum

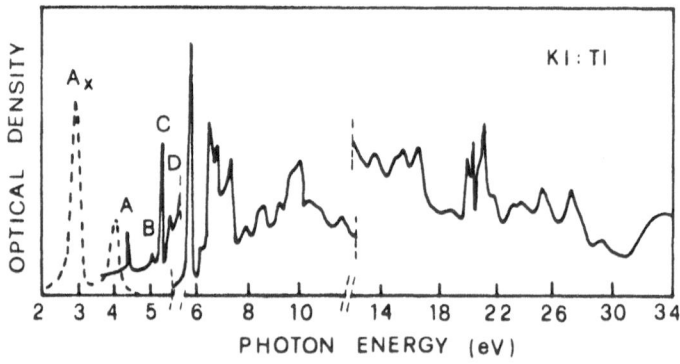

Fig. 20. Absorption spectrum of KI:Tl at 80 K. Above 5.6 eV the
absorption is due to the KI crystal (from Ref. 31 and
Ref. 36). Below 5.6 eV the Tl^+ ion absorption (from Ref.
32) and emission (broken line - from Ref. 33) bands are
shown. Note the changes of scale for both the optical
density and the photon energy.

shows mainly a broad peak around 18 eV and some subsidiary oscillations. The excitation spectra measured at LNT, Fig. 21b, for both
pure and doped crystals are alike. In this case, no energy transfer
towards a recombination center has been observed in the KI:Tl
samples. Thus energy transfer takes place only if the intrinsic
luminescence is quenched by the temperature. These results are
supported by the emission spectra measured for VUV exciting
photons. The emissions excited at 22.5 eV are presented for both
pure and doped KI and for RT and LNT in Fig. 22.[38] At RT the KI:Tl
sample shows the characteristic A_x emission. At LNT both types of
samples, even the most heavily doped ones (up to 0.003 mol %),
present only the intrinsic luminescence spectrum.

A problem that requires further investigations is related with
the minima around 20-22 eV in the excitation spectra of Fig. 21.
These minima coincide with the strong absorption maxima of the K^+
3p core excitons.[36] It is possible that the K^+ core excitons have
different channels for decaying. One way is via Auger emission, and
in this case an enhancement of the photoemitted electrons should be
found. A second way is through photodesorption, that can be
detected as an increase of the yield of the desorbed species.
Finally, the K^+ core excitons are well localized inside the unit
cell and may decay radiatively by direct recombination. In this
case the emission should occur in Stokes-shifted bands corresponding to the absorption bands.

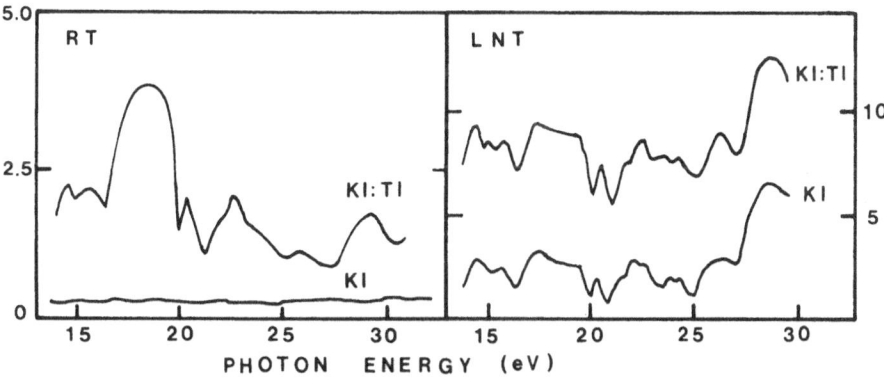

Fig. 21. a) Room temperature VUV excitation spectrum of KI and
KI:Tl. b) Liquid nitrogen temperature VUV excitation
spectra of the KI and KI:Tl emission at 3.44 eV. The pure
KI spectrum overlaps the KI:Tl one and it has been shifted
downwards for clarity (from. Ref. 38).

CONCLUSIONS

As we have discussed briefly in the first part of these notes, synchrotron radiation is a highly collimated, very intense, strongly polarized source with a continuous spectrum extending from the visible to the hard X-rays. It is also naturally pulsed, with high repetition rate and almost equal pulses. Some of these properties can be enhanced in special devices, such as wigglers and undulators. All these features gathered together in a single source make synchrotron radiation unique.

Several other sources are available in the vacuum ultraviolet. Continuous spectra are produced with low-pressure rare gas discharges or with the electron bremsstrahlung. However, they are rather weak, unpolarized, unstable and cover short energy intervals. Line sources extend in the far ultraviolet up to approximately 48 eV, with an intensity that in several cases exceeds that of the synchrotron radiation available in existing facilities. However, for a large number of investigations, it is crucial to go

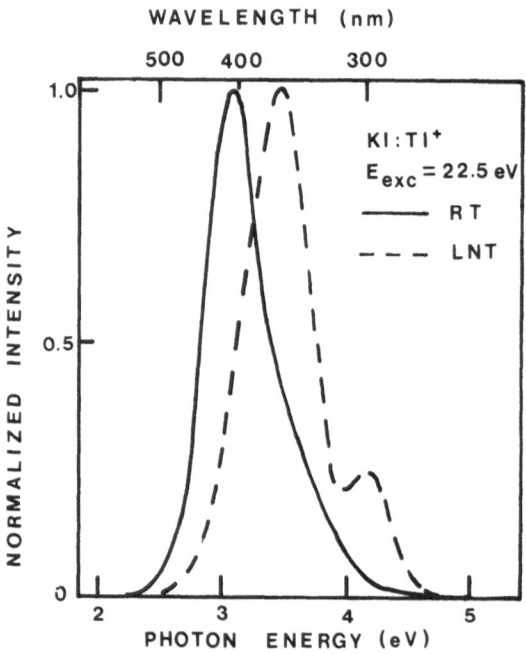

Fig. 22. Emission spectra of KI:Tl excited at 22.5 eV measured at 77 K and 300 K. The spectra have been normalized to the maximum intensity value (from Ref. 38).

to high photon energies and to have a continuous source. In discussing the applications of synchrotron radiation I selected a couple of examples emphasizing the great advantages of using synchrotron radiation. For instance, line sources are currently employed in photoemission for measuring EDC's, or for angle resolved experiments. But surface studies are better performed at higher photon energies then those available from line sources. Resonant photoemission, that is measured readily with the CIS method, does require the continuous spectrum. Similar arguments hold for luminescence spectroscopy too, for which also the time structure of synchrotron radiation becomes very important.

VUV lasers that are available nowadays or within a short time, are superior to synchrotron radiation in a number of experiments, in particular, those requiring spectral purity or very high intensity. The latter may reduce considerably measuring times, and this is almost essential in the case of samples deteriorating fast or for studying the dynamics of several phenomena during their evolution.

BIBLIOGRAPHY

Recent excellent books on synchrotron radiation and its applications are:

a. "Synchrotron Radiation", edited by C.Kunz (Springer, Berlin, 1978).
b. "Synchrotron Radiation Research", edited by H. Winick and S. Doniach (Plenum Press, New York, 1980).
c. "Handbook on Synchrotron Radiation", edited by E. E. Koch (North Holland, Amsterdam, 1983), Vol. 1a and Vol. 1b.

Specific volumes on photoemission and luminescence are:

d. "Photoemission and the Electronic Structure of Surfaces", edited by B. Feuerbacher, B. Fitton, and R. F. Willis (Wiley, New York, 1978).
e. "Photoemission in Solids", edited by M.Cardona and L.Ley (Springer, Topics in Applied Physics, Berlin), Vol. I (1978) and Vol. II (1979).

f. C. C. Klick and J. H. Shulman, "Solid State Physics", Vol. 5, pag. 97 (1957).

g. "Methods of experimental Physics", edited by E. Marton (Academic Press, New York 1959), Vol. 6B, pag. 293.

REFERENCES

1. For the hystorical news on synchrotron radiation see, e.g., E. E. Kock, D. E. Eastman, and Y. Farge, in "Handbook in Synchrotron Radiation", edited by E. E. Koch (North Holland, Amsterdam, 1983), Vol.1a, p.1, and references therein.

2. The derivation of the basic equations of synchrotron radiation emission can be found in several textbooks. Here we refer to: S. Krinsky, M. L. Perlman, and R. E. Watson, in "Handbook in Synchrotron Radiation", edited by E. E. Koch (North Holland, Amsterdam, 1983), Vol. 1a, p.65, and references therein.

3. D.H. Tomboulian and P.L. Hartman, Phys. Rev. $\underline{102}$, 1423 (1956).

4. Y. Farge, Appl. Opt. $\underline{19}$, 4021 (1980).

5. "Activity Report of the I.N.F.N. National Laboratories in Frascati", Report LNF-83/105 (1983), pag. 22 (in Italian).

6. D. J. Thompson and M. W. Poole, eds, "Europian Synchrotron Radiation Facility: The Machine." (The Europian Science Foundation, Strasbourg 1979).

7. R. Barbini and G. Vignola, Report LNF-80/12(R) (1980).

8. J. Freeouf, M. Erbudak, and D. E. Eastman, Solid State Commun. $\underline{13}$, 771 (1973).

9. I. Lindau, P. Pianetta, K. Y. Yu, and W. E. Spicer, Phys. Rev. B $\underline{13}$, 492 (1976).

10. P. Pianetta, I. Lindau, C. M. Garner, and W. E. Spicer, Phys. Rev. B $\underline{18}$, 2792 (1978).

11. U. Fano and J. W. Cooper, Rev. Mod. Phys. $\underline{40}$, 441 (1968).

12. S. T. Manson and J. W. Cooper, Phys. Rev. $\underline{165}$, 126 (1968).

13. C. Guillot, Y. Ballu, J. Paignì, J. Lecante, K. P. Jain, P. Thiry, R. Pinchaux, Y. Petroff, and L. M. Falicov, Phys. Rev. Lett. $\underline{39}$, 1632 (1979).

14. A. Fujimori and F. Minami, Phys. Rev. B $\underline{30}$, 957 (1984).

15. R. Brec, G. Ouvrard, A. Louisy, J. Rouxel, and A. LeMehaute, Solid State Ionics $\underline{6}$, 185 (1982).

16. J. Rouxel, P. Molinie, and L. H. Top, J. Power Sources $\underline{9}$, 345 (1983).

17. M. S. Whittingham, Progr. Solid State Chem. $\underline{12}$, 41 (1978).

18. A. LeMehaute, G. Ouvrard, R. Brec, and J. Rouxel, Mat. Res. Bull. $\underline{12}$, 1191 (1977).

19. W. Klingen, G. Eulenberger, and H. Hahn, Naturwissenschaften 57, 88 (1970).

20. F. S. Khumalo and H. P. Huges, Phys. Rev. B 23, 5375 (1981).

21. M. Piacentini, F. S. Khumalo, C. G. Olson, J. W. Anderegg, and D. W. Lynch, Chem. Phys. 65, 289 (1982).

22. M. Piacentini, F. S. Khumalo, G. Leveque, C. G. Olson, and D. W. Lynch, Chem. Phys. 72, 61 (1982).

23. G. K. Wertheim and S. Hufner, Phys. Rev. Lett. 28, 1028 (1972).

24. Y. Sakisaka, T. Ishii, and T. Sagawa, J. Phys. Jpn. 36, 1372 (1974).

25. M. Piacentini, V. Grasso, S. Santangelo, M. Fanfoni, S.Modesti, and A. Savoia, Nuovo Cimento D (1984) (in press).

26. M. Piacentini, V. Grasso, S. Santangelo, M. Fanfoni, S.Modesti, A. Savoia, Solid State Commun. 51, 467 (1984).

27. The MnPS$_3$ energy distribution curves are still unpublished. Those of FePS$_3$ and NiPS$_3$ are from Ref. 25.

28. F. Antonangeli, F. Bassani, F. Campolungo, A. Finazzi-Agrò, U. M. Grassano, E. Gratton, D. M. Jameson, M.Piacentini, N.Rosato, A. Savoia, G. Weber, and N. Zema, Report LNF-83/68(R) (1983).

29. R. D. Spencer and G. Weber, Ann. N. Y. Acad. Sci. 158, 361 (1969).

30. E. Gratton and R. Lopez-Delgado, Nuovo Cimento 56B, 110 (1980).

31. K. Teegarden and G. Baldini, Phys. Rev. 155, 896 (1967).

32. P. H. Yuster and C. J. Delbecq, J. Chem. Phys. 21, 892 (1953).

33. R. Edgerton and K. Teegarden, Phys. Rev. 129, 169 (1963).

34. J. M. Donahue and K. Teegarden, J. Phys. Chem. Solids 29, 2141 (1968).

35. Y. Farge and M. Fontana, "Electronic and Vibrational Properties of Point Defects in Ionic Crystals." (North Holland, Amsterdam, 1979), pag. 196.

36. A. Ejiri, M. Watanabe, H. Saito, H. Yamashita, T. Shibaguchi, H. Nishida, and S. Sato, "3rd International Conference on Vacuum Ultraviolet Radiation Physics", edited by Y. Nakai (Tokyo, 1971).

37. J. Ramamurti and K. Teegarden, Phys. Rev. 145, 698 (1966).

38. F. Antonangeli, F. Fermi, U. M. Grassano, M. Piacentini, A. Scacco, and N. Zema, Solid State Commun. 49, 323 (1984).

INJECTION LOCKING OF CW RING GAS LASERS

Jean Luc Boulnois

Quantel S.A., B.P. 23
91940 Les Ulis, France

The study of the non-linear characteritics of a regenerative oscillator under the influence of an external signal originates with the pioneering work of Adler[1], and since, the subject of "injection locking" has been refined by a number of authors[2-4]. Although the general analysis applies to almost any kind of self-sustained periodic oscillator, it has become of significant practical importance with the development of a wide variety of solid-state negative resistance microwave devices[5].

Basically, injecting a weak locking signal into a more powerful free-running oscillator stabilizes its oscillation frequency. However besides generating a synchronized fixed frequency signal, the technique can also be used for amplification and FM noise quieting, high-power generation by coherent summation, and also detection of FM signals.

As with microwave devices, feedback or regenerative techniques have become particulary appropriate in the efficient generation of frequency-stable high power laser sources. However besides being an extremely useful technique in laser stabilization, injection locking also plays a significant role in understanding mode-coupling effects within an oscillator, for example in the case of the response of ring laser gyroscopes and other related laser backscatter effects. This explains the present review of the basic principles of CW injection locking theory preceeding the discussion of recent experiments on single longitudinal mode CO_2 ring waveguide locked oscillators.

INTRODUCTION

Many laser applications, either of the "heterodyne type" where a Doppler beat signal is monitored such as in anemometry, velocimetry or with lidars, or of the "resonant interaction type" such as encoun-

tered in spectroscopy, pollution monitoring or off-center-line optical
pumping of lasers, make use of high power spectrally pure sources.
Usually the laser is required to operate with a single longitudinal
mode (SLM) at a particular frequency ; but the requirement of high
power implies a broadband spectrum in which several longitudinal modes
oscillate above threshold. In such cases mode selection is achieved by
introducing various intracavity dispersive elements such as etalons or
gratings, but these tend to become inefficient in high gain devices.
Thus very high power systems make use of oscillators-amplifier configurations where the optimum extraction efficiency is reached when the
oscillator saturates the amplifier ; these are large devices with
associated critical problems such as alignment and backscattering.

A competitive alternative is injection locking. One injects the
output of a narrow band, eventually frequency stable or frequency
tunable, low power master-oscillator (M.O.) into a broadband high-
power oscillator (P.O.) : under certain conditions on the recirculating power within the P.O., the device operates narrow band, SLM,
and is frequency-locked onto the injected signal.

The physical mechanisms responsible for the control of the
frequency behavior and the bandwith of the P.O. by the injected
signal are different, depending on the temporal regimes considered.

In the pulsed transient regime, the mechanism referred to is
"injection seeding". At this point, care should be exercised in the
use of concepts such as cavity modes which are basically steady-state
concepts whilst considering such transitory situations. The injected
signal can more accurately be viewed as providing an initial condition on one of the spatial and spectral modes of the pulsed oscillator, from which oscillation builds up preferentially as opposed to
building up from noise, with little continuing influence or control
by the injected signal beyond its setting of the initial field. The
mode with fastest growing amplitude saturates first, and for sufficiently long pump pulses it can deplete the gain of other modes via
cross-saturation. In other words, it can be said that the oscillating
mode keeps the amplitude information content of the injected field but
forgets about its phase content. In the case where the M.O. bandwidth
is narrow and the P.O. pulse sufficiently long, the frequency is generally "pulled" to the nearest P.O. cavity mode frequency. Examples of
pulsed injection locking can be found with dye lasers[7-8], CO_2-TEA
lasers[9], excimer lasers[10,11] and also Nd:YAG lasers[12].

In the steady-state regime, the amplitude of the injected field
plays the same role, in particular with regards to saturation where
the behavior of the injection-locked P.O. depends strongly in the
degree of mode coupling. But the role of the injected signal's phase
is most essential : the field from the M.O. adds to the recirculating
field within the P.O. and compensates for any net phase shift caused
by a difference between the recirculating frequency and the cavity

frequency. In other words the M.O. controls the phase and consequently the frequency of the P.O. Indeed oscillators can even be CW-locked off line-center within a given oscillating line (Conventional injection locking) or between completely different lines (hybrid injection locking). These techniques have been demonstrated with CW CO_2 lasers[13,14], dye lasers[15], argon lasers[16], in the long tail of gain switched TEA lasers[9] or with long pulse rare gas halide lasers[11].

RING LASER INJECTION LOCKING

A simple electronic model

It is of interest to derive the basic amplitude and phase injection locking equations by analysing a simple electrical lumped circuit[17] (Fig. 1) : a current generator $i_1(t)$, representing the external signal source is connected in parallel with a GLC circuit which includes a negative conductance $-G_n$ representing whatever existing negative-resistance (amplifying mechanism) that sustains the oscillations (transistor, vacuum tube, active element...). The "free running" oscillation frequency of this circuit is evidently $\omega_0^2 = 1/LC$. The basic question is : what is the time evolution of the voltage developed across the circuit ? The differential equation governing this circuit is :

$$C\frac{dv}{dt} + (G - G_n)v + \frac{1}{L}\int v(t)dt = i_1(t) \tag{1}$$

Upon assuming the injected signal from the current generator to have the form :

$$i_1(t) = I_1(t)\, e^{i[\omega_1 t + \phi_1(t)]} \tag{2}$$

one then writes the voltage as :

$$v(t) = V(t)\, e^{i[\omega_1 t + \phi(t)]} \tag{3}$$

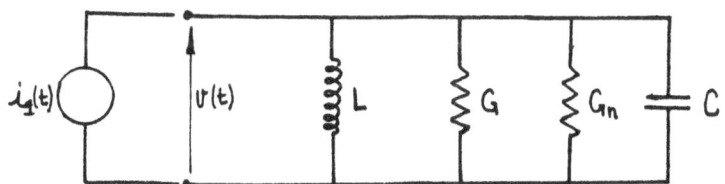

Fig. 1. Analog electronic circuit for injection locking.

In order to compute the integral in (1), one makes use of the slowly varying envelope approximation (SVEA) and drops the second order derivative and higher order terms in the integration for the phasor amplitude $\tilde{V} = V \exp(i\phi)$:

$$\int_{-\infty}^{t} \tilde{V}(t) e^{i\omega_1 t} dt = \frac{-i}{\omega_1}\tilde{V} e^{i\omega_1 t} + \frac{1}{\omega_1^2}\frac{d\tilde{V}}{dt} e^{i\omega_1 t} - \frac{1}{\omega_1^2}\int_{-\infty}^{t} \ddot{\tilde{V}} e^{i\omega_1 t} dt \tag{4}$$

319

Upon using this result, substituting into equation (1), separating into real and imaginary parts and using the resonance condition $\omega_1 \simeq \omega_0$, one deduces a set of two coupled equations for the voltage amplitude and phase :

$$\dot{V} + \frac{G-G_n}{2C} V = \frac{I_1(t)}{2C} \cos\left[\phi(t) - \phi_1(t)\right]$$
$$\dot{\phi} + \omega_1 - \omega_0 = -\frac{I_1(t)}{2C V(t)} \sin\left[\phi(t) - \phi_1(t)\right] \qquad (5)$$

Injection-locking can be completely analysed in the framework of equations (5) as done in the next section. At this point it suffices to mention that the phase equation is called Adler's equation[1].

Single-mode ring oscillator injection

Consider a single-mode ring laser cavity of perimeter p and an external field applied through a single input-output coupling mirror (Fig. 2). The main advantage of such a ring geometry is to effectively decouple the P.O. and ensure virtually no backscattering into the injecting M.O. Also, as with any other amplifier-oscillator it is then possible to optimize the mirror coupling coefficient (transmission T) in order to maximize the power extraction.

The coupled-mode equations are obtained by following the semiclassical formalism of Lamb[18-20], using Maxwell's equations whith the plane wave and SVEA approximations. The medium response is governed by the steady-state complex atomic susceptibility $\chi = \chi' + i\chi''$; it is assumed that the dispersive part simply produces a small pulling of the cavity frequency which is absorbed into the free-running oscillation frequency ω_0, whereas χ'' is responsible for gain and includes saturation ($\gamma_m = \omega \chi''$ is the laser growth rate).

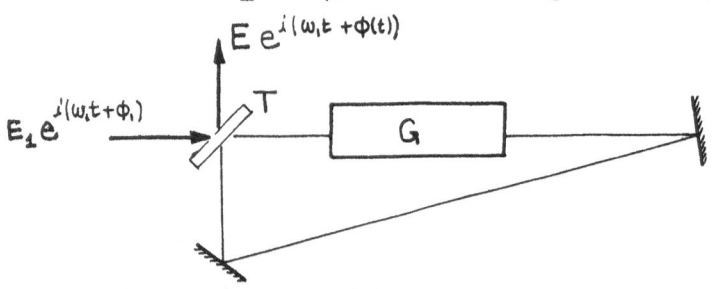

Fig. 2.   Schematic diagram of ring oscillator injection.

In these equations, the driving term due to injection is constructed by noticing that in each cavity roundtrip the injected signal at frequency ω_1, adds coherently an amount $\sqrt{T} E_1 \exp(i\phi_1(t))$ to the cavity field ; since the sum of the laser and injected fields experiences a 2π phase shift per cavity roundtrip time p/c, the laser field must be retarded in phase at a rate proportional to $-(c/p)E_1\sqrt{T}\sin\phi$ whereas the amplitude must increase at a rate $(c/p)E_1\sqrt{T}\cos(\phi-\phi_1)$ Upon measuring all fields in the same units, namely in terms of the wave amplitudes outside the ring cavity, the coupled equations are :

$$\dot{E} + \frac{1}{2}\left(\gamma_c - \omega\chi''\right)E = \gamma_e\, E_1(t)\, \cos\left[\phi(t) - \phi_1(t)\right] \tag{6}$$
$$\dot{\phi} + \omega_1 - \omega_o = -\gamma_e\, \frac{E_1(t)}{E(t)}\, \sin\left[\phi(t) - \phi_1(t)\right]$$

Here γ_c is the cavity energy decay rate (total cavity loss) whereas γ_e is the external energy decay rate (due to output coupling). It is readily seen that equations (5) and (6) have exactly the same form.

1- Steady-state solutions below threshold : amplifier

Let us assume steady-state output conditions[17] together with a constant sinusoidal input signal (E_1=constant, ϕ_1=constant) and a ring cavity operating below threshold ($\gamma_c > \gamma_m$) ; then eliminating the phase between the coupled equation yields :

$$\left(\frac{E}{E_1}\right)^2 = \frac{4\,\gamma_e^2}{(\gamma_c - \gamma_m)^2 + 4(\omega_1 - \omega_o)^2} \tag{7}$$

This is just the well-known power versus detuning relationship corresponding to the operation of the amplifier in the regenerative regime (Lorentzian lineshape) ; it presents the usual gain-bandwidth product which only depends on the external coupling rate :

$$g_o\, \Delta\omega_{1/2} = 2\,\gamma_e \tag{8}$$

Here g_o is the root gain at line center and $\Delta\omega_{1/2} = \gamma_c - \gamma_m$ is the full width at half maximum. Under these assumptions the phase equation becomes :

$$\sin\left(\phi - \phi_1\right) = \frac{-2(\omega_1 - \omega_o)}{\left[(\gamma_c - \gamma_m)^2 + 4(\omega_1 - \omega_o)^2\right]^{1/2}} \tag{9}$$

This equation clearly shows that the phase angle swings through 180° in the range $\Delta\omega_{1/2}$ about the resonance frequency. As the amplifier comes closer to threshold, the laser growth rate γ_m approaches the cavity loss rate γ_c and the bandwidth narrows : at threshold the phase suffers a complete discontinuity.

2- Steady-state solution above threshold : locked oscillator

Let us assume again steady-state output under constant injected conditions, ϕ_1 and E_1 constants, with very small injection signal ($E_1 \ll E$). Thus the oscillator is delivering an output amplitude E nearly equal to its free-running saturated output E_0.

Under these conditions, the amplitude equation does not bring determining informations on the injection locking process : it is automatically balanced since on the left side $\gamma_c \sim \gamma_m$, whereas the right side is infinitesimally small since $2\gamma_e E_1/E_0 \ll 1$.

The basic features of the locking process are obtained by

considering the phase equation which is now written :

$$\dot{\phi} + \omega_1 - \omega_0 = \Delta\omega_m \sin(\phi - \phi_1) \tag{10}$$

Here $\Delta\omega_m = \gamma_e E_1/E_0$. This equation is Adler's equation. It is clear that, for real values of the phase difference, the only possible steady-state solution is obtained when $|\sin\phi| < 1$; therefore :

$$|\omega_1 - \omega_0| \leq \Delta\omega_m \tag{11}$$

In other words, as the injected signal is tuned across the P.O. gain linewidth, there exists a range of frequencies where the ring oscillator is driven at the single frequency ω_1 or "injection-locked" by the M.O., while keeping essentially constant in amplitude. The maximum permissible excursion which the injecting frequency is allowed to make around the P.O. resonance frequency in order maintain this steady-state solution is precisely $2\Delta\omega_m$. It is known as the "locking range" : in terms of the mirror reflectivity R and the respective intensities I_1 of the injected signal, and I_0 of the free-running oscillator, the frequency locking range is :

$$\Delta\nu_{Lock} = \frac{c}{\rho}\,\frac{1-R}{\pi\sqrt{R}}\,\sqrt{\frac{I_1}{I_0}} \tag{12}$$

As the injected field frequency ω_1 is tuned across the locking range, the relative phase of the nearly constant output E shifts monotomically from $-\Delta\omega_m$ to $+\Delta\omega_m$: this quasi-linear variation (at least in the center portion of the locking range) is in fact the basis of demodulation techniques of FM signals.

3- Injection outside the locking range

Consider a free-running P.O. when a signal is suddenly injected at time t_0: as far as the amplitude is concerned, it is still reasonable to assume that the total laser amplitude E(t) is very close to its free running value E_0 since the amplitude equation should not be determinant as long as $E_1 \ll E_0$.

For constant amplitude of the injected signal, it is clear from equation (10) that whenever injection is performed outside the locking range namely when $|\omega_1 - \omega_0| > \Delta\omega_m$, there cannot exist a steady-state solution. In other words $\phi(t)$ is then a continuous function of time. Let $\Delta\omega = \omega_1 - \omega_0$, and let us define $K = \Delta\omega/\Delta\omega_m$; then the solution of equation (10) with initial conditions specifield at $t = t_0$ is[1] :

$$tg\,\frac{\phi(t) - \phi(t_0)}{2} = \frac{1}{K}\left[1 + \sqrt{K^2-1}\;tg\left[\Delta\omega_m\sqrt{K^2-1}\left(\frac{t-t_0}{2}\right)\right]\right] \tag{13}$$

Notice that the quantity $\Psi = \Delta\omega_m\sqrt{K^2-1}(t-t_0)/2$ is a linearly increasing function of time which takes values $\pi/2, 3\pi/2 \ldots$ where the argument of the tangent is infinite (successively positive and negative) ; thus $tg(\phi(t) - \phi(t_0))/2$ is also infinite for the same periodic values and consequently the relative phase $\phi(t)-\phi(t_0)$ must be a periodic function of time with a "beat frequency" ω_b such that $\Psi = \pi$. Therefore one deduces :

$$\omega_b = \sqrt{\Delta\omega^2 - \Delta\omega_m^2} \qquad (14)$$

A schematic plot of the variations of ω_b as a funtion of $\Delta\omega$ is given in Fig. 3.

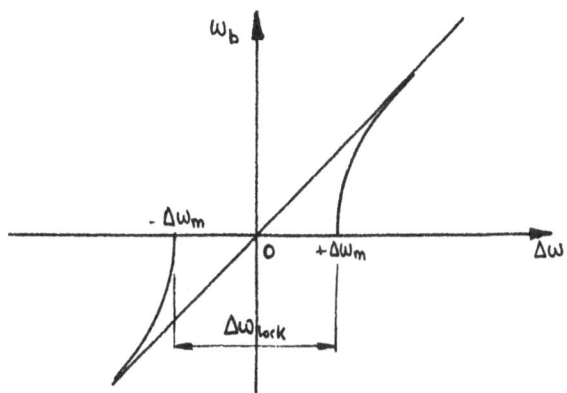

Fig. 3. Variations of the beat frequency as a function of detuning.

ω_b is the fundamental frequency of a complicate periodic variation of $\phi(t)$. In the total signal $E(t)\exp(i(\omega_i t +\phi(t))$, it then appears as a phase modulation about the carrier frequency with phase modulation sidebands $\omega_n = \omega_i + n\omega_b$. In particular the fundamental oscillation frequency $\omega_i - \omega_b$ is seen to be pushed away from the free-running frequency ω_0.

EXPERIMENTAL

A CW ring-waveguide oscillator amplifier injection-locked by a highly stable CO_2 waveguide master oscillator (M.O.) is presented for a laser radar transmitter application.

The folded legs ring-waveguide cavity includes a PZT for frequency tuning associated with a feedback stabilization loop consisting of a HgCdTe detector, a lock-in amplifier and an automatic electronic servo acting on the PZT. A tilted partially transmitting mirror is used for both input and output coupling. The ring contains a 30 cm long, internally water cooled, alumina block housing four DC excited discharges within 2 parallel capillary waveguide channels (Fig. 4) ; typical operation conditions are 8 mA per dischage at a pressure of 40 torr. As a single pass amplifier the ring is a high gain device $(G_0 = 5.4)$ and as an oscillator it has a low threshold and is thus capable of self-oscillation on a large number of CO_2 lines. Furthermore, a proper selection of the coupling mirror reflectivity R can be made in order to maximize the power extraction.

A tunable, single longitudinal mode, single line, low power CO_2 waveguide laser is injected through a careful mode matching technique imposing the same beam-waist for both lasers at the coupling mirror thereby ensuring quasi-EH_{11} fundamental mode emission of the ring. Because of homogeneous saturation together with a competition betwen driven oscillation and self-oscillation, at resonance, the drive at the injector frequency decreases the available gain, suppresses self-oscillation and completely locks the amplifier onto the injected signal ; under these conditions the injected ring waveguide operates single longitudinal mode (SLM).

Regenerative ring amplifier

In this configuration the CW waveguide ring amplifier operates below threshold with a mirror reflectivity at about 43% at 45° angle of incidence, and the PZT is tuned to resonance onto the injecting P(20) line at 944.2 cm^{-1}. Fig. 5 displays a typical P.O. oscillator response as a function of frequency : two successive single axial modes are observed with a free-spectral range of about 260 MHz. This data is completely consistent with the predictions of equation (7) : it demonstrates that the waveguide ring supports a single mode traveling wave at the injected frequency. With an injection power of 650 mW, the maximum power output obtained in this configuration is

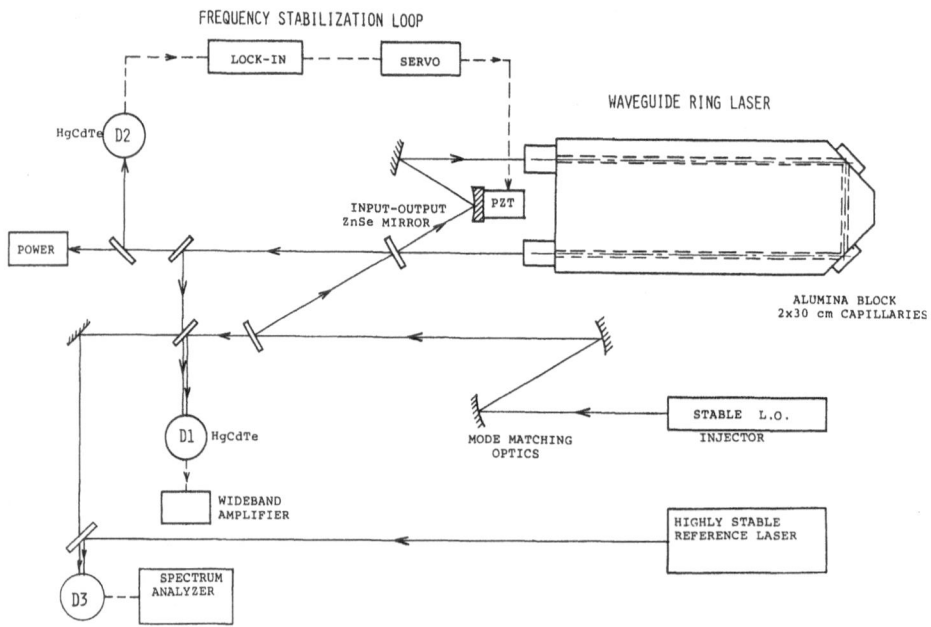

Fig. 4. Schematic diagram of injection locking experiment.

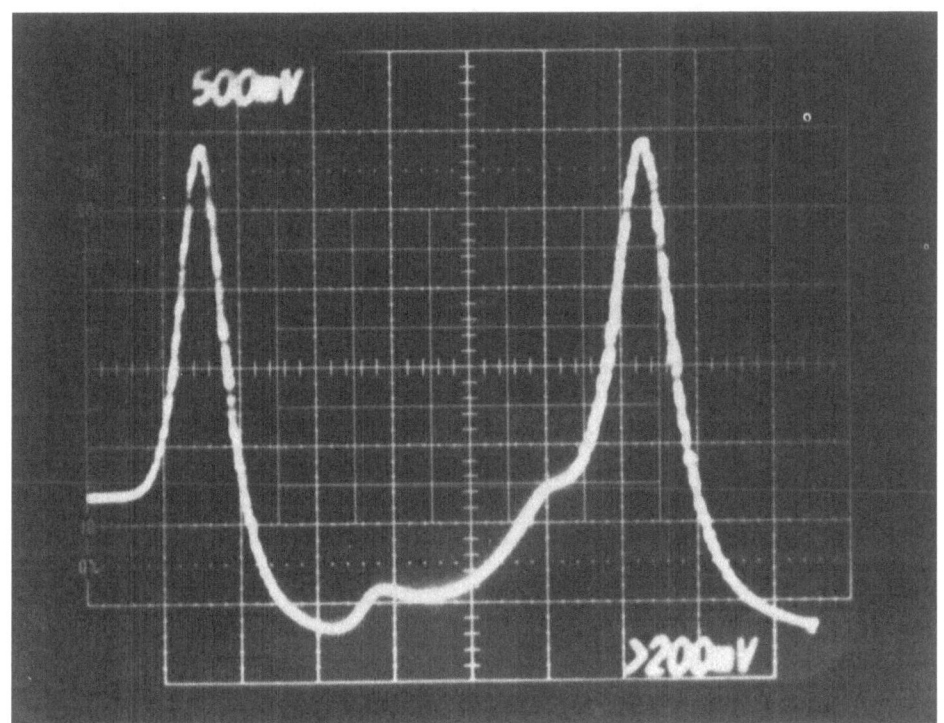

Fig. 5. Power versus detuning in the regenerative regime.

about 8 W ; the spatial mode distribution is excellent, with emission
in the fundamental EH_{11} mode and a half divergence angle of about
3 mrd. It is evident from Fig. 5 that the ring can easily be locked
in amplitude onto a peak center by using the automatic feedback loop.
Consequently the ring is brought to very frequency-stable emission,
with a jitter width of at most 1/1000 of the free spectral range.

2- Free-running ring oscillator

Since the ring waveguide is a high gain medium, above threshold
typical frequency sweeps show several CO_2 lines such as P(20),
R(20), P(26)... etc, sometimes overlapping as illustrated in Fig. 6.
However since the CO_2 laser medium is homogeneously broadened, only
one line, that with highest gain, oscillates at any one time. In order
to suppress most of the deleterious lines a 0.9 mm thick GaAs etalon
with anti-reflection coating on both faces is introduced into the ring
cavity. It may also be noted that by raising the current well above
maximum gain, one increases the losses by thermal effects and thereby
reduces the number of oscillating lines.

3 - Conventional injection locking

In this mode of injection, the M.O. and the P.O. ring both

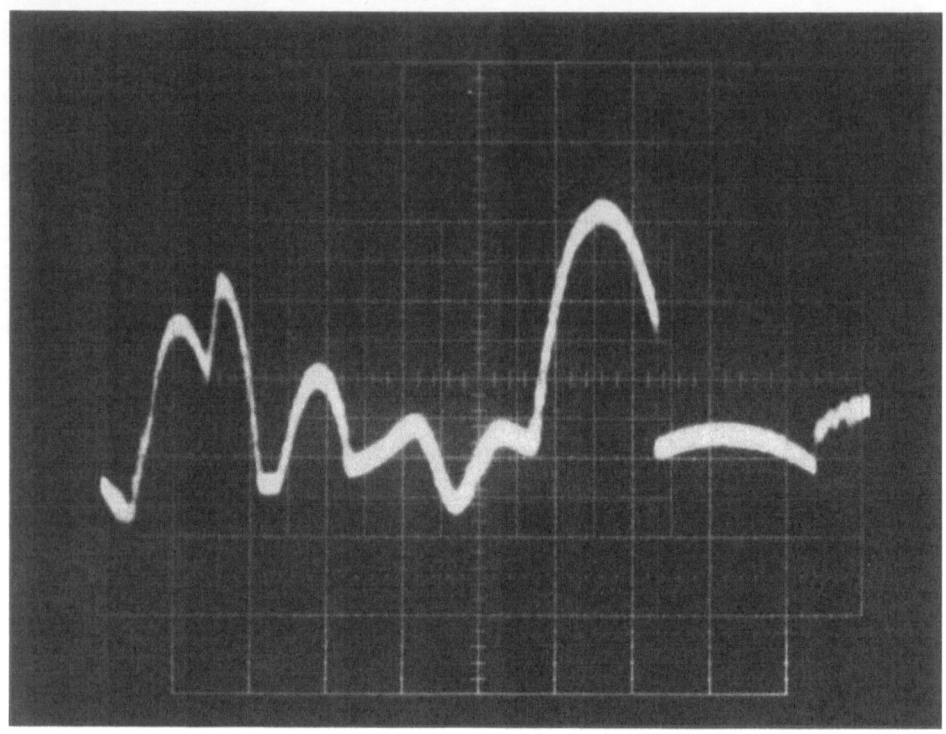

Fig. 6. Typical self-oscillating P.O. tuning curve.

operate on the same line within the P or R branches of CO_2. The
partial mirror reflectivity is now raised at about 75 %. Fig. 7
displays a typical power response when the free-running waveguide
ring is tuned across consecutive longitudinal modes within the P(20)
line. One distinctly notices the nearly parabolic gain linewith
typical of SLM operation together with a sharp spike at injection
resonance. In this particular experiment, the output power of the
ring is about 4 Watts and the spike culminates at about 8 W. The
reason is that the ring circulating power is far from being saturated and is thus enhanced in the locked region by the injected
power. The spike can be used as an amplitude discriminant in the
feedback loop in order to stabilize the power output : by raising
the discharge current, the gain can be raised for optimum extraction
and under such stabilized conditions the maximum achieved output
power is 13,5 W with 450 mW of injected signal, which is about 72 %
of the saturation power. Similar extraction efficiencies are
obtained on other CO_2 lines. Because of the proper spatial
matching, the output beam is in the fundamental EH_{11} mode with an
intensity of 250 W/cm^2 at the partially transmitting mirror.

This classical injection locking technique is useful to
amplitude stabilize and consequently frequency-stabilize the P.O. As
illustrated in Fig. 7, the injected signal significantly perturbs

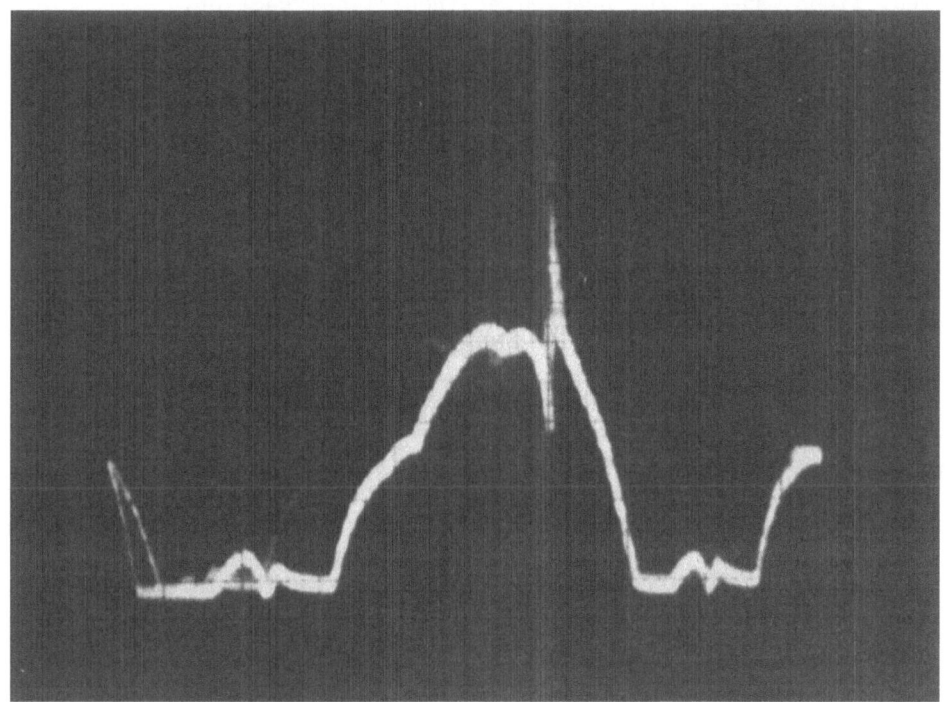

Fig. 7. Power versus tuning, conventional injection, P(20) line .

the gain of the active medium at low intracavity power, therefore it
is easy to monitor the peak location with the PZT and a high signal
to noise ratio (SNR) is available for the feedback loop. However as
the discharge current is raised, the background self-oscillation of
the P(20) line is increased to a high degree of saturation in which
the injection represents a marginal contribution : therefore the SNR
is small and the discrimination poor. This technique is consequently
sensitive to frequency fluctuations from thermal, mechanical or
acoustical origins and it is difficult to maintain the injection-
locked ring at a constant power level corresponding to the maximum
of the peak.

In order to explore the frequency characteristics of the P.O.
ring output, a heterodyne beat signal is monitored as the frequency
of the injected signal is swept within the gain linewith of the P.O.
The heterodyne signal resulting from mixing the outputs of both
oscillators is analyzed with a spectrum analyzer. As illustrated in
Fig. 8, it appears as a beat signal with two sidebands and an abrupt
absence of signal in a symmetric central region. In this locking
range the ring oscillator frequency follows that of the stable M.O.
The experimentally measured value of 2.5 MHz is within a few
percents of the theoretical formula (12). This spectral analysis
encompassing the presence of sidebands completely corroborates the

predictions made in the preceeding theoretical section.

4 - Hybrid injection locking

As previously indicated, the drawback of conventional injection locking for efficient high power frequency stabilized operation is that the SNR for amplitude discrimination decreases as the gain is raised. This problem is circumvented in hybrid injection[13] since this mode of operation consists in injecting a M.O. signal which corresponds to a line intentionally selected to be different from the P.O. self-oscillating line . If there exists a strong coupling of the molecular upper levels, such as the fast collisional relaxation within the CO_2 rotational manifold, then a homogeneous line competition ensues through which the injected signal effectively quenches the free-running P.O. and drives the high power oscillator at its own frequency. A clear demonstration of this technique is provided by the data of Fig. 9. Two consecutive single longitudinal modes of the P(20) self-oscillating line with peak power of about 13 Watts are clearly exhibited as the ring cavity is scanned ; when the latter is properly tuned with respect to the 350 mW injected signal on the P(18) line, strong homogeneous line competition occurs, and a sharp 14,5 W spike shows up at the cavity frequency resonant with the P(18) driving frequency. The power is about 78 % of the satu-

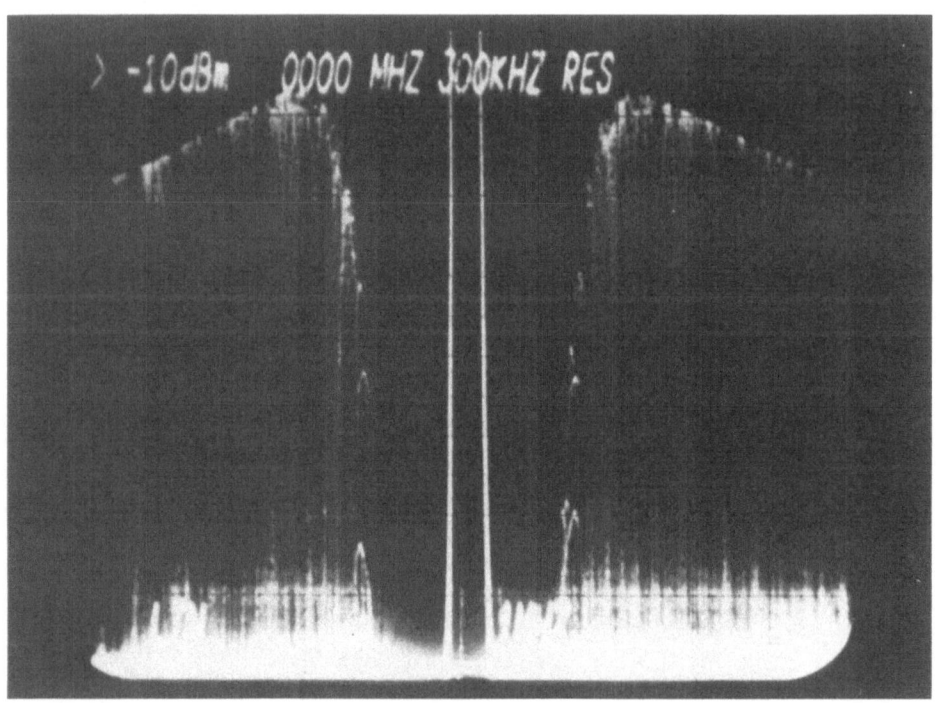

Fig. 8. Spectrum analyzer heterodyne signal with locking.

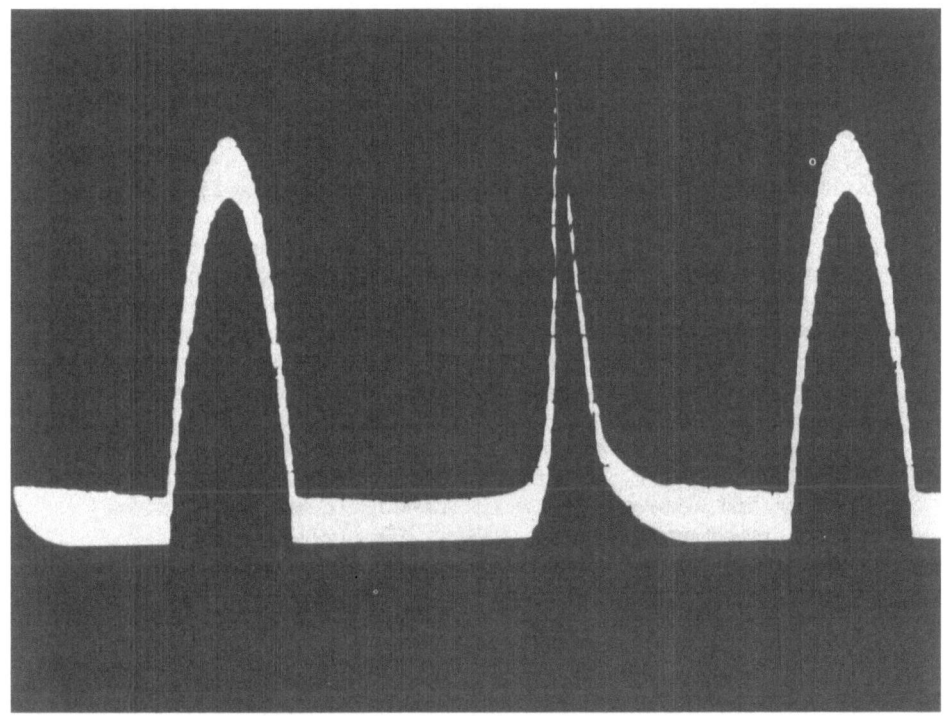

Fig. 9. Power versus tuning in hybrid injection locking,
self-oscillation on P(20) and injection on P(18).

ration power, and the EH_{11} fundamental mode output beam is of
excellent spatial quality. This technique, effectively separating
the driven and undriven components, offers a tremendous improvement
in the SNR for amplitude discrimination in the frequency stabilization feedback loop, since the contrast is now totally independant
of the P.O. gain. At very high power the hybrid injected waveguide
ring laser presents an excellent frequency stability and is much
less sensitive to cavity fluctuations : in a preliminary beat
experiment of the P.O. ring against a highly stable reference laser
(Fig. 1), frequency jitters of 200 kHz FWHM with a 100 us sweeping
time have been measured in this hybrid injection locking mode.

ACKNOWLEDGMENT

The author would like to thank A. VanLerberghe, P. Cottin and
P. Aubourg for their appreciated skills in running the experiment.

REFERENCES

I.   P. Adler, "A study of locking phenomena in oscillators," Proc.
     IRE, 34, 351 (1946) ; reprinted Proc. IEEE, 61, 1380 (1973).

2. R.D. Huntoon and A. Weiss, "Synchronization of oscillators," Proc. IRE, 35, 1415 (1947).

3. A. Blaquiere, "Les oscillateurs linéaires et le diagramme de Nyquist", J. Phys. Radium, 13, 527 (1952).

4. L.J. Paciorek, "Injection locking of oscillators", Proc. IEEE, 53, 1723 (1965).

5. K. Kurokawa, "Injection locking of microwave solid-state oscillators," Proc. IEEE, 61, 1386 (1973)

6. U.L. Stover and W.H. Steier, "Locking of laser oscillations by light injection," Appl. Phys. Lett., 8, 91 (1966).

7. J.J. Turner, E.I. Moses, and C.L. Tang, "Spectral narrowing and electrooptical tuning of a pulsed dye laser by injection locking to a CW dye laser," Appl. Phys. Lett., 27, 441 (1975).

8. J. Pinard and S. Liberman, "A frequency locked, single mode pulsed dye laser ; application to single frequency tunable UV generation," Opt. Commun., 20, 344 (1977).

9. J.L. Lachambre, P. Lavigne, G. Otis and M. Noel, "Injection locking and mode selection in TEA-CO_2 laser oscillators," IEEE J. Quantum Electron. QE-12, 756 (1976).

10. J. Goldhar et al, "Injection locking of a xenon fluoride laser," Appl. Phys. Lett. 31, 677 (1977).

11. I. Bigio and M. Slatkine, "Injection locking unstable resonator excimer lasers," IEEE J. Quantum Electron., QE-19, 1426 (1983)

12. Y.K. Park, G. Giulani and R.L. Byer, "Stable single-axial mode operation of an unstable-resonator Nd:YAG oscillator by injection locking," Opt. Lett., 5, 96 (1980)

13. C.J. Buczek and R.J. Freiberg, "Hybrid injection locking of higher power CO_2," IEEE J. Quantum Electron, QE-8, 641 (1972).

14. P. Cottin, A. Vanlerberghe et J.L. Boulnois, "High power injection-locked CO_2 ring waveguide laser", Conf. on Lasers and Electro-Optics, Annaheim, Paper ThL5, June 1984.

15. B. Couillaud, A. Ducasse, and E. Freys, "Injection locking of CW ring dye lasers, IEEE J. Quant. Electron, QE-20, 310 (1984).

16. C.N. Man and A. Brillet "Injection locking of argon-ion lasers," Opt. Lett. 9, 8, 333 (1984).

17. From A.E. Siegman, Lecture notes, E. Ginzton Laboratory, Standford University.

18. M. Sargent III, O. Scully, and W.E. Lamb, Jr., in "Laser Physics", Addison-Wesley, Reading, MA, 1974.

19. M.B. Spencer, and W.E. Lamb, Jr., "Laser with a transmitting window," Phys. Rev. A5, 884 (1972).

20. W.W. Chow, "Theory of line namwing and frequency selection in an injection locked laser," IEEE J. Quantum Electron. QE-19, 2, 243 (1983).

QUANTUM NOISE EFFECTS IN SINGLE-MODE INJECTION LASERS

Salvatore Piazzolla and Paolo Spano

Fondazione Ugo Bordoni
Viale di Trastevere 108
00153 Roma (Italy)

Recently,a great,increasingly growing interest has arisen in the field of optical coherent communication systems because of two main factors: first,the great development of the technology of semiconductor materials which has made commercially available injection lasers capable of stable operation in a single longitudinal and transverse mode; second,the well acquired capability of producing single-mode fibers with very low attenuation losses.

Coherent communication systems via optical fibers,both of the homodyne or heterodyne type,present,in fact,several advantages with respect to direct detection systems such as,for example,the ease of obtaining wavelength multiplexing with closely spaced channels,the possibility of employing optical amplifiers,a greater receiver sensitivity with a gain in the range of 10-20 dB depending on the adopted modulation system. For the above mentioned reasons,one immediately understands why a detailed knowledge of the noise properties of single-mode semiconductor lasers is absolutely needed,since these devices represent the most important part of the systems due to the possibility of employing them as transmitters,modulators, local oscillators and/or optical amplifiers. In fact,a detailed knowledge of the power spectral density of noise and,in particular,of the noise associated with the fluctuations of the instantaneous frequency of the emitted radiation,makes it possible to determine the correct shape and width of the emission line of the source,which is the most important characteristic of the device both from physical and applicative points of view.

To take into account the amplitude and frequency fluctuations of the laser field,the electric component of the radiation from a single-mode laser is written

$$\mathcal{E}(t) = E(t)\exp(i\,\omega(t)t) = (E_o + e(t))\exp(i(\omega_o t + \phi(t))) \quad (1).$$

Here $e(t) \ll E_o$, the deviation from the mean amplitude E_o, and $\phi(t)$, the random phase of the field such that $\dot{\phi}(t) = \omega(t) - \omega_o$ represents the deviation of the instantaneous frequency $\omega(t)$ from its mean value ω_o, are real quantities which give rise to amplitude and phase noise respectively.

In semiconductor lasers, which are characterized by a low Q cavity value, noise has to be actually associated with quantum processes, that is the spontaneous emissions of photons into the lasing modes, while in other kinds of lasers (solid state, gas, dye lasers) the most important causes of noise must be seeked for in the mechanical and thermal fluctuations of the cavity length. This can be immediately recognized by the following example. According to the Shalow-Townes' theory, the width, at half height, of the emission line of a single mode laser, evaluated taking into account only spontaneous emission processes, is given by

$$\Delta \nu_{sp} = \frac{h \omega_o \Delta \nu_c^2}{P} \qquad (2),$$

where $\Delta \nu_c$ represents the linewidth of the passive cavity and P is the emitted power. For an He-Ne laser with P=1mW, $\omega_o \cong 3 \cdot 10^{15} \text{rad} \cdot \text{s}^{-1}$, $\Delta \nu_c \cong 10^5 \text{ s}^{-1}$, we have $\Delta \nu_{sp} \cong 10^{-3} \text{ s}^{-1}$; in the case of an AlGaAs laser with P=1mW, $\omega_o \cong 2 \cdot 10^{15} \text{rad s}^{-1}$, $\Delta \nu_c \cong 10^{11} \text{ s}^{-1}$, so that $\Delta \nu_{sp} \cong 10^7 \text{s}^{-1}$ (the great difference of $\Delta \nu_c$ for the two lasers is due to the substantial differences in their mirror reflectivities and cavity lengths). Now, because the processes other than spontaneous emissions cause a further broadening of the line which can be easily kept as low as 10^4 s^{-1}, one immediately understands why in laser diodes only quantum noise must be pratically accounted for.

Intensity noise in semiconductor lasers has been fully analyzed both from an experimental and a theoretical point of view since many years[1,2] ; for what concerns phase and frequency noise, however, only in these last three years some of their peculiar characteristics have been observed and explained. The starting point was given by an experiment performed by Fleming and Mooradian[3] ; measuring the emission line of an AlGaAs injection laser they found, according to the Shalow-Townes' theory, it to be Lorentzian in shape, with a full width at half intensity which varied inversely with output power P, but about 50 times greater, at 300 K, than expected. The explanation of this unexpected fact is due to Henry[4]. He underlines as every spontaneous emission process gives rise to a total change in the phase of the field which is the sum of two different contributions: the first is the one considered usually, associated with the random phase of the spontaneously emitted photon, while the second has to be attributed to the change in the field intensity, subsequent to any emission process, which determines a variation of the refractive index of the cavity. Considering the two above contributions, Henry was able to state that the emission line had to be Lorentzian in form but that its width had to be enhanced by a factor $(1 + \alpha^2)$, when compared to that predicted by conventional theories; α, the so-called "linewidth enhancement factor", is the ratio between the derivatives of the real and imaginary part of the refractive-index in the active region with respect to the carrier density,and,while being negligible in other kinds of lasers,is greater than unity in injection lasers as to substantially determine their emission line-width (values in the range 2-6 have been found,or estimated,depending on the composition and the physical structure of the laser itself[4-6]). This last fact can be understood considering that semiconductor lasers act as detuned cavity oscillators [4].

The above theory,however,even if accounts for the excessive anomalous broadening,is valid only when an equilibrium or,to be more precise,a quasi-equilibrium condition exists. In fact,it takes into account only changes of the instantaneous phase of the electromagnetic field over time intervals,that in injection lasers are of the order of 1 ns,during which the relaxation oscillations undergone by the system after any spontaneous emission process have completely died out. That is equivalent to consider in the power spectrum of $\dot{\phi}(t)$,the instantaneous frequency deviation of the field,only that region of frequencies $\Omega < \tau_r$, τ_r being the relaxation oscillation damping time. In order to have a more complete information about the shape of the emission line,one must also take into account the effect of the relaxation oscillations consequent to any spontaneous emission event. In fact,one expects some form of correlation in the laser noise over time intervals shorter than τ_r ,and hence a substantial difference in the shape of the power spectrum of the instantaneous frequency deviation at $\Omega > \tau_r$ with respect to the completely flat behaviour one would have if the emission lineshape were fully Lorentzian.

Phase noise measurements in the high frequency region have been performed using the experimental set-up shown in fig.1[7,8] .

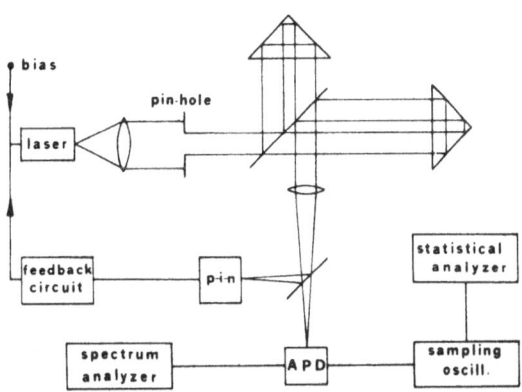

Fig.1-The experimental set-up.

It consists essentially of a Michelson interferometer whose mirrors.

are substituted with two right angle prisms in order to avoid
undesired reflections which could induce instabilities in the source,
a single-mode semiconductor laser stabilized in temperature within
0.1 °C. The output from the interferometer is spatially filtered and
detected by a fast APD whose signal is sent to a spectrum analyzer
or to a sampling oscilloscope followed by a statistical analyzer.
In so doing, one can obtain both the power spectral density and the
probability density function of the signal from the APD. A slow feedback circuit, driven by a fraction of the light output from the interferometer detected by a p-i-n photodiode, is used to keep the mean
wavelength of the laser radiation locked to the interferometer.
It is easy to see that if the interferometer is adjusted in quadra-
ture, that is $\omega_o\tau = \pi/2$, where τ is the time delay between the beams
travelling in its two arms, and τ is small enough that $|\Delta_\tau\phi(t)| = |\phi(t+\tau) - \phi(t)| \ll \pi/2$, the power spectra $S_{\Delta\tau\phi}(\Omega)$ and $S_{\dot\phi}(\Omega)$, associa-
ted to the phase shift $\Delta_\tau\phi(t)$ and to the instantaneous frequency
deviation $\dot\phi(t)$ respectively, are related to the power spectrum
$S_i(\Omega)$ of the signal from the APD by the relation

$$S_{\Delta\tau\phi}(\Omega) = \frac{S_{\dot\phi}(\Omega)}{\Omega^2} (1-\cos\Omega\tau) = S_i(\Omega) / (i_1^2 R/4) \qquad (3).$$

Here R is the input impedance of the spectrum analyzer and i_1 is the
maximum excursion of the fringe pattern revealed from the APD.

   Figure 2 shows $S_{\dot\phi}(\Omega)$ for different lasers measured at different
operating conditions. At low frequencies these spectra exhibit a
flat behaviour just as predicted by conventional theories; in the
high frequency region, however, they show the presence of a sharp resonant peak at nearly the same frequency of the already known peak
in the power spectrum of intensity noise[1,2] (for completeness, fig.3
shows the measured power spectra of intensity noise of two lasers
of fig.2). This fact suggests that the above peak in $S_{\dot\phi}(\Omega)$ must be
related to the change of the refractive-index of the active medium
consequent to the spontaneous emission processes, and its most important consequence is that the emission lineshape is not Lorentzian.
In fact, being $\dot\phi(t)$ a stationary ergodic normally distributed
process[8,9], the lineshape can be shown to be

$$S_\mathcal{E}(\omega) \propto \int_{-\infty}^{+\infty} \exp(i(\omega - \omega_o)\tau) \exp(-\sigma^2(\tau)/2) \, d\tau \qquad (4),$$

where

$$\sigma^2(\tau) = \langle (\Delta_\tau\phi(t))^2 \rangle = (2/\pi) \int_0^\infty S_{\dot\phi}(\Omega)(1-\cos\Omega\tau)/\Omega^2 \, d\Omega \qquad (5)$$

is the variance of $\Delta_\tau\phi(t)$ which, in turn, is a zero mean, ergodic and
normally distributed process, too[8,9]. The emission lineshape would
be Lorentzian only in the case in which $\sigma^2(\tau) \propto |\tau|$. Figure 4 shows
$\sigma^2(\tau)$ evaluated through eq.(5) for two different lasers; as it re-
sults immediately, $\sigma^2(\tau)$ is proportional to $|\tau|$ only at values of $|\tau|$
high enough that the relaxation oscillations after any spontaneous

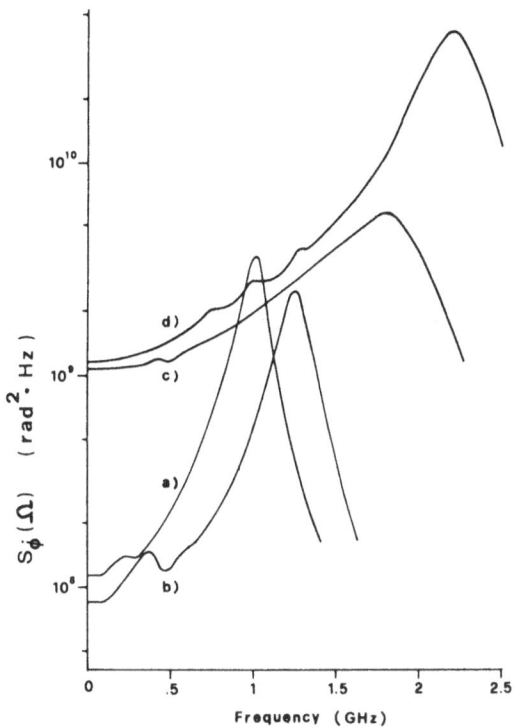

Fig.2- Two-sided spectral density $S_{\dot{\phi}}(\Omega)$ vs.frequency. Curve a and b : ITTLS7709 laser at P=4mW and 7mW respectively.
Curve c: RCA C86014 at P=1.6mW.
Curve d : Laser Diode LCW10 at P=3mW.

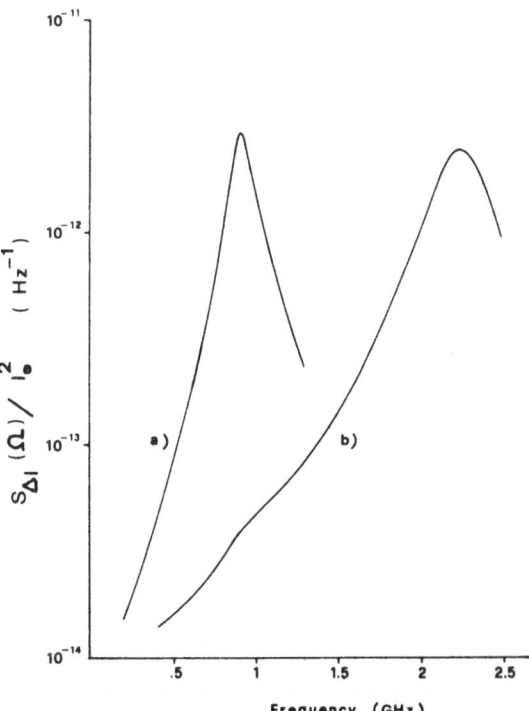

Fig.3- Two-sided normalized intensity-noise spectra $S_I(\Omega)$ vs. frequency.
Curve **b** : Laser Diode LCW10 at P=3mW.
Curve a : ITTLS7709 at P=4mW.

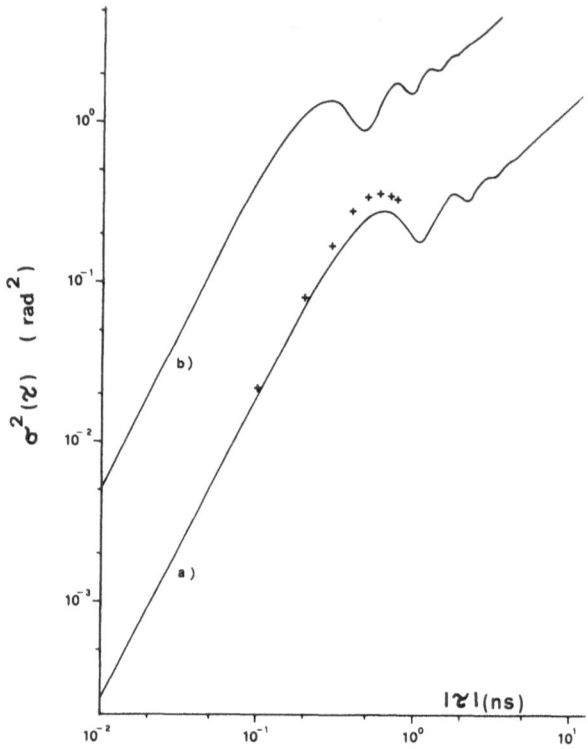

Fig.4- Variance $\sigma^2(\tau)$ versus τ for the ITTLS7709 laser
(curve a) and Laser Diode LCW10 laser operating
in the same conditions of fig.3.

emission has died out. At low values of $|\tau|$, the curves show at first
a quadratic dependence on $|\tau|$ followed by an oscillatory one; this
is due to the mentioned relaxation phenomena of the cavity refractive-index which alter the spectrum of the noise and introduce a non
zero correlation time in the phase fluctuations. Crosses in fig.4
represent values of $\sigma^2(\tau)$ derived directly from measurements of the
probability density function of $\Delta_\tau \phi(t)$ obtained through the statistical analyzer. For all the employed lasers the distribution of $\Delta_\tau \phi(t)$
turned out to be Gaussian.

Figure 5a shows the normalized lineshape of one of the employed
lasers, evaluated through eq.(4) by using the measured $S_{\dot\phi}(\Omega)$, while
fig.5b shows the lineshape of the same laser obtained directly by
means of a Fabry-Perot scanning interferometer. The lineshape deviates from a Lorentzian form because of the appearance of the two
satellite peaks, due to the presence of the oscillating part in $\sigma^2(\tau)$,
not predicted by conventional theories.

Recently, papers from different authors, which face the problem
of phase and frequency noise from a theoretical point of view, have
been published, giving results which fully explain all the experimental results reported above [11-14]. These theoretical approaches hinge

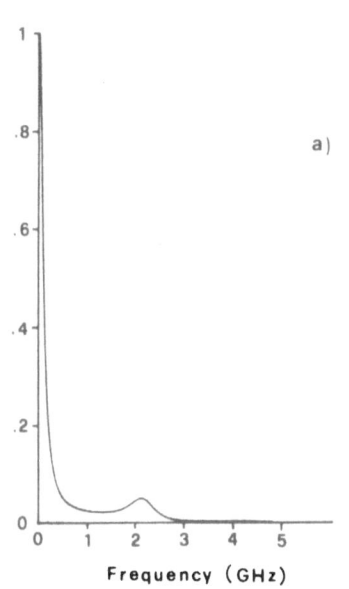

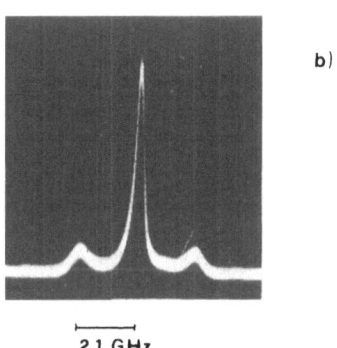

Fig.5- Emission lineshape of the Laser Diode LCW10 at P=3mW : a) computed through the measured values of $S_\phi(\Omega)$; b) measured with a Fabry-Perot interferometer.

Frequency (GHz)

2.1 GHz

upon the use of a set of equations which describe the time evolution of the electric field of the cavity mode[15] and of the injected carrier density[16], together with eq.(1). Every spontaneous emission process gives rise to small deviations of the amplitude and phase of the field and of the carrier density around their stationary values, so that a first order approximation is sufficient to describe accurately the time evolution of the physical system after any perturbation has taken place. The corresponding equations, written in terms of the new variables $N(t)=n(t)\Gamma V$ and $I(t)=|\mathcal{E}(t)|^2 V$, which represent respectively the number of carriers and the number of photons belonging to the mode in the cavity (V is the mode volume, Γ the filling factor and n(t) the density of the injected carriers), read

$$\begin{cases} \Delta\dot{I}(t) = (G_n I_0 / \ \Gamma V) \ \Delta N(t) & (6a) \\ \dot{\phi}(t) = (G_n \ \alpha / \ 2\Gamma V \ ) \Delta N(t) & (6b) \\ \Delta\dot{N}(t) = -\gamma_e \Delta N(t) - \Gamma G_0 \Delta I(t) & (6c), \end{cases}$$

where G_0 and I_0 are the stationary values of gain and number of photons, G_n the derivative of gain with respect to the carrier density, $\gamma_e = \gamma + (G_n I_0/V)$, γ being the inverse of the spontaneous lifetime of the excited carriers and $\alpha = -(2\omega_0 \eta_n v_g / G_n c)$ is the linewidth enhancement factor (η_n being the derivative of the cavity refractive-index η with respect to n and v_g the group velocity of the light).

The set of equations (6) shows how $\Delta I(t)$, $\Delta N(t)$ and $\dot{\phi}(t)$, the deviations of the number of photons and excited carriers, and of

the frequency from their stationary values,respectively,are coupled
among themselves. In order to describe noise in the system under study,we must add to eqs.(6) Langevin terms,which take into account the
randomness of the spontaneous emission events. For eqs.(6a) and (6b)
these terms can be modeled as

$$
\begin{cases}
F_{\Delta I}(t) = \sum_i (2I_o^{1/2}\cos\theta_i + 1)\delta(t - t_i) & (7a) \\
F_\phi(t) = \sum_i I_o^{-1/2} \sin\theta_i \, \delta(t - t_i) & (7b),
\end{cases}
$$

where θ_i is the phase of the i-th photon,emitted at a time t_i,relati-
ve to the phase of the field just before t_i,and $\delta(t)$,the delta function,accounts for the discontinuous changes in intensity and phase
during spontaneous emission events.
The Langevin term for eq.(6c) is

$$
F_\Delta(t) = -\sum_i \delta(t - t_i) \tag{7c},
$$

because any emission causes a unitary decrease in the number of carriers. Because the complete set of equations,that is eqs.(6) to which
the corresponding Langevin terms have been added,is linear,their solutions,which can be found by means of standard mathematical techniques,
will be expressed as the sum of the responces of the unperturbed system to any i-th term in eqs.(7). Furthermore,since the phase θ_i is
independent from the time t_i of the emission and uniformly distributed,the autocorrelation functions,and hence the power spectral densi-
ties of $\Delta I(t)$, $\Delta N(t)$,and $\dot\phi(t)$ can be simply derived [13]. The power
spectra of intensity and frequency noise read

$$
S_{\Delta I}(\Omega) = R_s \frac{((G_n I_o/\Gamma V)(G_n I_o/\Gamma V - 2\gamma_e)+(2I_o+1)(\gamma_e^2 + \Omega^2))}{((\Omega_R^2 - \Omega^2)^2 + \gamma_e^2 \, \Omega^2)} \tag{8}
$$

$$
S_\phi(\Omega) = \frac{R_s}{2I_o} \left(\frac{\alpha^4 \Omega_R^4 (2I_o + 1 + \Omega^2/(\Gamma G_o)^2)}{2I_o((\Omega^2 - \Omega_R^2)^2 + \gamma_e^2 \, \Omega^2)} + 1 \right) \tag{9},
$$

where R_s is the average rate of spontaneous emissions in the mode
and $\Omega_R = (G_n G_o I_o/V)^{1/2}$.
Equation (8),which,as stressed before,was previously derived [1,2],
and (9) represent the complete solution of noise problem in single-
mode semiconductor lasers.The unitary term into the brackets in
eq.(9) is the term usually considered and is due to the randomness
of the relative phase of any spontaneously emitted photon. The other
term,which is proportional to α^2,and therefore strictly bound to the
refractive-index changes induced by carrier density fluctuations,is
responsible both for the appearance of the satellite peaks and for

the excess broadening of the emission line. At low frequencies,in fact,that is when $\Omega \ll \Omega_R$,the term into brackets is well approximated by $(1+\alpha^2)$,in agreement with Henry's results[4]. It must be stressed that an increase in the emitted power,and then in I_o,determines both a reduction of low frequency noise content proportional to I_o^{-1} and an increase in the frequency of the resonant peak proportional to $I_o^{1/2}$. Figures 6a and 6b show $S_\phi^*(\Omega)$ and $S_{\Delta I}(\Omega)/I_o^2$ for an ITTLS7709 gain guided laser operating at λ=865.5nm with P=4mW ; the continuous curves are computed through eqs.(8) and (9) choosing appropriate values for the physical parameters[13],and the dashed ones are obtained experimentally.

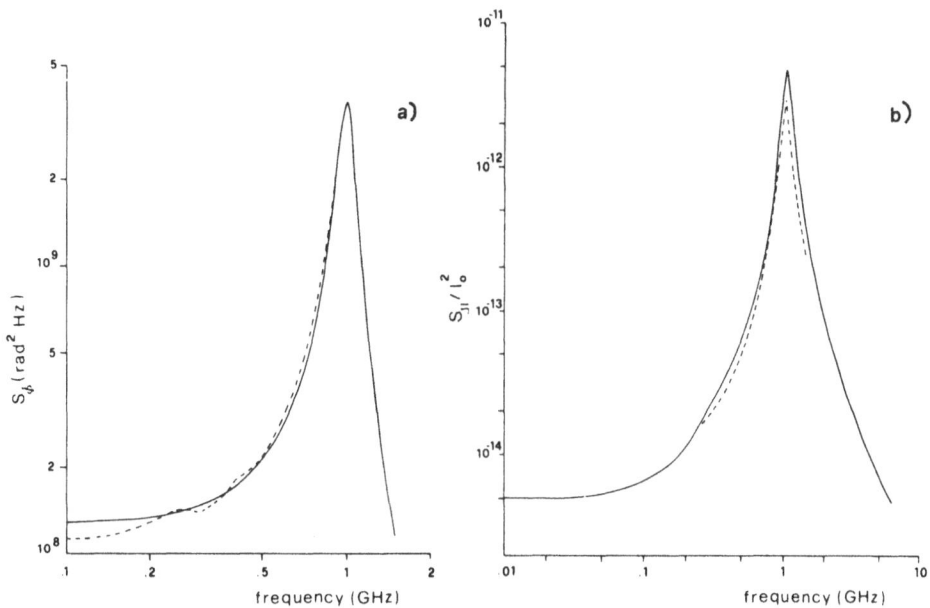

Fig.6- Two sided normalized power spectral densities of
frequency (a) and intensity noise (b) for an
ITTLS7709 laser emitting, at λ=865.5nm, a power
P=4mW.The continuous curves are theoretical,
while the dashed ones are experimental.

Substituting the right side of eq.(9) into eq.(5),one can directly evaluate $\sigma^2(\tau)$[17]. The complete resulting expression,which is rather long and involved,can be substantially simplified as follows

$$\sigma^2(\tau) = R_s/(2I_o)((1+\alpha^2)|\tau| + \alpha^2(1/\gamma_e - \exp(-\gamma_e|\tau|/2)$$

$$(\cos\Omega_o \tau/\gamma_e + 3\sin\Omega_o|\tau|/(2\Omega_o)))) \qquad (10),$$

339

where $\Omega_0 = (4\Omega_R^2 - \gamma_e^2)^{1/2}/2$ is the relaxation oscillation frequency. Equation (10) can be simply shown to account for the behaviour of $\sigma^2(\tau)$ as a function of $|\tau|$, as previously described. Equivalent expressions for $\sigma^2(\tau)$ have been found by other authors[12,14]. Substituting eq.(10) into eq.(4) allows to evaluate the normalized lineshape function which reads

$$G(\omega) = (1 - \frac{R_s \alpha^2}{2I_0\gamma_e}) \frac{2\,\Delta\Omega_1}{(\omega - \omega_0)^2 + \Delta\Omega_1^2} +$$

$$\frac{R_s}{4I_0} \left(\frac{2/\gamma_e(\,\Delta\Omega_1 + \gamma_e/2) + 3/\Omega_0(\omega - \omega_0 + \Omega_0)}{(\omega - \omega_0 + \Omega_0)^2 + (\,\Delta\Omega_1 + \gamma_e/2)^2} + \right.$$

$$\left. \frac{2/\gamma_e(\,\Delta\Omega_1 + \gamma_e/2) + 3/\Omega_0(\omega - \omega_0 - \Omega_0)}{(\omega - \omega_0 - \Omega_0)^2 + (\,\Delta\Omega_1 + \gamma_e/2)^2} \right) \tag{11},$$

where $\Delta\Omega_1 = R_s(1+\alpha^2)/(2I_0)$.

By inspecting eq.(11), it is possible to recognize the characteristic Lorentzian behaviour of the central lobe (which is essentially due to the flat low-frequency content of $S_\phi(\Omega)$), and the presence of the sideband peaks shifted in frequency by an amount Ω_0, in agreement with the experimental observations.

As a conclusion, it is worth while to stress that the presented results, besides showing and explaining some interesting physical phenomena such as the excessive linewidth and the presence of satellite peaks in the emission line of single-mode injection lasers, have an immediate pratical application in the design of optical coherent communication systems due to the strong dependence of their performances (for example, in such systems the error probability is an increasing function of $\sigma^2(\tau)$[18]) on the characteristics of the above mentioned devices.

References

1. D.E.Cumber,"Intensity fluctuations in the utput of CW laser oscillators I",Phys.Rev. 141:306 (1966).
2. H.Jackel and H.Melchior,"Fundamental limits of the light intensity fluctuations of semiconductor lasers with dielectric transverse mode confinement",5th ECOC,Amsterdam (1979).

3. M.W.Fleming and A.Mooradian,"Fundamental line broadening of single mode GaAlAs diode lasers",App.Phys.Lett. 38:511 (1981).

4. C.H.Henry,"Theory of the linewidth of semiconductor lasers", IEEE J.Quantum Electron. QE-18:259 (1982).

5. C.Harder,K.Vahala and A.Yariv,"Measurement of the linewidth enhancement factor α of semiconductor lasers",App.Phys.Lett. 42:328 (1983).

6. K.Kikuchi,T.Okoshi and T.Kawai,"Estimation of linewidth enhancement factor of CPS-type AlGaAs lasers from measured correlation between AM and FM noises",Electron.Lett. 20:450 (1984).

7. S.Piazzolla,P.Spano and M.Tamburrini,"Characterization of phase noise in semiconductor lasers",App.Phys.Lett. 41:695 (1982).

8. B.Daino,P.Spano,M.Tamburrini and S.Piazzolla,"Phase noise and spectral line shape in semiconductor lasers",IEEE J.Quantum Electron. QE-19:266 (1983).

9. A.Papoulis,"Probability,Random Variables,and Stochastic Processes",Mc Graw-Hill,New York (1965).

10. H.E.Rowe,"Signal and Noise in Communication Systems",Van Nostrand Reinhold,New York (1965).

11. R.Schimpe and W.Harth,"Theory of FM noise in single-mode injection lasers",Electr.Lett. 19:136 (1983).

12. K.Vahala and A.Yariv,"Semiclassical theory of noise in semiconductor lasers-Part I",IEEE J.Quantum Electron. QE-19:1096 (1983);"Semiclassical theory of noise in semiconductor lasers-Part II",ibid.:1102 (1983).

13. P.Spano,S.Piazzolla and M.Tamburrini,"Phase noise in semiconductor lasers:a theoretical approach",IEEE J.Quantum Electron. QE-19:1195 (1983).

14. C.H.Henry,"Theory of the phase noise and power spectrum of a single mode injection laser",IEEE J.Quantum Electron. QE-19: 1391 (1983).

15. W.E.Lamb,Jr.,"Theory of optical masers",Phys.Rev. 134:A1424 (1964).

16. G.H.B.Thompson,"Physics of Semiconductor Laser Devices", J.Wiley & Sons,London (1980).

17. S.Piazzolla and P.Spano,"Analytical evaluation of the line shape of single-mode semiconductor lasers",Opt.Commun.,to be published.

18. Y.Yamamoto,"Receiver performance evaluation of various digital optical modulation-demodulation systems in the 0.5-10 μm wavelength region",IEEE J.Quantum Electron. QE-16:1251 (1980).

INJECTION-LOCKED SINGLE MODE TEA CO$_2$ LASERS:

SOME RECENT MEASUREMENTS AND INNOVATIONS

D.M. Tratt, A.K. Kar, and R.G. Harrison

Physics Department
Heriot-Watt University
Riccarton, Edinburgh, U.K.

INTRODUCTION

The purpose of this article is to present an updated state-of-the-art report on injection-locking as applied to TEA CO$_2$ laser systems. As such, this brief overview reflects the continuing sustained interest in the operational behaviour of such systems and their applicability to a wide spectrum of tasks in a variety of research areas.

Since its first demonstration in 1966[1], laser injection-locking has been particularly favoured over other methods[2,3] for the purpose of extracting single-frequency output from intrinsic-ally broadband systems, such as solid state[4,5], semiconductor[6,7], excimer[8,9], dye[10,11] and even free-electron[12] lasers.

In the specific realm of TEA CO$_2$ laser applications, injection locked systems have been used as MIR[13] and FIR[14] OPML pump sources, differential absorption lidar (DIAL) transmitters[15], for uranium isotope separation by two-frequency multiphoton absorption[16], non-linear optical studies in gases[17] and semiconductors[18], etc. One such system has also been incorporated into a 10 kJ inertial confinement fusion (ICF) driver[19].

It is not the purpose of this tutorial to give a detailed theoretical picture of the injection-locked TEA CO$_2$ laser, but rather to concentrate on some recently obtained measurements which throw some light on the mode selection process as it is more commonly formulated[20,21], but with more emphasis placed on the action of intrapulse frequency sweeping, or chirp, and the polarization of the injected signal. Consideration will also be given to practical aspects of injection technique.

While keeping this rationale in mind, it is nevertheless instructive to study first a simplified general model of a laser under the influence of an externally-derived injection signal.

THE LASER WITH INJECTED DRIVE SIGNAL: A QUALITATIVE TREATMENT

In this analysis we consider a generalised conception of the laser oscillator injected with a cw signal described in the plane-wave approximation by:

$$E_i(t) = A_i e^{j\omega_i t} \tag{1}$$

in which A_i and ω_i assume their usual meanings.

The evolution of the cavity mode $E(t)$ with frequency ω lying closest to that of the injected signal may then be described by the driven damped oscillator equation:

$$\ddot{E}(t) - \gamma \dot{E}(t) + \omega^2 E(t) = E_i(t) \tag{2}$$

where $E(t) = A(t) \exp(j\omega t)$ and γ represents the system gain, written here as a negative damping term.   When evaluated, equation (2) becomes:

$$[\ddot{A}(t) + (2j\omega-\gamma) \dot{A}(t) - j\omega\gamma A(t)]e^{j\omega t} = A_i e^{j\omega_i t} \tag{3}$$

If we now assume $A(t)$, the cavity field amplitude, to be slowly varying and recognise that for a typical optical system the gain term is negligible in relation to the field frequency, then we may assume $\ddot{A}(t)$ to be negligable and that $\omega \gg \gamma$.   Thus, equation (3) reduces to:

$$[2j\omega\dot{A}(t) - j\omega\gamma A(t)] e^{j\omega t} = A_i e^{j\omega_i t} \tag{4}$$

which we rearrange as:

$$\dot{A}(t) - \frac{\gamma}{2} A(t) = \frac{A_i}{2j\omega} e^{j(\omega_i-\omega)t} \tag{5}$$

The general solution of this differential equation can be shown to be:

$$A(t) = \left[\frac{A_i/2j\omega}{j(\omega_i-\omega)-(\gamma/2)}\right] e^{j(\omega_i-\omega)t} + \left[\frac{A_i/2j\omega}{(\omega_i-\omega)^2+(\gamma^2/4)}\right] \gamma e^{\gamma t/2}$$

$$\cdots\cdots\cdots \tag{6}$$

Thus:

$$E(t) = \left[\frac{A_i/2j\omega}{j(\omega_i-\omega)-(\gamma/2)} \right] e^{j\omega_i t} + \left[\frac{A_i/2j\omega}{(\omega_i-\omega)^2+(\gamma^2/4)} \right] \gamma \, e^{\gamma t/2} e^{j\omega t}$$

$$\ldots\ldots \quad (7)$$

This is a usefully revealing result. It is clear from equation (7) that in the circumstance where the laser is operating at oscillation threshold (i.e. $\gamma = 0$), the latter term on the RHS vanishes and the driven cavity field appears as the regeneratively amplified injection frequency.

This is true injection-locking, as observed in cw systems[22]. However, when the gain is "switched" to a level well above threshold, we notice that the natural oscillation frequency of the laser will very rapidly dominate over the drive frequency so that the laser operates on the driven cavity mode, rather than the injected frequency. This is precisely the behaviour observed with injected TEA CO_2 systems[20].

Siegman has pointed out that in such circumstances use of the term *injection-locking* is misleading and incorrect, and that *injection-seeding* would constitute a more appropriate usage[23]. The analysis propounded here shows that both operational regimes, although in effect quite distinct, can nevertheless be success-fully incorporated into a single unified descriptive treatment.

SOME NOVEL PRACTICAL ASPECTS OF INJECTION-LOCKING

A variety of different techniques have been evolved for the purpose of coupling an injected signal into a TEA CO_2 laser[24], but while most of these techniques provide quite efficient coupling of the injection signal to the driven laser, they also possess the ability to couple back a proportion of the TEA output to the master oscillator. This has a marked destabilizing effect on the master oscillator performance[25,26], so that some means of alleviating this problem becomes desirable, if not essential.

Our studies into polarization coupling phenomena[27] have re-vealed a means by which this detrimental optical feedback may be largely suppressed. The method is based on the fact that even with an injected signal of nominally orthogonal polorization (relative to that preferentially supported by the driven cavity) stable, reproducible single mode output prevails. In itself, this result is not particularly remarkable, since although in theory there should be no coupling between the two oscillators, there are in practice numerous processes which might cause a proportion of the injected signal to be scattered into the preferred plane of polarization of the slave cavity. The powers required for injection-locking are so minute ($\sim \mu W$ [24,28]) that even quite weak effects may suffice.

Much more significant is the injected power dependence of the mode selection zone, given for the case of our compact system[27] by fig. 1. For the case where the injected polarization is identical to that of the TEA output (labelled 0 in fig. 1), we note the width of the mode selection zone increases monotonically with injected power, consistent with existing experimental evidence[21]. In marked contrast, however, the case of orthogonal injection (labelled 90 in fig. 1) shows a definite peak in the mode selection range. This was a surprising and unexpected result, since it testifies to the existence of a finite injection power for which the mode selection efficiency appears optimal. The source of this somewhat anomalous behaviour remains as yet unidentified, but nevertheless affords an elegant and conceptually radical means of achieving large injection coupling efficiencies whilst simultaneously providing a high degree of optical isolation between the driven laser and the master oscillator. Furthermore, the

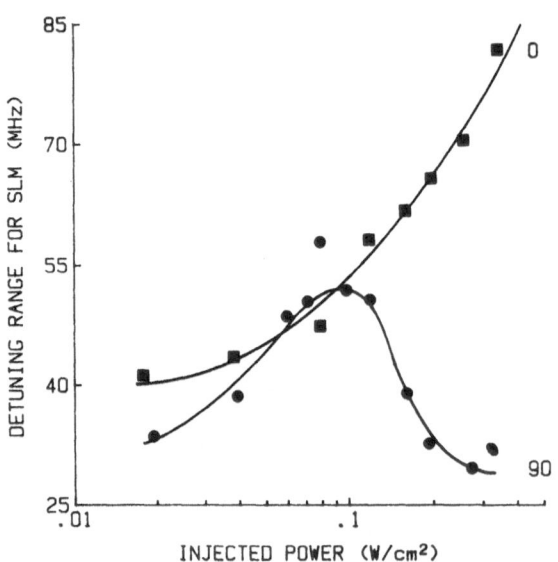

Figure 1.   Variation of mode selection range with injected power for injected polarizations parallel (0°) and orthogonal (90°) to that defined by the TEA cavity.

application of this refinement imposes no inherent constraints on the realization of the full output potential of the TEA laser, since no external beam splitters or coupling holes, etc. are required.

The experimental configuration used to demonstrate these principles is shown in fig. 2.

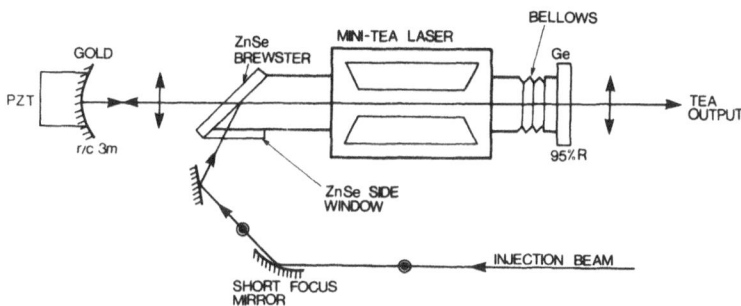

Figure 2.   Novel coupling geometry for TEA CO_2 injection locking.

Here, the intracavity Brewster window is itself used as the injection coupling element.   By injecting an orthogonal polarization we in effect access a regime of high reflectivity from the Brewster window without affecting its transparency with respect to the TEA CO_2 radiation.

Before concluding this section we note that the cavity system of fig. 2 contains no dispersive element for rotational line selection.   It was found that this configuration would only generate single mode output when injected with lines in the immediate vicinity of the 10P-band gain maximum (i.e. 10P(18) or 10P(20)).   Seeded with 10P(12) radiation, for example, the TEA laser output consisted of an initial SLM pulse at the injected wavelength followed by a secondary multimode pulse at the 10P(18) wavelength (see fig. 3).   Similar behaviour was observed with the injected source tuned to the 10R-branch and has been described by a number of authors, also operating TEA cavities with no means of rotational line control.

The implication of this observation is that injection alone is insufficient to suppress oscillation on the lines with highest gain.   Some workers have sought to alleviate this problem by the insertion of an intracavity SF_6 cell as a dispersive element.

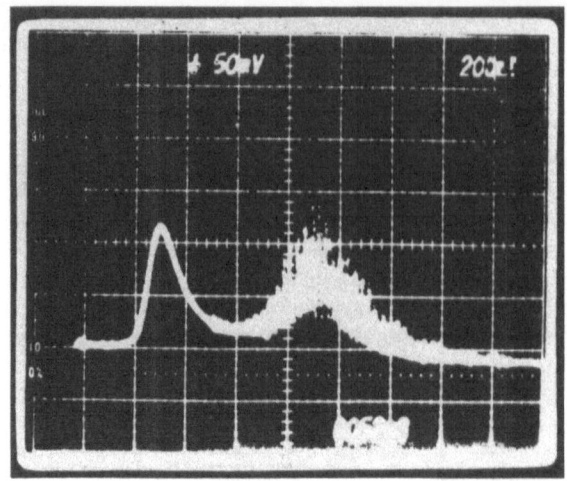

Figure 3.   Output from cavity of fig. 2 when injected
with 1OP(12) radiation.   The initial single
mode pulse occurs at the injected wavelength
while the subsequent multimode emission occurs
at the 1OP(18) line.

Since this solution may not always be practicable, it must be con-
cluded that single line SLM operation may best be guaranteed by
the use of a grating-tuned cavity.

ANALYSIS OF FREQUENCY SWEEPING EFFECTS

Intrapulse frequency sweeping was first invoked in the context of TEA CO_2 injection-locking in 1977[29], although no systematic
study was conducted at that time.   More recently, it was shown
theoretically how frequency sweeping acting over the TEA gain
build-up interval effectively broadens the mode selection zone,
with an accompanying asymmetry about resonance whose sense is
dependent on that of the chirp-rate[30].

In the conventional description of injection-locking[20,21] the
detuning parameter ϕ is assumed constant, and expressed by the
identity:

$$\phi = 2\pi\tau.\Delta\nu \quad , \tag{8}$$

in which τ is the TEA cavity round-trip transverse time and $\Delta\nu$ is the frequency mismatch between the injected signal and the TEA cavity mode to which it couples.

In our analysis we shall incorporate intrapulse chirp phenomena in this formulation by allowing ϕ to assume a time dependence of the form:

$$\phi(t) = \phi_0 - 2\pi\tau\Delta\nu(t) , \qquad (9)$$

where ϕ_0 is the initial detuning and $\Delta\nu(t)$ now represents the chirp. The standard expression describing the laser field evolution within the driven cavity thus becomes:

$$E(t + \tau) = R^{\frac{1}{2}} E(t) \exp[G(t) + j \phi(t)] + E_i \qquad (10)$$

in which R is the output coupler *power* reflectance, E_i is the injected amplitude and $G(t)$ is the functional form which describes the single-pass power gain at any given time.

The dominant chirp mechanism prevailing over the gain build-up period is due to electron plasma effects occurring during the TEA discharge current pulse. By measuring the time dependence of this discharge current it is possible to predict with some confidence the expected distribution of the intrapulse frequency sweeping[31]. For a range of values for the injected power the mode selection zone was determined by evaluating the range of detuning over which the dominant-to-residual mode intensity ratio exceeded 40 dB (this criterion having been chosen because a 1% modulation depth in the experimentally observed output signal is practically the limit of subjective judgement techniques).

The result of this theoretical exercise is reproduced in fig. 4. We note that the effect of intrapulse chirping is to considerably broaden the mode selection zone (relative to the case of zero chirp - fig. 4, stippled region) and to impart to it an asymmetry about resonance (i.e. nominal zero detuning), consistent with the previous work[30].

To test this theory using our own compact system, it was necessary to directly measure the frequency of the TEA CO_2 output using the apparatus depicted by fig. 5. No separate local oscillator has been used; instead the TEA output was heterodyned against the master oscillator signal, so that the observed beat frequency represented a direct measure of the detuning between the injected signal and the TEA cavity.

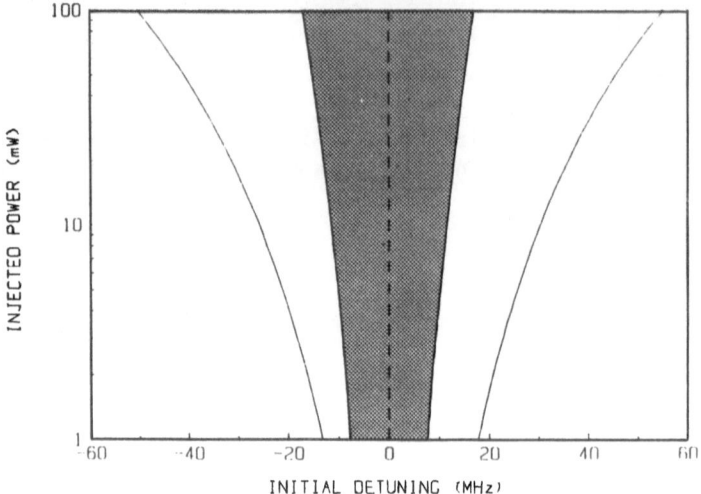

Figure 4. Variation of mode selection range with injected
power. The stippled region denotes the zero chirp
case.

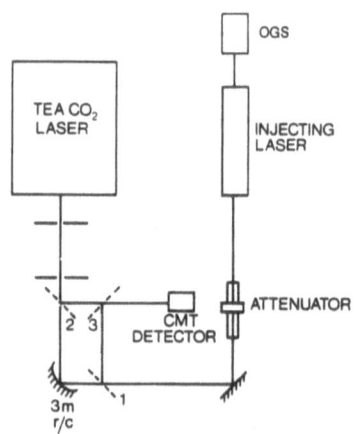

1. ZnSe BEAM SPLITTER
2. NaCl BEAM SPLITTER
3. Ge COATED ZnSe BEAM SPLITTER

OGS OPTO GALVANIC STABILIZER

Figure 5. Heterodyne frequency measurement apparatus. The
arrangement of beam splitters shown ensures that
the two beams reaching the CMT detector are of
comparable magnitude.

Fig. 6 shows a typical experimental run in which the TEA cavity
length was scanned through the mode selection zone at a constant
injected power of ∿ 1.5 mW. Superimposed on the experimental
data is the equivalent theoretical result, illustrating that the
model suggested here is quite capable of predicting both the
magnitude and sense of the injection-locked detuning range
asymmetry about resonance, as determined from the experimental
evidence.

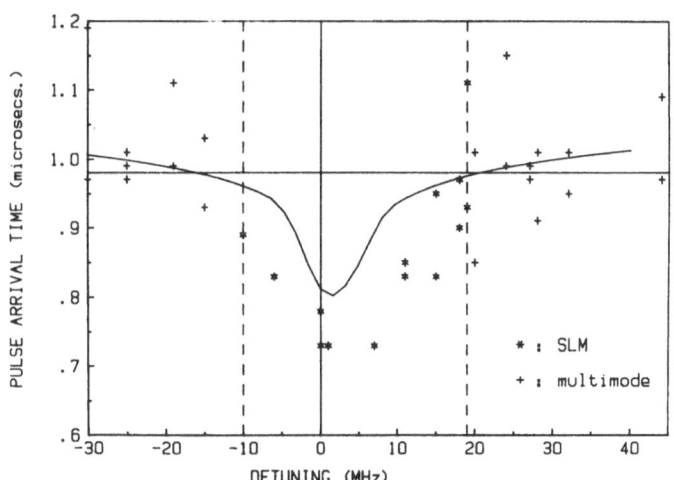

Figure 6. Dependence of pulse arrival time on initial
detuning, showing the experimentally derived mode
selection zone (broken lines). The intersection
of the horizontal line with the theoretical curve
denotes the range over which the residual mode
rejection ratio exceeds 40 dB.

Moreover, this feature is seen to shift the optimum injection
efficiency operating point (pulse arrival time minimum) to a
frequency slightly lower than that of the injected signal; a fact
which may assume importance in some practical systems.

CONCLUSION

The performance of injection-locked single-mode TEA CO_2
laser systems has been investigated with respect to polarization
and intrapulse frequency sweeping. It has been demonstrated how
these two particular aspects in turn affect the mode selection
process with the consequent appearance of several interesting and
unexpected properties, certain of which have a direct bearing on
the construction and operation of practical systems.

ACKNOWLEDGEMENT

This article has been abstracted from material arising out of a study carried out under financial support by the Procurement Executive of the UK Ministry of Defence, sponsored by DCVD.

REFERENCES

1. H.L. Stover and W.H. Steier,
    Appl. Phys. Lett, 8 (1966) 91.
2. P.W. Smith, Proc. IEEE, 60 (1972) 422.
3. S.L. Chin, Opt. Laser Technol., 12 (1980) 85.
4. Y.K. Park, G.Giuliani and R.L. Byer, Opt. Lett. 5 (1980) 96.
5. Y.K. Park, G. Giuliani and R.L. Byer, IEEE J. Quant. Electron QE-20 (1984) 326.
6. R. Lang, IEEE J. Quant. Electron., QE-18 (1982) 976.
7. T. Anderson, S. Lundquist and S.T. Eng, Appl. Phys. Lett. 41, (1982) 14.
8. O.L. Bourne and A.J. Alcock, Appl. Phys. Lett., 42 (1983) 777.
9. I.J. Bigio and M. Slatkine, IEEE J. Quant. Electron, QE-19 (1983) 1426.
10. U. Gamiel, A. Hardy and D. Treves, IEEE J. Quant. Electron QE-12 (1976) 704.
11. P. Flamant and G. Megie, IEEE J. Quant. Electron, QE-16 (1980) 653.
12. J.J. Yeh and A.B. Budgor, Proc. SPIE, 453 (1984) 41.
13. C.R. Jones, J.M. Telle and M.I. Buchwald, Technical Digest, 10th Int. Conf. on Quant. Electron, Atlanta, Ga. (1978) p.671.
14. A. Semet, L.C. Johnson and D.K. Mansfield, Int. J. Infrared Millim. Waves, 4 (1983) 231.
15. R.T. Menzies and G. Megie, Technical Digest, Topical Meeting on Coherent Laser Radar for Atmospheric Sensing, Aspen, Co. (1980), paper ThC8.
16. G. Koren, J. Appl. Phys., 54 (1983) 2827.
17. R.G. Harrison, W.J. Firth and I.A. Al-Saidi, Phys. Rev. Lett. 52 (1984), 15 July (to appear).
18. A. K. Kar, J.G.H. Mathew, S.D. Smith, B. Davies and W. Prettl Appl. Phys. Lett. 42 (1983) 334.
19. C. Yamanaka, S. Nakai, M. Matoba, H. Fujita, Y. Kawamura, H. Daidi, M. Inove, F. Fukemara and K. Terai, IEEE J. Quant. Electron, QE-17 (1981) 1678.
20. J. L. Lachambre, P. Lavigne, G. Otis and M. Noel, IEEE J. Quant. Electron, QE-12 (1976) 756.
21. H. Tashiro, T. Shimada, K. Toyoda and S. Namba, IEEE J. Quant. Electron, QE-20 (1984) 159.
22. C.J. Buczek, R.J. Freiberg and M.L. Skolnick, Proc. IEEE, 61 (1973) 1411.
23. A.E. Siegman, Lasers (University Science Books, Mill Valley, Ca, USA, 1983).

24. G. McClelland, S.D. Smith, R.G. Harrison and D. Tratt, Int. J. Infrared Millim. Waves, 2 (1981) 571.

25. J. L. Lachambre, P. Lavigne, M. Verreault and G. Otis, IEEE J. Quant. Electron, QE-14 (1978) 170.

26. P.W. Pace and J.M. Cruickshank, IEEE J. Quant. Electron, QE-16 (1980) 937.

27. A.K. Kar, R.G. Harrison, D.M. Tratt and C.A. Emshary, Appl. Phys. Lett., 42 (1983) 12.

28. U.P.Oppenheim, R.T. Menzies, and M.J. Kavaya, IEEE J. Quant. Electron, QE-18 (1982) 1332.

29. J.R. Izatt, C.J. Budhiraja and P. Mathieu, IEEE J. Quant. Electron, QE-13 (1977) 396.

30. N.R. Heckenberg, B.J. Renton, S.T. Shanahan and W. Wright, Appl. Phys. B, 29 (1982) 67.

31. D.V. Willetts and M.R. Harris, J. Phys. D: Appl. Phys. 15 (1982) 51.

AN Er^{3+}: GLASS EYE-SAFE MILITARY LASER RANGEFINDER

C. Tarenzi and M. Zorgno

FIAR - Fabbrica Italiana Apparecchiature
Radioelettriche
Milano (Italy)

ABSTRACT

A military eye-safe rangefinder based on a Er^{3+}: glass
Q - switched laser is described. The output energy of the laser
is up to 50 mJ with a pulse duration of $\sim$ 25 ns. Under typical
operating conditions the expected range of the system may be up to
$\sim$ 5 Km.

The application of laser technology in the military field has
been continuously growing in the last decade. The main tactical
systems so far developed and, sometimes, also deploied include
(Fig. 1):(i)Rangefinders. They measure the distance from a given
target from the measurement of the roundtrip time of a short laser
pulse. (ii)Target designators. They allow intelligent weapons to
be guided toward the given target. (iii)Laser radars. They give information about position and speed of the target. Another important application is provided by laser simulators. These systems
may take different forms all with the purpose of simulating firing
in a real battlefield situation and thus training the military personnel.
The choice of the laser for each of the above applications
has always been made by taking into account the status of development of both laser beam transmitter and receiver as well as the
transparency properties of the atmosphere. Thus, rangefinders incorporating ruby, Nd:glass, Nd:YAG and CO$_2$ lasers, designators
using Nd:YAG lasers, radar systems using etherodyne detection whith
CO$_2$ lasers and simulators employing semiconductor lasers have mostly
been developed. The increasing number of laser sources which may
thus be present in a modern battlefield is however enphatizing the

problem of eye hazard of the personnel. In fact, at visible and near IR wavelengths, serious ocular damage may be caused to the retina due to the relatively high transmission of the ocular media. Safety values for the emitted energy (for Class I-exempt) range between 0.2 (for ruby) to 2 uJ (for Nd:YAG) in a 7 mm diameter beam (Fig. 2). At the CO_2 laser wavelength, however, the beam is strongly absorbed by the epythelium. In this case corneal damage is the limiting phenomenon and the safety value (again for Class I-exempt) raises to 79 uJ in a 1 mm diameter beam.

At the wavelength of 1.54 um, however, corneal damage threshold is considerably increased and retinal damage is prevented by the absorption of the ocular media and by the reduced focussing capability of both the corneal and the crystalline lens. This results in an increase of the allowed energy of about two orders of magnitude compared to the case of a CO_2 laser (see Fig. 2). At this last wavelength two types of lasers seem to be more suitable: the Er^{3+}: glass laser and the Nd:YAG laser whose wavelength is Raman shifted in a high pressure cell containing a suitable gas (e.g. methane).

For low repetition rate rangefinders, Er^{3+}: glass laser seems to be a suitable choice for the following two main reasons: (i)Cost effectivity, semplicity and roughness of the laser structure. (ii)Compatibility with the well experimented Nd:glass technology. An Er^{3+}: glass oscillator is indeed similar in design and operation to a Nd:glass laser. The main problem arises from the poor thermal conductivity of glass so that operation seems to be limited to low repetition rate ($\sim$ 1 pulse/sec). It should finally be noticed that both lasers, besides being eye-safe, present the following other two main advantages: (i)Improved atmospheric propagation, due to reduced scattering, compared to that of a 1 um beam. (ii)Good sensitivity of both Ge and InGaAs detectors. New developments at these wavalengths are also expected (e.g. avalanche photodiodes) for applications in the field of optical fiber communications.

Two types of Er^{3+}: glass Q-switched lasers have been considered and developed (or in development): the first one makes use of a rotating prism while the second one uses a polarizer-Pockel cell combination as Q-switching element. The interesting possibility of a saturable absorber has also been considered. In all cases the laser rod is pumped by a single linear flashlamp in a close-coupled configuration. The lamp is energized by a simple pulse-forming-network made of a single capacitor and an inductance. Triggering is provided by a parallel system. The schematic drawings of the laser head are shown in Fig. 4 and 5. Overall cavity length is smaller than 10 cm. The energy output vs. energy input characteristic for both normal laser action and Q-switched operation are shown in Fig. 6 and 7 respectively. Pulse duration in Q-switching operation is about 25 ns and beam divergence is less than 5 mrad.

Based on a laser of this kind an eye-safe laser rangefinder is in the phase of development. The predicted range capability

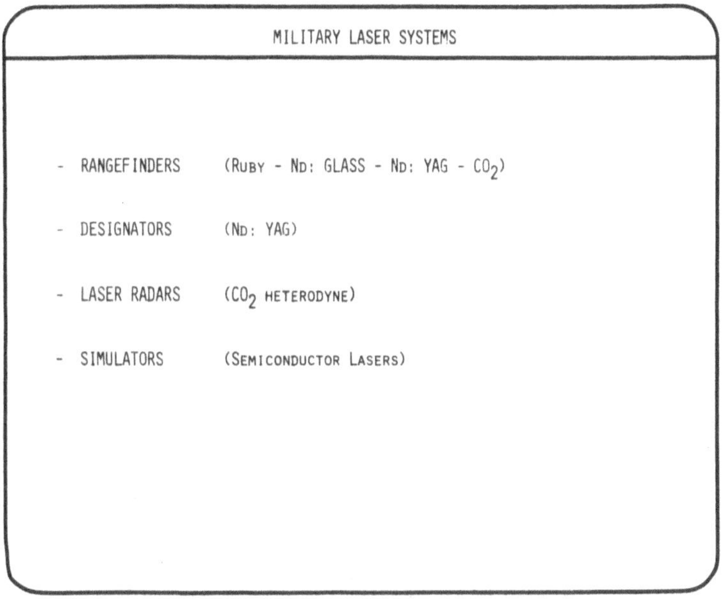

MILITARY LASER SYSTEMS
- RANGEFINDERS (RUBY - ND: GLASS - ND: YAG - CO_2)
- DESIGNATORS (ND: YAG)
- LASER RADARS (CO_2 HETERODYNE)
- SIMULATORS (SEMICONDUCTOR LASERS)

Fig. 1.

MAX. ENERGY PER SHOT (CLASS I - EXEMPT) [μJ]
7900
790
790
790
2
0,5
0,2
RUBY
GA-AS
ND: YAG
ER: GLASS
ER: YLF
HO: YLF
CO_2
0,69
0,9
1,06
1,54
1,73
2,06
10,6
λ [μM]

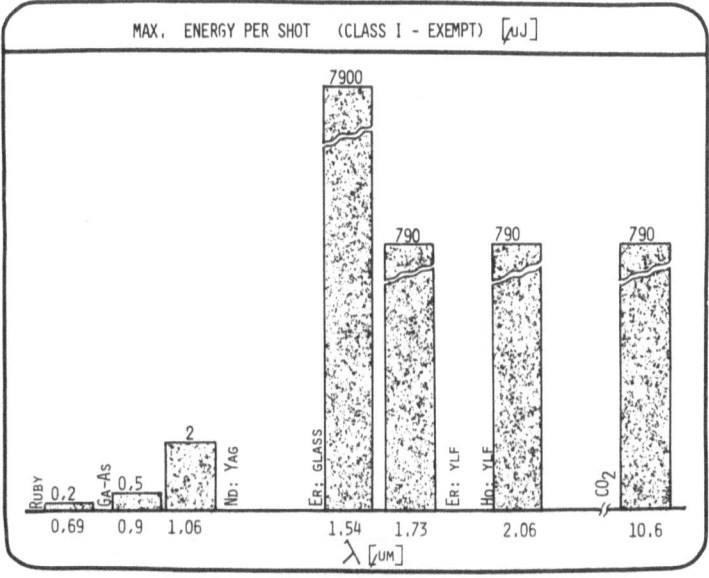

Fig. 2.

```
ERBIUM : GLASS LASER  -  FEATURES


    - TOTAL EYE-SAFETY

    - COMPATIBILITY WITH Nᴅ: GLASS LASER

    - GOOD SENSITIVITY OF DETECTORS AT  λ=1.54 µM

    - GOOD ATMOSPHERIC PROPAGATION

    - COST EFFECTIVITY, SIMPLICITY AND  ROUGHNESS  OF THE LASER STRUCTURE
```

Fig. 3.

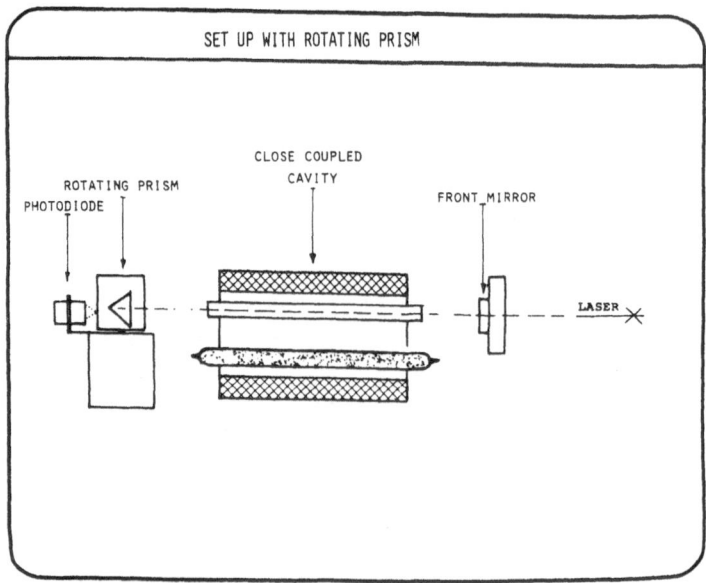

Fig. 4.

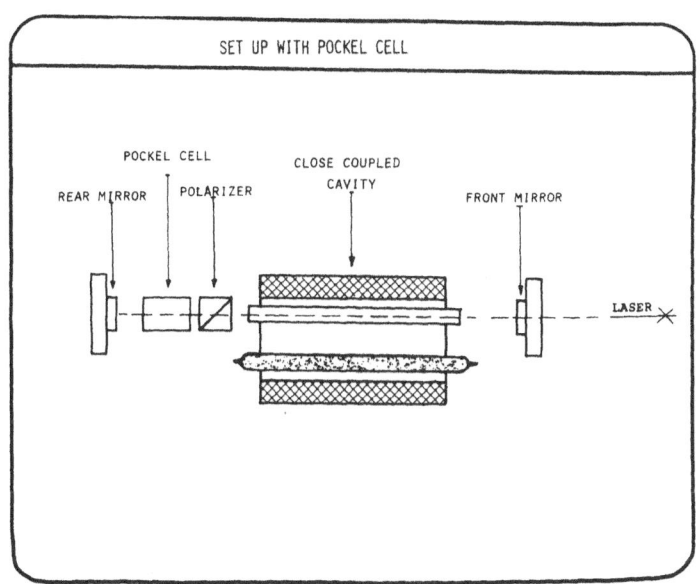

Fig. 5.

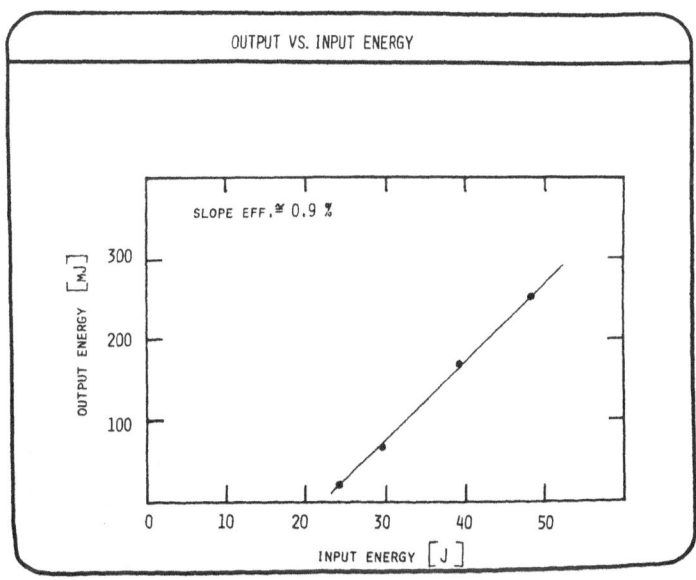

Fig. 6.

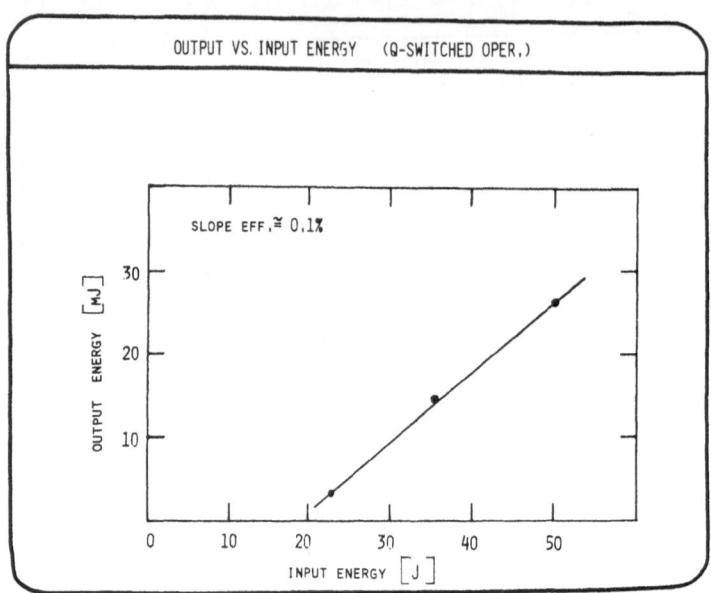

Fig. 7.

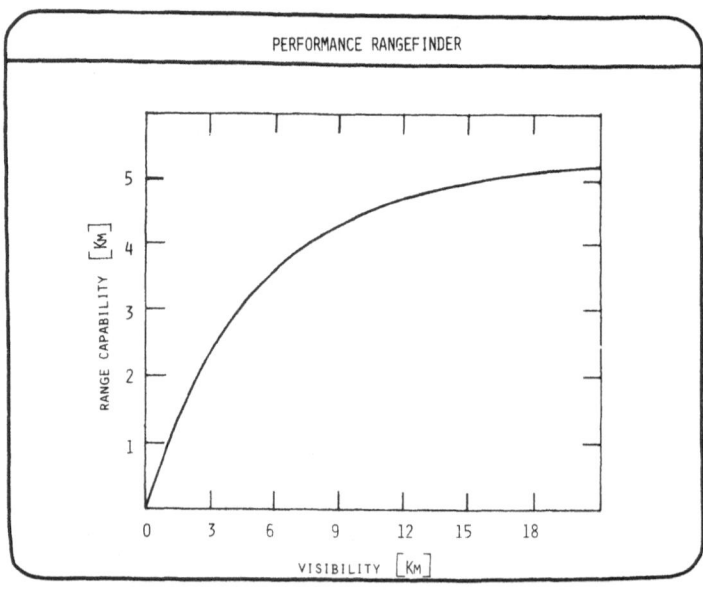

Fig. 8.

of this instrument as a function of atmospheric visibility is shown in Fig. 8. For a visibility of $\sim$ 20 Km, the expected range is seen to be $\sim$ 5 Km. This rangefinder is expected to have wide field of application for all cases in which a system of medium range and low repetition rate is of interest.

EYE-SAFE LASERS FOR MILITARY APPLICATIONS

G. Alessio

SELENIA S.p.A. v. Castelli Romani 2

00040 Pomezia, Roma

Abstract   Applications of eye-safe lasers in military fields are described with particular reference to tank telemetry. Experimental results for a Raman-Shifted Nd-YAG laser are given.

INTRODUCTION

The measure of the distance between a tank and a target is normally performed by a Q-Switched Nd-YAG laser which emits short pulses towards the target. A suitable optical reciever is used to detect the back-scattered echo and thus the distance is obtained from the round trip-time.

Normal requirements for this kind of systems are:

max range to be measured	10000 m
accuracy	1 - 5 m
measure rep. frequency	12 p.p.m.

From the laser point of view these become:

pulse energy max	50-60 mj
duration FWHM	8-10 nsec
divergence	.5 mrad

EYE-SAFE REQUIREMENTS

Eye-safe lasers are important in military tank-rangefinders because they can be used in training of troops (in which the eye safety is a key requirement). Furthermore a big industrial effort is being given to the development of hand-held eye-safe rangefinders (for infantry use).

The 1.06 µm wavelength emitted from Nd-YAG laser is not safe for the human eye because of the high transmittance of the ocular media and the appreciable retinal absorption. Moreover at 1.06 µm the focusing power of the eye lens is quite good, meaning that the laser radiation is well focused on the retina where the produced hot-spot can cause permanent damage.

Longer wavelengths (10 µm) are equally dangerous because they cause permanent damage on the surface of the cornea: it is in fact not transparent to the radiation and so the optical power is absorbed in a very thin layer thus causing local surface burns on the cornea.

Safe radiations (the safety has to be defined) are those whose wavelengths are not well transmitted within the ocular media and are poorly absorbed by the retina. Furthermore the focusing power of the eye lens should be very low.

Safety is defined via the MPE (Maximum Permissible Exposure) which is found to be [1]:

| MPE | 1.06 µm | $5\mu j/cm2$ |
| MPE | 1.54 µm | $1000mj/cm2$ |

at these two reference wavelengths.

The 1.54 µm wavelength is not the only one possible, but is perhaps the most convenient from the telemetry point of view. In fact the choice of a wavelength has to be traded-off among several factors: the feasibilitiy of its generation, the atmospheric transmittance, the detectability etc.

For instance the Er:Glass laser emits directly at 1.54 µm, but it has some drawbacks. Firstly it is a three-level system and thus needs strong optical pumping (typically 50 J of electrical energy are necessary to obtain 25 mj of Q-Switched optical output). Furthermore the Q-switched operation is quite difficult (since a 1.54 µm saturable absorber, at present, is unavailable and a Pockels cell is expensive and cumbersome).

Another way to obtain the required wavelength is to shift a shorter wavelength (easy to generate) via a non linear interaction.

From this point of view the Stimulated Raman Scattering (S.R.S.) is one of the most promising and reliable techniques as far as minimum use of new

technologies and materials is concerned; moreover a
good conversion efficiency and a full compatibility
with 1.06 μm existing sources make it the best
industrial choice. The gaseous methane features a
vibrational scattering of 2.900 cm-1 and thus shifts
the Nd-YAG 1.06 μm wavelength to 1.54 μm with a
theoretical quantum efficiency of 69%. Experimental
energy conversion efficiencies of 20% have been
reported in literature [2], [3].

From the physical standpoint the S.R.S. is a non
linear process in which a "big" photon's energy is
shared between a phonon and a "smaller" photon (longer
wavelength). This process is managed by a spontaneous
scattering cross-section $d\sigma/d\Omega$, and only if the pump
fluency is higher than a certain threshold value does
it behave as a stimulated process. The first Stokes
intesity is given by: [4]

$$I_s(L) = I_s(0)*exp[(g_s*I_p-\gamma)*L]$$

where:

I_s	first Stokes intensity
γ	loss coefficient
g_s	differential Raman gain
L	interaction length
I_p	pump intensity

In this equation g_s is proportional to $d\sigma/d\Omega$, and γ/g_s
is the threshold intesity.

EXPERIMENTAL RESULTS

A breadboard has been set up in order to
demonstrate the feasiibility of the S.R.S. technique.
It consists of a Nd-YAG passively Q-Switched laser
module, a pressure gas cell with A.R. coated windows
and all the suitable optics to focus and recollimate
the pump beam.

In order to detect the 1.54 μm output a wide-band
pyroelectric detector has been used, and the
unconverted 1.06 μm pump beam has been filtered via a
silicon window and optical filters.

The optimum methane pressure was found to be ~70
Kg/cm$_2$ at which the efficiency curve saturates. This
is due to the broadening of the vibrational line that,
above a certain value of the pressure, counterbalances
the growth of the number of the molecules (due to the
pressure). In this condition and with the given pump
beam (50 mj/10nsec pulses from a ϕ= 5 mm laser rod) the
optimum focal length was found to be 120 mm (focusing

lens). Further shortening causes arcing within the cell and a fluctuating output at 1.54 μm.

FUTURE DEVELOPMENTS

The goal is to design of a compact, rugged, reliable pressure cell (100 mm max length) with focusing windows (to minimize the number of the optical elements) and to obtain a conversion efficiency of 20% minimum with stable 1.54 μm output. This device will be integrated in a 1.06 μm telemeter in order to achieve rangefinding operation at the 1.54 μm eye-safe wavelength.

REFERENCES

1. Sliney & Wolbarsht: Saftey with lasers and other optical sources.

2. D.G. Bruns, H.W. Bruesselbach, D.A. Rockwell: A Nd-YAF- based eye-safe Raman laser. Proc Int. Conf. on Lasers 1980.

3. D.G. Bruns, H.W. Bruesselback, D.A. Rockwell: Scalable visible Nd-YAG pumped Raman laser source. IEEE J.Q.E. vol. 18, no. 8, Aug 1982.

4. A. Grasiuk & Zubarev: High power tunable IR Raman lasers. Appl. Phys. 17, 1978.

SELECTED TOPICS IN LASER SPECTROSCOPY

R. G. Brewer, R. G. DeVoe, A. Schenzle, M. Mitsunaga,
J. Mlynek, N. C. Wong and E. S. Kintzer

IBM Research Laboratory
San Jose, California 95193

Microscopic Theory of Optical Line Narrowing of a Coherently Driven Solid[1,2]

The optical Bloch equations which incorporate the phenomenological population (T_1) and dipole dephasing (T_2) times have been tested recently by optical free induction decay (FID) measurements on an impurity-ion crystal $Pr^{3+}:LaF_3$ at 1.6°K. At low optical fields, the observed Pr^{3+} optical linewidth is dominated by magnetic fluctuations arising from pairs of fluorine nuclear flip-flops where the condition $T_1 >> T_2$ prevails. At elevated fields, this nuclear broadening mechanism is quenched and the Bloch equations are violated with $T_2 \rightarrow T_1$. A microscopic theory appropriate for a low temperature impurity solid is developed which reveals the above features both for optical and radio frequencies, and a simple physical interpretation of this line narrowing phenomenon is given. Modified Bloch equations of a novel form are derived to second order and yield analytic FID solutions over the entire range of optical field strength. A discussion of the earlier NMR theories will be given pointing out similarities and differences.

Raman Heterodyne Detection of Nuclear Magnetic Resonance[3-5]

We report a new way of detecting nuclear magnetic resonance in solids and gases using a coherent optical and radio frequency induced Raman effect. The technique, which employs heterodyne detection, is capable of monitoring nuclear resonances under cw conditions or coherent spin transients, both in the ground and excited electronic states, with high sensitivity and precision. We have demonstrated the versatility of the method in sodium vapor and the dilute rare earth impurity ion crystals $Pr^{3+}:LaF_3$ and $Pr^{3+}:YAlO_3$. Raman detected Pr^{3+} spin echoes of nuclear quadrupole transitions have been observed not only in the $^3H_4$ ground state but also for the first time in the $^1D_2$ excited electronic state. From the cw spectrum, the Pr^{3+} hyperfine splittings and the magnetically broadened inhomogeneous linewidths are determined with kilohertz precision, about a fivefold improvement over earlier measurements.

In addition, a novel Raman interference effect has been discovered in $Pr^{3+}:YAlO_3$. Experiments confirm theoretical arguments that the two inequivalent Pr nuclear sites in the host lattice generate coherent Raman signals of opposite sign that interfere. A necessary requirement, which can now be examined in detail, is that the principal nuclear quadrupole Z axis of Pr be incongruent in the ground and optically excited electronic states. This effect may be only one of a class of Raman coherence phenomena. For example, in addition to site interference, Raman detected nuclear Zeeman transitions of a single site can exhibit interference also.

It is clear that a broad range of cw and pulsed NMR experiments can now be explored in ground and optically excited states of low temperature impurity ion solids or even gases using Raman heterodyne detection.

Laser Frequency Division and Stabilization[6]

An optical interferometric technique is proposed for stabilizing and measuring a laser frequency in terms of a radio frequency (rf) standard. In preliminary studies, a sensitive optical dual frequency modulation scheme (DFM) allows locking a laser to an optical cavity and the cavity in turn to a radio frequency reference with a noise level of 60×10^{-3} Hertz or 2 parts in 10^{10}. In principle, the laser frequency ω_0 can acquire the stability of the rf standard and being locked to a high order multiple n of the rf frequency ω_1 facilitates the optical-rf division $\omega_0/n = \omega_1$.

References

1. R. G. DeVoe and R. G. Brewer, Phys. Rev. Lett. 50, 1269 (1983).
2. A. Schenzle, M. Mitsunaga, R. G. DeVoe and R. G. Brewer, Phys. Rev. A 30, 325 (1984).
3. J. Mlynek, N. C. Wong, R. G. DeVoe, E. S. Kintzer and R. G. Brewer, Phys. Rev. Lett. 50, 993 (1983).
4. N. C. Wong, E. S. Kintzer, J. Mlynek, R. G. DeVoe and R. G. Brewer, Phys. Rev. B 28, 4993 (1983).
5. M. Mitsunaga, E. S. Kintzer and R. G. Brewer, Phys. Rev. Lett. 52, 1484 (1984).
6. R. G. DeVoe and R. G. Brewer, Phys. Rev. A (October 1984).

INDUSTRIAL APPLICATION OF HOLOGRAPHIC INTERFEROMETRY

Hans Steinbichler

Postfach 312 8207 Endorf / Germany

Rapid development of high-power lasers allows the application of holograhic interferometric methods in industrial environments. Holographic interferometry offers some essential advantages (1,2,3) over other kinds of testing and measuring methods:

- Non contact, inertialess measurement
- Object is not affected
- Two-dimensional image information
- Testing under practice related loading
- Relevance between holographic display and actual flaws

1. Holographic non destructive testing

The principle of holographic, non-destructive testing is based on the fact, that defects inside a component cause a typical surface deformation which can be recognized as fault-specific deformation. A weak, defective area generally deforms in a different way than faultless zones. Even if these fault-typical deformations are only in the micrometer range, they can be clearly identified on account of the measuring accuracy of holograhic interferometry.

Another advantage of holographic non-destructive testing is, that a component can be subjected to a realistic stress during the test. So, as an example, a high pressure tube is stressed by internal pressure (4). A weak, defective spot is more deformed than the surrounding areas. This deformation, which is typical for flaws, can be identified by holographic interferometry.
This realistic stress assures a certain relevance between holographic display and actual damage.

weak spot

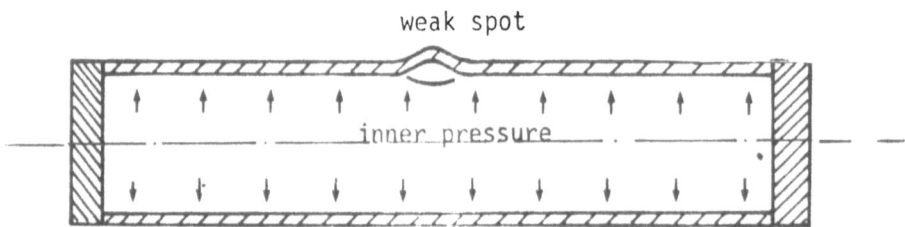

fig. 1: Section of an inner pressurised tube with a flaw

On figure 2 the holographic investigation of a fibre reinforced plastic tube is shown. These tubes are designed for pressures of more than 1000 b. The tool machine operated in bad condition. The glass fibres, which were winded helically, got a flaw at each winding period.

fig. 2: Holographic investigation of fibre reinforced plastic tubes. Flaws at each winding period

As a comparison the deformation of a faultness tube is shown in fig. 3.

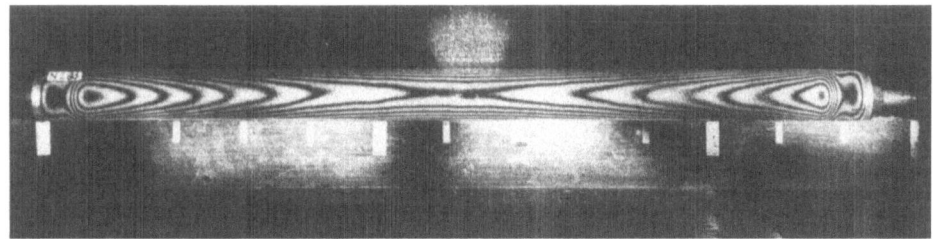

fig. 3 : Holographic investigation of fibre reinforced plastic tubes; tube without flaws

The growing use of fibre reinforced plastics in the aerospace activities entails additionel applications. In composite materials holographic testing has already proved its worth.

Some helicopter rotorblades are made of glass or carbon fibre reinforced plastics. A shell of fibre reinforced plastics (fig.4) is filled with polyurethene foam. By changing the surrounding pressure an inner flaw, which is usually connected with an air pocket, causes a surface deformation.

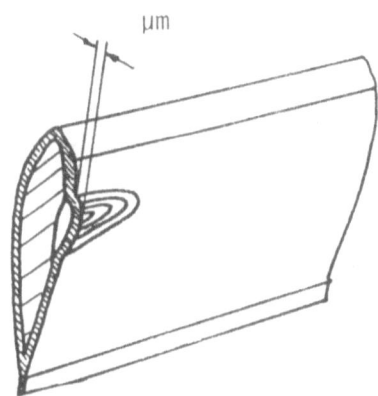

µm

fig. 4: Principal of holographic testing of rotor blades

371

The rotorblades are put in a chamber which is slightly evacuated for the test.

A pressure difference of 20 mb is already sufficient for a clear identification of the flaws (fig.5). The rotorblade is illuminated and observed thru a window in the chamber. Flaws can be identified by elliptical or annular interference fringes. The size of the flaws is equivalent to the size of the local fringe pattern. The depth of the flaw can be calculated by the number of fringes.

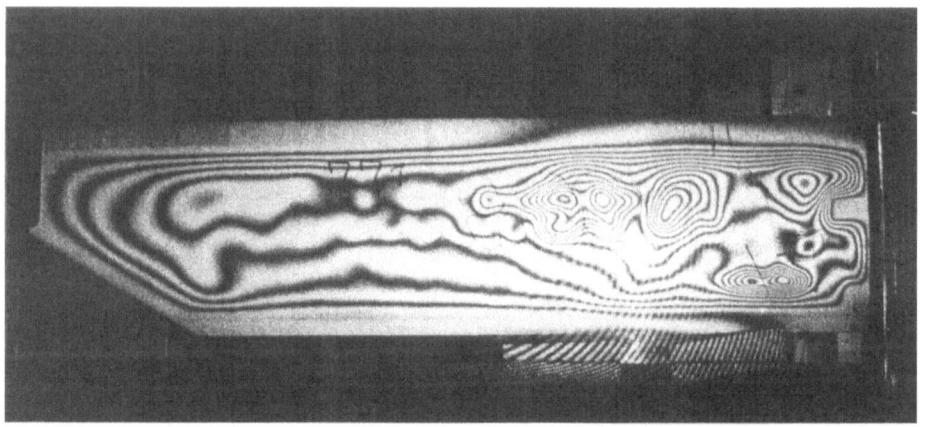

fig. 5: Holographic test of helicopter rotorblades made of fibre reinforced plastics

The same test principal is used for the testing of adhesive connections, e.g. of rubber and metal (fig. 6).

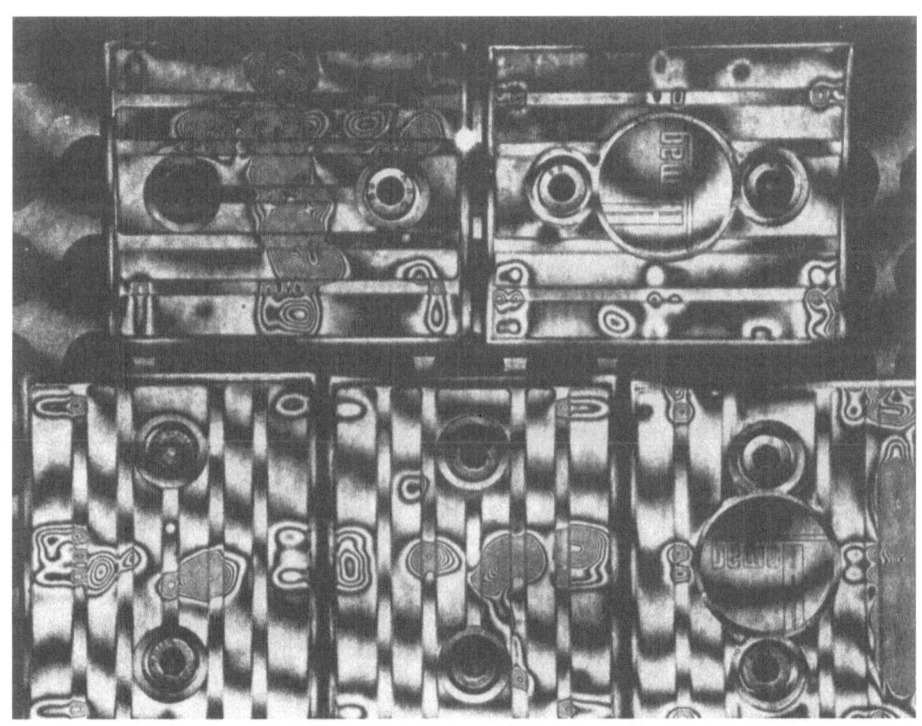

fig. 6: Holographic Test of a adhesive connection of rubber and metal

The demonstration of the test of turbine rotaries concludes the holographic non-destructive testing (4). An annular solder seam, which connects the center and the turbine blades, is to be tested. (fig. 7)
The turbine wheel is fixed on a chuck and loaded in the centre. In-homogeneities in the interference pattern indicate faults in the solder seam.

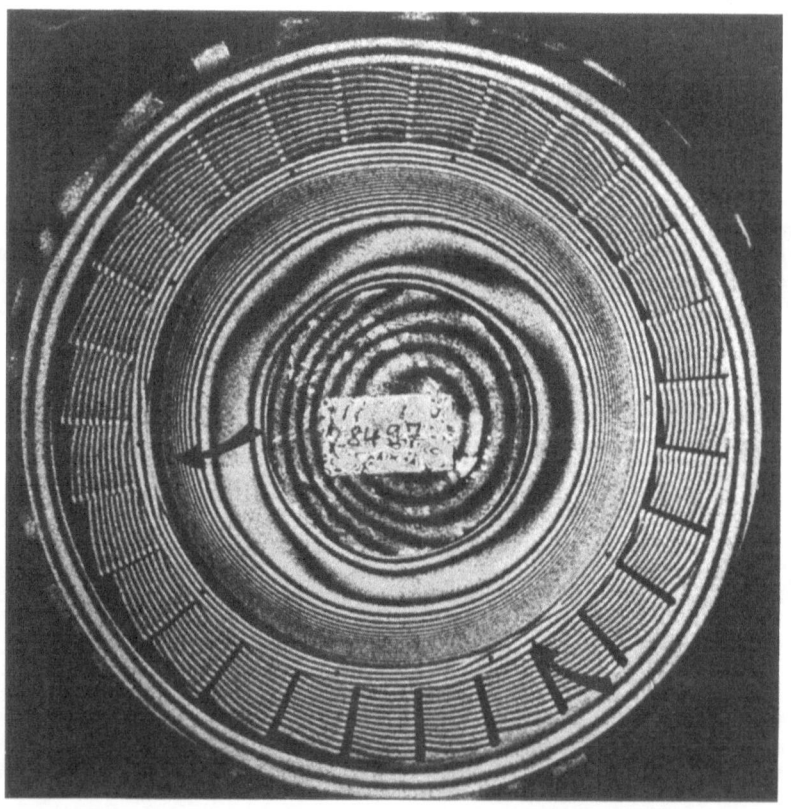

fig. 7: Holographic testing of turbine wheels

2. Evaluation of Design Characteristics

Holographic Interferometry makes the diagrammatic measurement
of deformations possible. Thus it is, simular to photoelasti-
city, an aid in the evaluation of design characteristics. The
spectrum of application stretches from static cases of loading
(2,3,5) to the investigation of oscillation (3,4,5) e.g. investigation of sound transmission in car components.

Holographic interferometry can also be used in extremely non
laboratory like environment, e.g. under space simulated conditions. The thermal distortion of an antenna reflector was
measured under spaced simulated condition (4). Thermal distortions of the reflector change the transfer characteristics,
therefore the distortion has to be measured in advance. The
antenna was fixed in a space simulation chamber. It was illuminated and observed through windows in the chamber. Fig. 8 shows
the hologram of the deformed antenna.

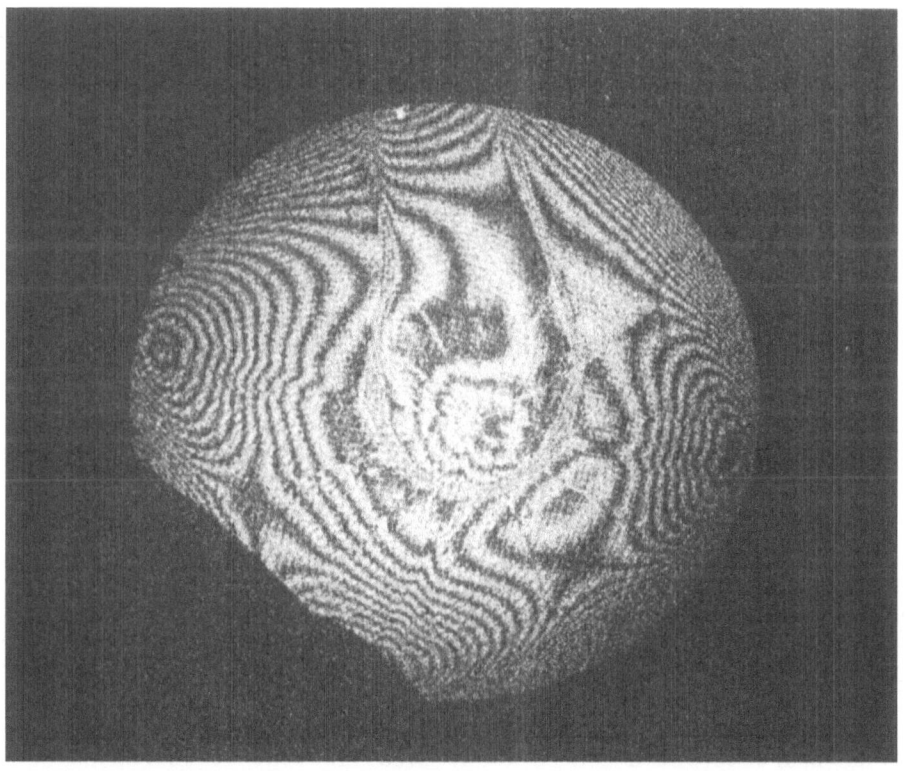

fig. 8: Thermal deformations of an antenna under space simulated
conditions

The evaluation of holograms, taken during the cool-down from
- 35,6°C to - 119°C, is shown in fig. 9.

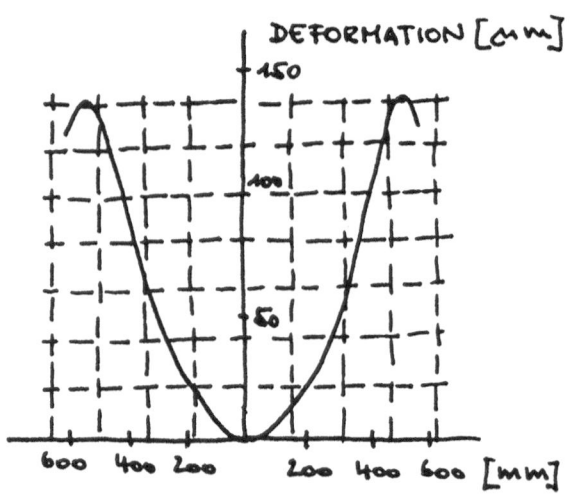

fig. 9: Thermal deformation of an antenna during cool-down
- 35,6°C to - 119°C
(in cooperation with Dr. Frey, IABG, Ottobrunn)

Capabilities of pulsed laser holography is suited very well for
determining amplitude distribution on dynamically loaded parts,
i.e. also on running engines. Between two laser pulses, the
change in amplitude can be determined on the object under consideration. By means of appropiate triggering of the laser pulses at
certain frequencies, the effect of these frequencies on the vibration characteristic can be identified. In fig. 10 is represented the amplitude distribution of a running engine, with determination refering to the 2nd order deflection of the whole assembly,
related to the engine speed.

376

The possibility of analyzing non-stationary, shock-excited vibrational states makes the holographic interferometry useful in the field of ballistic research (15). So vibrational modes of guns (fig.11) can be ascertained. When the gun is fired, the barrel is vibrating and this vibration influences the trace of the bullet.

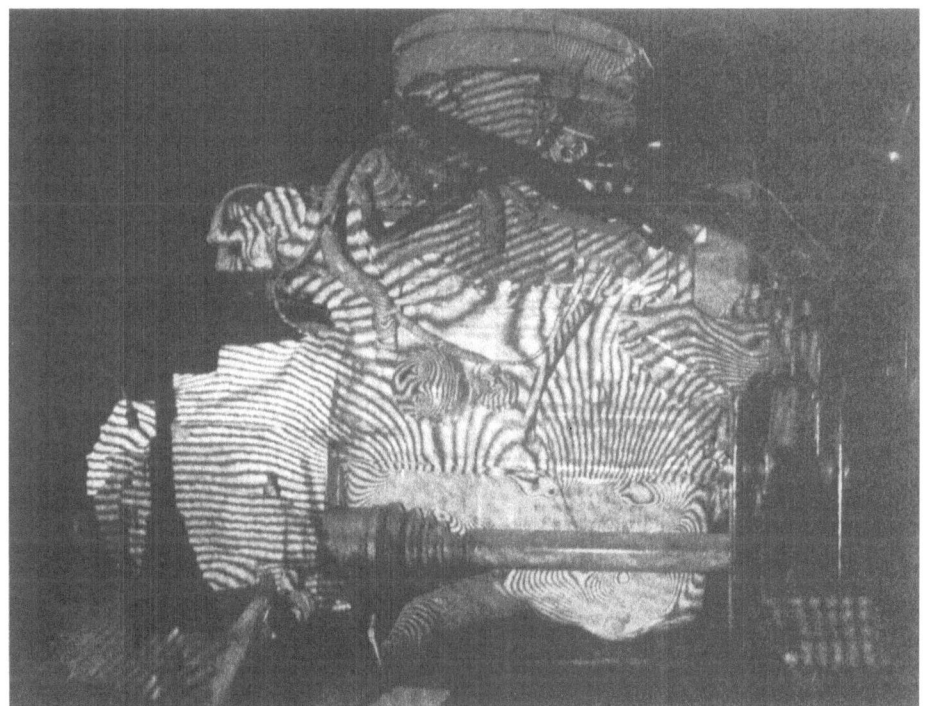

fig. 10: Second order deflection of a running engine-gear assembly relate. (By courtesy of Ford Werke AG, Köln)

The quantitative evaluation of the interferograms obtained from such investigations is extremely difficult. Therefore one tried to stimulate the interferograms by calculation the interference field using simple assumptions with respect to the possible vibrational state of the weapon. The result of these calculations were plotted by a table-top computer. In fig. 22 an example of these calculations is shown side by side with an interferogram obtained directly from the barrel. The dominant pattern of the calculated interferogram is very similar to the experimental result.

By following this procedure one made use of the main advantage of the the holographic analysis, i.e. the possibility to show whether a theoretical model represents the characteristic features of the vibrational state of the object under investigations. Thus a specific model can be checked rather easily, even if some additional disturbances would have to be taken into account for a detailed analysis.

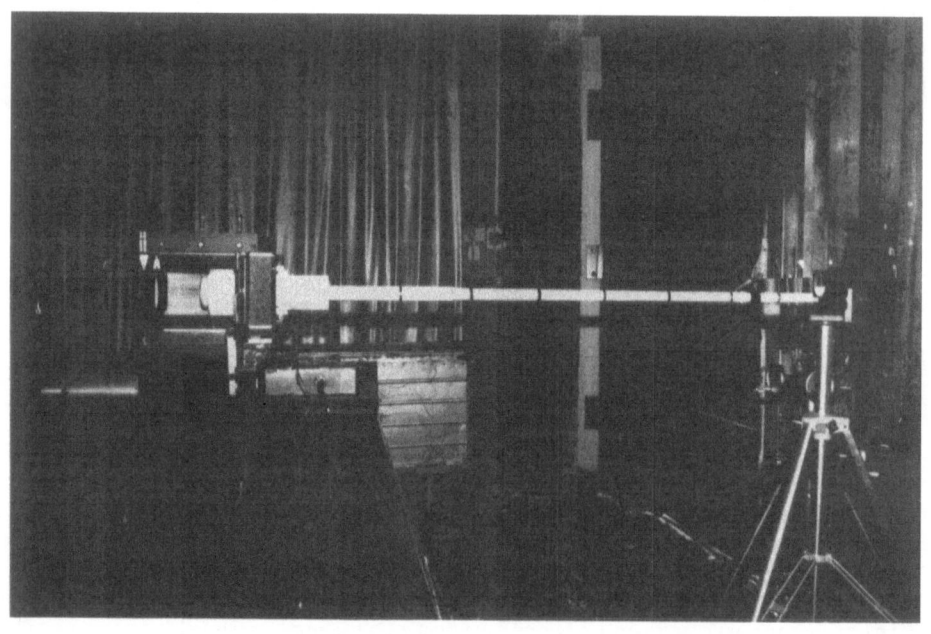

fig. 11: Experimental arrangement for vibration studies of a gun (cal. 20 mm) (Zwingel, Fa. Diehl, Nürnberg)

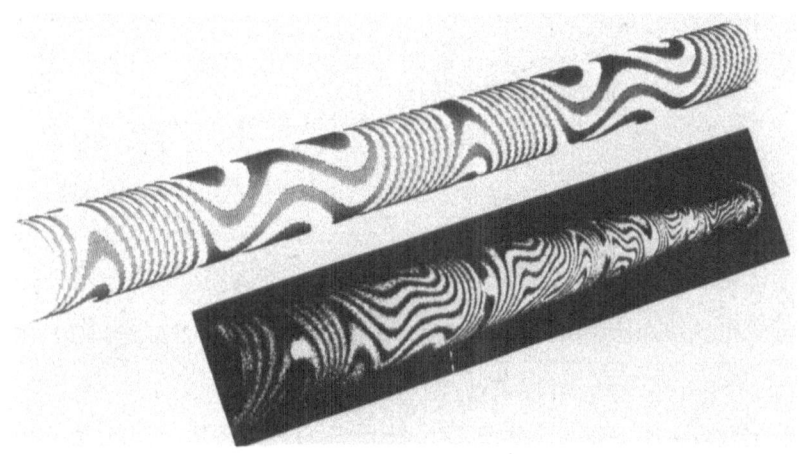

fig. 12: Calculated interferogram compared with the interferogram
obtained directly from the barrel
(Zwingel, Fa. Diehl, Nürnberg)

References

(1) Francon, M.:   Holographic (1972)
                         Springer Verlag Berlin

(2) Robertson, E.:   The Engineering Uses of Holography
     Harvey, J. M.:   Cambridge University Press 1970

(3) Vest, C. M.:   Holographic Interferometer (1979)
                        John Wiley and sons, Inc., New York

(4) Steinbichler, H.: Holography in Europe (1982)
                        SPIE-Proceedings: Industrial and Commer-
                        cial Applications of Holography, San Diego

(5) Felske, A.:   Doppelpulsholografie als Hilfsmittel für
     Happe, A.:   Untersuchungen von Schwingungserregung und
                        Schwingungsfortpflanzung
                        Vortrag auf der Frühjahrstagung der Deutschen
                        Physikalischen Gesellschaft e.V. am
                        23.03.1972 in der Universität Kiel

(6) Zwingel, D.:   Vibration Analysis by Triple Pulse Holography,
     Steinbichler, H.: Proceedings: Laser and Electrooptics 1981

PHASE CONJUGATION: TECHNIQUES AND APPLICATIONS

Concetto R. Giuliano and David A. Rockwell

Hughes Research Laboratories
3011 Malibu Canyon Road
Malibu, California 90265

I.   INTRODUCTION

One of the most interesting research areas accessed with intense laser sources is that of nonlinear optics. Shortly after the first lasers became available, a variety of nonlinear phenomena were found to arise in optical media. Many of these phenomena, such as second harmonic generation and stimulated Raman scattering, have been widely exploited to produce intense coherent optical beams at wavelengths far removed from those of existing sources, thereby opening up new methods of research into the physics of optical materials, as well as providing laser sources for a variety of new applications. These nonlinear wavelength conversion techniques are reviewed elsewhere in this volume. The present article reviews nonlinear optical phase conjugation, a phenomenon that is receiving intense interest not because of the production of new wavelengths, but because it offers realistic solutions to several difficult problem areas arising in a variety of applications. This article is not intended to review the entire field of nonlinear optical phase conjugation, since several comprehensive, current reviews are already available.[1,2]   Instead, we intend to discuss the basic physical concepts of phase conjugate devices, and summarize recent results in a variety of important applications areas.

The term "phase conjugation" is derived from the mathematical description of electromagnetic radiation in which the electric field is written as the real part of a complex expression:

381

$$E(x,y,z,t) = Re\{\psi(x,y,z)e^{i\omega t}\} \tag{1}$$

where

$$\psi(x,y,z) = A(x,y)e^{i[-kz-\phi(x,y,z)]} \tag{2}$$

These two equations describe a wave with a frequency ω moving in the positive z-direction with a wavelength $\lambda=2\pi/k$. The real amplitude function $A(x,y)$ describes the transverse spatial variation of the wave, and the phase function $\phi(x,y,z)$ represents the transverse variation of the phase at the field point z. The phase conjugate of E is defined as

$$E_{conj} = Re[\psi*(x,y,z)e^{i\omega t}]$$

$$= Re[A(x,y)e^{i[kz+\phi(x,y,z)]}]e^{i\omega t} \tag{3}$$

and represents a wave moving in the negative z-direction with the sign of the phase term $\phi(x,y,z)$ reversed relative to that of the original wave. Note that the phase conjugate wave is found by taking the complex conjugate of only the spatial part of E, leaving the temporal part unchanged. We can also think of the process as a reflection with a phase reversal; this perspective is implied in the Soviet literature in which the process is described more physically as wavefront reversal. Since an equivalent mathematical description of the conjugate field is achieved by leaving the spatial part of E unchanged and reversing the sign of the time, t, we often refer to this process as "time reversal."

To put this general mathematical description on a more physical basis, we consider the specific example of a wave radiating from a point source, as is indicated in Figure 1. If z represents the axial distance from the source, the spatial variation of the phase is

$$\phi(x,y,z) = \frac{\pi(x^2+y^2)}{\lambda z} \tag{4}$$

(This expression neglects diffraction effects, and is not valid near z=0). According to Eqs. (1) and (2) the spatial dependence of the electric field wavefront is proportional to

$$\exp[i(-kz - \frac{\pi(x^2+y^2)}{\lambda z})] \tag{5}$$

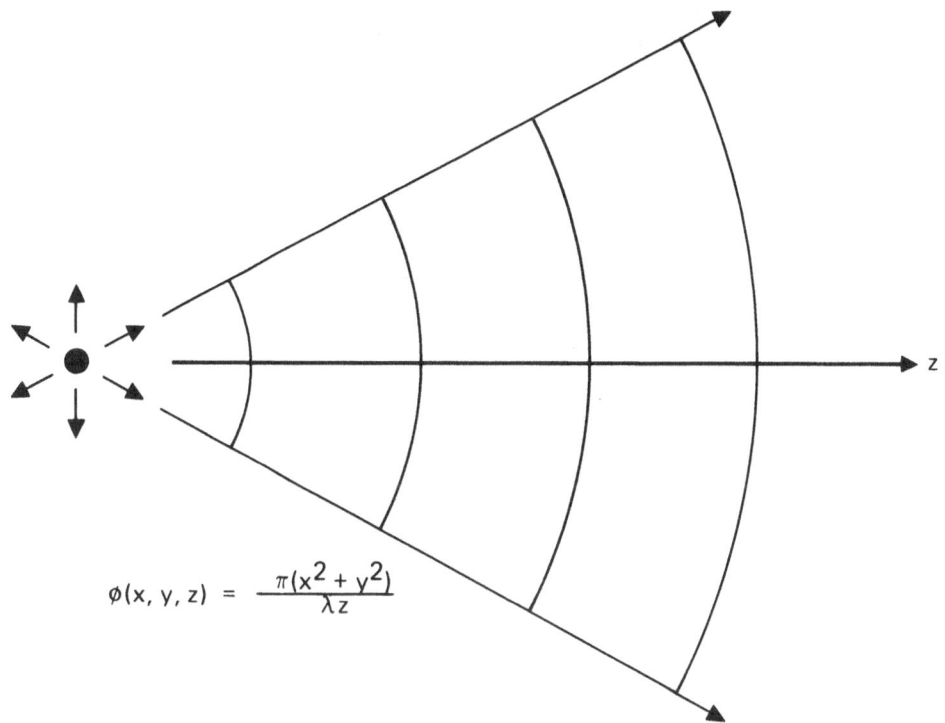

$$\phi(x, y, z) = \frac{\pi(x^2 + y^2)}{\lambda z}$$

Figure 1. Schematic diagram showing the transverse phase distribution of a spherical wave diverging from a point source.

In this case for any value of z, ϕ specifies how the phase must increase off axis to represent the diverging wave propagating from left to right. According to Eq. (3), the spatial dependence of the conjugate electric field wavefront is proportional to

$$\exp\left[i\left(kz + \frac{\pi(x^2 + y^2)}{\lambda z}\right)\right] \tag{6}$$

In this case, ϕ specifies how the phase must vary off axis to represent a converging wave propagating from right to left.

These ideas are illustrated in Figure 2 where we compare reflection of a point source from an ordinary mirror with that from a phase conjugate mirror. For the ordinary mirror, we observe the usual specular reflection in which the diverging wave continues to diverge. In contrast, the same diverging wave striking the phase conjugate mirror becomes a converging wave that retraces its path back to the source.

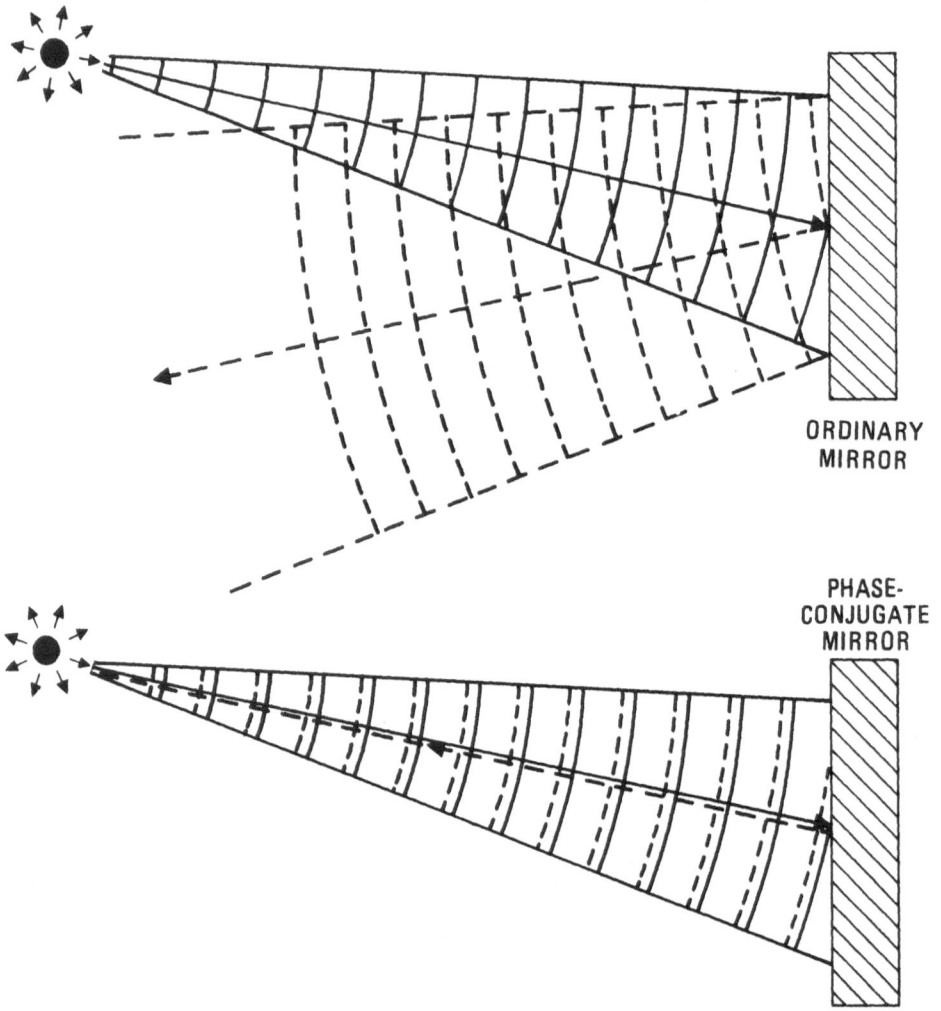

Figure 2. Comparison of reflections of a diverging wave from an ordinary mirror and a phase-conjugate mirror.

As a more interesting example, consider the case in which the wave incident on the conjugator was originally a uniform plane wave, but was aberrated by a distorting medium such as a piece of bottle glass, a turbulent atmosphere, a severely strained optical element, or simply a glass plate with a bump in it (see Figure 3). Such a wave incident on a conjugator would result in a new wave that would be as severely aberrated as the input wave. However, because of the wavefront reversal, when the output wave passes back through the aberrator it will emerge completely free of distortion. This is in contrast to what occurs if the conjugate mirror were replaced with an ordinary mirror, in which case double-passing the aberrator doubles the wavefront distortion. It is just this type of demonstration that was the subject of the first published observations in 1971 and 1972 of optical phase conjugation by researchers in the Soviet Union. Zel'dovich and coworkers[3] and Nosach and coworkers[4] observed wave-front conjugation resulting from the nonlinear optical phenomenon, stimulated Brillouin scattering (SBS). Stepanov and coworkers[5] observed conjugation while experimenting with real-time holography. This phenomenon is similar to degenerate four-wave mixing (DFWM), a nonlinear phenomenon that Hellwarth[6] recognized is a conjugation process; subsequent experimental demonstration was performed by Bloom and Bjorklund[7].

With these basic ideas in mind, it is now appropriate to review in the following section some theoretical principles relating to phase conjugation arising from a variety of nonlinear processes. Specifically, we will consider degenerate four-wave mixing, stimulated Brillouin scattering, and the photorefractive effect. We will then continue with an overview of some promising applications of nonlinear optical phase conjugation.

II. THEORETICAL CONCEPTS

Phase conjugate reflections occur in many materials by virtue of a variety of nonlinear interactions. With a few exceptions they all share one common principle: the presence of the electromagnetic radiation induces a spatial variation of the optical properties of the medium such that light is scattered. This spatial variation in the medium is often called a "grating" because its effect is often similar to that of an ordinary diffraction grating. The grating can be created by one pump beam in

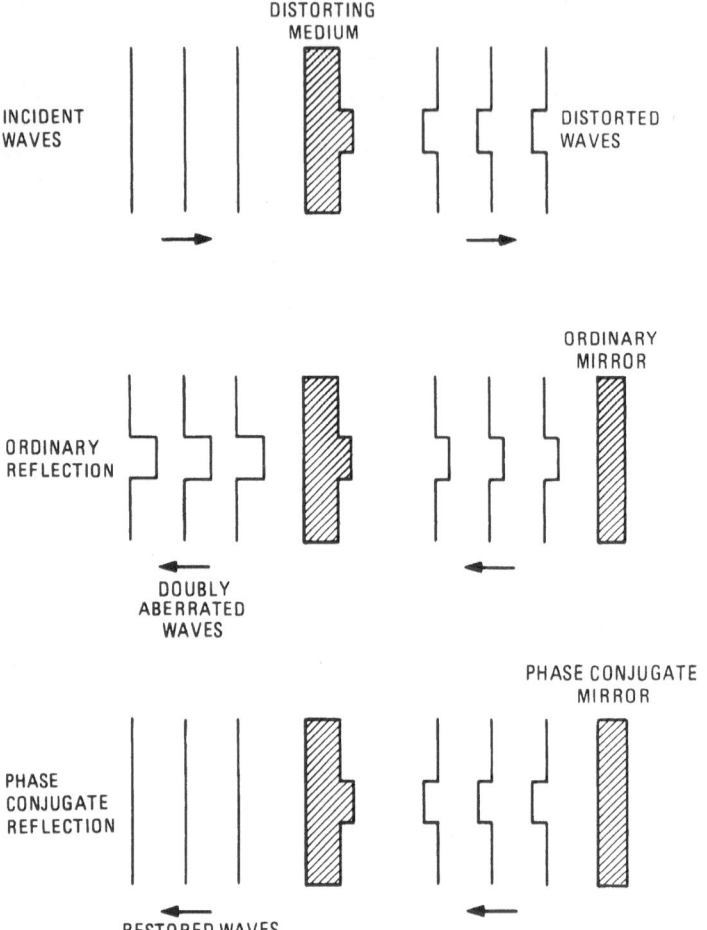

Figure 3. Schematic diagram showing compensation for an optical distortion using a phase conjugate mirror.

the presence of spontaneously generated noise radiation, as in stimulated scattering, or several beams, as in four-wave mixing. In either case, theoretical arguments and experimental measurements agree that the scattered light is the phase conjugate of (one of) the beam(s). Since most of the specific processes are treated elsewhere,[1,2] we will review here only three representative examples: degenerate four-wave mixing, stimulated Brillouin scattering, and the photorefractive effect. In each case, we will describe the nature of the "grating" and show how it arises from the presence of the optical beams.

Degenerate Four-Wave Mixing

Four-wave mixing is a nonlinear process in which three input waves mix to yield a fourth (output) wave. The three input waves consist of two planar counterpropagating pump waves, labeled E_f and E_b (f for forward and b for backward), and a probe wave E_p, entering at an arbitrary angle to the pump waves. All three couple through the third-order susceptibility, $\chi^{(3)}$, to yield a fourth wave, E_s, the signal wave which is proportional to the spatial complex conjugate of E_p.

The conjugate wave is produced by the third-order polarization $P_{nl} = \chi^{(3)} E_f E_b E_p$; P_{nl} is proportional to the product of the amplitudes of the three input waves. More specifically, the nonlinear polarization that yields the conjugate of E_p can be shown to arise (for isotropic media) from the contributions of three separate terms:

$$\vec{P}_{nl} = A(\theta)(\vec{E}_f \cdot \vec{E}_p^*)\vec{E}_b + A(\pi-\theta)(\vec{E}_b \cdot \vec{E}_p^*)\vec{E}_f + B(E_f \cdot E_b)E_p^* \quad (7)$$

(The more complex situation for anistropic media is not discussed here). The first two terms are responsible for the analogy between the degenerate mixing and holography. Each contains a scalar product corresponding to the interference between one of the pump waves and the probe wave; the product is multiplied by the field of the other pump wave. Thus, each term corresponds to the creation of a hologram from one of the pumps fields and the probe while simultaneously reading it out with the other pump. This is illustrated in a simple way in Figure 4, which shows the holographic (or dual grating) picture of degenerate four-wave mixing. The formation and readout processes are shown separately here, although they actually take place at the same time. The formation process is shown as the generation of two overlapping grating structures (fringe patterns), also shown separately for simplicity. Each one consists of a series of planes, with normals in the directions $\vec{k}_f - \vec{k}_p$ and $\vec{k}_b - \vec{k}_p$; the separation of the planes is given by

$$D = \lambda/(2\sin\theta/2) \quad (8)$$

(a) SCHEMATIC CONFIGURATION

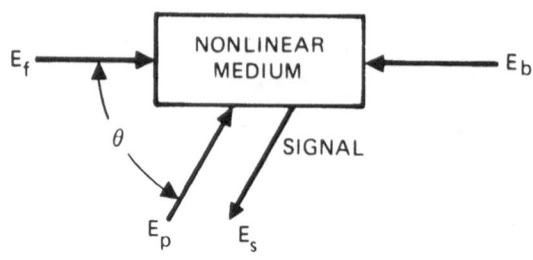

(b) FORMATION   8187-3

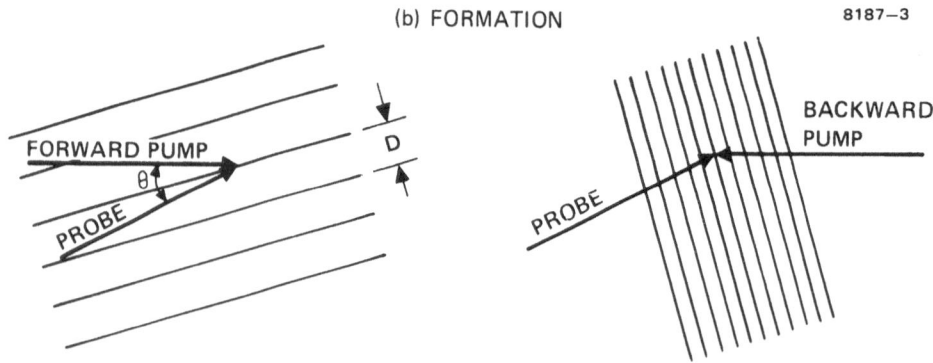

(c) READOUT

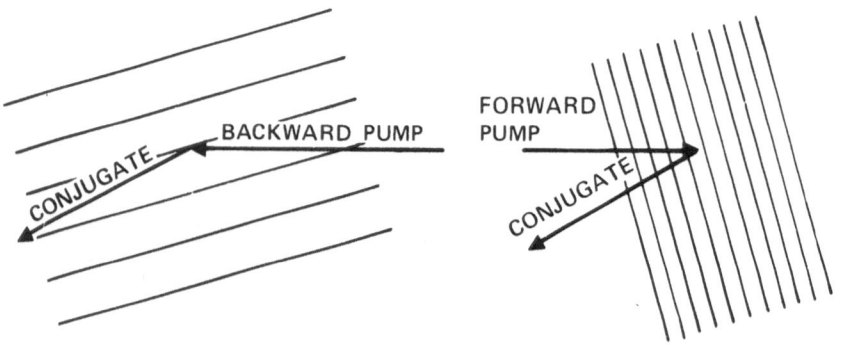

GRATING SPACING $D = \lambda/2 \sin \theta/2$

Figure 4. Diagram representing four-wave mixing: (a)
schematic arrangement showing the four interacting waves;
(b) formation and (c) readout of the "grating."

388

One refers to the pattern arising from the interference between forward pump and probe as the large-spaced grating; the one between the backward pump and probe is the small-spaced grating.

The readout or playback process occurs when the backward pump scatters from the large-spaced grating and the forward pump scatters from the small-spaced grating, yielding the conjugate wave. Thus, the phenomenon described by the first two terms contributing to $\vec{P}_{nl}$ can be viewed as one in which the refractive index of the nonlinear material is spatially modulated as a result of the interference between pump and probe. This is followed by scattering by the other pump.

The term $B(\vec{E}_f \cdot \vec{E}_b)\vec{E}^*$ has no holographic analog. The scalar product of E_f and E_b corresponds to a nonlinear index which has no spatial modulation but which oscillates at a frequency 2ω. The probe wave interacting with this driven coherent excitation at 2ω creates a polarization that results in the generation of a conjugate wave.

The relative magnitudes of the coefficients A and B depend strongly on the properties of the nonlinear medium chosen for the four-wave interaction. In particular, if the nonlinear medium has an optical resonance for a single quantum transition at a frequency near ω (the wave frequency used in the experiment), large enhancements of the four-wave mixing signals arising from the first two terms are possible compared with that obtainable from a nonresonant system.

For example, consider a nonlinear medium consisting of an ensemble of two-level atoms. The near-resonant contribution to the nonlinear index will manifest itself as a spatial modulation of the populations of the lower state relative to that of the upper state. The gratings formed as a result of interference between the pumps and the probe would be "population gratings," that is, if one were to walk in the direction $\vec{k}_f - \vec{k}_p$ in the nonlinear medium, one would notice that the population of atoms in the excited state relative to the ground state varies sinusoidally with a period $D = \lambda/(2\sin\theta/2)$. Alternatively, if the medium possesses a pair of energy levels of the same parity, allowing them to couple coherently through an interaction involving two quanta, then the third term in $\vec{P}_{nl}$ may be dominant in contributing to the four-wave signal.

Degenerate four-wave mixing allows for conjugate reflectivities in excess of unity. That is,

$$I_{signal}/I_{probe} > 1. \tag{9}$$

Many such examples have been observed experimentally (see below). This fact has practical implications for four-wave mixing.

Another characteristic of the mixing process that can have practical implications involves its behavior for waves of different polarization. In fact, experiments in which the polarization of pump and probe waves have been manipulated to achieve the desired result have enhanced our understanding of the four-wave mixing.

Note that each of the terms in the equation for $\vec{P}_{nl}$ involves the scalar product of two fields multiplied by a third field. Thus, a given term will contribute to the nonlinear polarization only if the scalar-product term is nonzero—that is, only if the fields in the scalar product have polarization components along a common direction. This fact can be exploited to explore the fundamental properties of four-wave mixing. For example by performing an experiment in which $\vec{E}_f$ and $\vec{E}$ are linearly copolarized while $\vec{E}_b$ is cross polarized, one is examining the contribution of only the first of the three terms in $\vec{P}_{nl}$, that is, the large-spaced grating. Hence, by appropriate selection of co- and cross-polarized combinations, the contributions of the different terms in $\vec{P}_{nl}$ can be examined for various nonlinear materials.

We can carry this one step further by examining what might be expected in the following situation: We have a gaseous, nonlinear medium that consists of atoms or molecules having a two-level, optically-allowed transition in the vicinity of the frequency of our four-wave-mixing experiment. Now, because of the resonant enhancement, we would expect the contribution from the first two terms—the "population grating" terms—in $\vec{P}_{nl}$ to be orders of magnitude greater than that from the third term.

The interference patterns between pump and probe waves are fixed in space; if the atoms were stationary, the population gratings would also be fixed in space. The depth of modulation of the population gratings (and hence the four-wave reflectivity) would depend both on the light intensity—which determines how fast atoms are promoted to the upper level—and on the lifetime of the

390

upper level prior to decay back to the ground state.
Now, if the atoms can move an appreciable distance--say,
an appreciable fraction of a grating period (or fringe
spacing) during an excited-state lifetime--we have
another mechanism by which the four-wave signal can be
degraded: a grating "wash-out" effect.

How long does it take for an average atom to move
the length of a grating period? That depends on how big
the grating period is. In fact, it suggests a nice
experiment. Remember, the grating spacing,
$D=\lambda/(2\sin\theta/2)$, can vary all the way from $\lambda/2$ to infinity,
depending on θ. This means that we would expect the
large-spaced grating to suffer less from degradation due
to atomic motion than the small-spaced grating. By using
polarization tricks, we can choose to look at the effects
of one grating at a time, comparing the results with what
we would expect from a theory that takes the effects of
atomic motion into account. This, in fact, has been
done: the agreement between theory and experiment is
excellent.[9,10]

A contrasting feature of our expression for $\vec{P}_{nl}$ is
that the third term--the 2ω coherent term--is not
expected to degrade in any way due to motion effects, a
characteristic which has practical implications.

Stimulated Brillouin Scattering

Stimulated Brillouin scattering involves the
generation of a coherent acoustic wave when an intense
optical wave interacts with a nonlinear medium. The
mechanism generally acts through electrostriction--that
is, the tendency of the medium's density to increase in
proportion the electric field intensity. Above some
threshold intensity a light wave allowed to propagate in
a nonlinear medium produces an intense back-scattered
optical wave, and its frequency is down-shifted by an
amount equal to the acoustic frequency.

The collinear process, which involves
counterpropagating incident and scattered light waves,
has the highest gain of all the possible scattering
processes and is the only one observed in stimulated
Brillouin scattering. The acoustic wave propagates in
the same direction as the incident wave; it can be
thought of as a moving mirror or stack of dielectric
plates, from which the incident wave reflects to generate
a Doppler-shifted scattered wave. In fact, the

wavelength of the acoustic wave is half that of the optical wave in the medium; thus the acoustic wave serves as a moving half-wave dielectric stack.

When thinking of stimulated Brillouin scattering in terms of electrostriction, note that the two counterpropagating optical waves form a moving interference pattern. The speed of the pattern, given by $V=(\omega_1-\omega_2)/(k_1+k_2)$, is equal to the speed of sound in the medium. In fact, it is this condition that allows for the buildup of both the acoustic wave and the scattered optical wave, at the expense of the incident optical wave.

Analysis indicates that, under certain conditions, the process for which the Brillouin gain is greatest is the one in which the scattered wave is the conjugate of the incident wave.[3] Consequently, an aberrated input wave generates an equally aberrated acoustic wave through stimulated Brillouin scattering, with a phase surface that matches it exactly. One can think of the nonlinear process as creating (in the medium) a deformable mirror whose surface is just right to reverse the phase of the reflected wave from that of the incident wave. Thus, when the reflected wave retraces the incident path, the medium removes from it whatever phase errors were introduced in the first pass. If it were possible to take a moving picture of the incident wave the complete behavior of the conjugate wave would be portrayed by running the film backwards.

Stimulated Brillouin scattering can be made to occur in a highly controlled manner and with efficiencies approaching unity. This is especially true under the conditions where it works best as a conjugator, in multimode optical-waveguide configurations. It has been observed over a wide range of wavelengths and under both continuous and pulsed conditions.

The Photorefractive Effect

Photorefractive materials have recently become the subject of intensive research[11] because of their great potential applicability as a phase conjugate device. Although these materials are generally utilized in configurations involving four-wave mixing it is treated here as a separate topic because the materials have such interesting physical behavior.

Photorefractive materials are some of the most sensitive materials that have been shown to perform optical phase conjugation; the sensitivity is comparable to that of optical film. These materials include $BaTiO_3$, $Bi_{12}SiO_{20}$ (BSO) and $KaTa_{1-x}Nb_xO_3$ (KTN).

The photorefractive effect first appeared in the literature as "index damage" in $LiNbO_3$ used to frequency double Nd:YAG radiation. Exposure to the intense 532 nm beam induced long term refractive index changes in the material, thereby spoiling the phase matching condition required to achieve efficient second harmonic generation. Within just a few years, it was recognized that this "problem" could be used to advantage; Chen and co-workers[12] demonstrated that the light-induced index change could be used to store high-quality holographic images in $LiNbO_3$ and $LiTaO_3$. This discovery was the beginning of the active research into the photorefractive effect.

This spatially varying refractive index, or grating, is induced by two interfering laser beams in the following way. First, the light causes charges to migrate from bright to dark regions within the crystalline material. The charges are believed to arise from low-lying traps formed by impurity or defect sites in the crystal. Second, the separation of the positive and negative charges induces a strong electrostatic field within the crystal (the field strength is on the order of 10^5 V/m). Third, the electrostatic field induces refractive index changes through the linear electro-optic effect.

Some unique features of this process are worthy of note. First, as is indicated in Figure 5, the index grating may have a phase shift with respect to the spatial variation of the light intensity. This fact leads to the possible coupling of energy from one beam to another as the two beams propagate through the photorefractive material.[13] Second, the magnitude of the induced index change is not dependent on the total intensity of the two beams, but only on their ratio. This arises because the steady-state charge distribution is determined by two factors: the periodic spatial variation in the intensity pattern tends to force the charges into a periodic distribution, while the uniform component tends to randomize the charge locations and erase the periodic distribution. The periodic spatial variation is greatest when the two beams are equal.

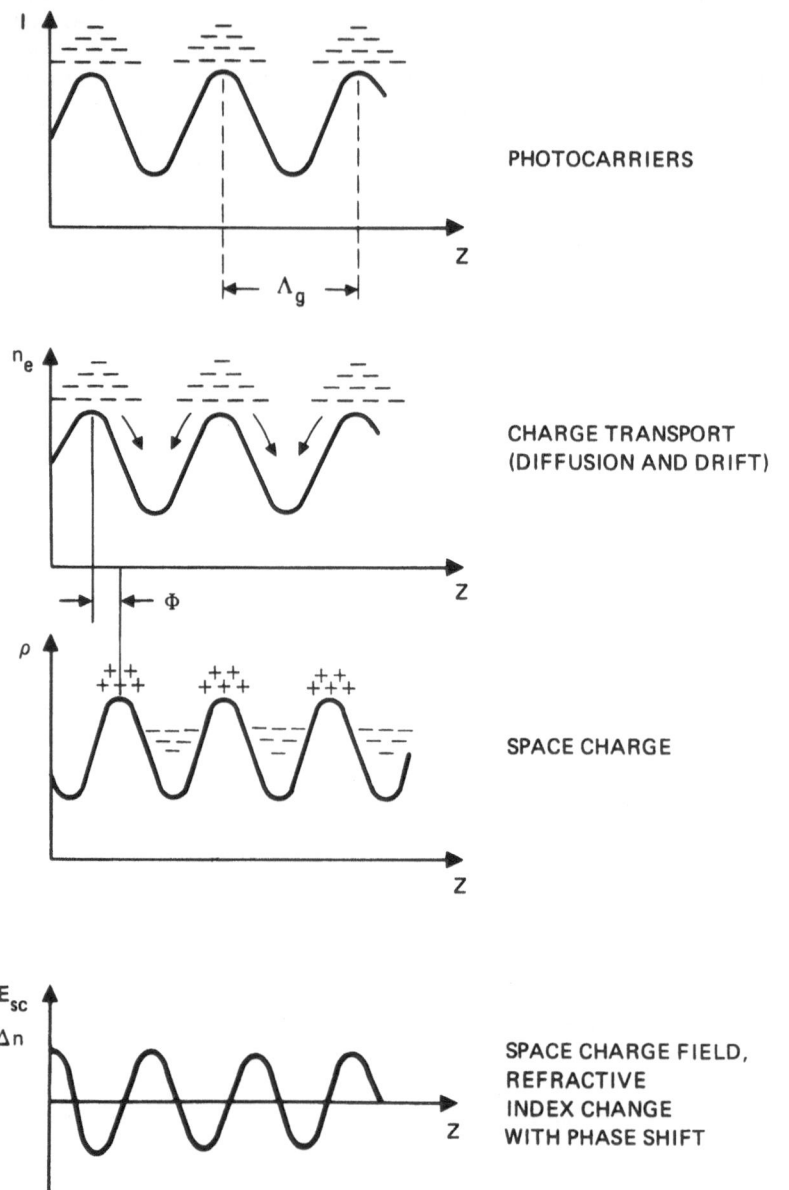

Figure 5. Schematic diagram representing the photorefractive effect. Photocarriers, generated by the nonuniform optical radiation, diffuse within the crystal to create a nonuniform charge distribution which modulates the refractive index via the electro-optical effect.

One of the most exciting demonstrations involving photorefractive materials is the self-pumped phase conjugator.[14] This demonstration relies on the beam coupling referred to above to generate two counterpropagating beams from a single incident beam. This concept is finding a variety of applications, such as phase conjugate resonators (reviewed below).

III. APPLICATIONS

Phase conjugation devices are rapidly becoming major elements in a wide range of applications. Before reviewing major areas, we consider how good the fidelity of the aberration correction can be. The answer is, perfect--within the limits of measurement employed so far. Evidence of the ability of a conjugator to correct for optical distortion is depicted in Figure 6. These far-field photographs of a laser beam show the unperturbed beam (a), the beam after passing through a random aberrator (an etched glass plate) (b), and the profile after the aberrated beam was conjugated and passed back through the etched glass plate (c). When the detailed profile in (a) is compared with that in (c) using a multiple-exposure photographic technique, they are seen to be identical within experimental measurement error.[15]

These measurements extend from the center of the beam out into the wings, where the intensity is as low as 10^{-5} of the axial peak intensity. The most severe random aberration used in this work degraded the unperturbed, diffraction-limited beam to about 35 times the diffraction limit. This degradation corresponds to reduction of the axial intensity to about 1/1000 of the unperturbed condition. Such compensation effects have been demonstrated for both stimulated Brillouin scattering and degenerate four-wave mixing.[15]

The only apparent limitation to the fidelity of the conjugation process is the effective aperture of the conjugator itself. This is equivalent to saying that the laws of diffraction still hold. In other words, if a conjugator having finite aperture a is located a distance L away from a random-phase plate, the smallest transverse scale of aberrations that will be compensated will be of the order L/a. Smaller-scale transverse bumps (aberrations) in the phase fronts emanating from the phase plate will not be compensated when the conjugate wave passes back through.

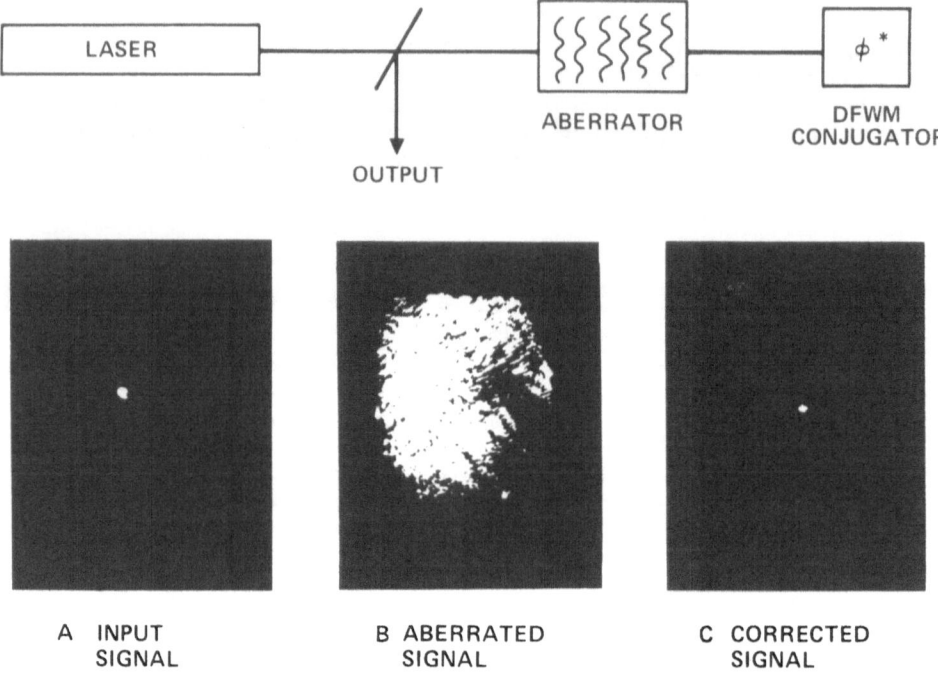

Figure 6. Aberration compensation using phase
conjugation. Using the experimental arrangement at the
top, the input far-field intensity distribution A is
aberrated to produce the pattern B, which is phase
conjugated and sent back through the aberrator to produce
the pattern C.

Another way of saying this is that light coming from
small-scale regions (smaller than L/a) will diffract by
the time it reaches the conjugator so that an appreciable
fraction will spill over the collecting aperture; and if
you do not collect all the light, you cannot conjugate it
completely.

It wasn't until several years after the first
observations of phase conjugation that a potential
application to adaptive optics was first described.[16]
The ability of phase conjugators to restore severely
aberrated waves to their original state on passing

through the distorting medium twice suggests a potential application to adaptive optics. The concept is illustrated in Figure 7. The goal is to deliver energy from a laser system to a target or receiver, while minimizing the effects of distortions that tend to spread the beam and reduce the energy density at the target. A practical situation where this might be applied is laser fusion, where energy is to be deposited onto a target pellet.

The first step is the generation of a reference. This is done, for example, by illuminating the target with a pulsed laser source having a wavelength within the gain bandwidth of the system's amplifiers. Some of the light reflected from the pellet is captured in the aperture of the focusing element (shown in the figure as a lens) and enters the optical system. Note that in a real system containing many optical elements and possibly a propagation medium, the reference wave accumulates phase distortions; these can cause its wavefront to deviate substantially from what it would be if it propagated in free space and encountered perfect optical components.

The second step is amplification of the distorted reference wave; the third is conjugation. After conjugation, the wave undergoes a second amplification and propagates back through the optical train; this time, because of the conjugation process, the phase distortions accumulated on the first pass are eliminated in a reverse sequence, as the wave makes the second pass through the system. The result is delivery to the target of an intense pulse of light that is virtually diffraction limited. More precisely, the pulse wavefront incident on the target is a replica of the reference wavefront that radiated from the target in the first place.

The beauty of the scheme is that once the reference is created everything else follows automatically. It has particular value in laser-fusion systems that irradiate the target from several directions. Here, the problem of beam alignment, pointing, and focusing is extremely complex in the conventional technology, demanding that the target be precisely located within a narrowly defined field of view and requiring sophisticated sensor/servo systems. This is to say nothing of the large number of optical elements, turning mirrors and so forth within the optical train, each of which is a source of optical distortion.

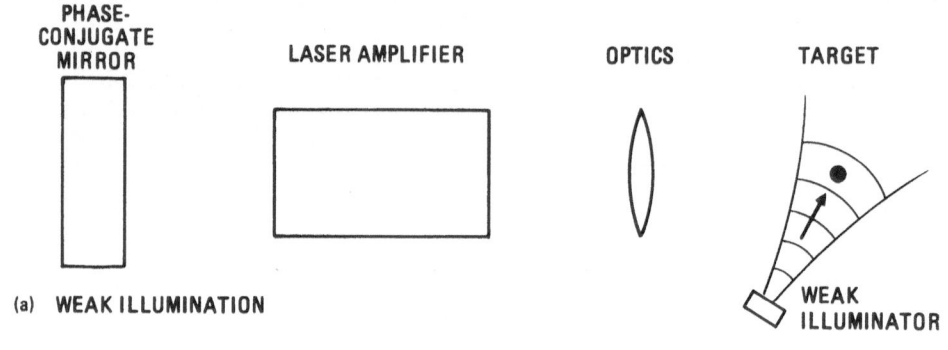

(a) WEAK ILLUMINATION

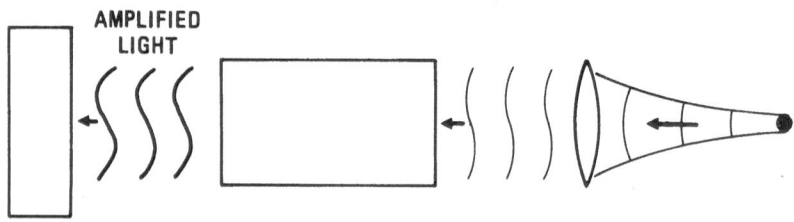

(b) ABERRATION (LASER + OPTICS)

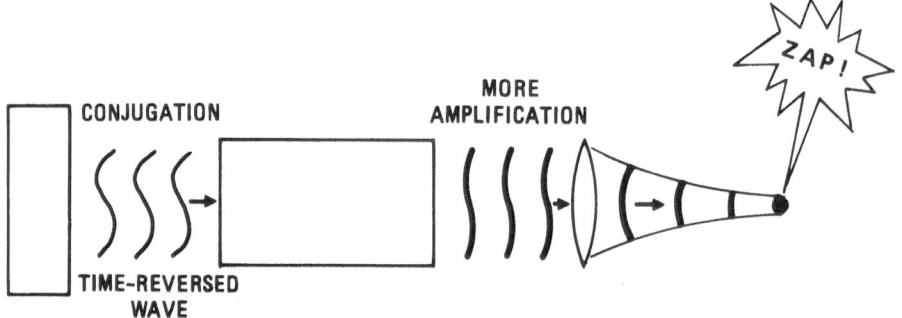

(c) COMPENSATION + INTENSE ILLUMINATION

Figure 7. Phase conjugate adaptive optics, as applied,
for example, to laser fusion. The sequence of events is
(a) illumination of the pellet with a probe beam; (b)
reflection and amplification, resulting in a distorted
wave; (c) conjugation; and (d) a second pass through the
distorting medium, producing coherent illumination of the
pellet.

Phase conjugators in such multiple-arm systems have the potential for eliminating this complexity. A single reference pulse illuminating the target to initiate the process can result in precise delivery of energy from a multiple-arm system. Pointing and focusing are provided automatically, no matter where the pellet is located in the target chamber (within reason). Simultaneous arrival of the intense pulses can be ensured by making all arms the same length to fairly loose tolerances—no more than a small fraction of the pulse length (1 nsec or 30 cm, approximately).

Similar considerations apply to the delivery of laser energy to a remote receiver or target through a turbulent atmospheric path. Quite often, the energy must be delivered in a beam that has minimal distortion. In "conventional" adaptive optics, one measures in some way the atmospheric turbulence along the propagation path; one then uses deformable mirrors to predistort the outgoing laser beam in a way that compensates exactly for the atmospheric distortions to be encountered along the propagation path to the target. The atmospheric path errors can be sensed by measuring the phase-front of a reference wave that originates from a bright glint or a beacon located at the target. The reference wave from the target glint is received at the aperture of the laser transmitter. This aperture is divided into a number of subapertures; a local wavefront tilt for each of these is measured—with, for example, Hartmann sensors or shearing interferometers. The size of each subaperture is chosen to be consistent with the scale of atmospheric turbulence expected at the particular operating conditions. The wavefront-tilt information is used to drive actuators on a deformable mirror, from which the outgoing laser beam reflects on its way out to the target. Through this sequence of wavefront measurement, error signal generation, and mirror deformation, the outgoing wave is transmitted as the phase conjugate of the incoming reference wave.

Here we see the essential contrasting features between conventional adaptive optics and nonlinear adaptive optics. In the conventional case, the reference wave is measured and discarded, and the results of the measurement used to obtain the necessary settings on the deformable-mirror actuators. This is followed by transmission of a beam from a different source, the laser (presumably, one relatively free of distortion), through the optical system. In the nonlinear case, the reference wave is not discarded, nor is it measured in the usual

sense of the word. It is amplified, conjugated, and
retransmitted.

One distinct advantage of the nonlinear over the
conventional approach to wavefront correction of laser
beams is that compensation will still occur even if the
reference wave has substantial amplitude variation over
the wavefront, as will be the case in the event of severe
turbulence; the conventional approach does not compensate
for amplitude variations—only phase variations.

In the conventional, adaptive-optics approach the
reference wave is measured and then discarded and it
therefore does not need to have the same wavelength as
the wave that is ultimately transmitted. It can be
derived from a completely different source that is in no
way related to the laser. Of course, if the reference
wave and transmitted wave have vastly different
wavelengths, the compensation process may suffer because
of dispersive effects. This fact, nevertheless
represents a potential advantage over the nonlinear
approach in which the reference wave length must be
compatible with the amplification process and the
conjugation process. If the conjugator is based on
degenerate four-wave mixing then the reference wave must
satisfy specific coherence requirements relative to the
pump waves, consistent with the response time of the
nonlinear medium—namely that the difference between the
pump and reference frequencies be no greater than the
reciprocal of the response time.

Optical Resonators

An interesting and potentially valuable application
for phase conjugate optics is the use of a phase
conjugator as an element in an optical resonator. A
phase-conjugate resonator is an optical resonator in
which one (or both) of the conventional mirrors is
replaced by a conjugate mirror. Several papers[17-20] have
predicted the properties of these devices, and a few of
these properties have been demonstrated.[21-23] Practical
considerations have also been studied.[24]

There are several unique properties that phase-
conjugate resonators are expected to exhibit. One is
that such a resonator will not possess longitudinal modes
that depend on the cavity length. An ordinary optical
resonator possesses longitudinal modes separated in
frequency by $c/2L$, where L is the cavity length. This

results from the boundary condition that after one round trip (two reflections) the wave that corresponds to a resonant mode must constructively interfere with itself. This requires that the net accumulated phase after one round trip must be an integral multiple of 2π, in other words, the only waves that "fit" in the resonator are those for which $n\lambda=2L$ (n an integer).

In a phase conjugate resonator on the other hand, the phase that is accumulated as the wave propagates from the ordinary mirror to the conjugate mirror is subtracted by the same amount on the way back to the ordinary mirror; in one round trip, the net accumulated phase is thus alway zero. Consequently, a phase conjugate resonator of length L can support any wavelength consistent with the bandwidth of the gain medium and the conjugator itself. Moreover, a conjugate resonator oscillating at a particular wavelength will continue to oscillate at that wavelength, independent of variations of the cavity length. This is in contrast to an ordinary resonator, whose spectral output will exhibit "mode hopping" and frequency drift if the cavity length is allowed to vary.

Another property of a phase-conjugate resonator makes it highly attractive for application to high-power oscillators: It can compensate for intracavity optical distortion. One can show that, when light is extracted from the "ordinary-mirror" end of the resonator, the transverse phase of the wave only depends on the output mirror's figure (its detailed surface shape); the phase will not depend on any other sources of distortion within the body of the resonator.[17] This feature has been demonstrated qualitatively in laboratory experiments.[21]

Power and Image Transmission

One intriguing potential application of phase conjugation is in the transmission of high power electromagnetic radiation from a space-borne power generating station to a terrestrial site. Many concepts have appeared over the years that involve conversion of solar energy to electrical energy in a space station, followed by beaming of the energy to earth via coherent optical or microwave radiation. NASA is also interested in the possibility of direct conversion of solar energy for pumping high power space-borne lasers. An intriguing possibility arises from the concern for safety, in requiring a provision for accurate and reliable pointing of the multi-megawatt beams from the space station to the earth station.

Phase conjugation suggests a unique approach for solving this potential problem. The space-based power-generating station consists of a power source (solar radiation), a coherent oscillator (high-energy laser), and four-wave mixing medium. The space-borne high-energy laser provides the counterpropagating pumps for the four-wave mixing medium; in one configuration, the medium could be contained within the laser's resonant cavity. Ideally, the pump power and the nonlinear medium would be chosen to provide coupling from the laser resonator through the medium, via a pilot beam originating at the terrestrial receiving station. The conjugation process returns the earth-directed output beam along the same path on which the input beam is received. This ensures that the only energy delivered to earth is along the pilot-beam path. Moreover, the energy circulating inside the space-borne laser will be coupled out only when the pilot beam is present.

Optical phase conjugation can be useful for a number of applications requiring delivery of a "special" field distribution from one point to another. An example of a special field distribution is an image. One of the first such applications suggested has to do with the transmission of images along optical fibers. A problem in transmitting an image along an optical fiber and recoverning it at the other end arises from modal dispersion in the fiber, this causes a scrambling of spatial information, which can seriously degrade the image. By sending the image through a length of fiber, into a conjugator and finally through another fiber identical to the first, with the conjugator at the midpoint of the path, one could offset the effects of modal dispersion and recover the original image. The real question here is whether or not it would be possible to find lengths of fiber with "identical" modal-dispersion characteristics, an issue yet to be addressed experimentally.

Another area involving imaging and optical phase conjugation is in photolithography. Projection of complex patterns onto photo-resist layers is of great technological importance in the microelectronics industry. Projection systems using conventional optical techniques are very complex, due to the needs for near-diffraction-limited, low f-number performance.

One approach to a lensless 1:1 projection system employing optical phase conjugation is shown in Figure 8 The object, in this case a mask or transparency, is illuminated from the back with a laser. The image is formed on the photoresist surface of a substrate, after being reflected from the conjugator by the way of the beam splitter. The advantage of this scheme is that a diffraction-limited performance can be achieved without using expensive optical components. (The only element requiring high optical quality is the beam-splitter). This approach achieves the same goal as contact photolithography, without placing the mask in direct contact with the sample (a step that may be highly undesirable). Projection of high-quality images using such a technique has recently been demonstrated.[25]

A variation on this application is one where a special field distribution is to be delivered to a target plane at intensities that are high enough to damage a mask or transparency. This scheme is also shown in Figure 8 but this time including the amplifier. In this case, the mask is illuminated with a low-power beam; this is amplified, conjugated, reamplified, and delivered to the target as a high-intensity beam having a special field distribution. The scheme could be applied, for example, to situations involving laser annealing in cases where only specific samples areas are to be irradiated and no others. Another possibility is laser fusion, where it may be necessary to illuminate the target pellet with other than nearly uniform illumination. Still another potential application is irradiation of a wire with a high-power laser to create an intense, linear x-ray source.

Pump Manipulation

Four-wave mixing is more complicated experimentally than stimulated Brillouin scattering because it requires the use of auxiliary pump waves. This very fact, however, gives it greater flexibility. The reason is that the pump parameters can be modified to yield an output wave which, in addition to being the spatial conjugate of the input wave, has some other desirable property that can be exploited for a specific application. Several examples indicate the possibilities.

Pointing control by pump "misalignment". Suppose the target in the laser-fusion example in Figure 5 were

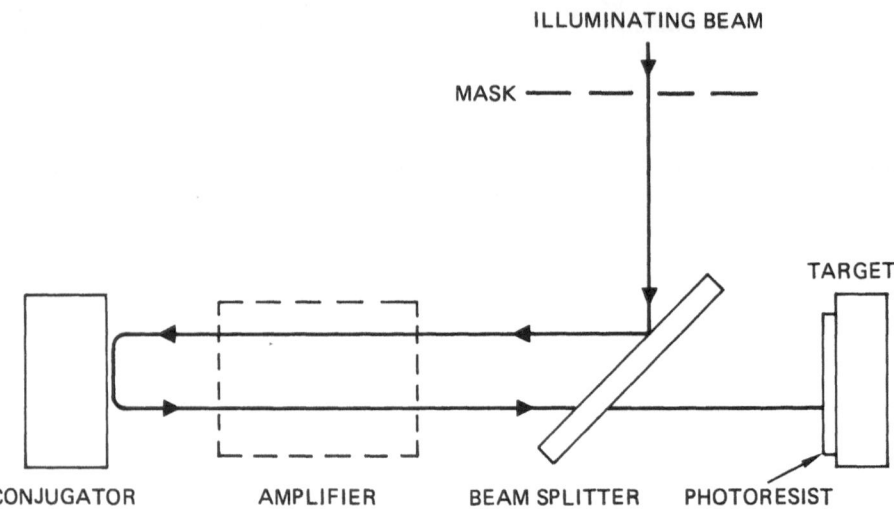

Figure 8. Application of phase conjugation to photolithography. A light beam is used to form an image of the mask onto a target. The ability of a phase-conjugate "mirror" to provide an undistorted image is particularly useful if the mask could be damaged by the high intensities used to form the image; the conjugate mirror restores the distortions introduced by the amplifier.

moving transverse to the direction of the beam. If its speed is sufficiently large, it will have moved an appreciable fraction of a beam diameter in the time it takes light to travel from the target to the conjugator and back to the target again. In this case, delivery of the laser energy to the target requires that it be possible to override the tendency of a good conjugator to produce a true time-reversed beam. We can deviate from the precise "retro" behavior by intentionally misaligning the pumps from the counterpropagating geometry. The degree to which the direction of the conjugate wave deviates from the refrence wave, as the pumps are misaligned, can be calculated in a straightforward manner from the phase-matching conditions; it is a function of the pump-misalignment angle and the pump-probe angle. The conjugate-wave reflectivity drops off as the pumps are misaligned—but only a few percent for misalignments of the order of a few milliradians.[26]

Thus by controlling the precise propagation direction of the pump waves one can control the direction of the conjugate wave. It is important to remember, of course, that compensation for optical-path errors will be less perfect if the conjugate wave is not allowed to retrace exactly the path taken by the reference wave. The extent to which the compensation is degraded depends upon the detailed spatial structure of the propagation path errors, and will ultimately limit the amount of point-ahead angle (or lead angle) that can be tolerated in an adaptive optical system.

Pump polarization manipulation. As discussed earlier, it is possible to generate a conjugate wave whose polarization is orthogonal to that of the reference wave. This fact allows the physical separation of the two waves through the use of polarization splitters and opens the door to a variety of applications. One is repointing the conjugate wave away from the backward direction, another is refocusing to a plane other than that from which the reference originates. Figure 9 shows schematically how such a polarization-manipulation scheme might work.

Pump temporal modulation. Another, similar, application of four-wave mixing is in covert optical communications that require information to be conveyed to one or more remotely located, mobile receiver sites from an air- or space-born platform, without broadcasting over a wide area. Here again, the four-wave mixer is situated in the transmitter and the receivers are equipped with interrogators which illuminate the remotely located transmitter with lasers tuned to a predetermined operating frequency. For this example, the interrogating beam and one of the pump beams operate continuously and the other pump is pulse-modulated in an appropriate fashion. The information is transmitted back to the interrogation site as a modulated conjugate wave. Only those sites possessing proper interrogating capability can obtain the information. An added advantage is that the total power required to operate the transmitter can be many orders of magnitude smaller than that required for a broadcasting system.

Several other pump parameters can be varied, with potentially interesting results. The imposition of phase variations on the pump waves result in a transfer to the signal wave, and has the potential for imposing focus on other phase information on the conjugate wave. This could be important in certain applications of adaptive optics.

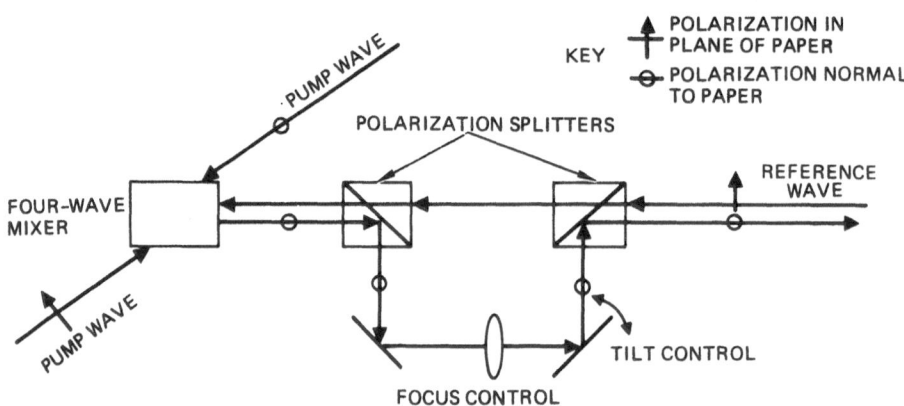

Figure 9. Modulation of polarization to produce a conjugate wave whose direction or focus can be changed. Because the conjugate and probe beams are polarized orthogonally, the conjugate beam can be separated and thus refocussed or realigned.

Correlation and Convolution

The fact that the four-wave signal is proportional to the product of the three input fields suggests an application that, though having little to do with the conjugation property of four-wave mixing, is sufficiently interesting to mention in passing. One can readily obtain the spatial Fourier transform of a field through the appropriate use of lenses. By multiplying the Fourier transforms of probe and pump fields having transversely varying phase or amplitude, one can obtain a resultant signal field that has the properties of a correlation or convolution between the input fields.

Thus, to obtain correlation between two of the input fields, the probe wave and one of the pump waves are chosen to have transverse information and the other pump wave to be a point source.[27]   To perform a convolution, information is placed on the pump waves and the probe wave is made a point source. Autocorrelation and autoconvolution as well as other more complex operations are also possible with this approach. Considering the potential for high-speed, real-time information processing for spatially complex optical fields, such four-wave-mixing applications have exciting possibilities.

IV.   CONCLUSION

After relatively few years in the young field, most of the fundamental physics of optical phase conjugation is well understood. Many predictions of the details of degenerate four-wave mixing have been experimentally confirmed, with good agreement; deviations from theoretical predictions are at least qualitatively understood. Many demonstrations have been made under both pulsed and cw condtiions, over a wide range of wavelengths, and for many different materials. These are summarized in Reference 2, which includes pertinent comments (percent reflectivities, power densities, conjugator longitudinal dimensions) along with references. A number of impressively high reflectivities have been observed at modest power densities, in samples of modest dimensions.

This is only the beginning; much remains to be done, especially in finding new nonlinear materials tailored for specific applications. The only demonstrations carried out so far have been on a laboratory scale using well behaved, low-average-power lasers. None have yet been done at high average powers with conjugators larger than about a centimeter in cross section. In addition, we have yet to experiment by taking a weak input wave, amplifying it and conjugating it, with sufficient overall gain to produce an output wave with adequate energy for practical applications. And there are several other issues to be considered: the fidelity of conjugation via four-wave mixing under very high reflectivity conditions, the extent to which other nonlinear phenomena compete with the desired conjugation process, how a conjugator behaves under extremely weak signal conditions, and how the four-wave-mixing reflectivity degrades as the probe-pump ratio approaches unity.

REFERENCES

1. Robert A. Fisher, ed. Optical Phase Conjugation (Academic, New York, 1983).
2. David M. Pepper, "Nonlinear Optical Phase Conjugation," The Laser Handbook, Vol. 4, edited by M. Bass and M. Stitch.
3. B. Ya.Zel'dovich, V. I. Popovichev, V. V. Ragul'skii, F. S. Faizullov, Sov. Phys. JETP 15, 109 (1972).
4. O. Yu.Nosach, V. I. Popovichev, V. V. Ragul'skii, F. S. Faizullov, Sov. Phys. JETP 16, 46 (1971).
5. B. I. Stepanov, E. V. Ivakin, A. S. Rubanov, Sov. Phys. Doklady 16, 46 (1971).
6. R. W. Hellwarth, J. Opt. Soc. Amer. 67, 1 (1977).
7. D. M. Bloom, G. C. Bjorklund, Appl. Phys. Lett. 31, 592 (1977).
8. R. L. Abrams, R. C. Lind, Opt. Lett. 2, 94 (1978); Opt. Lett. 3, 205 (1978).
9. D. G. Steel, R. C. Lind, J. F. Lam, C. R. Giuliano, Appl. Phys. Lett. 35, 376 (1976).
10. S. M. Wandzura, Opt. Lett. 4, 208 (1979).
11. Jack Feinberg, "Optical Phase Conjugation in Photorefractive Materials," Optical Phase Conjugation, Robert A. Fisher ed., (Academic, New York, 1983), p. 417.
12. F. S. Chen, J. T. LaMacchia, and D. B. Fraser, Appl. Phys. Lett. 13, 223 (1968).
13. Jack Feinberg, ibid., p. 431.
14. J. O. White, M. Cronin-Golomb, B. Fischer, and A. Yariv, Appl. Phys. Lett. 40, 450 (1982).
15. R. C. Lind, C. R. Giuliano, Conf. on Laser Engineering and Applications, Washington, D.C., June 1979 (unpublished).
16. V. Wang, C. R. Giuliano, Opt. Lett. 2, 4 (1978).
17. J. F. Lam, W. P. Brown, Opt. Lett. 5, 61 (1980).
18. J. M. Bel'dyugin, M. G. Galushkin, E. M. Zemskov, Sov. J. Quantum Electron. 9, 20 (1979).
19. J. AuYeung, D. Fakete, D. M. Pepper, A. Yariv, IEEE J. Quantum Electron. QE-15, 1180 (1979).
20. P. A. Belanger, A. Hardy, A. E. Siegman, Appl. Opt. 19, 602 (1980).
21. D. M. Pepper, D. Fakete, A. Yariv, Appl. Phys. Lett. 33, 41 (1978).
22. J. Feinberg and R. W. Hellwarth, unpublished.
23. R. C. Lind and D. G. Steel, unpublished.
24. C. R. Giuliano, R. C. Lind, T. R. O'Meara, and G. C. Valley, Laser Focus, February, 1983, p. 55.
25. M. D. Levenson, Opt. Lett. 5, 182 (1980).
26. R. C. Lind, T. R. O'Meara, R. K. Jain, D. G. Steel,

R. A. McFarlane, Hughes Research Laboratories preprints.
27. P. Avizonis, F. A. Hopf, W. D. Bomberger, S. F. Jacobs, A. Tomita, K. H. Womack, Appl. Phys. Lett. <u>31</u>, 435 (1977).

WAVELENGTH CONVERSION BY STIMULATED RAMAN SCATTERING

David A. Rockwell and Hans W. Bruesselbach

Hughes Research Laboratories
3011 Malibu Canyon Road
Malibu, California 90265

INTRODUCTION

In considering a broad subject area such as the physics of new laser sources, one must necessarily review recent progress in the particular area of wavelength conversion via nonlinear optical processes. From a fundamental perspective, the field of nonlinear optics offers researchers a unique opportunity to study basic optical properties of materials in the presence of intense electromagnetic radiation in which the electric field strength might be comparable to that of the field binding the valence electrons to an atom. From an applications perspective, nonlinear optical devices offer a practical method to generate coherent radiation at wavelengths significantly removed from those of existing laser sources, thereby obviating the necessity for developing a fundamentally new laser source in every wavelength range that might be of interest. This article reviews the physics and applications of the specific nonlinear process of stimulated Raman scattering. Recent research shows that Raman devices offer great promise for producing multiple wavelengths from a single suitable pump laser, with projected average output powers well in excess of 10 watts. With a realistic energy conversion efficiency $\gtrsim 50$ percent, Raman devices are becoming established as a major factor in the physics of new laser sources.

The spontaneous Raman scattering process first received active attention during the late 1920s and early

1930s. The Raman effect is an inelastic light scattering
process; that is, a photon is scattered by a medium, and,
in the process, imparts some of its energy to the
scattering medium. This process is schematically
indicated in Figure 1(a). An incident photon with an
energy $h\nu_i$ scatters from the medium which is in some
initial state (often the ground state), and leaves with
an energy $h\nu_f$. A quantum of energy $h\nu_s = h(\nu_i - \nu_f)$ is left
in the medium. The quantity ν_s is called a Stokes
frequency of the medium, and the process is called a
Stokes process. Figure 1(b) schematically indicates the
anti-Stokes process in which a medium initially in an
excited state is left in a lower energy state after the
scattering process, with the scattered photon having an
energy greater than the initial photon energy by an
amount $h\nu_s$. In general, analysis of the Raman-scattered
light reveals many spectral lines, each one corresponding
to a transition between specific energy levels of the
material system. Often the lines correspond to
vibrational or rotational excitations (in the case of
molecules) or optical phonons (in a solid), but
electronic excitations can also be observed. By
carefully measuring the spectrum of the scattered light,
it is possible to obtain detailed information on the
energy level structure of the scattering medium.

The physics of the spontaneous Raman process has
been well understood for quite some time, and is
thoroughly discussed in textbooks[1]. It has been very
useful in elucidating the structure of complex molecules
in the infrared spectral region. Although Raman
spectroscopy yields the same energy level values as one
obtains from measurements of the infrared spectrum, the
process is fundamentally different. The Raman effect
depends on the polarizability of the molecule (and its
variation with the particular quantum state in which the
molecule happens to be), but is entirely independent of
the presence of a permanent dipole moment. Thus a Raman
spectrum often appears in molecules that have no infrared
spectrum; in this way Raman spectroscopy complements
infrared spectroscopy.

The stimulated Raman effect was first observed[2]
shortly after the demonstration of the first ruby laser.
While using a nitrobenzene Kerr cell as a Q-switch a loss
was observed. Careful investigation revealed radiation
at an invisible wavelength longer than that of the ruby
laser. The effect was correctly interpreted[3] as
originating from a stimulated Raman process in the
nitrobenzene.

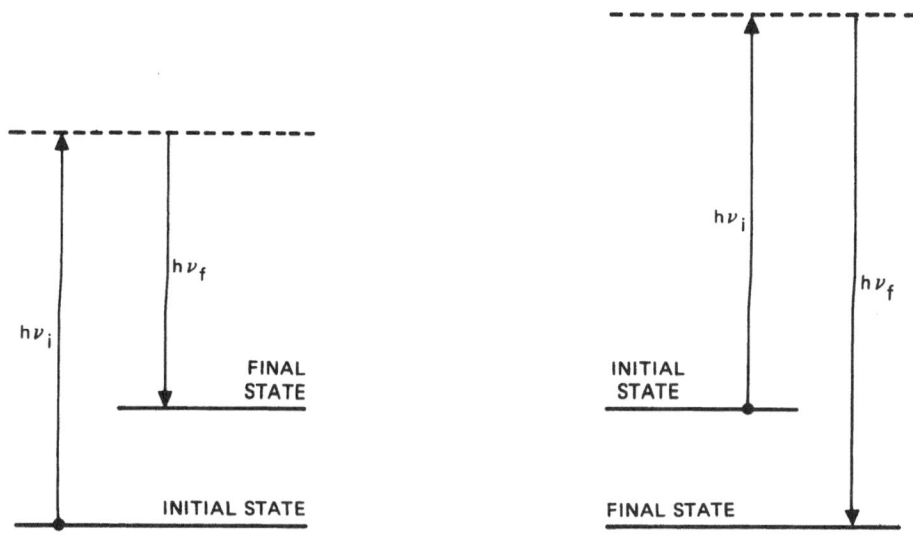

(a) STOKES PROCESS
$\nu_f = \nu_i - \nu_s$

(b) ANTI-STOKES PROCESS
$\nu_f = \nu_i + \nu_s$

Figure 1.   Schematic diagram showing transitions involved
in Raman scattering.   For a Stokes process (a) the final
photon has less energy than the initial photon, and the
scattering medium absorbs a quantum of energy.   For an
anti-Stokes process (b) the final photon has more energy
than the initial photon, and the scattering medium loses
a quantum of energy.

The theory of stimulated Raman scattering is reviewed in
the next section.   The semi-classical approach taken
provides considerable insight into the fundamental
physical mechanisms responsible for the phenomonon.   The
following section briefly mentions a variety of
applications of Raman technology, and then explores in
considerable detail the significant factors involved in
achieving efficient wavelength conversion using a
practical Raman oscillator/amplifier configuration
capable of producing output energy >1 Joule.

413

THE THEORY OF STIMULATED RAMAN SCATTERING

Although the Raman process is inherently quantum mechanical[4], more physical insight is gained if it is described using a semi-classical approach in which the photons are represented by the appropriate electromagnetic fields, and the Raman medium is described in quantum-mechanical terms. In this way, the rigorous quantum mechanical details of the Raman process are not necessary, and an appreciation of the basic physical interaction mechanisms is possible. In this spirit, this discussion begins by using Maxwell's equations to describe the propagation of electromagnetic radiation in a medium characterized by a non-vanishing polarizability. The departure from ordinary linear propagation is made by allowing the polarization to have a nonlinear dependence on the electric field. By assuming the nonlinearities are small enough that the field amplitudes are slowly varying, we can establish a set of linearized differential equations which contain the essential physics of the Raman process. Further physical understanding of the nature of Raman scattering is then achievable by observing that the nonlinear polarization can be related to, for example, the vibrational excitations of a molecule. Finally, we will show how this description of Raman scattering yields an expression for the medium-dependent Raman gain, the parameter of most practical significance in Raman devices.

The Nonlinear Wave Equation

The starting point for this derivation of the nonlinear wave equation is Maxwell's equations. They show how a nonlinear polarization can generate an electromagnetic wave at a frequency different than that of the waves already present. We defer until later the question of how the nonlinear polarization is generated by the electric fields. For a homogeneous, non-magnetic, non-conducting material containing no free charges, the four Maxwell equations are

$$\nabla \cdot \vec{D} = 0 \tag{1}$$

$$\nabla \cdot \vec{B} = 0 \tag{2}$$

$$\nabla \times \vec{B} = \frac{1}{c} \frac{\partial \vec{D}}{\partial t} \tag{3}$$

$$\nabla \times \vec{E} = \frac{1}{c} \frac{\partial \vec{B}}{\partial t} \tag{4}$$

where, as usual, $\vec{E}$ is the electric field, $\vec{B}$ is the magnetic field, $\vec{D}$ is the displacement field

$$D = \vec{E} + 4\pi\vec{P} \tag{5}$$

and $\vec{P}$ is the polarization in the medium. The polarization is assumed to have two explicit components.

$$\vec{P} = \vec{P}_o + \vec{P}_N \tag{6}$$

where $\vec{P}_o$ is the ordinary linear component and $\vec{P}_N$ is the nonlinear component. The wave equation is found by taking the curl of Eq. (4), and setting that equal to $-\frac{1}{c}\frac{\partial}{\partial t}$ operating on Eq. (3). Using the result that

$$\nabla \times (\nabla \times \vec{E}) = \nabla (\nabla \cdot \vec{E}) - \nabla^2 \vec{E} \tag{7}$$

and assuming the fields are polarized perpendicular to their respective propagation directions so $\nabla \cdot \vec{E} = 0$, we have

$$\nabla^2 \vec{E} = \frac{1}{c^2} \frac{\partial^2 \vec{D}}{\partial t^2} \tag{8}$$

Using Eqs. (5) and (6), we have

$$\nabla^2 \vec{E} - \left(\frac{n}{c}\right)^2 \frac{\partial^2 \vec{E}}{\partial t^2} = \frac{4\pi}{c^2} \frac{\partial^2 \vec{P}_N}{\partial t^2} \tag{9}$$

In obtaining this last equation, we have used the usual definition $D = \epsilon \vec{E}$ for the <u>linear</u> permittivity of the medium ϵ and ordinary <u>linear</u> refractive index $n^2 = \epsilon$. In the absence of nonlinear effects, the right-hand side of this equation is zero; the solutions to the <u>linear</u> equation are fields of the form

$$\vec{E}(\vec{r},t) = \frac{1}{2} \vec{A}(\vec{r},t)\, e^{-i(\vec{k}\cdot\vec{r} - \omega t)} + c.c. \tag{10}$$

where $\vec{k}$ is the wavevector and ω is the frequency of the electromagnetic wave. These latter two quantities satisfy the dispersion relation

$$k = \frac{n\omega}{c} \tag{11}$$

where k is the magnitude of the vector $\vec{k}$. For the nonlinear case, we shall see that $\vec{P}_N$ can have frequency components that differ from those of the waves already present in the medium, thereby generating waves at new frequencies.

We now represent the nonlinear polarization in the form

$$\vec{P}_N = \frac{1}{2} \vec{\phi}_N(\vec{r},t)\ e^{-i(\vec{k}\cdot\vec{r}-\omega t)} + c.c. \tag{12}$$

In general, other terms also exist with different frequencies or wave vectors, but they will not be in phase synchronism with the electric field of Eq. (10) and, hence, they will have no average effect. We now substitute the fields representing the solutions to the linear equation, Eq. (10), and the expression for the nonlinear polarization, Eq. (12) into Eq. (9), the wave equation. The left-hand side spatial derivatives are

$$\nabla^2\vec{E} = \frac{1}{2}\ (\nabla^2\vec{A} - 2i\vec{k}\cdot\vec{\nabla}\ \vec{A} - k^2\vec{A})e^{-i(\vec{k}\cdot\vec{r}-\omega t)} + c.c. \tag{13}$$

It is now useful to make the "slowly varying envelope approximation" that the amplitude $\vec{A}$ increases only very slightly in a distance of one wavelength. As a result the following inequalities hold:

$$|k^2\vec{A}| \gg |\vec{k}\cdot\nabla\vec{A}| \gg |\nabla^2\vec{A}| \tag{14}$$

Similar inequalities apply when $\frac{\partial}{\partial t}$ is substituted for ∇, and ω is substituted for k. With this approximation, the wave equation becomes

$$\left\{\left[2i\vec{k}\cdot\nabla + 2i\omega\left(\frac{n}{c}\right)^2\frac{\partial}{\partial t}\right] + \left[k^2 - \left(\frac{n\omega}{c}\right)^2\right]\right\}\vec{A} = \frac{4\pi\omega^2}{c^2}\vec{\phi}_N \tag{15}$$

The second term on the left-hand side is zero according to the dispersion relation Eq. (11), and we are left with

$$\left(\frac{\partial}{\partial z} + \frac{n}{c}\frac{\partial}{\partial t}\right)\vec{A} = -i\ \frac{2\pi k}{n^2}\vec{\phi}_N \tag{16}$$

where we have defined the z-axis as the direction along which the assumed plane wave propagates. Equation (16) shows how a nonlinear polarization can effect an electric field having the same frequency and wave vector.

The Nonlinear Polarization

This electromagnetic treatment of Raman scattering continues using the model of Yariv[5] in which the Raman medium is taken as a collection of N harmonic oscillators per unit volume, each representing one molecule. For simplicity, we assume a one-dimensional system and neglect x-and y-derivatives. The normal vibrational coordinate is denoted by $X(z,t)$. The equation of motion for a single oscillator is

$$\frac{d^2 X}{dt^2} + \Gamma \frac{dX}{dt} + \omega_s^2 X = \frac{F(z,t)}{m} \tag{17}$$

where Γ is the damping constant corresponding to the spontaneous Raman scattering linewidth $\Delta\nu_R = \Gamma/2\pi$, ω_s is the (undamped) resonance frequency, $F(z,t)$ is the driving force, and m is the mass.

The driving term relates the molecular vibration to the electric fields. It is found by starting with the expression for the electromagnetic energy U in an isotropic medium composed of such molecules

$$U = \frac{1}{2} \vec{D} \cdot \vec{E} \tag{18}$$

or, using the definition of the displacement field, Eq. (5), and the definition of the polarizability α of an individual molecule

$$\vec{P} \equiv N\alpha\vec{E} \tag{19}$$

we have

$$U = \frac{1}{2} (1+4\pi N\alpha) E^2 \tag{20}$$

In ordinary linear optics, α is simply a constant that depends on the medium selected. We are interested here in the case where the polarizability might depend on the vibrational coordinate, or

$$\alpha = \alpha_o + \left(\frac{\partial\alpha}{\partial X}\right) X + \dots \tag{21}$$

417

We shall see in what follows that this dependence leads to interesting nonlinear properties. Using Eq. (21) in Eq. (20) we find the electromagnetic energy to be

$$U = \frac{1}{2} [1 + 4\pi N(\alpha_o + (\frac{\partial \alpha}{\partial X}) X)]E^2 \tag{22}$$

The force per unit volume has a magnitude $\partial U/\partial X$; taking the derivative and dividing by N yields the force per oscillator

$$F(z,t) = 2\pi (\frac{\partial \alpha}{\partial X}) \overline{E^2}(z,t) \tag{23}$$

where the bar indicates an average over a few optical periods. This averaging is required because the molecules are unable to respond to optical frequencies. Equation (23) shows how an electric field can couple to molecular vibrations through the nonvanishing differential polarizability $(\partial \alpha/\partial X)$.

This field-induced excitation of the molecular vibrations reacts back on the electromagnetic fields. Equations (19) and (21) state that the molecular vibration at the frequency ω_s causes a modulation of the dielectric properties of the medium at ω_s. This, in turn, can lead to an energy exchange between electromagnetic fields separated in frequency by multiples of ω_s. In particular, we now consider the interaction of a laser field at a frequency ω_1 and a Stokes field at a frequency $\omega_2 = (\omega_1 - \omega_s)$. This Stokes field is always present in the medium, at least at a level corresponding to the "noise" due to spontaneous Raman scattering of the laser radiation. The total field is the sum of the two fields

$$E(z,t) = \frac{1}{2} A_1(z)e^{i\omega_1 t} + \frac{1}{2} A_2(z)e^{i\omega_2 t} + c.c. \tag{24}$$

so that

$$\overline{E^2}(z,t) = \frac{1}{4} A_1(z)A_2^*(z)e^{i(\omega_1-\omega_2)t} + c.c. \tag{25}$$

Substituting Eq. (23) and Eq. (24) into Eq. (17), the molecular equation of motion, gives

$$\frac{1}{2}(-\omega^2+i\Gamma\omega+\omega_s^2)X(z)e^{i\omega t} = \frac{\pi}{2m} (\frac{\partial \alpha}{\partial X})A_1 A_2^* e^{i(\omega_1-\omega_2)t} \tag{26}$$

where

$$X(z,t) = \frac{1}{2} X(z) e^{i\omega t} + c.c. \tag{27}$$

It follows from Eq. (26) that the molecular vibration is driven at a frequency $\omega = \omega_1 - \omega_2$ with a complex amplitude

$$X(z) = \frac{\pi}{m} \left(\frac{\partial \alpha}{\partial X}\right) \frac{A_1(z) A_2^*(z)}{[(\omega_s^2 - (\omega_1 - \omega_2)^2 + i\Gamma(\omega_1 - \omega_2)]} \tag{28}$$

This expression for the amplitude of the molecular vibrations can be related back to the polarization of the medium to derive an expression for the nonlinear part P_N introduced in Eq. (6). From Eqs. (19) and (21), we have

$$P = N[\alpha_o + \left(\frac{\partial \alpha}{\partial X}\right) X(z,t)] \, E(z,t) \tag{29}$$

The first term involving α_o gives the ordinary linear polarization P_o. The nonlinear polarization is found using the second term with Eq. (28)

$$P_N(z,t) = N\left(\frac{\partial \alpha}{\partial X}\right) \{\frac{\pi}{m} \left(\frac{\partial \alpha}{\partial X}\right) \frac{A_1 A_2^* e^{i(\omega_1 - \omega_2)t}}{(\omega_s^2 - (\omega_1 - \omega_2)^2 + i\Gamma(\omega_1 - \omega_2))} + c.c.\}$$

$$\times \, [A_1(z) e^{i\omega_1 t} + A_2(z) e^{i\omega_2 t} + c.c.] \tag{30}$$

This expression contains terms oscillating at several frequencies: ω_1, ω_2, $(2\omega_1 - \omega_2)$, and $(2\omega_2 - \omega_1)$. Consider first the term oscillating with the frequency ω_2 that generates the Stokes wave; it may be written

$$P_N(\omega_2; z,t) = \frac{1}{2} \phi_N(\omega_2; z) e^{i\omega_2 t} + c.c. \tag{31}$$

where

$$\phi_N(\omega_2; z) = \frac{2\pi}{m} N \left(\frac{\partial \alpha}{\partial X}\right)^2 \frac{|A_1|^2}{(\omega_s^2 - (\omega_1 - \omega_2)^2 - i\Gamma(\omega_1 - \omega_2))} A_2(z) \tag{32}$$

The Nonlinear Susceptibility and Raman Gain

The coefficient that relates an induced polarization to the inducing electric field is called the susceptibility χ. For linear optics, χ is independent of the electric field. In the present case we can derive an expression for χ_R, the complex Raman nonlinear susceptibility, which is defined by the relation

$$\phi_N(\omega_2; z) \equiv \chi_R |A_1(z)|^2 A_2(z) \tag{33}$$

From Eq. (32), which showed the relationship between the nonlinear polarization and the fields coupling to the molecular vibrations, we have

$$\chi_R(\omega_2) = \frac{2\pi}{m} N \left(\frac{\partial\alpha}{\partial X}\right)^2 [\omega_s^2 - (\omega_1 - \omega_2)^2 - i\Gamma(\omega_1 - \omega_2)]^{-1} \tag{34}$$

In most cases of practical interest the damping coefficient is small, $\Gamma/\omega_s \ll 1$. This means that the term $(\omega_1 - \omega_2)$ differs from ω_s by small factors of order Γ/ω_s. Hence, we may simplify the denominator of Eq. (34) in the following way. First we note that

$$\omega_s^2 - (\omega_1 - \omega_2)^2 = [\omega_s + (\omega_1 - \omega_2)][\omega_s - (\omega_1 - \omega_2)] \tag{35}$$

$$\simeq 2\omega_s [\omega_s - (\omega_1 - \omega_2)] \tag{36}$$

except for small factors. Also,

$$\Gamma(\omega_1 - \omega_2) \simeq \Gamma\omega_s \tag{37}$$

except for small factors. Hence, Eq. (34) may be rewritten

$$\chi_R(\omega_2) \simeq \frac{\frac{\pi}{m} N\left(\frac{\partial\alpha}{\partial X}\right)^2}{\omega_s [\omega_s - (\omega_1 - \omega_2) - i\ \Gamma/2]} \tag{38}$$

If we explicity define the real and imaginary parts of χ_R by

$$\chi_R(\omega_2) \equiv \chi_R'(\omega_2) - i\chi_R''(\omega_2) \tag{39}$$

we have

$$\chi_R'(\omega_2) \simeq \frac{\pi}{m\omega_s} N\left(\frac{\partial\alpha}{\partial X}\right)^2 \frac{[\omega_s - (\omega_1 - \omega_2)]}{\{[\omega_s - (\omega_1 - \omega_2)]^2 + \Gamma^2/4\}} \tag{40}$$

and

$$\chi_R''(\omega_2) \simeq -\frac{\pi}{m\omega_s} N\left(\frac{\partial\alpha}{\partial X}\right)^2 \frac{\Gamma/2}{\{[\omega_s - (\omega_1 - \omega_2)]^2 + \Gamma^2/4\}} \tag{41}$$

We see that the Raman susceptibility is a Lorentzian, as is its linear counterpart[6]. Both the real and imaginary parts are plotted in Figure 2.

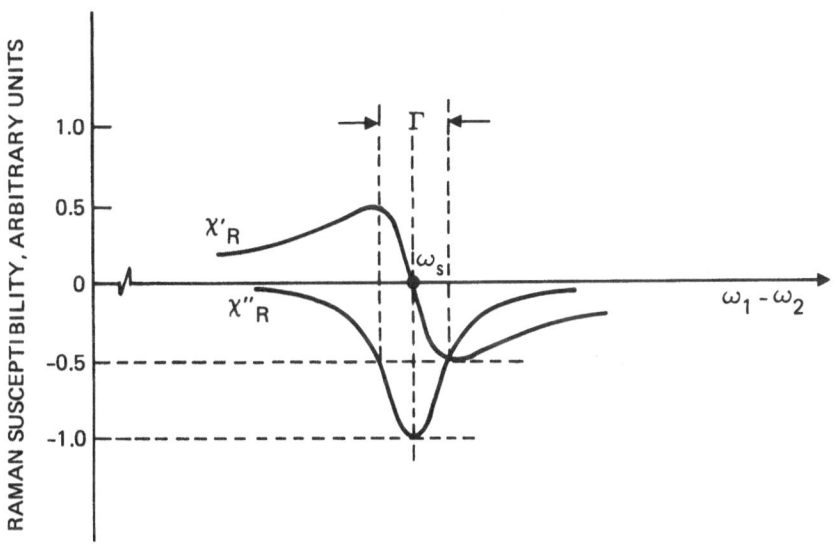

Figure 2. Real and imaginary parts of the Raman nonlinear susceptibility χ_R as a function of the frequency difference $(\omega_1 - \omega_2)$ of the pump and Stokes beams, respectively. (Taken from Reference 5).

The fact that $\chi_R''(\omega_2)$ is negative leads to gain at the frequency ω_2. This can be seen by using the steady-state form of Eq. (16), the differential equation relating the field amplitude to the nonlinear polarization. In the steady state, the time derivative vanishes. Substituting Eq. (33) for the nonlinear polarization, and Eq. (39) for the Raman susceptibility into Eq. (16), we have

$$\frac{dA_2}{dz} = - [ik_2 \frac{2\pi}{n^2} \chi_R' |A_1|^2 + \frac{2\pi k_2}{n^2} 2 \chi_R'' |A_1|^2] A_2 \qquad (42)$$

Thus, the z-dependence of A_2 has two parts. The first part depends on χ_R' and represents a change in the real refractive index in the medium. The second part has z-dependence given by

$$\exp[- \frac{2\pi k_2}{n^2} \chi_R'' |A_1|^2 z] \qquad (43)$$

and represents gain whenever $\chi_R'' < 0$. If we write this as $\exp(\beta z)$, defining a gain constant β, we have from Eqs. (43) and (41)

$$\beta = k_2 \{ \frac{\pi^2 (\frac{\partial \alpha}{\partial X})^2 \Gamma}{n^2 m \omega_s \{ [\omega_s - (\omega_1 - \omega_2)]^2 + \Gamma^2/4 \}} \} N |A_1|^2 \qquad (44)$$

All the factors in the large brackets are properties of the Raman medium. We conclude that the Raman gain coefficient at the Stokes frequency is proportional to the product of a medium-dependent factor, the molecular number density, and the pump laser intensity.

Anti-Stokes Waves

In the discussion following Eq. (30) it was mentioned that several frequency components were contained in the nonlinear polarization. We have just seen that the terms involving ω_2 lead to the Stokes Raman gain at that frequency. Gain can also exist at the anti-Stokes frequency $\omega_3 = 2\omega_1 - \omega_2 = \omega_1 + \omega_s$ if certain phase matching conditions are met.[1] From Eq. (30) we see that the term involving ω_3 can be written in the form

$$P_N(\omega_3; z, t) = \frac{1}{2} \phi_N(\omega_3; z) e^{i\omega_3 t} + c.c. \qquad (45)$$

where, now,

$$\phi_N(\omega_3; z) = \frac{2\pi}{m} N (\frac{\partial \alpha}{\partial X})^2 \frac{A_1 A_1 A_2^*}{[\omega_s^2 - (\omega_1 - \omega_2)^2 + i\Gamma(\omega_1 - \omega_2)]} \qquad (46)$$

422

It is of interest to show explicitly the z-dependence of this term. Substituting terms of the form

$$\vec{A}_j(\omega_j;\vec{r}) = \vec{A}_j(\omega_j)e^{-i\vec{k}_j\cdot\vec{r}} \tag{47}$$

the rapid spatial dependence on the right-hand side of Eq. (46) is proportional to

$$\exp[-i(2\vec{k}_1-\vec{k}_2)\cdot\vec{r}] \tag{48}$$

Hence, anti-Stokes radiation will be emitted in any direction $\vec{k}_3$ that satisfies the phase matching condition

$$\vec{k}_3 = 2\vec{k}_1-\vec{k}_2 \tag{49}$$

Because of dispersion within the medium, the anti-Stokes radiation is emitted in a cone centered about the pump laser axis. Figure 3 schematically shows how this angle is determined. A straightforward calculation for hydrogen ($\nu_s=4155$ cm^{-1}) with a pump laser at $\lambda=532$ nm yields $\theta_{PM} \cong 0.8\sqrt{\rho}$ mrad where the density ρ is measured in amagats. Forward anti-Stokes can reduce the gain of the forward-going Raman radiation, and can lead to a dominance of backward-Stokes radiation under the proper experimental conditions[7].

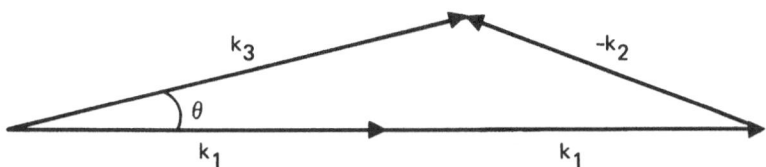

Figure 3. Phase matching diagram for the creation of an anti-Stokes photon, k_3 via four-wave mixing among two pump photons, k_1, and a Stokes photon k_2. Due to dispersion in the Raman medium, the process is most efficient for a specific phase match angle θ.

RAMAN DEVICES

The theoretical concepts outlined above have been
utilized in a wide variety of fundamental studies and
device applications. An excellent review of many
fundamental studies is available in a recent book by
Weber[8], and the interested reader is referred there for
information of that nature. Several articles[9,10] are
also available which review a variety of applications.
The remainder of the present article will mention some
recent device demonstrations, and then review important
considerations involved in scaling Raman devices to
medium energies ($\gtrsim$1 Joule) with a pump-to-Stokes energy
conversion efficiency $\gtrsim$50 percent. This energy scaling
involves a configuration in which a "seed" pulse from a
Raman oscillator is used to stimulate conversion of pump
radiation to a longer wavelength in a Raman amplifier.
Interest in such a device arises not only from the fact
that a new wavelength is produced, but also because this
process represents a method to "clean-up" an energetic
pump beam: under appropriate conditions, the output
Stokes beam can have a much higher irradiance than that
of the pump beam[11,12].

Eyesafe Laser

A variety of laser device applications are presently
being met using a Nd:YAG laser operating at a wavelength
of 1.06 μm. This laser is preferred because of its
advanced state of engineering development. However,
because of the serious potential for eye damage, such
laser devices have, in the past, been primarily
restricted to military and laboratory environments.
Research into the wavelength dependence of laser eye
damage shows that the transparency of the cornea
gradually falls to a very low value for wavelengths
longer than ~1.4 μm; such wavelengths are absorbed in the
cornea and are not focused on the retina. This fact
signifies that wavelengths $\gtrsim$1.4 μm are nearly three
orders of magnitude safer than shorter wavelengths. With[13]
this in mind, a program was undertaken to demonstrate
an eyesafe laser rangefinder using a Q-switched Nd:YAG
laser pumping a methane Raman cell to produce a
wavelength of 1.54 μm. This eyesafe wavelength falls
within a good atmospheric transmission window[14] from
about 1.5 μm to 1.8 μm, and also coincides with the peak
sensitivity range of a germanium photodiode[15] receiver.
The device was intended for use as an eyesafe cloud
height indicator at civilian airports.

Pulse Repetition Frequency Scaling

High pulse repetition frequencies (PRFs) at moderate average powers and efficiencies have been primary objectives of recent investigations of wavelength-agile laser sources. Nonlinear frequency conversion using stimulated Raman scattering has been used with Nd:YAG lasers operating at 1064 nm to produce wavelengths from 1106 nm to 1907 nm[16]. By using the fourth harmonic of this pump laser at 266 nm, Stokes shifted wavelengths as short as 270 nm are produced[17]. To generate shorter wavelengths from a given pump laser, either inefficient anti-Stokes Raman shifts[18] are required, or nonlinear mixing of the pump with the Stokes shifted wavelengths may be used[19]. Scaling the average power by increasing the energy per pulse is most practical in gases[20], where nearly 100 percent quantum efficiency has been demonstrated[21,22]. Because the heat deposited in the medium by the difference in energies of the incident and scattered photons produces optical distortion[23], however, the PRF is limited to very low values by the thermal time constants of the Raman medium. In methane gas, for example, Raman laser operation produces a severely distorted output beam when a simple high-pressure cell is used at PRFs exceeding ~5 Hz. A PRF of 60 Hz has been reliably achieved[24] using a compact Raman resonator incorporating a circulating gas Raman cell. By removing the heated gas in the time interval between successive pump laser pulses, the Raman process can always occur in an optically homogeneous volume of gas. This PRF scaling demonstration, coupled with the potential for wavelength agility, represents a significant step in the development of a practical, tunable Raman laser scalable to high average power.

Raman Oscillator/Amplifier

The remainder of this section discusses a Raman oscillator/amplifier. Where specific numerical results are given they apply to a Raman process involving the vibrational transition in pressurized hydrogen (ν_s =4155 cm^{-1}) pumped at a wavelength of 532 nm. This discussion ignores any detailed analysis of the microscopic interactions causing the Raman scattering. Instead, we simply treat the medium as a system that will amplify light at a frequency ω_2 in the presence of a pump laser at a frequency $\omega_1 = \omega_2 + \omega_s$, where ω_s is the Stokes frequency shift. (Refer to the discussion leading up to Eq. (44)). With a slight change to notation more compatible with experimental parameters, we simply write the expression for the increase in the Stokes intensity I_s as

$$\frac{dI_s}{dz} = g \, I_s(z) \, I_o(z) \tag{50}$$

where z is the propagation distance, g is the medium-dependent Raman gain coefficient, and I_o is the pump laser intensity. Since the probability of scattering is proportional to the number of scatterers present and inversely proportional to the linewidth (see Eq. (44)) g is functionally dependent on the density of the gas. It is also related to the Raman cross-section of the molecules. Equation (50) does not include spontaneous Raman scattering, which is obviously necessary to initiate the stimulated process, since the above equation predicts no additional Stokes radiation if there is none initially (I_s=0). We will defer until later the important discussion of what spontaneous scattering levels can reasonably be expected; for the present, we simply solve the equation as it stands.

Returning to Equation (50), then, we may remove one of the variables by using the conservation of energy or photons:

$$\lambda_p I_p = \lambda_p I_o - \lambda_s I_s \quad, \tag{51}$$

where I_o is the input pump intensity, and λ_s and λ_p are the output (Stokes) and pump wavelengths, respectively. The ratio λ_p/λ_s appears often in these analyses; we shall define it as Q. It is always less than one for Stokes processes, and represents the ratio of the energy of the Stokes to the pump photon; it is the maximum possible energy conversion efficiency according to the Manley-Rowe relationship[25]. The energy difference is simply the energy imparted to the vibrating molecules. Equation (50) now becomes

$$\frac{dI_s}{dz} = gI_s(I_o - I_s/Q) \quad, \tag{52}$$

which is readily integrated to give the analytical solution[26],

$$I_s(out) = \frac{QI_o}{1 + (QI_o/I_s(in) - 1)\exp(-gI_oZ)} \quad, \tag{53}$$

where I_s(out) is the Stokes intensity coming out of the
gain region of length z, and I_s(in) is the injected
Stokes seed intensity. Plotting this equation for the
output intensity as a function of the gain exponent,
gI_oz, gives the performance of an ideal Raman device. We
have plotted in Figure 4 a family of these curves for
various values of the injected intensity ratio,
I_s(in)/I_o.

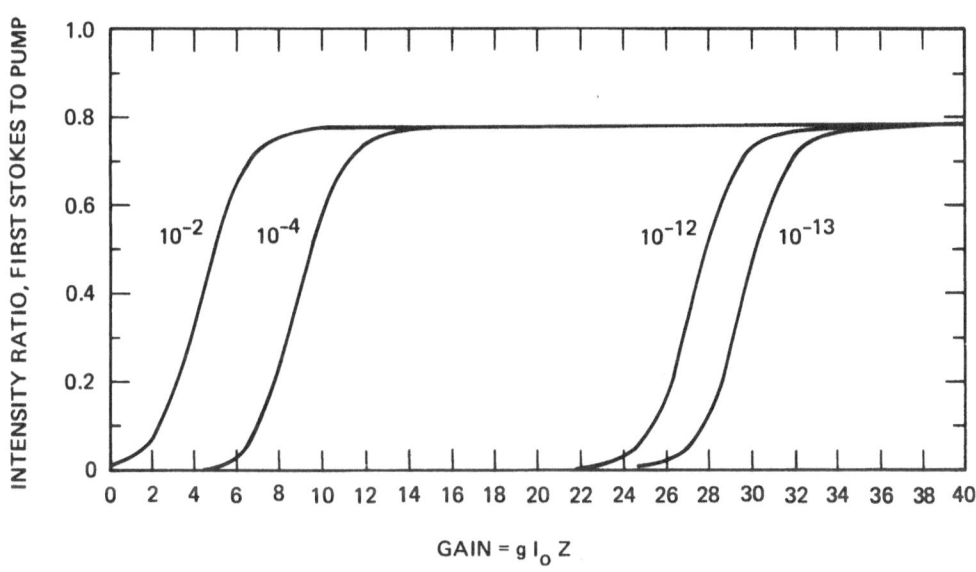

Figure 4.   Solutions to the plane wave Raman
amplification rate equation showing first-Stokes
production as a function of small-signal Raman gain for
several values of the injected Stokes intensity ratio,
I_S(in)/I_o.

    Several features are apparent in the figure, the
details of which will be discussed below.

o   The curves are identical in shape but are
           displaced according to the value of the
           injected Stokes intensity ratio. This brings
           up the issue of what amounts of spontaneous
           Stokes intensity are inherently injected, and,
           hence, at what gain levels the device can be
           expected to become an oscillator. It will be
           shown below that other processes not yet

considered are relevant to this issue. Figures
for an appropriate level of injection for a
practical Raman amplifier can then be
calculated.

o   The output intensity saturates at 100% quantum
efficiency and remains there. It thus appears
that extremely good (Manley-Rowe limited)
conversion could easily be obtained. However,
these curves are calculated under the
"plane-wave" assumption that 100% of the pump
is at a single intensity. This assumption is
inappropriate for real lasers having spatial
and temporal intensity variations; calculations
for real lasers produce greatly modified
amplification curves, as will be detailed
below.

o   The Raman gain parameter, g, appears only in
the exponent, multiplied by I_o and the gain
region length, z. This allows a tradeoff of
these parameters in a design.

These observations motivate the following discussion of
how the equations must be modified and what precautions
must be taken in the design of a practical Raman
amplifier or oscillator. We will proceed by developing a
set of design equations that realistically describes the
available experimental data. We will see that practical
complications can arise that make it difficult to achieve
the very high conversion efficiency indicated above.

Effect of Injected and Spontaneous Stokes

As the injected Stokes intensity is made smaller, it
is evident from Figure 4 that the onset of a rapid
increase in Stokes output requires a proportionately
larger value of $gI_o z$. For very small ratios of injected
Stokes-to-pump intensity, Eq. (53) may be approximated by
$I_s(\text{out})/I_s(\text{in}) = \exp[gI_o z]$. This equation can be
inverted to define a threshold value for the small signal
gain ($gI_o z$) at which rapid conversion from the pump to
the Stokes wavelength ensues. At this gain, $I_s(\text{out}) \simeq I_o$
and $gI_o z \simeq \ln(I_o/I_s(\text{in}))$. This formulation suggests that
weak spontaneous scattering can thus be incorporated into
the Raman amplification model by simply representing it
as a specific level of injected Stokes. This new model
will also make the theory applicable to a Raman
oscillator starting from spontaneous scattering. We

must, therefore, calculate the spontaneous scattering
into the pump beam geometry and determine the gain level
at which significant amplification of spontaneous Stokes
would be expected.

The spontaneous starting level may be calculated by
use of the measured ordinary Raman scattering cross-
section for the various gases. The definition of the
differential scattering cross-section[27] states that

$$\frac{P_{sc}}{P_p} = N \frac{d\sigma}{d\Omega} \Delta\Omega z \quad , \tag{54}$$

where P_{sc} is the scattered power, P_p is the incident laser
power, N is the molecular density, $d\sigma/d\Omega$ is the
differential scattering cross-section per molecule, $\Delta\Omega$ is
the solid angle into which the scattering occurs, and z is
the beam propagation distance. We will assume $\Delta\Omega$ to be
the solid angle of a diffracting Gaussian beam which has a
full angle divergence of $2\lambda/\pi w_o$. The solid angle is
therefore $\lambda^2/\pi w_o^2$. If we also make the assumption that
the scattering volume is that which is occupied by a
Gaussian beam before its area expands by a factor of two,
then we also have the relationship, $z=2z_o=2\pi w_o^2/\lambda$, where
we have used the usual definition of the confocal
parameter, z_o,[28] as the characteristic waist length for a
Gaussian beam[28]. Inserting these into Equation (54)
results in

$$\frac{P_{sc}}{P_p} = 2N\lambda \left(\frac{d\sigma}{d\Omega}\right) \quad . \tag{55}$$

If we calculate this number for hydrogen at
60 atmospheres, for which the scattering cross-section[29]
is 5.28×10^{-30} cm^2/sr-molecule, we obtain 1.2×10^{-12}. Thus,
the spontaneous scattering corresponds to an injected
signal into the propagating beam that is proportional to
the molecular density and is approximately twelve orders
of magnitude less than the pump laser. For this to be
amplified to the level of the pump beam, a small-signal
gain corresponding to $gI_o z$ equal to 28 is required. This
injected Stokes level of 10^{-12} was indicated in Figure 4.

The theoretical threshold gain requirement of $\sim 10^{12}$
(or $\sim e^{28}$) is extremely high - much higher, in fact, than
observed experimental values for the oscillation
threshold for stimulated Raman scattering. Historically,
there has been considerable controversy over this issue

for liquids, which exhibit other complicating intensity-enhancing nonlinear effects such as self-focusing. In hydrogen such complicating effects are minimal, and it has been shown[30] that the principle reason for the occurrance of stimulated scattering before gains of e^{30} are achieved is feedback from reflections near the Raman medium. By assuming that these reflecting surfaces form a laser resonator, Avizonis et al.[30] quantified this effect by using the simple equation for the oscillation threshold in a laser resonator: $R_1 R_2 \exp[2gI_o z]=1$, where R_1 and R_2 are the reflectivities of the resonator end mirrors; we have assumed two passes of Raman gain in the exponent. Considering the possibility that the windows of the Raman cell are the resonator mirrors, and that they are antireflection-coated to give them a reflectivity of 0.2%, we find that oscillation will ensue when $gI_o z$ equals only 6.2. Improvement of the coatings to 0.1% reflectivity only allows $gI_o z$ to reach 6.9. The possibility of the Raman cell windows acting as resonator mirrors is removed in practice by tilting the windows sufficiently. When uncoated windows are used, they are typically placed at or near Brewster's angle[31]. However, since even extremely small amounts of feedback can evidently cause premature appearance of successive Stokes, this same threshold equation may be used heuristically to estimate the influence of other light scattered back into the gain region. Note that the threshold equation states that if the stray feedback from either end of the cell is greater than $\exp[-gI_o z]$, then oscillation will occur prematurely, and the interval of $gI_o z$ over which conversion to a specific desired Stokes order occurs will then be narrowed. For hydrogen this implies any feedback greater than 10^{-12}. This could conceivably arise from many sources, including dust or small particles in the gas, scratches, digs, or pits in windows.

In typical experimental situations, then, much lower Raman thresholds occur than would be predicted by theory. For example, in recent work Fulghum et al.[32] report $gI_o z=10$, and Chang and Djeu[33] report amplified spontaneous emission at values of $gI_o z$ over 6. Komine[34] conjectures that parasitic oscillation problems may occur for injection levels of 10^{-4}; i.e., $gI_o z$ equal to 9. For the remainder of this discussion, therefore, we assume that $gI_o z=8$ is a realistic value. This means that the feedback into the gain region from stray reflections, scratches and digs, scattering from dust, etc. must be less than 3×10^{-4}, a value that may be realized in practice by the use of high quality windows and coatings, tilting the windows and other nearby optical surfaces, and general care in the placement of all optics.

Higher Stokes Generation

Figure 4 only considered conversion to the first Stokes. For the same reason that the first Stokes appears, higher Stokes can also appear. That is, once the intensity at the first Stokes is comparable to that of the pump, it becomes a pump for second Stokes. If no energy is injected at any Stokes order, all orders must build up from the spontaneous level according to equations analagous to Equation (53). This produces a family of curves, as shown in Figure 5 which was originally derived by Shen and Bloembergen[35]. In subsequent figures we have not shown the rapid transition interval in detail since it is narrow as compared to the flat regions. In fact, this transition is even more rapid than indicated by our equations because, for simplicity, we have so far not included the effects of parametric processes[36] discussed below.

We will first discuss the ramifications of Figure 5. The calculated intensity ratios, denoted by K, include the effects of linear loss. Two sets of curves are drawn, one for a loss of 10^{-4} cm^{-1}, which is the minimum possible from consideration of Rayleigh scattering in the gas, and one for a loss of 3×10^{-3} cm^{-1}, a reasonable figure for typical situations. Note that the linear loss can significantly reduce the conversion efficiency, especially if long cells are used. Reference 35 assumed gain lengths of only 50 cm; such a length has 96% transmission with the larger loss; a two meter cell with the smaller loss would have 98% transmission.

Note that Figure 5 shows a gain increment of $\sim e^{30}$ between successive Stokes orders. If one includes the lowered oscillation threshold due to the scattered light effects discussed above, the curves are changed, as shown in Figure 6. The region of dominance for each Stokes order is smaller, having a width corresponding to the assumed increment in $gI_{o}z$ of 8, which is required for each Stokes to appear. No cascading beyond the fourth Stokes takes place because the fourth-Stokes photons have less energy than the lowest hydrogen vibrational state.

Another cause for the premature appearance of higher Stokes is their generation via parametric processes[37]. Higher Stokes are generated by four-wave-mixing of the

431

immediately lower Stokes or pump. This type of process
was indicated earlier as a way of generating anti-Stokes
radiation. Phase matching is required for these
processes to become significant. Because of dispersion
in the medium, phase matching only occurs when the beams
are at certain angles to each other (the phase match
angle is ~10 mrad for the hydrogen at ~1000 psi). The
phase-match angles vary as the square root of the
density. Hence phase matching is easier, and therefore
the problem of eliminating higher-Stokes generation is
aggravated, when working at lower densities. High-
density operation and the use of well collimated beams
minimize the possibility of phase matching and higher
Stokes competition.

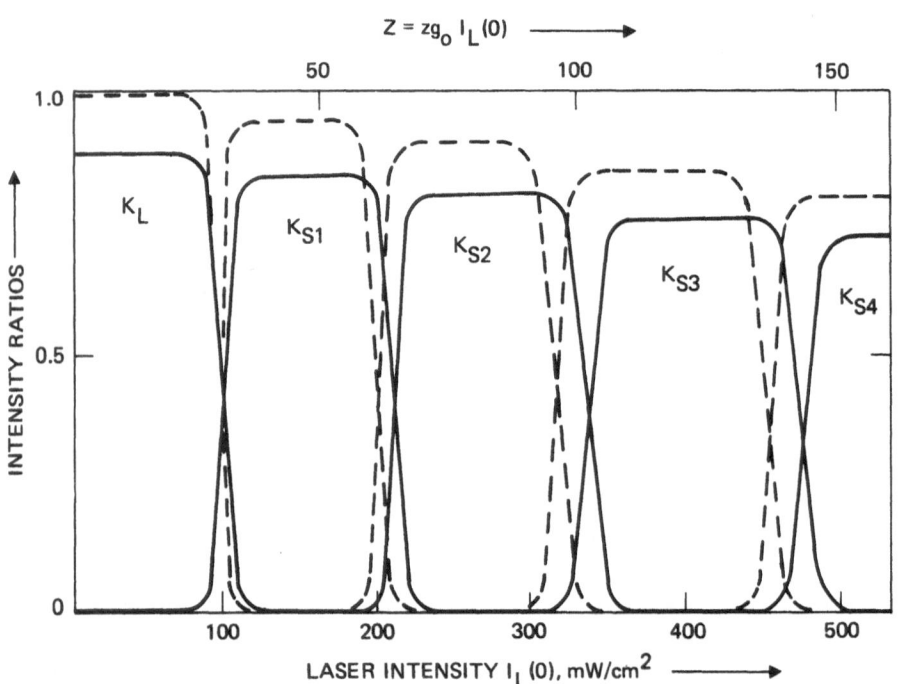

Figure 5. Calculated intensity ratios of laser and
Stokes light, reproduced from reference 35. Incoming
laser intensity is assumed to be 300 MW/cm^2. Ratios (K)
for the laser (L) and successive Stokes (S1,S2,...) are
plotted as a function of intensity on the lower scale or
equivalently small gain (gI$_L$z) on the upper scale. In
this situation g=3.0 cm/GW(CS_2); the Stokes shift was
only 658 cm^{-1}, using a Ruby laser at 14,320 cm^{-1}.

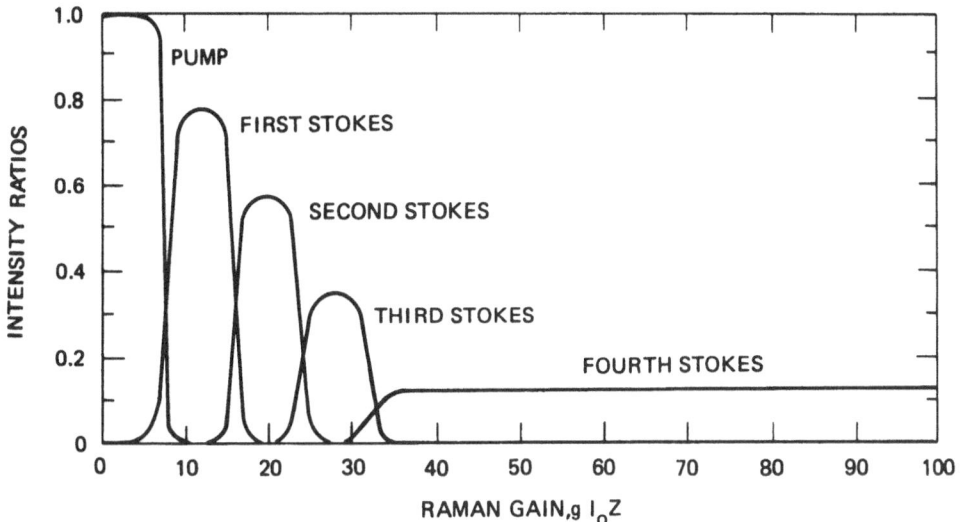

Figure 6. Solutions to the plane wave Raman rate equations, assuming threshold at typical experimentally observed level.

To review, the curves, as they are drawn in Figure 6, may represent a Raman oscillator starting from spontaneously generated Stokes noise. Conceptually, an amplifier differs from this only in that Stokes radiation generated elsewhere is deliberately injected. It is important to remember, however, that even in a Raman cell intended to be used as an amplifier, spontaneously generated output will appear at approximately the gain level increments indicated in Figure 6.

By using proper injected Stokes levels, a Raman amplifier can be optimized for higher Stokes orders. This has been demonstrated by Komine[34], who used this theory in his scaling studies of Raman converters. We reproduce his theoretical Figure 13 in Figure 7. He was able to obtain significant conversion to the second Stokes. Similar design curves are drawn by Holliday[38]; however, as we have discussed earlier, the widths of $gI_oz = \sim 30-40$ that he uses for the region where second Stokes will dominate are unrealistically high.

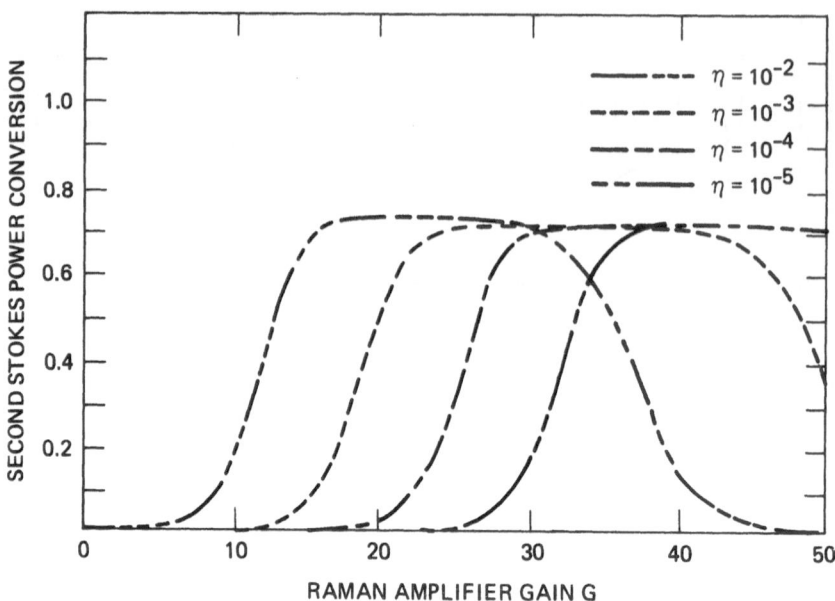

Figure 7. Theoretical second-Stokes power conversion vs
Raman amplifier gain for various Stokes injection levels
Conversion is from 353 nm to 500 nm via second-Stokes in
hydrogen. η is the ratio of injected seed to pump.
(Reproduced from reference 34.)

The Effect of Pump Beam Intensity Characteristics

The above theory is applicable only to situations in
which 100% of the laser beam is at a single intensity.
This ideal situation allows a straightforward solution to
the coupled nonlinear differential equations describing
the interactions. This assumption is never valid for
real beams which have nonuniform spatial and temporal
intensity distributions; that is to say, the output of a
real laser beam is distributed over a range of
intensities. This obvious and very important fact must
be incorporated into a theory in order for it to
correctly predict Raman amplifier performance. Similar
considerations have been found to be important for other
nonlinear processes such as efficient frequency tripling
of Nd:Glass radiation[39,40].

We will first calculate how the laser output is distributed over a range of intensities. This will lead to a discussion of how this distribution affects the shape of the curves in Figures 4 through 7. We will then review how well experimental data in the literature fit this more complete theory.

The beam that most closely approximates the ideal assumption of a single uniform intensity is one with a "top hat" intensity distribution; i.e., the beam is square in both space and time. Such a beam is depicted schematically in Figure 8(a). Figure 8(b) plots the distribution of beam energy as a function of intensity. For the "top hat" spatial intensity distribution, all the pulse at any instant in time is at the same intensity, I_o; it is a delta function at I_o. If the temporal shape of our example were not square, however, the output distribution at any instant is still a delta function, but at an intensity that would uniformly rise and fall as the pulse evolved temporally. Therefore, the energy distribution would not be a delta function, but rather some broader distribution that depends on the shape of the pulse. This distribution is shown in Figure 8(c) for a Gaussian temporal profile.

For the "top hat" pulse, deducing the intensity distributions was obvious. For other beams the calculation is usually straightforward. The Gaussian beam, in both space and time, is an important example and the one most commonly used for theoretical calculations. Consider the general Gaussian beam:

$$I = \exp[-r^2 - t^2] \quad . \tag{56}$$

For simplicity, we have normalized all parameters, so that the peak intensity is 1.0, the width in space is 1.0, and the width in time is 1.0. The usual radial coordinate is r and time is t. We now ask the question: what fraction of this beam's energy resides in any given intensity interval? If we make the variable substitution $R^2 = r^2 + t^2$, Eq. (56) becomes $I = \exp[-R^2]$. Thus, it is now easily visualized that the spatial and temporal parts of the beam having constant intensity lie on the sphere defined in x,y,t space by R=constant. Energy is the product of the intensity, area, and time. The differential volumes on the sphere have units of area multiplied by time. The differential dE, of the total output energy, E, at this intensity is simply the intensity times the differential volume. The

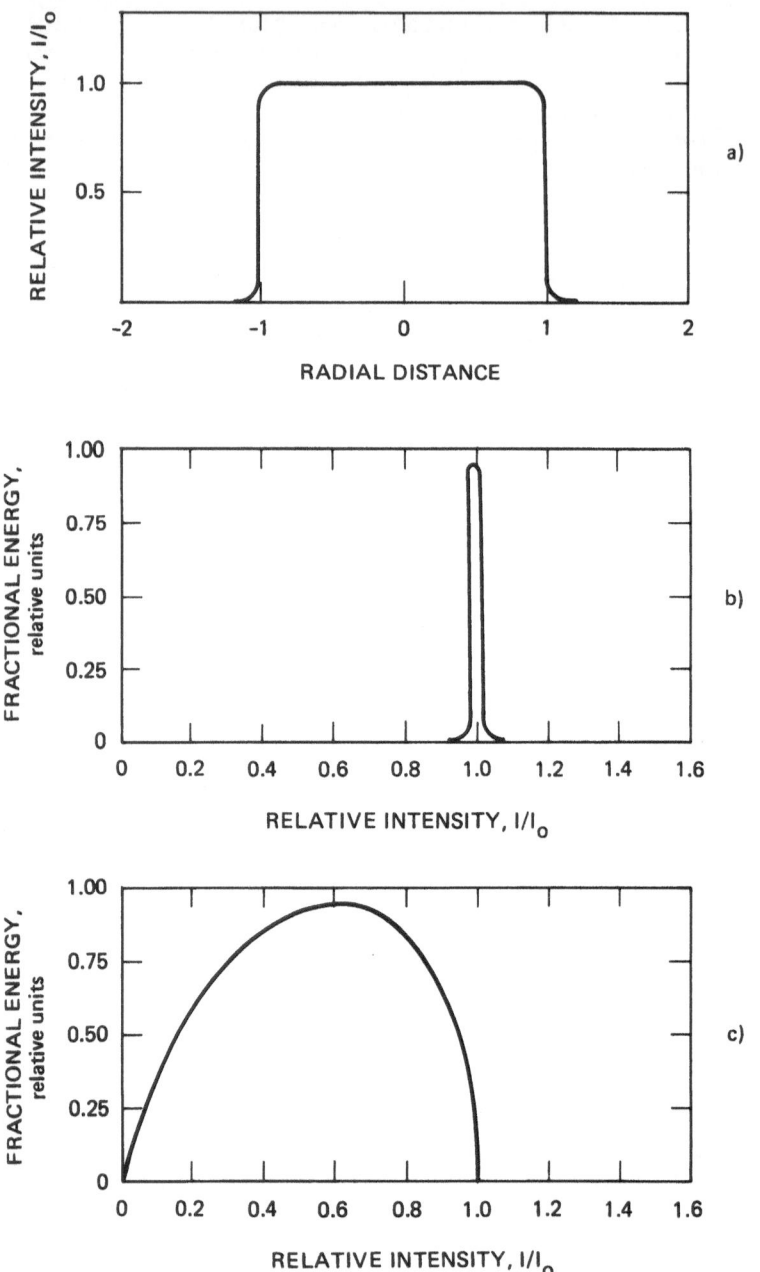

Figure 8. Energy distribution in a "top hat" beam: (a) spatial intensity distribution; (b) all the energy is distributed at one intensity if the temporal shape is also square; (c) the energy is distributed over a range of intensities for a "top hat" spatial profile and a Gaussian temporal profile.

differential dE, of the total output energy, E, at this
intensity is simply the intensity times the differential
volume. The differential volume, dV, is the surface area
of the sphere times dR. Hence $dE=4\pi IR^2dR$. The fraction
of the total energy, E, that is at this intensity is
dE/E. The fraction of the energy that is within the
fraction dI/I of the intensity is therefore equal to
IdE/EdI. Using the above expression for dE, we have
$IdE/EdI=4\pi I^2R^2dR/EdI$. E is a constant obtained by
integrating the differential energy over all R; it equals
$2/\pi$. Using the Gaussian relationship between I and R,
our expression for the distribution of intensity in a
Gaussian beam finally becomes

$$\frac{IdE}{EdI} = I \sqrt{-\ln I/\pi} \quad . \tag{57}$$

This equation is plotted in Figure 9(c). The ordinate is
a measure of the fraction of the total output at each
intensity value. As expected, there is essentially no
output at zero intensity. There is also essentially no
output with exactly the peak intensity, since this peak
intensity appears at only one infinitesimal area during
one instant of time. It is important to observe that
most of the energy is spread fairly uniformly over the
entire possible range.

Let us now examine how this intensity distribution
will affect the theory for Raman amplification. This can
be visualized by imagining the realistic situation of an
amplifier or Raman cell where the length (z) and gain
coefficient (g) are kept fixed and the intensity of the
pump beam (I_o) is slowly increased; i.e., the number gI_oz
is being increased. If the beam had all its energy at
the same intensity, curves such as those in Figure 5
would be generated. However, if the beam had a
distribution of intensities, as all real beams do, the
parts of the beam at higher intensity would be operating
at the higher Stokes orders, while the parts that were at
a lower intensity would still be at lower Stokes, or
perhaps not even above threshold. Since a Gaussian beam
has approximately equal energy at all intensities, in the
limit of high intensity, approximately equal fractions of
the energy in the beam would be operating at all points
on the abscissa of Figure 5 or 6; since the ranges of
intensities over which particular Stokes dominate are
approximately equal, essentially equal energy would be
obtained at all possible Stokes orders. To put the same
statement in terms of quantum conversion efficiency, it

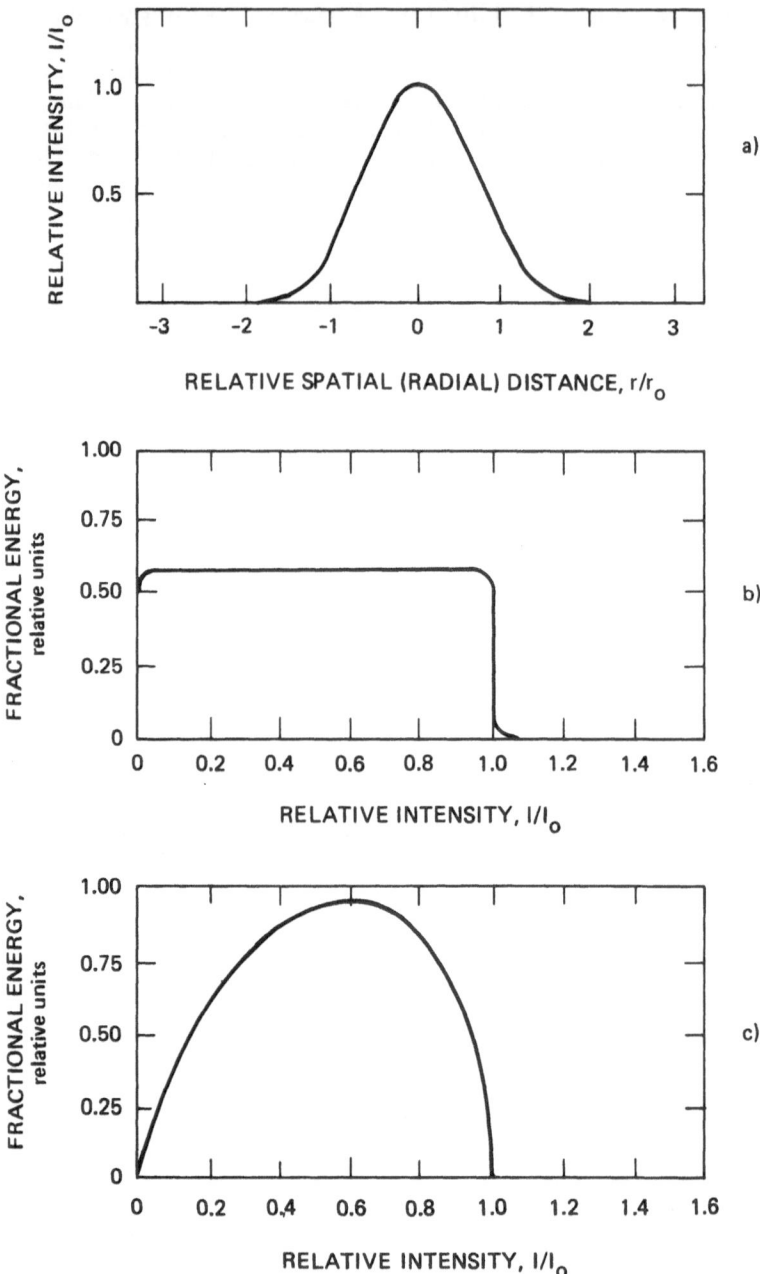

Figure 9. Energy distribution of a Gaussian beam with respect to intensity: (a) spatial intensity distribution; (b) assuming a square temporal shape, energy is distributed uniformly at all intensities less than peak; (c) assuming a Gaussian temporal shape, energy is distributed as calculated in the text.

can be seen that with a Gaussian beam the best conversion
that can be obtained to the first Stokes is 50%, to the
second Stokes, 33%, to the third Stokes, 25%, to the
fourth, 20%, etc. This fact represents a limitation to
Raman conversion schemes that rely on consecutively
higher Stokes order shifts.

A formulation of the Raman gain equations including
this consideration was done by von der Linde et al.[41] in
1969. We reproduce their Figure 2 in Figure 10. They
considered the beam to be Gaussian in space only. For
this less physical, but easier to analyze, case an
analysis similar to that above shows that the energy is
rigorously divided equally among all possible
intensities – this was shown in Figure 9(b). Hence, it
will be observed that their quantum conversion
efficiencies are nearly correct as estimated above.
Other investigators[32] have also calculated this theory
and report good agreement with experiment.

We have seen that for a beam with the character
usually encountered in pump lasers, the conversion
efficiency is limited by the fact that the beam has a
spread of intensities. There are two ways in which the
conversion can be improved. One way is to maximize the
fraction of the pump laser intensity range that
corresponds to the desired Stokes order. For example,
injecting a first-Stokes seed into the Raman amplifier
reduces the $(gI_o z)$ range required before first-Stokes
radiation begins to deplete the pump, thereby allowing
more of the pump laser intensity range to participate in
first-Stokes production. This effect is indicated in
Figure 11, which is to be compared with Figure 6 where no
finite injected Stokes was assumed (i.e., the Stokes
radiation builds up from noise). If the peak $gI_o z$ for
the pump beam is made ~10, and the injected first Stokes
is made 0.01 times the pump, then all but approximately
10 to 15% of the intensity range is dominated by first
Stokes, meaning that for a beam with a Gaussian intensity
distribution 85 to 90% quantum conversion can be
expected. For reference we note that if we wish to
optimize the second Stokes, we must allow a range of $gI_o z$
over which first Stokes dominates; thus the fraction of
the range over which the second Stokes dominates is
proportionally smaller. The maximum possible quantum
conversion for a Gaussian beam in that situation is
approximately 70%, independent of the assumptions about
the onset of higher Stokes. Another way to overcome this
problem is to design the pump laser to be closer to a
"top hat" intensity distribution. It is not always

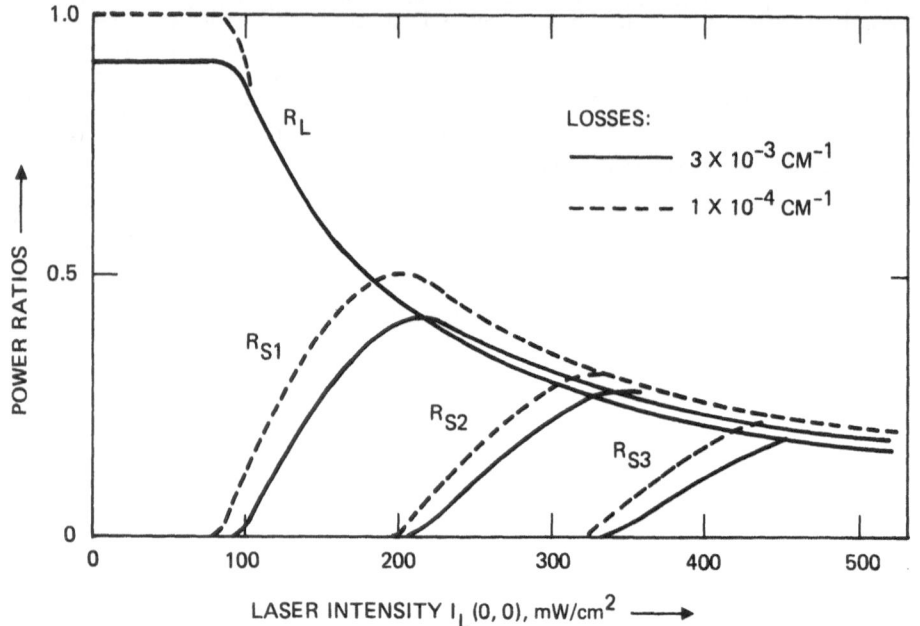

Figure 10. Calculated power ratios of laser and Stokes
light, reproduced from reference 41. A Gaussian spatial
profile was assumed. The temporal variation of intensity
was not considered.

achieved in experimental practice, however, because one
can not consistently place the entire Raman device in the
near-field of the pump laser (even the diffraction
occurring in a distance of one meter appreciably alters
the top hat distribution).

Cell Length and Intensity Trade-Offs

For a given material and pressure, the gain
parameter (g) is fixed. For example, several
investigators have measured the Raman gain in hydrogen,
dating as far back as 1965, with most using a ruby laser
as the pump[42-45], but more recently other pump lasers
have been employed[46-48]. As shown in Eq. (44), the gain
varies inversely as the wavelength of the converted
light[47,48]. The Raman gain for hydrogen at the 532 nm
pump wavelength is ~2.2 cm/GW. The other two parameters

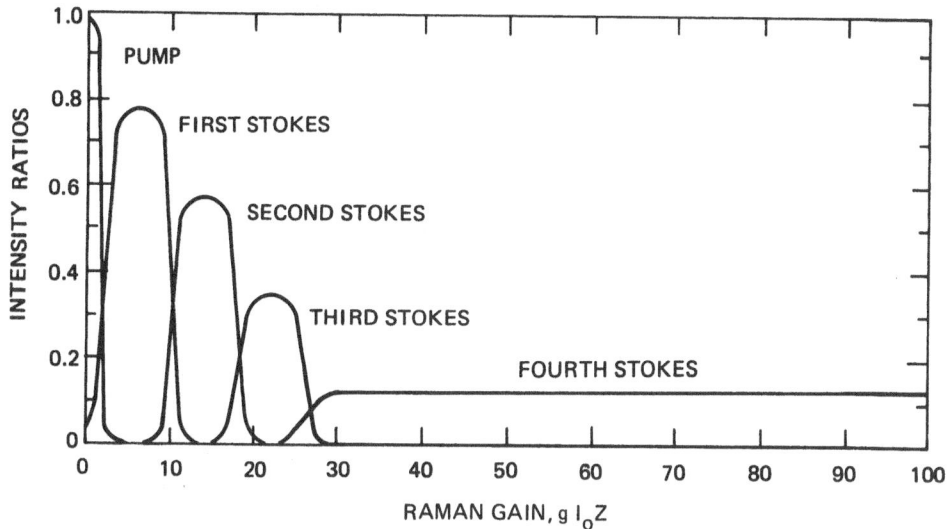

Figure 11. Solutions to Raman rate equations assuming a first-Stokes seed injection level of 1 percent of the pump.

in the exponent in the denominator of Eq. (53) can be traded off against each other in an amplifier design. For example, the use of a Q-switched neodymium laser for the pump gives ~15 nsec pulses. This provides very high peak intensities, and allows the use of cells much shorter than those appropriate for discharge-pumped lasers for which pulse lengths typically vary from 100 nsec to 1 μsec. Shorter pulse durations allow the use of proportionately shorter cells, all other parameters being equal. A similar trade-off exists between beam diameter and cell length. In order to achieve the shortest possible amplifier in the context of this trade-off, one makes the spot size as small as possible, consistent with window damage considerations.

Pump Laser Coherence Effects

The coherence properties of the pump and the seed lasers affect the performance of a Raman oscillator/amplifier. By coherence properties we mean those properties relating to the number and spectral width of modes, their spacing, and their relative phase. These may be specified for any pump laser with varying degrees of precision. The simplest representation is to state the coherence length or the linewidth of the laser. The most precise characterization would include specification of the amplitude, phase, and linewidth of every transverse and axial mode of the laser. In practice, such detailed knowledge is usually only possible for a single-mode laser.

According to Raman laser theory the output phase front of a Raman amplifier is exactly determined by the phase front of the seed laser (the oscillator). This can be seen from Eq. (42) in which any phase information in the pump field is absent. However, this is only true if the pump laser operates single-mode. With a multi-mode pump laser the situation becomes more complicated, because nonlinear mixing between the several pump laser and seed laser modes is possible in the Raman medium. Hence, the output beam can adopt some of the phase characteristics of the pump beam. If the seed beam is also multi-mode, then which seed-pump mode interactions build up most quickly or dominate is affected by the degree to which the pump and seed modes involved are phase- and spectrally-matched. The above issues affect the performance, the design, and the scaling of an oscillator/amplifier in several ways.

One of the most easily observed manifestations of these issues occurring with the use of a multi-mode pump laser is the "phase healing length".[49] When the pump and seed beams are not temporally correlated, the relative phases of the many modes are not conducive to the occurance of coherently additive four-wave mixing between them, and low amplifier gain results. Temporal correlation (and with it phase correlation) of the modes and the resulting high gain are achieved by superimposing the temporal fine structure of the pump and Stokes beams in the amplifier cell. The phase correlation can be maintained in the amplifier so long as the linewidths are sufficiently narrow that significant dispersion does not occur.

The situation is much simpler in the single-mode case. No intermode four-wave mixing occurs, so no phase correlations are necessary. Maximum Raman gain is obtained so long as there is reasonable overlap between the temporal envelopes of the pump and seed pulses.

CONCLUSION

The principles and ideas reviewed in this paper are being used in a variety of applied research programs throughout the world. Raman devices have proven to be consistent with a wide range of systems applications constraints, such as requirements for small size, high efficiency, and high reliability. By carefully considering the physics of Raman devices, and how Raman conversion depends on pump laser characteristics such as spectral purity and nonuniform intensity distributions, Raman oscillator/amplifier configurations can be optimized to produce >50 percent energy conversion from the pump to the first-Stokes wavelength. With demonstrated pulse repetition frequency scaling to $\gtrsim 100$ Hz, and energy scaling to >1 Joule, Raman lasers offer a potential for average output power performance that is presently limited by pump laser capabilities.

The authors gratefully acknowledge helpful discussions on the subject of stimulated Raman scattering with S.M. Wandzura and G.C. Valley.

REFERENCES

1. See, for example, G. Herzberg, Molecular Spectra and Molecular Structure, Vol. I. Spectra of Diatomic Molecules (Van Nostrand Reinhold, New York, 1950).
2. E.J. Woodbury and W.K. Ng, Proc. IRE 50, 2347 (1962).
3. R.W. Hellwarth, Phys. Rev. 130, 1850 (1963).
4. For a quantum mechanical description of stimulated Raman scattering see, for example, D. Marcuse, Principles of Quantum Electronics (Academic, New York, 1980).
5. A. Yariv, Quantum Electronics (Wiley, New York, 1975), p.484.
6. F. Zernike and J.E. Midwinter, Applied Nonlinear Optics (Wiley, New York, 1973), p.4.
7. H.W. Bruesselbach, D.A. Rockwell, S.M. Wandzura, and G.C. Valley, "Efficient Wavelength Conversion with a Backward-Stokes Raman Laser," presented at OSA meeting, New Orleans, Oct. 1983.
8. A. Weber, ed. Raman Spectroscopy of Gases and Liquids (Springer-Verlag, New York, 1979).
9. A.Z. Grasyuk, Sov. J. Quant. Electron. 4, 269 (1974).

10. A.Z. Grasiuk and I.G. Zubarev, Appl. Phys. 17, 211 (1978).

11. N.F. Andreev, V.I. Bespalov, A.M. Kiselev, and G.A. Pasmanik, Sov. J. Quant. Electron. 9, 585 (1979).

12. G.C. Valley, IEEE J. Quant. Electron. QE-18, 1370 (1982).

13. D.G. Bruns, H.W. Bruesselbach, and D.A. Rockwell, Proc Int. Conf. on Lasers'80, 406 (1980).

14. H.W. Yates and J.H. Taylor, "Infrared Transmission of the Atmosphere," NRL Report 5453, U.S. Naval Research Laboratory, Washington, D.C. (1960).

15. Judson Infrared, Inc., Ft. Washington, PA, Series J-16 Germanium photodiode.

16. W.R. Trutna and R.L. Byer, Appl. Opt. 19, 301 (1980).

17. J. Paisner and S. Hargrove, "A Tunable Laser System for the Ultraviolet, Visible, and Infrared Regions," Energy Technology Review, Lawrence Livermore Laboratories, UCRL Report 52000-79-3, March 1979.

18. Quanta Ray, Mountain View, CA 94043, now makes a dye laser pumped Raman laser tunable to 190 nm using anti-Stokes Raman shifts, but at a generally lower efficiency than the Stokes shifted wavelengths.

19. J.G. Meadors and M.A. Poirier, IEEE J. Quant. Electron. QE-8, 427 (1972).

20. A.J. Glass, IEEE J. Quant. Electron. QE-3, 516 (1967).

21. N. Djeu and R. Burnham, Appl. Phys. Lett. 30, 473 (1977).

22. P. Rabinowitz, A. Stein, R. Brickman, and A. Kaldor, Appl. Phys. Lett. 35, 739 (1979).

23. E. Wild and M. Maier, J. Appl. Phys. 51, 3078 (1980).

24. D.G. Bruns, H.W. Bruesselbach, H.D. Stovall, and D.A. Rockwell, IEEE J. Quant. Electron. QE-18, 1246 (1982).

25. J.M. Manley and H.E. Rowe, Proc. IRE 47, 2115 (1959).

26. W.H. Culver and E.J. Seppi, J. Appl. Phys. 35, 3421 (1964).

27. A. Yariv, ibid, p.476.

28. A. Yariv, ibid. p.111.

29. W.R. Fenner, H.A. Hyatt, J.M. Kellman, and S.P.S. Porto, J. Opt. Soc. Am. 63, 73 (1973).

30. P.V. Avizonis, K.C. Jungling, A.H. Guenther, and R.M. Heimlich, J. Appl. Phys. 39, 1752 (1968).

31. H. Komine, E.A. Stappaerts, S.J. Brosnan, and J.B. West, Appl. Phys. Lett. 40, 551 (1982).

32. S.F. Fulghum, D.W. Trainor, C. Duzy, and H.A. Hyman,

Topical Meeting on Excimer Lasers (Incline Village, NE, 1983).

33. R.S.F. Chang and N. Djeu, Opt. Lett. $\underline{8}$, 139 (1983).

34. H. Komine, Scaling Studies of Efficient Raman Converters, Technical Report AD-110159, (1981) p.16.

35. Y.R. Shen and N. Bloembergen, Phys. Rev. $\underline{137}$, A1787 (1965).

36. M. Sparks, Phys. Rev. Lett. $\underline{32}$, 450 (1974).

37. J.H. Newton and G.M. Schindler, Opt. Lett. $\underline{6}$, 125 (1981).

38. J.N. Holliday, Opt. Lett. $\underline{8}$, 12 (1983).

39. W. Seka, S.D. Jacobs, J.E. Rizzo, R. Boni, and R.S. Craxton,
Opt. Comm. $\underline{34}$, 469 (1980).

40. R.S. Craxton, Opt. Comm. $\underline{34}$, 474 (1980).

41. D. von der Lind, M. Maier, and W. Kaiser, Phys. Rev. $\underline{178}$, 11 (1969).

42. E.E. Hagenlocker and W.G. Rado, Appl. Phys. Lett. $\underline{7}$, 236 (1965).

43. E.E. Hagenlocker, R.W. Minck, and W.G. Rado, Phys. Rev. $\underline{154}$, 226 (1967).

44. N. Bloembergen, G.G. Bret, P. Lallemand, A. Pine, and P. Simova, IEEE J. Quant. Elect. $\underline{QE-3}$, 197 (1967).

45. N. Bloembergen, Am. J. Phys. $\underline{35}$, 989 (1967).

46. P. Rabinowitz, A. Stein, L.R. Brickman, and A. Kaldor, Opt. Lett. $\underline{3}$, 147 (1978).

47. W.R. Trutna, Y.K. Park, and R.L. Byer, IEEE J. Quant. Elect. $\underline{QE-15}$, 648 (1979).

48. W.K. Bischel, Laser Focus, (October, 1983), p.156.

49. E.A. Stappaerts, W.H. Long, Jr., and H. Komine, Opt. Lett. $\underline{5}$, 4 (1980).

OPTICAL FIBRE SYSTEM AND RELATED SOURCES

Bruno Costa - Emilio Vezzoni

CSELT - Centro Studi e Laboratori Telecomunicazioni

Via G. Reiss Romoli, 274 - 10148 - Torina, Italy

1. Introduction

Optical fibres are finding increasingly widespread use in all telecommunication applications. Industrially developed countries have plans for installation of thousands of kilometers of optical cables in the near future. Technical advantages of optical fibres are the justification for this success: low attenuation, high bandwidth, low weight and small size, insensitiveness to e.m. disturbances, suitability for a large range of applications, easy upgradability. Applications include intra building and intra vehicle communications, local area distribution (links between computers), subscriber loop, links between telephone exchanges, inter city, long distance communication, submarine cables.

Correspondingly a range of optical sources with characteristics appropriate to the specific needs of the various applications is required.

In the following an overview of optical fibres and systems will be given, followed by an examination of related optical sources.

2. Optical fibres

The structure of modern optical fibres for telecommunications is illustrated in fig. 1. The inner core region, with a higher refractive index (n_o) with respect to the cladding (n_1)

447

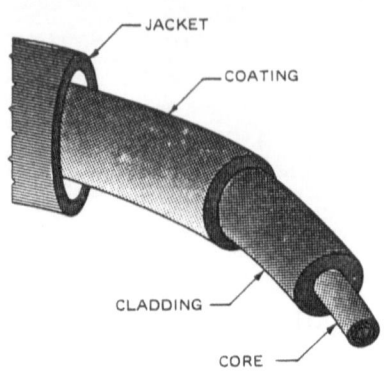

Fig. 1 - Structure of a modern fibre for telecommunications.

is mainly based on Silica, codoped, in general, with Germania and Phosporus Oxide for increasing the index. The present trend is to use only Germania, as it gives better resistance to loss degradation due to hydrogen (H_2) contamination. The cladding region is pure silica, at least in the outmost part.

Around the core an addition of fluorine, that decreases the refractive index of silica, is often employed.

For mechanical protection the fibre is coated with some plastic material (typically acrylates or silicone) with thicknesses ranging from tens to hundreds of microns.

For insertion in a cable a further jacket is applied either in loose or tight form.

For telecommunication purposes two types of optical fibres are now in use: multimode, graded index fibres and monomode fibres.

2.1 Multimode fibres

Multimode fibres support many propagating modes; this is due to the fact that they have a core diameter 2a, which is large compared to the wavelength, λ, of the radiation (about 50 μm against 0.8 - 1.5 μm) and a comparatively large numerical aperture (N.A. = $n^2_0 - n^2_1 \simeq 0.2$).

The total number of supported modes in a graded index fibre is $N = V^2/4$, where $V = (2\pi a/\lambda) \cdot (N.A.)^2$. The main consequence of multimode propagation is a relatively large time dispersion produced in fibre on a propagating pulse; the reason is that each mode travels at a different speed in the fibre, so that an originally narrow pulse launched into the various modes spreads out, and the final output pulse is simply the superposition of all the pulses carried by the different modes, arriving at different times.

This fact, called modal dispersion, limits the maximum repetition rate that can be transmitted along a fibre, as the time period between successive pulses cannot be smaller than the pulse width.

Time dispersion, defined for instance as the r.m.s. of the impulse response of the fibre, may range, for a 1 km length, from 0.2 to 1 ns in best fibres.

A complementary representation, useful for system design, may be obtained in the frequency, rather than the time, domain; in this case the fibre is characterized by a transfer function, i.e. in terms of its response to the varying frequency of a sinusoidal signal; the frequency at which the response falls by 3 dB with respect to zero frequency is called the fibre "bandwidth".

In order to improve the bandwidth performance it is necessary to equalize, as much as possible, the time of flight of the various modes; this may be accomplished by suitably shaping the index distribution; it has been found that a nearly parabolic law of refractive index change as a function of radius is quite satisfactory. For this reason multimode optical fibres for telecommunications are of the "graded index" type; bandwidths of such fibres may range from 200 to 1500 MHz·km.

Concerning attenuation the remarkable progress in fabrication techniques has led to the routine production of fibres with attenuations very close to the intrinsic limits of silica, i.e. 2-2.5 dB/km at 0.85 μm wavelength, and 0.5 - 0.8 dB/km at 1.3 μm.

Multimode fibres are now a sufficiently well established industrial product. This is reflected in the existence of international standards, issued since 1980, concerning the main characteristics of graded index fibres for telecommunications. These are summarized in table I.

Table 1 - Physical characteristcs of 50/125 μm graded index optical fibres.

PARAMETER	VALUE	UNIT
Core Diameter	50 ± 3	μm
Cladding diameter	125 ± 3	μm
Concentricity	< 6%	Referred to core diameter
Core non circularity (Difference between longest and shortest chord through the core centre)	< 6%	"
Cladding non circularity (Difference between longest and shortest chord through the cladding center)	< 2%	Referred to cladding diameter
Refractive index profile	Near parabolic	
Maximum N.A. (a range is defined)	0.19-0.24 ± 10%	

2.2 Monomode fibres

Fibres that carry only one mode are called monomode fibres. Single mode operation is achieved in fibres with sufficiently low V value, as in this case all modes but the fundamental one are beyond cut-off. Low V values correspond to small core radius and small N.A., that in single mode fibres attain values of 3-5 μm and 0.1-0.14 respectively.

The greatest advantage of using single mode fibres is the elimination of modal dispersion, leading to very large bandwidth. The bandwidth limitation is only due to what is called "chromatic dispersion": it is caused by the fact that the speed of the optical signal in the waveguide depends on the frequency, or wavelength, both due to the dispersion properties of the material (the refractive index is a function of wavelength) and to the dependence of the mode group velocity on the wavelength (waveguide dispersion).

Chromatic dispersion is conveniently given in terms of time delay per unit wavelength difference, per unit length traversed; specifically in ps/nm·km. A plot of tipical dispersion in a single mode fibre is given in fig. 2.

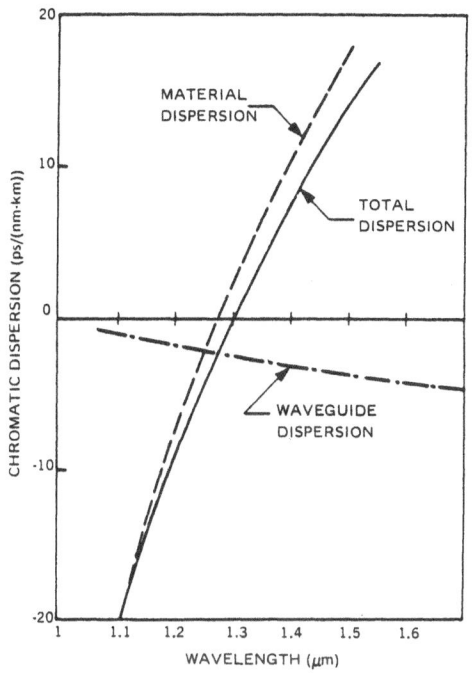

Fig. 2 –   Chromatic dispersion in a monomode fibre, decomposed in material and waveguide contribution.

It is interesting to note that a wavelength exists where the chromatic dispersion becomes zero; clearly at that wavelength the maximum possible bandwidth is achieved; in addition it has to be observed that the bandwidth limitation is not produced only by the fibre, but comes from a combination of fibre and source properties; in fact the smaller the source spectral width, the larger the available bandwidth. Transmission rates of more than 2 Gbit/s over 130 km of optical fibres have been demonstrated, corresponding to bandwidths larger than 200 GHz·km.

Use of single mode fibres is also beneficial from the attenuation point of view; due to the lower refractive index difference required, a smaller amount of dopant is added to silica, which results in smaller intrinsic losses; in fig. 3 a comparison of estimated intrinsic attenuation limits for multimode and monomode fibres is shown. At the wavelengths of interest, namely 1.3 and 1.55 μm, losses of 0.4-0.6 and 0.2-0.4 dB/km can be routinely achieved.

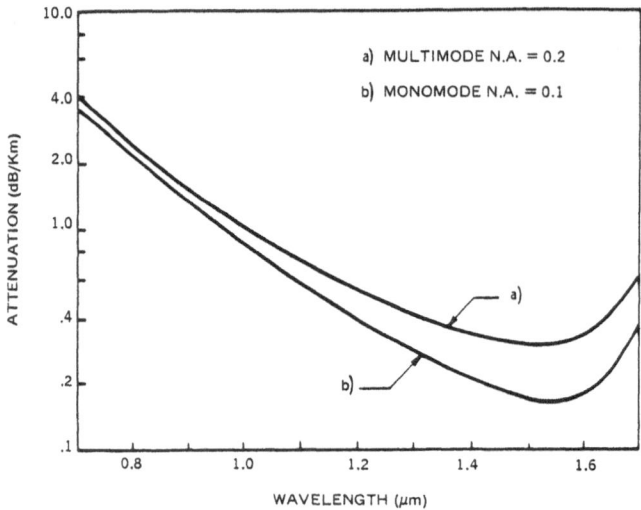

Fig. 3 -   Curves of minimum expected attenuation (intrinsic loss) of multimode and single mode fibres.

Table 2 - Characteristics of monomode fibres at 1.3 μm .

PARAMETER	VALUE	UNIT
Mode field diameter	9 ÷ 10 (±1)	μm
Cladding diameter	125 (±3)	μm
Concentricity error	0.5 ÷ 3	μm
Cladding non circularity	< 2%	
Cut-off wavelength	1100÷1250 (or 1280)	nm
Attenuation coefficient	< 1.5	dB/km
Dispersion	< 6	ps/(nm·km)

Monomode fibres have reached the commercial stage too, and in table II a recently proposed (1984) standardization of the main characteristics is reported.

3. Fabrication technique

As already mentioned present methods allow the fabrication of optical fibres with attenuations very close to intrinsic limits. Losses in fibres are basically caused by absorption and scattering phenomena. A spectral region exists (0.8 - 1.7 μm) where absorption due to the material itself gives only a minor contribution to total attenuation, the main contribution being Rayleigh scattering, caused by unavoidable material density fluctuations. Attenuation caused by Rayleigh scattering goes as the inverse of the fourth power of wavelength; thus use of optical fibres at longest possible wavelengths is more advantageous. A large absorption may be caused by metallic impurities and water radicals (OH-); in order to produce low loss fibres such contaminants are to be kept at very low levels, say less than 1 part per billion.

Techniques that have successfully achieved that goal are based on synthetic formation of the fibre material via chemical vapour deposition. The method consists basically in activating a reaction between halogenides of the involved elements (Si; Ge, P) in vapor phase and oxygen, giving, as a final product, oxydes of the quoted elements which form the glass material. A very high temperature is required to activate the reaction.

The possibility of having very pure starting reactants, in liquid form, the further purification obtained by the vaporization, use of a protected environment for the process allow an extremely pure material to be obtained and consequently very low-loss fibres to be fabricated.

A brief description of the MCVD (Modified Chemical Vapor Deposition) is given in the following.

The process consists of two main steps: fabrication of a preform and drawing of the corresponding fibre from the preform. The preform is a glass rod, 1.5-3 cm diameter, with a composition exactly similar to that of the final fibre (a cladding, a core, a proper refractive index profile); the fibre is simply obtained by heating the preform at sufficiently high temperature (>2000°C) so as to reduce its viscosity to a level low enough that it can be pulled with a small force; pulling speed is adjusted to get the right diameter for the resulting fibre. During the pulling process, the protective plastic coating is applied.

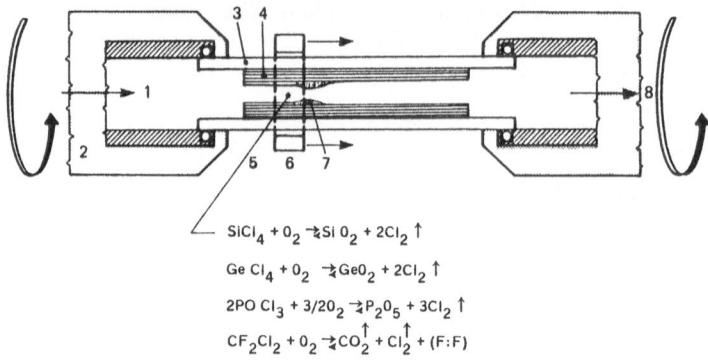

$$SiCl_4 + O_2 \rightarrow SiO_2 + 2Cl_2 \uparrow$$

$$GeCl_4 + O_2 \rightarrow GeO_2 + 2Cl_2 \uparrow$$

$$2POCl_3 + 3/2O_2 \rightarrow P_2O_5 + 3Cl_2 \uparrow$$

$$CF_2Cl_2 + O_2 \rightarrow CO_2 \uparrow + Cl_2 \uparrow + (F:F)$$

Fig. 4 –   Schematic representation of MCVD process.
            1: inlet of reactants ($SiCl_4$, $GeCl_4$, $POCl_3$, CF_2Cl_2, O_2)
            2: synchronously rotating chucks
            3: support silica tube
            4: layers of doped silica, deposited and consolidated
            5: reaction zone
            6: burner
            7: dispersion of timy glass powder
            8: gases outlet.

The main features of preform fabrication are schematically shown in fig. 4. The starting element is a silica tube, held in the chucks of a glass lathe and rotating around its axis. Reactants, in gaseous form, are made to flow along the tube, in the required proportion; an oxy-hydrogen flame, provided by a burner, heats a zone of the tube at temperatures around 1500-1700°C; when reactants cross the high temperature region oxidation reactions as indicated in the figure occur; the resulting particles are made, by the heat distribution, to adhere to the tube walls, and are consolidated into glass; the burner traverses at low speed the whole tube length and, as a result, a thin (~30-50 μm), very pure glass layer is deposited on the inner wall of the tube; the process can be repeated many times, possibly changing the composition at each step with a predetermined law, until the designed structure is obtained. At the end of the process the tube is not completely filled; in order to close the inner hole the temperature is further raised, so that the surface tension forces cause the softened tube gradually to shrink and finally completely collapse.

Based on the same principle different techniques have been developed; it is possible to grow the preform starting from a supporting rod (e.g. a ceramic one) and depositing the synthetic particles around the rod; glass particles are formed directly in the burner flame, letting the reactants to flow together with oxygen and hydrogen; this process is called OVPO (Outside Vapour Phase Deposition); if the deposition is in the axial direction of the starting rod, we have VAD (Vapour Axial Deposition). A different source of energy, i.e. microwave power, can be used in the inside CVD process: in this case it is called PCVD (Plasma CVD).

All this processes have allowed fabrication of very low loss optical fibres, both in laboratories and on an industrial basis. In Fig. 5 the attenuation spectral curve of a state of the art single mode fibre, produced at CSELT, is shown. The minimum loss value is 0.2 dB/km at 1.55 μm wavelength.

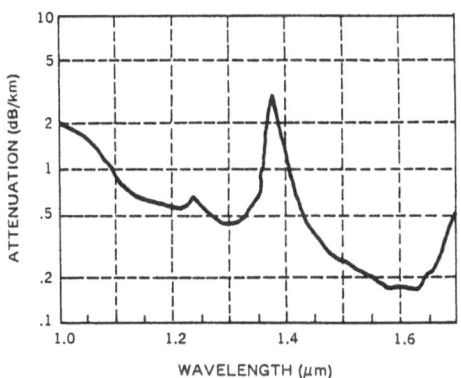

Fig. 5 -   Attenuation characteristics of a state-of the art single mode fibre in the long wavelength region.

Outside processes (OVPO, VAD) give a higher yield, as a larger amount of material per unit time can be deposited. Inside processes (MCVD, PCVD) show perhaps more flexibility in structure design. In any case optimization of the processes for efficient industrial production is still to be performed.

Lower losses than those presently attainable may be obtained by using non oxide glasses. Remembering that fundamental scattering loss decreases as λ^{-4}, losses as low as $10^{-2} \div 10^{-3}$ dB/km should be reached in the 2-6 μm wavelength region.

Suitable materials are glasses based on fluorides or chalcogenides. Expected spectral loss curves for some compositions are shown in Fig. 6.

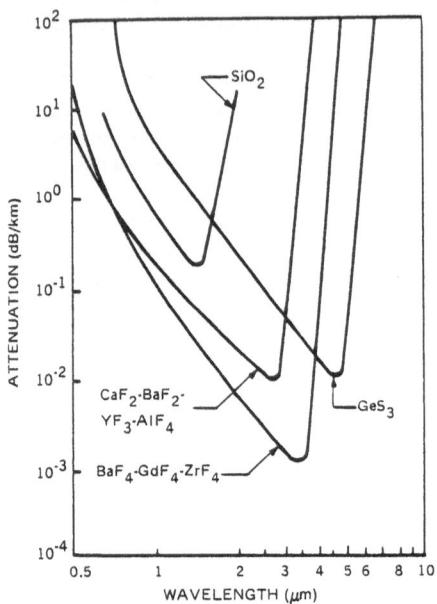

Fig. 6 – Theoretical intrinsic loss of various non-oxide glasses, compared to silica (SiO_2), in the infrered region.

4. Systems and Applications

Optical fibres may be employed in any kind of application requiring transmission of signals; in particular they can replace all the existing systems based on metal cables. The question whether it can be done economically is the only point to be considered for the choice.

Transmission systems can be broadly divided into analog and digital ones. While the fibre itself can transmit equally well both types of signal, the characteristics of available sources (insufficient linearity, as it will be better discussed later) tend to favour the use in digital system. In addition, and perhaps more important, the general trend of telecommunications is to evolve towards an all digital network.

Hierarchical levels for digital transmission systems in Europe include bit rates of 2, 8, 34, 140, 565 Mbit/s. As already mentioned optical fibres may find use in a variety of applications: communications inside buildings or vehicles (ships, planes) where small size, small weight, flexibility, insensitiveness to e.m. disturbances are of importance; local area networks (e.g. links between computers); subscriber loop, i.e. distribution of wideband services to both business and home subscribers; interoffice links, in urban area; trunk telecommunications, submarine links.

In view of the afore said applications a choice has to be made between multimode and single mode fibres, as well as between various possible wavelengths of operation. On this last point the basic elements to be considered are the spectral behaviour of attenuation and bandwidth of the fibre, the availability of appropriate sources and detectors and their specific features. Traditionally three wavelength regions of interest are considered, i.e. those around 0.85, 1.3 and 1.55 μm; they are often referred to as first, second and third window. Originally the reason for this distinction was due to the presence of large absorption peaks due to water, centered at 0.95, 1.25, 1.38 μm, that left low-loss windows at the quoted wavelengths. Progress in fabrication processes has virtually eliminated such peaks, with some residual effect at 1.38 μm (see also Fig. 5). Therefore the main interest for 0.85 μm is the availability of well established optical sources and, above all, of excellent detectors (Si photodiodes), with very low noise.

The 1.3 μm region exhibits lower attenuation and zero chromatic dispersion, which is interesting particularly for single mode fibres or for multimode fibres used in connection with broadband sources (LED's). At 1.55 μm the minimum attenuation values are found, particularly in single mode fibres; if also dispersion is reduced, either by using very narrow linewidth sources or by fabricating dispersion shifted fibres, the utmost in terms of length capacity product can be obtained.

Regarding the choice between multimode and monomode fibres the following considerations apply. Multimode fibres, thanks to comparatively large core size and NA, allow easier and more efficient coupling to optical sources, especially to incoherent ones, and, similarly, pose less stringent requirements on jointing and connectorization. Their main limitation in terms of transmission performance is dispersion: all systems from 140 Mbit/s and above are essentially bandwidth-limited, i.e. the maximum repeater spacing is set by bandwidth limitations rather than attenuation ones. Moreover, the bandwidth being a function of wavelength, there is a certain lack of flexibility, in that optimum operation at a certain wavelength requires optimizatin of fibre profile for that wavelength.

An additional difficulty in designing systems based on multimode fibres arises from the fact that the behaviour of the bandwidth versus the link length is quite unpredictable; in fact the total bandwidth depends on such factors as systematic and random irregularities of the index profile, change of index profile from one fibre to another, the amount of exchange of power among modes, called mode conversion caused, for instance, by microbending, mode conversion due to splices, the original mode distribution excited by the optical source. On the other hand multimode fibres are not less expensive than the more performant single mode fibres (in fact they are, in principle, more expensive).

In conclusion their use seems appropriate in systems where component cost (optical sources, detectors, connectors and splices) is a dominant factor, which means all kind of short distance application mentioned above, LAN and subscriber loop distribution, as long as the demand for bandwidth does not become too large.

Concerning single mode fibres, this transmission properties are good enough to adequately cover all kind of today foreseeable applications. The region of operation of single mode fibre based systems, in terms of repeater spacing against bit rate, under a few reasonable assumptions on source and detector proper-

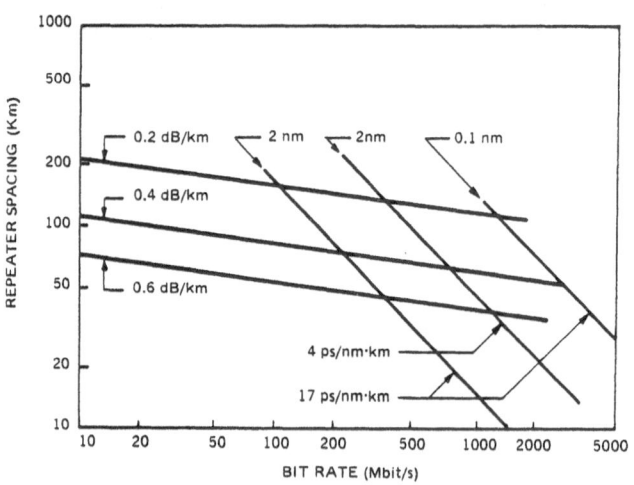

Fig. 7 –   Limiting curves for maximum repeater spacing vs. bit rate. Parameters are:
a) fibre attenuation (dB/km);
b) fibre dispersion (ps/nm·km);
c) source spectral width (nm).

ties as well as on signal coding, is shown in Fig. 7. As already mentioned monomode fibres are, in principle, less expensive than multimode fibres, essentially because they have a smaller core and therefore longer lengths can be obtained from one preform.

System design is easier because both dispersion and attenuation follow a simple addition law in single mode fibres. On the other hand monomode fibres require good quality sources (single mode lasers) for efficient optical power coupling; splices and connectors call for more sophisticated, and consequently more expensive, techniques and devices.

Therefore applications are presently more concentrated on long distance communications. Moreover single mode fibres offer quite interesting prospects for future developments. One reason is that they may be naturally interconnected to integrated optical devices, that are based on single mode propagation in planar waveguides. Useful integrated optical devices are: modulators, switches, isolators, filters. A second reason is that monomode fibres maintain the coherence properties of the source; availability of a coherent optical beam allows the development of coherent detection schemes, of the homodyne or heterodyne type. Present day systems are based on direct detection of the intensity of the optical signal; in coherent systems the optical signal from the fibre is mixed with a local oscillator; in this case thermal noise of the receiving set-up is avoided and only the fundamental limitation given by quantum noise is left. Consequently sensitivity enhancement of 10-20 dB with respect to intensity detection is ideally attainable. Such systems require however extremely stable, narrow line width sources that are not presently available.

In addition the polarization of the signal must be the same as that of the local oscillator beam. As a consequence either polarization maintaining fibres have to be used (research is under way, but the task appears very difficult) or active devices at the output end are to be employed in order to properly change the polarization of the incoming beam.

In general, both for multimode and monomode fibre based links, it is possible to increase the transmission capacity, or multiplex different signals, or to achieve bidirectional transmission, by using wavelength division multiplexing (WDM). This consists in sending a plurality of signals from sources emitting at different wavelengths into an optical fibre, by means of an appropriate mixing device. At the output end of the fibre a proper wavelength selective device allows the separation of the different channels. For bidirectional transmission at each end a device is needed which is able to separate the ingoing and the outgoing beams, that have different wavelengths. Two representative schemes of WDM transmission are shown in Fig. 8.

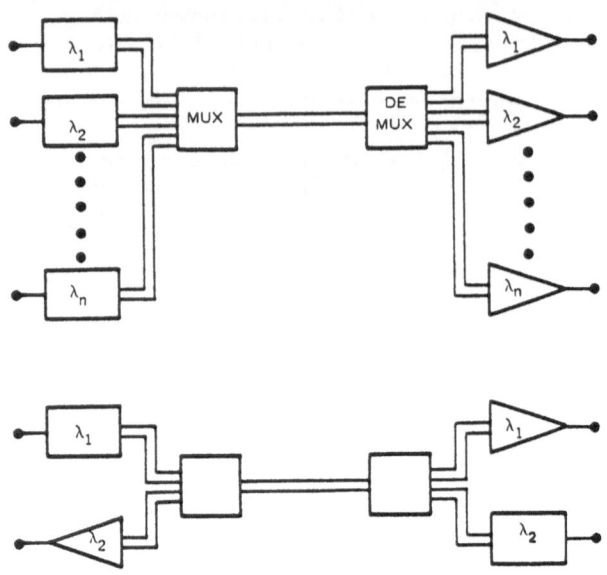

Fig. 8 – Schematic picture of WDM systems

As already pointed out all the various applications so far described call for sources with specific properties. The remainder of the paper deals with the subject of optical source properties and their relationship to system requirements.

5. Sources for optical fibre systems: general requirements

To take advantage of the many qualities of optical fibres, efficient sources of radiation must be employed, matching the specific requirements set by the lightguide transmission characteristics and the more general needs of reliable telecommunication systems.

The spectral transmission characteristics of glass-based optical fibres set a first strict requirement on the emission wavelength of the sources, that must fall in one of the low attenuation ranges, that is around 0.85, 1.3 and 1.55 µm; the 1.3 µm range is moreover favoured by the absence of chromatic dispersion in the fibres.

The small core size (in the range 10-100 µm) and the limited acceptance angle (in the range 5°-20°) of the fibres call, furthermore, for high radiance sources, i.e. sources concentrating much power on a restricted area, and being as directive as possible.

To exploit the large bandwidths of multimode and mainly single mode fibres, the sources of practical interest should withstand modulation in the range 10 MHz - 10 GHz.

More general system considerations set a number of further requirements. The sources should be preferably directly modulated, e.g. through the bias current, to avoid the need of additional external modulators; they should be very efficient as well, to reduce electrical power consumption; they should be rugged, to allow easy mounting and maintenance operations.

The source reliability should meet specifications that, for some kind of systems, like the submarine ones, can be very strict and hard to satisfy.

Finally, the cost of these devices should be compatible with the system cost reduction, which is a main target of optical fibre telecommunications.

None of the conventional sources meet all these requirements toghether; incandescent or gas discharge lamps have not radiance enough, nor can be directly modulated; gas or solid state lasers have higher radiance but require external modulation, have low efficiency and relatively low reliability.

The only way to meet all the above mentioned requirements is to turn to electroluminescence as can be obtained by current injection across a forward biased p-n junction. The recombination of electrons and holes around the metallurgical junction of a semiconductor diode can be a very efficient source of near infrared radiation.

6. Light emitting diodes and laser diodes

Two different kinds of sources can be obtained: the light emitting diode (LED), when only spontaneous emission occurs; and the more sophisticated laser diode (LD), relying upon stimulated emission mechanisms. At present, LED's and LD's are the only widely used sources for optical fibre systems.

The operation at the desired wavelength, mainly depending on the bandgap of the semiconductor material, can be obtained by compound materials of groups III and V.

AlGaAs can emit radiation in the range 0.8-0.9 μm, by adjusting the Al content. Similarly, InGaAsP can be tuned at any wavelength between 1 and 1.6 μm, varying the atomic fraction of the different elements.

To achieve laser action in a diode, two general conditions are needed. First, the carrier density must be high enough to reach the population inversion condition in the active region where radiative recombinations take place, thus realizing the amplifying medium. Second, an optical feedback must be supplied, using two opposite chip facets, mirror-like cleaved, that result in a Fabry Perot resonator. When the injection current is increased, a threshold level is found, where the total cavity losses are balanced by the material gain; above this level, stimulated emission takes place, and the laser output increases linearly with the current. The spectrum shows one or more lines, corresponding to the resonance frequencies of the Fabry-Perot cavity, with a FWHM envelope width of a few nanometers (Fig. 9a).

Light emitting diodes have a simpler behaviour, showing the output power increasing more or less linearly with the current, with a saturation at high injection levels. LED spectra are approximately gaussian, with FWHM widths of 30-40 nm and 80-120 nm for AlGaAs (or Ga As) and InGaAsP devices respectively (Fig. 9b).

The first laser diode was demonstrated in 1962; these early devices, simple homojunctions made of GaAs, were not able to operate CW at room temperature, which is a fundamental condition for system applications, due to the poor confinement of the injection carriers at the p-n junction. The obstacle was overcome in 1968-1970 by the demonstration of the double heterostructure laser diode (DH-LD), basically made of a low Al content active layer, surrounded by two higher Al content confining layers.

Due to the potential barriers at the layers interfaces this structure results in a strong carrier confinement in the active (recombination) layer, while the refractive index change, through a waveguiding mechanism, introduces additional confinement to the generated photons (Fig. 10).

As a result, DH-LD's can stably and safely operate CW at room (and higher) temperature. To further reduce the CW operating current, a stripe geometry contact is used, narrower than 20 μm (Fig. 11). All modern laser diodes make use of similar structures although a little more complicated by further requirements.

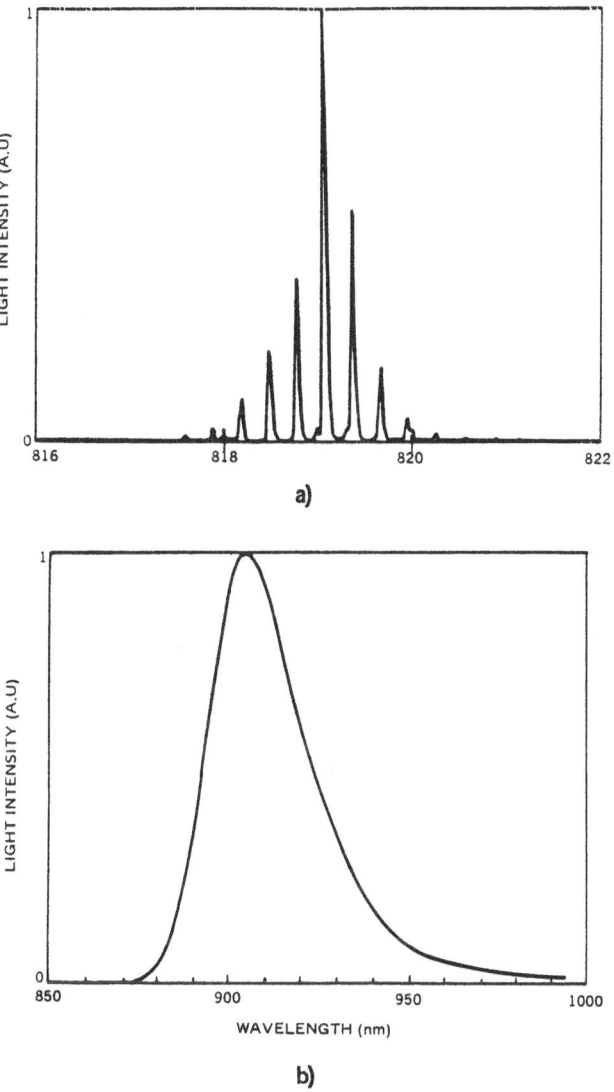

Fig. 9 –   Typical emission spectra of: a) AlGaAs laser diode;
b) GaAs LED

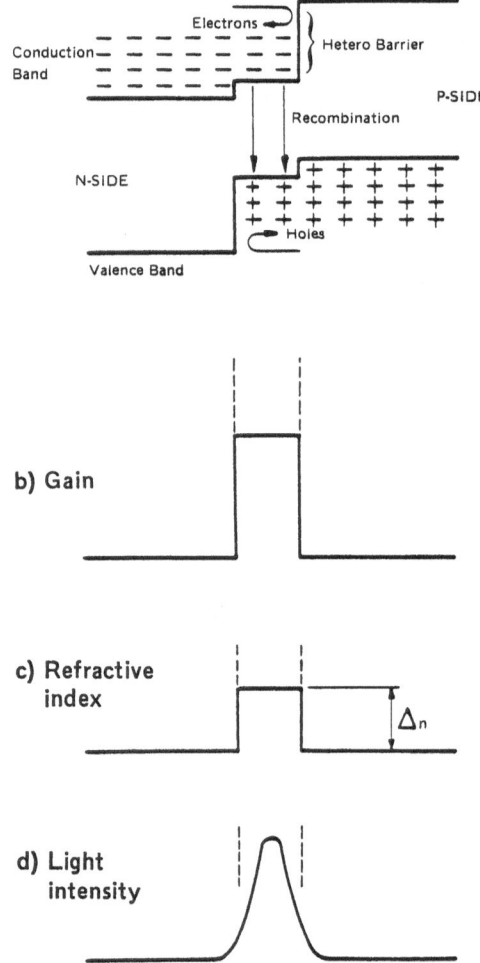

a) Carrier confinement

Electrons
Conduction—
Band
Hetero Barrier

P-SIDE

Recombination

N-SIDE

$+ + + + + +$
$+ + + + + +$
$+ + + + + +$
$+ + + + + +$

Holes

Valence Band

b) Gain

c) Refractive index

Δ_n

d) Light intensity

Fig. 10 – Operation principle of the double heterostructure laser diode.

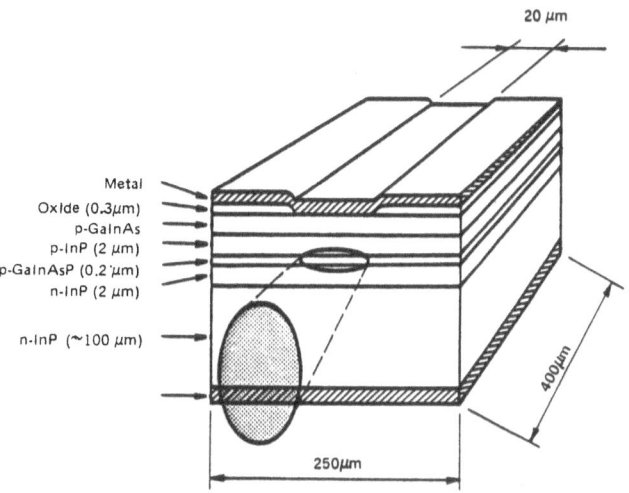

Fig. 11 - Schematic of an InGaAs/InP double heterostructure laser
diode, with oxide insulated stripe contract.

To grow such multi-layer crystal heterostructures epitaxial growth techniques are used, like liquid-phase epitaxy (LPE)
and vapour-phase-epitaxy (VPE), starting from GaAs and InP
substrates, for AlGaAs and InGaAsP components respectively. A very
promising technique for mass production of optoelectronic devices,
which is at present broadly investigated, is the metallorganic
chemical vapour deposition (MOCVD), that allows an excellent
control of the layer thickness over large wafer areas.

The key point, in all this processes, is the careful
control of the growth parameters, in order to obtain a good
matching between the lattice constants of the different layers,
thus avoiding the formation of dislocation networks and other lattice defects, that adversely affect the reliability of the
resulting devices.

Once the wafers are prepared and processed, they are
cut, and the two mirror-like facets are obtained, generally by
cleavage. Dielectric layers can be deposited on the facets, to
passivate them, or change their reflectivity.

Typical laser diodes have cavity lengths of a few
hundreds μm; the light spot on the output mirror is more or less
elliptical, with a transverse size of a few microns; the output
beam is not circularly symmetrical, showing a higher divergence in
the plane perdendicular to the junction, of a few tens of degrees.

CW operating currents are lower than 200 mA, with optical output powers that can reach some tens of mW, limited by the damage threshold of the facets. For reliability considerations, LD's are generally operated below 10 mW (Fig. 12).

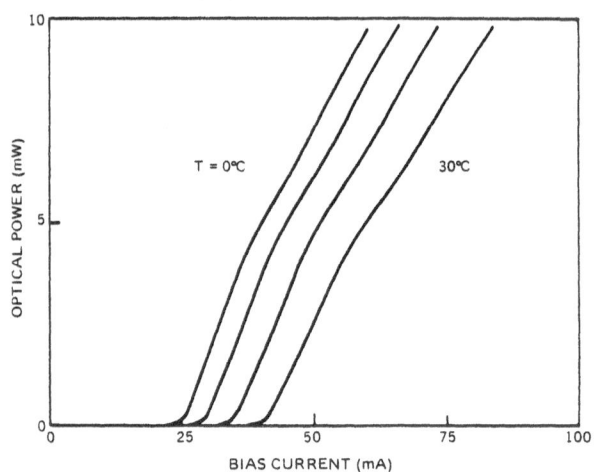

Fig. 12 - Power vs. current characteristics of a laser diode, at different operating temperatures.

The coupling efficiency into optical fibres is good, and can be increased by means of suitable lens-systems. More than 1 mW of optical power can be coupled into single mode fibres; 2-3 mW can be reached in standard multimode graded index fibres.

LED's can be manufactured according to two different geometries: surface emitting (Fig. 13a) and edge emitting (Fig. 13b). In the former case the radiation is emitted perpendicularly to the plane of the junction, while in the latter the output is on a plane parallel to the junction.

Double heterostructures are commonly used for LED's as well, although efficient GaAs homojunctions are still produced.

Surface emitters have circular emitting areas with 30-50 μm diameters, and the emission pattern is far less directional than the one of laser diodes. This reduces the coupling efficiency into optical fibres to few percents. Optical powers of 20-100 μW can be coupled into graded index multimode fibres. Single mode fibres are not used with such emitters. The output power can be increased at the expense of bandwidth; therefore, the

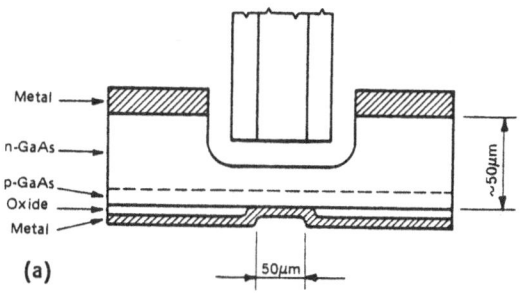

(a)

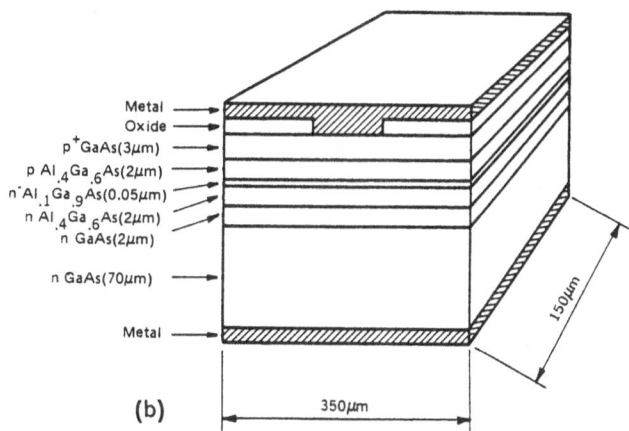

(b)

Fig. 13 - LED structures: a) surface emitter (homojunction),
b) edge emitter (heterojunction).

devices can be tailored to reach the best power/bandwidth tradeoff for a given application. Commercial sources have bandwidths in the range 30-100 MHz.

Edge emitters have a geometry very similar to that of laser diodes. Yet, an appropriate design prevents stimulated emission. The output characteristics are nearly similar to those of surface emitters.

In the following we are going to examine the specific features of the sources which are required according to the particular system application.

7. Sources for digital systems

The preferred use of optical fibres is, at present and much more in the future, for digital systems, due to the general trend of transmission systems to rely on digital technology; LD sources for these applications must present a number of important features. First of all, they must be fast enough to undergo modulation at bit rates that are rapidly increasing: 34, 140, 560 Mb/s and even 1 Gb/s or more. High quality commercial laser diodes can be modulated up to 1 Gb/s without much effort. For higher rates, care must be taken to reduce the capacitance of the chip and the parasitic elements of the package; in any case, an upper limitation to the repetition rate is set by the so called "pattern effect", consisting in a strong dependence of the mark (logic "1") level on the preceding bits (Fig. 14), resulting in a dramatic intersymbolic interference.

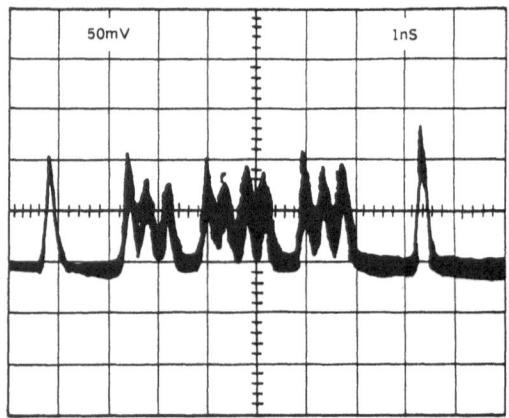

Fig. 14 - Response of a laser diode to a data sequence at the repetition rate of 2.5 Gb/s: the "pattern effect" is clearly visible.

This effect is intrinsically tied with the interaction mechanism between injected carriers and photons inside the active cavity, and is likely to limit the bit rate of direct modulation to a few Gb/s.

Another important feature of LD sources is the need of operation in the fundamental spatial mode. Like in any resonant cavity, in laser diodes, too, the electromagnetic field can be spatially distributed in characteristic configurations, the "modes" of the cavity, which are progressively activated by increasing the injection current above the threshold.

As the appearance of higher order modes after the fundamental one is accompanied by undesired effects, like non linearities in the power vs. current characteristic (the so called "kinks", as can be seen in the output characteristics of Fig. 12, around 5mW) noise, strong ringing in the pulse response and changes in the geometrical characteristics of the output beam, a well controlled lateral guiding mechanism must be introduced, shifting the onset of higher order modes towards higher output powers, where the LD is not supposed to work. This lateral waveguiding must be good enough to ensure that the fundamental mode be very stable, to avoid non linearities and other unwanted effects similar to the above mentioned ones.

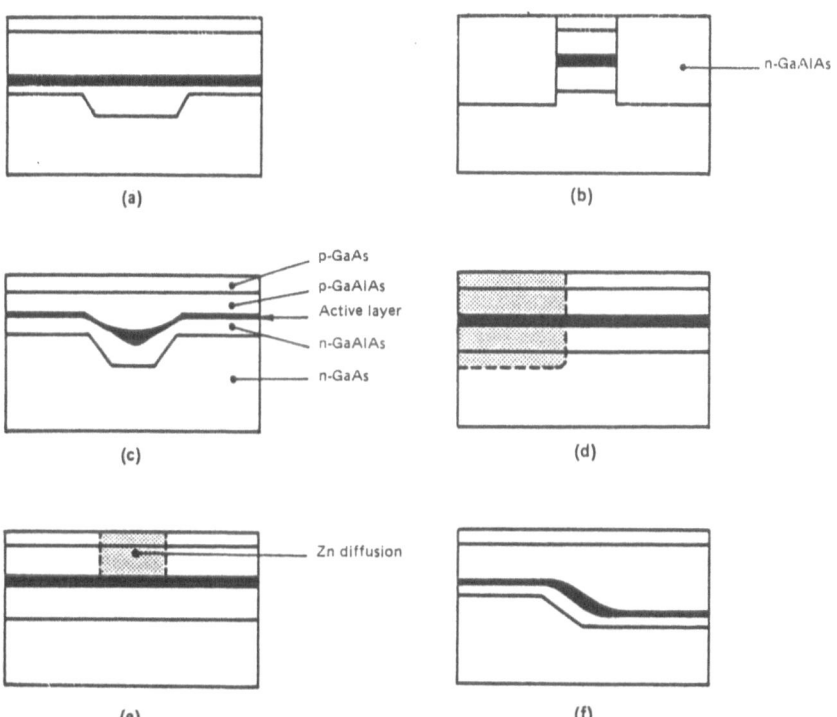

Fig. 15 - Schematic structures of GaAlAs/GaAs index guide lasers:
a) channeled substrate planar;
b) buried heterostructure;
c) embedded structure;
d) transverse junction stripe;
e) deep diffusion;
f) terraced substrate.

Such a mode control can se obtained basically in two
different ways: by "gain guiding" or by "index guiding". In the
former case, the injected current and hence the gain is confined
in a very narrow region by reducing the size of the contact stripe
to a few microns; in the latter, a true waveguiding mechanism is
introduced varying in some way the effective refractive index
along the plane of the junction. Many index guide structures are
schematically shown in Fig. 15 (AlGaAs/GaAs) and Fig. 16
(InGaAsP/InP). A tipical index guide structure, for instance, is
the "chanelled-substrate-planar laser diode" (CSP-LD), shown in
Fig. 15a, where a channel in the substrate introduces the desired
waveguiding effect. An even more effective lateral confinement is
obtained by the "buried heterostructure" (BH-LD), shown in Fig.
15b and 16a, which is produced burying the active layer between
walls of different composition. Beside the optical confinement,
this structure presents a very efficient lateral current confinement, thus resulting in the lowest possible threshold currents
(10-20 mA at room temperature).

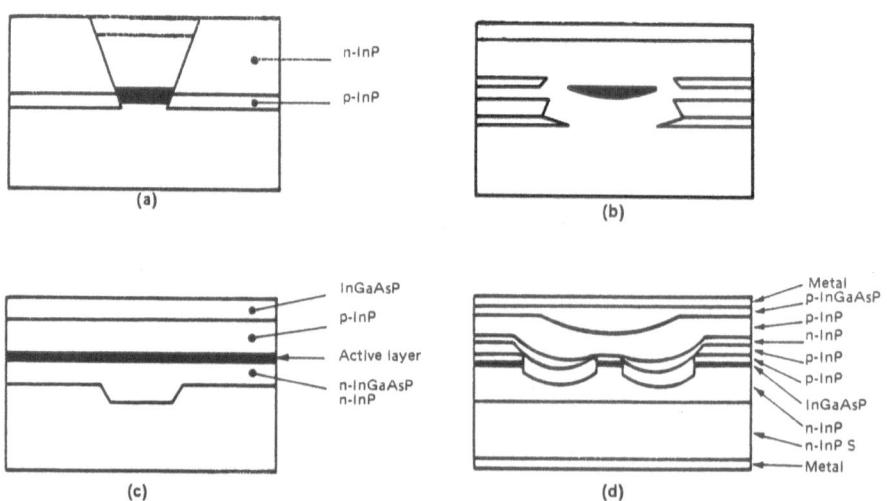

Fig. 16 - Schematic structures of InGaAsP/InP index guide lasers:
a) buried heterostructure
b) buried crescent
c) plano convex waveguide
d) double channel planar buried heterostructure.

Such a threshold current reduction is another significant goal, from the system point of view; as the threshold current

increases with the operating temperature, it is important to keep the injection current as low as possible, to avoid the enhancement of such effect, due to internal heating of the laser chip. This is particularly true for InGaAsP laser diodes, that show a stronger drift of threshold current with temperature in comparison with AlGaAs devices. InGaAsP BH-LD can still operate at temperatures exceeding 100°C (see Fig. 12), and this can avoid the inclusion of a Peltier-effect thermo-electric cooler in the laser package, resulting in a simpler, cheaper and more reliable device.

8. Sources for analog systems

For some applications, mainly TV programs distribution in broadband networks, the analog transmission can be still attractive, due to its lower bandwidth requirement. Beside the characteristics above described, which are common to all kind of systems, the optical sources for analog systems must present the additional features of high linearity and low intrinsic noise.

The linearity of above threshold LDs operating in the fundamental lateral mode, as can be deduced from CW power vs. current characteristics, is apparently satisfactory, (see for instance the curves of Fig. 12, in the 1÷5 mW range) especially for index-guided LDs. Under modulation, reflections from pig tail ends, connectors, couplers, can seriously influence the linearity.

As far as noise is concerned, the output of the LD is affected by noise, generally referred to as "relative intensity noise" (R.I.N.), arising from different causes: fluctuations intrinsic to the statistical nature of the carrier recombination process ("intrinsic noise"), reflections from discontinuinties of the refractive index in the optical transmission path or backscattered light ("reflection noise"), spatial filtering of the laser beam. The presence of more longitudinal modes, in reciprocal competition, can add further noise ("partition noise"). Tipical values of total R.I.N. can be as high as -120 dB/Hz, and can be critical for high quality transmission.

Several expedients can be divised to limit these effects. For instance, intrinsic noise can be reduced using low threshold current LDs while reflection noise can be avoided employing anti reflection coatings on the critical surfaces, tilted end connectors, optical isolators. Partition noise is reduced fixing the operating temperature of the LD at a value corresponding to low modal competition. As a general rule, better performances can be obtained using FM or PFM instead of baseband modulation and, finally, the injection of a high frequency signal (several hundreds MHz), by scrambling the longitudinal modes of the cavity, can be effective, too.

Whenever the length of the link is short enough, LEDs sources can be used for both analog and digital systems instead of LDs with obvious economical advantages, the LED being intrinsically cheaper and its coupling to the fibre easier. The linearity is poor, but compensation techniques can be easily implemented; furthermore, the LED is not affected by reflections, nor partition noise.

The broad spectral width, combined with the chromatic dispersion of the fibre, prevents the use of the LED on long links and at high frequencies.

9. Sources for specific applications

9.1 Long distance systems

For long distance transmission through single mode fibres at very high bit rates the chromatic dispersion penalty can be avoided using narrow spectral width laser diodes, possibly operating on a single longitudinal mode, as near as possible to the zero dispersion wavelength of the fibre.

Many structures have been developed operating on the single longitudinal mode in CW conditions; most of the spatially stabilized index guided LD's have this property. Unfortunately, whenever the LD is modulated, its originally single mode spectrum broadens to many modes, due to the transient behaviour of electrons and photons in the active region.

A survey of proposed solutions to the problem of dynamic single longitudinal mode laser diodes, is reported in Fig. 17. It is possible, for instance, to use LDs with very short cavity length to broaden the longitudinal mode spacing; external resonators like mirrors or gratings can be used as well. More efficient and stable solutions require different laser structures. The "cleaved coupled cavity" LD (C^3 - LD), for instance, is obtained simply by closely butt coupling two laser chips, one of which is used as an additional external resonator; it has been shown that stable fundamental mode oscillation can be achieved in this way. Other solutions have been proposed, all relying on multiple resonator configurations.

A different way to achieve the same result makes use of the "injection locking" technique; a master LD, operating CW on the single longitudinal mode, injects its radiation into a second LD, undergoing modulation, whose emission line is therefore "locked" to the master one, and is prevented from broadening.

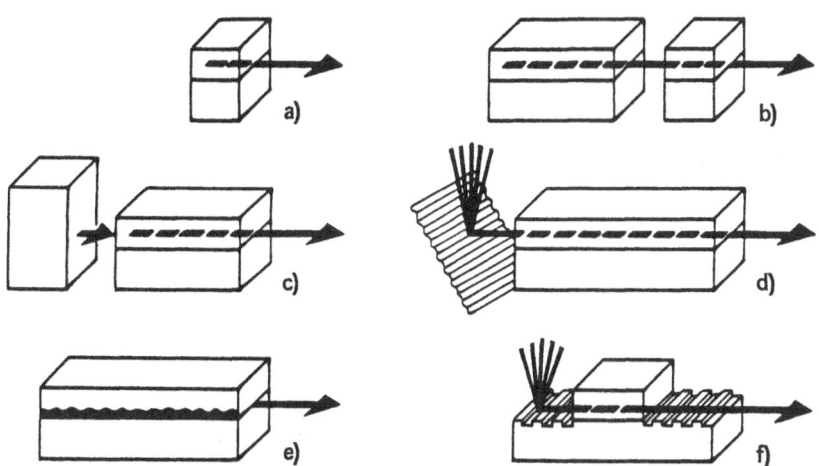

Fig. 17 - Dynamic single longitudinal mode laser diodes:
a) short cavity
b) cleaved coupled cavity
c) injection locked
d) external grating
e) distributed feedback
f) distributed Bragg reflector.

The most promising solution, yet, that avoids the need of external cavities and related aligments, is the "distributed feedback" laser diode (DFB-LD); this source has basically the same structure of the already discussed DH-LD, with the exception of the optical feedback, which is no longer supplied by the two mirror-facets, but is obtained through a grating, directly built in the active layer or in the adjacent layers. If suitable measures are taken to prevent the onset of Fabry-Perot cavity nodes and if appropriate phase conditions are provided at the cavity boundaries, this structure operates on a single longitudinal mode, with a rejection of more than 30 dB over spurious modes, even when modulated in the Gb/s range. This is an exciting feature, in view of systems operating at 1.55 μm, where the fibre attenuation has the absolute minimum (0.2 dB/km). Altough the DFB-LD operates on a single longitudinal mode, still there is a residual line broadening, which ultimately affects the spectrum when the device undergoes modulation; it is a "chirping" effect, due to the dependence of the refractive index of the semiconductor on the carrier density, that causes wavelength variations during the evolution of a current pulse. This can result in linewidths of the order of a few Angstroms. Therefore, whether systems at very high bit rates (>5 Gb/s) will be developed, over long optical lines, they will probably need external modulators, to overcome both speed and spectral limitation of LDs.

9.2 Medium and short distance systems

For transmission at moderate bit rates over medium distances, multimode fibre are used, with LD sources. In this case another problem arises, due to the so called "modal noise".

This kind of noise is due to spatial filtering, occuring for instance at misaligned joints, in the presence of laser speckle variations in the fibre core. The speckle is produced by the interference between the different propagating modes of the fibre, and varies due to temperature changes and modulation of the sources, that slightly shifts the operating wavelength, or due to vibrations and deformations of the fibre cable. The larger is the laser spectrum (i.e. the shorter its coherence length), the less visible is the speckle at a given distance from the source and the lower are the effects of spatial filtering. In this case broad spectrum, multimode LDs are preferred, provided that chromatic dispersion is no longer significant, due to the limited length of the link. If the link is short enough, the choice of an LED would completely eliminate modal noise problems.

For transmission of high bit rates over very short distances, (e.g. a few hundreds meters), like intra office links, multimode fibres and high speed LEDs can be used, operating at 1.3 μm to contain the effects of chromatic dispersion. Fast surface LEDs are currently under development, able to operate at 1 Gb/s or more.

A different choice could be stimulated by the envisaged mass production of GaAlAs lasers for optical disk reading, that should bring to very low cost devices. Operation at short wavelength (0.85 μm) should not be a major obstacle if limited link lengths are considered (in the order of 1-5 km).

9.3 Broadband distribution networks

For wideband distribution to subscribers a main factor is terminal equipment cost and reliability. In this case, the development of LEDs with large bandwiths is a possible solution. As the distribution of advanced services can imply the introduction of several channels from the exchange to the subscribers, beside one channel in the opposite way, multiplexing techniques are required. A very promising technique, although still not clearly economical, is the "wavelength division multiplexing" (WDM), based on the use of optical filters which sum different wavelengths at the input of a fibre and separate them at the output.

The use of LED sources, with their broad spectrum, pratically limits the number of multiplexable channels at two or three; when a higher number of channels is required, LD sources must be used, which allow a closer WDM channels spacing.

10. Sources for coherent detection

Finally, future transmission systems will employ, beside conventional direct detection, advanced coherent detection techniques like heterodyne or homodyne, which allow, theoretically, an improvement of the receiver sensitivity up to 20 dB. In this case, beside the transmitter laser, a second one, the local oscillator, must be used at the receiver end (Fig. 18). Both sources need extremely stable and narrow spectral linewidths, of the order of 1 MHz or less. DFB-LDs are the most probable candidates as sources for coherent detection, showing at present, in the best case, linewidths of a few MHz.

Much work has still to be done to well understand the noise causes affecting the LD linewidth and to devise appropriate solutions to this problem.

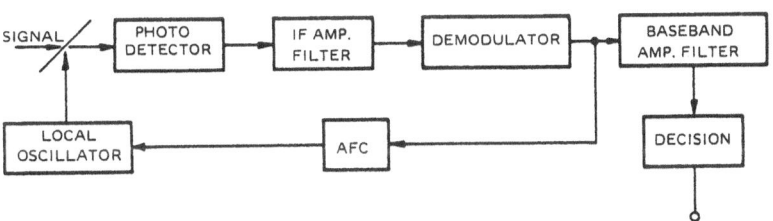

Fig. 18 - Schematic of heterodyne detection receiver.

11. Sources for mid-infrared fibres

In the further future, whether mid-infrared glass fibres (in the 3-6 µm range) for telecommunication will become available, with their predicted ultra-low attenuation, this will call for adequate sources.

At present lead-salt laser diodes operating in the mid-IR are only produced for spectroscopy applications, operating CW only at cryostatic temperatures.

If the 3-6 µm transmission band will become practicable, the development of suitable sources, meeting all the requirements of telecommunication systems, will be a challenging activity for the future.

TWO POSSIBLE MEASUREMENTS OF THE EFFECTIVE MASS OF ELECTRONS INJECTED INTO A LIQUID

G. Ascarelli

Physics Department Purdue University
West Lafayette In.47907,U.S.A.

The experiments that are proposed aim at the measurement of the effective mass of an electron injected into an insulating liquid by either studying the modulation of the photoemission from a semiconductor into a liquid when a large magnetic field is applied to the photocathode or the absorption of f.i.r. radiation by electrons trapped in a potential well near the surface of the liquid. In the former case the modulation should be periodic in 1/B and, depending on the initial electron states in the photocathode, should appear as either a series of steps or a series of maxima in the photoelectric yield. From the separation of these steps the effective mass of the electrons can be calculated. In the latter type of measurement the electrons are trapped in a potential well near the surface by a combination of their image charges and an external electric field. Both the depth of the well and the energy levels in the well depend on both the external electric field and the electron effective mass.

I. Magneto photoemission

The novel technique proposed for this experiment has as its aim the measurement of the effective mass of electrons in a liquid.This is to be accomplished by combining ideas that have independently evolved in different sub fields of physics:the de Haas-Van Alphen effect in metals, the photo emission from metals and semiconductors into liquids (used e.g. in the measurement of the energy of the conduction band minimum of a liquid) and photoemission from semiconductors into vacuum.

The idea is to illuminate a piece of intrinsic semiconductor (e.g. Ge) with light about 5 or 10 meV the above threshold for photoemission into the liquid. The experiment should attempt to detect changes in the magnitude of the photo current as a function of the applied magnetic field. These changes should be periodic in 1/B. Like in the case of the de Haas-Van Alphen effect their period is a measure of the effective mass of an electron in the conduction band. These steps should resemble those found in the I-V characteristics of tunnel diodes at low temperature.

The energy of the light that is necessary, in the case of pure Ge ($\sim 10^{10}B/cm^3$) in Xe, is between 4.1 and 4.2 eV, since the bottom of the conduction band of Xe is between 0.6 and 0.7 eV below vacuum[1] and the threshold for photo emission from Ge into vacuum is 4.8 eV[2].

In the geometry being envisaged here, the applied magnetic field is perpendicular to the surface of the semiconductor. Each time a Landau level of the conduction band of the liquid crosses the level corresponding to the maximum of the kinetic energy of the electrons emitted from the solid there will be a change of the photoelectric yield in a way similar to what happens in a metal-insulator-metal junction[3] or when a Landau level crosses the Fermi surface in the de Haas-Van Alphen effect (Fig 1).

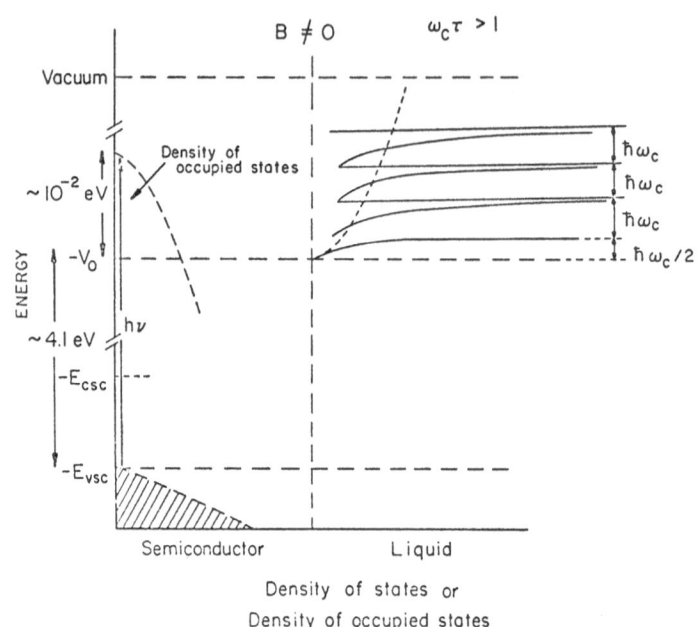

Fig.1 Position of the Landau levels and the corresponding density of states in the liquid. It is assumed that the Landau level structure in the excited states of the semiconductor is completely washed out in order to simplify the drawing of the picture. The energy scale used to indicate the Landau level structure is expanded in comparison with the energy used to indicate the photon energy.

It is legitimate to inquire if a traditional cyclotron resonance experiment would not be more suitable, more easely interpreted and more easely carried out than our proposed experiment.

Electrons in liquid Kr and liquid Xe near their triple points have mobilities[4] respectively of 1600 cm^2/Vsec and 2000 cm^2/Vsec, therefore the condition $\omega_c \tau > 1$ is satisfied respectively for fields of 6.2T and 5T. From exciton spectra in the solid the exciton reduced mass[5] in Xe is estimated to be 0.34 m_0 while that in Kr is 0.43 m_0. In the absence of better information the same values will be assumed to hold for the liquid.

A traditional cyclotron resonance experiment would have to be carried out at fields $\simeq$10T and the source of e.m. radiation could be an HCN laser. However, optimistically, it would be difficult to produce more than $\simeq 10^{11}$ electrons/cc using traditional radiation chemistry techniques, specially since the fringing field of the magnet (usually a superconducting solenoid) would make the steering of the electron beam from the accelerator rather difficult.

Without complicated and cumbersome interferometric techniques (that may be unrealizable in this case) it will be difficult to detect the power absorbed by much less than 10^{14} carriers. As a result the s/n ratio attainable in such measurement is insufficient.

To understand more clearly the physical ideas underlaying the proposed measurement we must consider the photoemission process. According to Spicer[6] this can be divided in 3 steps:

1) photon absorption in the photocathode;
2) transport of the electron from the interior of the photocathode to the surface;
3) transmission through or over a surface barrier.

Gobeli and Allen[7] using the results of Kane's theory[8] suggested that in Ge step 1) corresponds to an indirect absorption, the "missing" crystal momentum being provided by either phonons and/or by imperfections in the bulk near the surface. In the case of step 2), when the photon energy is sufficiently close to threshold, an electron that experiences any inelastic collision (mainly optical phonon emission) will have its energy associated with the momentum normal to the surface decreased below the threshold for photo emission and will not emerge from the solid[6,9]. Therefore if the energy of the incident photon differs from the threshold energy by less than an optical phonon the distribution of electrons reaching the surface will reflect the occupied density of states of the top of the valence band and the final density of states in the semiconductor (Fig.1).

The current from the photocathode is strongly influenced by the position of the Landau levels in the liquid (Fig.1).[10] As an example, in the simple case of a rectangular barrier the photo current can be written:

$$J(B)=\int_0^\Delta \sum_{n_l} g_l(E_l,n_l)\sum_{n_s} f(E_s,n_s)D(E_s,n_s,S)dE_s\,dE_l$$

The subscripts s and l refer respectively to the semi conductor and the liquid; E_s, E_l are the corresponding energies associated with the motion along the magnetic field. For each value of spin S the sum over n_s runs from 0 to a maximum such that $E_s + \hbar\omega_{cs}(n_s+1/2) + g\beta BS = \Delta$. This value corresponds to the maximum energy above the minimum of the conduction band of the liquid that a photoexcited electron in the semiconductor can have. When the photoexcited electrons originate from the top of the valence band of the semiconductor Δ will depend on magnetic field as it reflects the position of the first Landau level in the valence band of the semiconductor. The density of states in the liquid that is associated with a given Landau level n_l is $g(E_l,n_l)$. The density of photoexcited electrons in the semiconductor that have energy E_s and are in a Landau level n_s sufficient to escape from the solid is $f(E_s,n_s,S_s)$. The above integrand must be furthermore multiplied by suitable delta functions that reflect the selection rules associated with the transmission through the barrier. Examples are spin and energy conservation. The zero of energy is taken to be the photo-emission threshold and Δ is the maximum energy of an injected electron in the liquid. The quantity $D(E_s,n_s,S)$ is the probability of transmission of an electron through or over the surface barrier. As an example, in the case of a rectangular barrier of width W and height V_b above the photo emission threshold, we have using the WKB approximation[10].

$$D = \exp{-2W(2m/h^2[V_b - E_s - \hbar\omega_{cs}(n_s + 1/2) - g\beta BS])^{1/2}\delta(S_l, S_s)}$$

What is important to notice is that even if the Landau levels in the semiconductor side of the barrier are broadened out of recognition the integral over E_l and the sum over n_l will give rise to steps in the current, each step corresponding to the case of an n_l such that $(n_l + 1/2)\hbar\omega_{cl} + g\beta BS = \Delta$ (Fig. 2). The reason is that each time a Landau level of the liquid goes above Δ, there is a drastic decrease of the number of states in the liquid into which an electron can be injected from the photocathode.

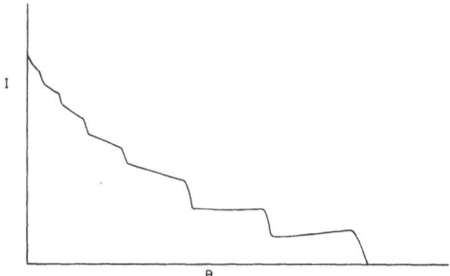

Fig.2 Expected form of the photocurrent as a function of magnetic field when the Landau levels in the semiconductor are not taken into account. If the electrons are instead excited from the top of the first Landau level of the valence band of the semiconductor, and there is not an excessive lifetime broadening, the above steps are substituted by maxima. It is always assumed that no surface states are present in the forbidden gap of the semiconductor.

The magnetic field induced variation of the density of states in the liquid is not, however, the only source of oscillations of the photoelectric yield. Oscillations of the absorption coefficient associated with transitions from the top of the valence band to the final state of the semiconductor from which electrons are emitted is also important insofar as it varies the position below the surface where the excited electron is created. At least one of the resulting periods should be easely recognizable because it involves the light hole mass $(m_{lh}^* = 0.042m_o)$ and should be important at relatively low magnetic fields. With a maximum value of $\Delta \simeq 10~meV$, the oscillations of the yield arising from the Landau levels associated with the light hole will disappear beyond $\Delta = \hbar\omega_{cl}/2$ i.e. $B \simeq 8 T$. From deviations of the low field oscillations of the yield from what is expected from the light hole mass the effective mass in the excited state of the semiconductor can be calculated. The results can be checked at high magnetic fields where the oscillatory magneto absorption associated with the heavy hole $(m_{hh}^* = 0.28m_o)$ should become visible. However, from band structure calculations[11], there is a very large density of states of Ge $\simeq 4.1 eV$ above the conduction band maximum (Fig.3). The effective mass along the [111] direction is much larger than a free electron mass and the density of states along Λ_3 is much larger that at any other point of the Brillouin zone that can contribute to photoemission into the liquid. Electrons arising from these states should give an overwhelming contribution to photoemission.

Finally two other sources of oscillations of the yield should appear in correspondence to the magnetic field for which a Landau level on either the semiconductor or on the liquid side of the barrier rises above Δ. Only the latter gives information on states of the liquid.

480

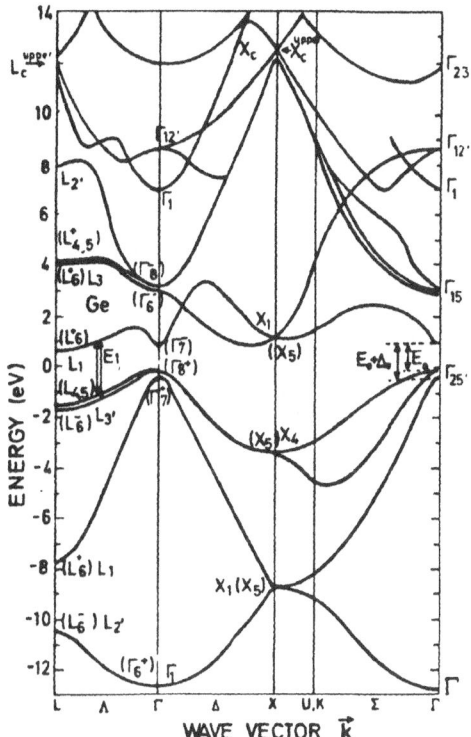

Fig. 3 Band structure calculation of Ge [11]. The letters in parenthesis denote double group symmetries, those without parenthesis the single group. Notice the very flat band near the point L ≃4ev above the valence band maximum. These states should dominate the tunneling from Ge into liquid Xe.

They should be recognized by the fact that their period cannot be calculated from the semiconductor masses considered above.

The succession of steps mentioned in the previous paragraphs could resemble a series of maxima if, on account of some combination of band structure, impurity states, surface states or electron relaxation, the population of excited electrons in the semiconductor could be concentrated above the photoemissive threshold in a range of energies small compared with the separation of the Landau levels. If the relaxation time of the electrons on the semiconductor side of the barrier is sufficiently long there should always be a maximum of the electron density at Δ. This reflects the large density of states at the first Landau level of the valence band of the semiconductor from where the electrons are optically excited.

Another way to peak the density of excited electrons near Δ is to use a level deep in the forbidden gap of the semiconductor as the initial state from where the electrons are excited. In the case of Ge, Zn whose level is ≃0.1 eV above the valence band maximum[12] and whose solubility is over 10^{18} cm^{-3} could provide the localized state. These impurities could be easely introduced e.g. by ion implantation. It is however unclear what effect this would have on surface states.

An order of magnitude of the total photoemission can be estimated e.g., in the case of Ge, from the results of Gobeli and Allen[2]. From their data we estimate that the photoemission 10 meV above threshold has quantum efficiency 10^{-8}. If e.g. 10 mW of chopped light, from a suitably frequency doubled dye laser (final energy $\simeq 4.1$ eV), is incident on the sample one would observe a current $\simeq 10^{-11}$ A. This is 1000 times larger than the noise of a commercial current amplifier for a phase sensitive detector[13]. This estimate of the yield may be too small on account of the large density of states that is calculated 4.1 eV above the conduction band maximum. There is no comparable density of states that can influence the photo emission into vacuum.

A more attractive alternative is to use a pulsed light source and a charge sensitive amplifier similar to those commonly used in nuclear spectroscopy, e.g. in conjunction with Si-Li detectors. These amplifiers have $s/n=1$ corresponding to about 1000 carriers collected on a capacity $\simeq 20$ pF. A pulsed light source delivering about 10 mJ e.g. over 100 ns would give rise to a burst of 10^8 electrons. Such sources can be obtained either starting from a Yag or an excimer laser.

A problem that has not been considered and that should be taken into account is that, in order to have a sharp cutoff of the density of states, there cannot be any appreciable band bending due to surface states. Thus the density of states in the band gap of the semiconductor, that are localized near the surface, should be small. These considerations imply that the dewar where the experiment is to be carried out must be pumped to at least 10^{-9} or 10^{-10} torr (this is also desirable to avoid sample contamination) and, in order to observe electrons originating from the valence band, the semiconductor must be cleaved while submerged in the rare gas liquid. The maintenance of a clean surface should not be as difficult as in the case of photo emission into vacuum because both the semiconductor and the container walls are at temperatures ~160K and the semiconductor surface is shielded from contaminants by the rare gas liquid. In the case when the electrons originate from deep impurity levels in the semiconductor (e.g Zn) the impurities must be more or less uniformly distributed to a depth of the order of $10\mu m$ to assure that under illumination there will be no significant band bending over a distance of the order of the escape depth of the photoelectrons i.e ~$10 nm$. If the impurities are introduced by ion implantation, laser annealing of the accompanying damage can be easely accomplished using the same laser used for the rest of the experiment.

II. FIR of electrons trapped in a surface well

As pointed out above, a traditional cyclotron resonance measurement is probably unfeasible on account of the almost unavoidable interaction between the necessarily large magnetic field ($>10T$) produced by some sort of solenoid and the accelerator beam used to produce ionization in the liquid. However, a FIR analog of the measurement of the effective mass of ions in liquid He, developed by Poitrenaud and Williams[14], can be adapted to measure the effective mass of electrons in the conduction band of a liquid.

An electron in a liquid is repelled from the surface by an image charge[14,15] that gives rise to a potential

$$V = \frac{e(\epsilon_l - \epsilon_g)}{16\pi\epsilon_l(\epsilon_l + \epsilon_g)x}$$

where x is the distance of the electron from the surface of the liquid ϵ_l and ϵ_g are respectively the dielectric constants of the liquid and the vapor.

If a uniform electric field (F) forces the injected electrons towards the surface, each carrier is confined to a potential well

$$V(x) = \frac{A}{x} + eFx$$

where

$$A = \frac{e^2(\epsilon_l - \epsilon_g)}{16\pi\epsilon_l(\epsilon_l + \epsilon_g)}$$

The states near the bottom of the well can be approximated by an harmonic oscillator whose frequency is:

$$\omega_o^2 = \frac{2(eF)^{3/2}}{m^* A^{1/2}}$$

This is very similar to what happens when electrons are localized in the channel structure of an FET[16].

Like in the case of the magneto photo emission a resolved resonance absorption requires $\omega_o \tau \gtrsim 1$. In the case of liquid Xe, on the basis of TOF mobility measurements this corresponds to $\omega_o \geq 2.6 x 10^{12}$ i.e. $1/\lambda \geq 14 cm^{-1}$ (assuming $m^* = 0.34 m_o$). In the case of Ar $(m^* = 0.5 m_o)$ $1/\lambda \geq 37 cm^{-1}$.

An indication of the experimental parameters calculated assuming a hypothetical value of the resonance of $100 cm^{-1}$ are given in table I.

Table I. Some parameters of a potential well in which the electrons would absorb $100 cm^{-1}$ radiation in the case of liquid Ar and liquid Xe near the triple point. The relative dielectric constant of the vapor is taken equal to 1.

Liquid	Ar	Xe
Field	2.3 10^4 V/cm	1.83 10^4 V/cm
X_o	46$\mathring{A}$	55$\mathring{A}$
depth of well	2.1 10^{-2} eV	1.87 10^{-2} eV
temperature	85 K (7.8 10^{-3} eV)	162 K (1.39 10^{-2} eV)
diel.const.	1.53	1.61
eff. mass	0.5	0.34
$\omega_o \tau$	2.7	7.1

Contrary to the case of the FIR spectra of electrons trapped in an FET channel[16] the absorption cannot be detected as an attenuation of the incident beam (whose polarization must be perpendicular to the surface of the liquid) because there are too few electrons ($\simeq 10^8$) to absorb a detectable fraction of the incident power. Another detection mechanism must be devised.

Despite the fact that we do not know the scattering processes that determine the mobility of an electron in a liquid, there must be mechanisms that give rise to the dissipation of the energy that the electron gains from an external electric field. At very large fields there is breakdown $(F \simeq 10^6 V/cm)$. In the case of liquid Ar near the triple point the time of flight (TOF) velocity is nearly saturated with fields of $\simeq 10^4 V/cm$. The corresponding fields in the case of liquid Xe are much smaller. Although TOF measurements do not necessarily reflect the microscopic mobility[17-19] fields.

If the field associated with the FIR radiation is sufficiently intense ($\gtrsim 10^4 V/cm$), we may expect a decrease of the d.c. conductivity that reflects the power the FIR transfers to the electrons. This effect, known in gasses at radio frequencies as the Luxembourg effect, has been exploited to carry out cyclotron resonance measurements in insulators[20]. Similar effects are well known in gas lasers when the DC current of the discharge is changed by the introduction into the laser cavity of a species that absorbs at the laser frequency.

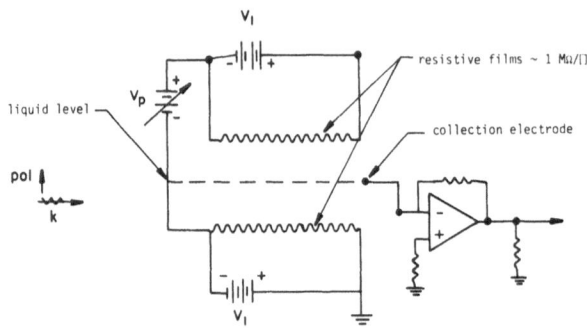

Fig.4 Experimental configuration for the proposed FIR optical absorption by electrons trapped on the surface of a liquid. The voltage source V_p produces a field $\simeq 10^4 V/cm$ that is used both to sweep the electrons towards the surface of the liquid and to create the po- tential well. The resistive plates instead, create the longitudinal field that imparts a drift velocity to the electrons parallel to the surface of the liquid.

A schematic of the intended experiment is given in Fig.4. The voltage V_p provides the field necessary to both sweep the electrons produced by the ionizing radiation towards the liquid vapor interface and to create one of the "walls" of the well. The two resistive plates instead have as a function to accelerate the electrons towards the collecting electrode. The longitudinal field created by these plates should be relatively small so as to be near the threshold of the region where the non ohmic behaviour is noticed ($\simeq 20 V/cm$ in the case of Xe and $\simeq 500 V/cm$ in the case of Ar). The expected dependence of the current on time is indicated in Fig.5. The magnitude of the dip depends on how close the FIR frequency is to ω_0.

The power absorbed by the electrons from the laser is proportional to σF^2 where

$$\sigma = \frac{\sigma_0}{1 + (\omega - \omega_0)^2 \tau^2}$$

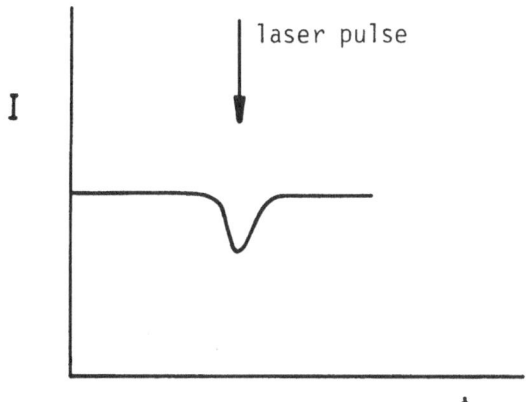

Fig.5 Expected time dependence of the electron current current following a short pulse of ionizing radiation. "Heating" of the electrons by the laser field gives rise to a decrease of the electron mobility that is detected by a dip in the electron current. The ionizing radiation reaches only a small fraction of the sample.

In this expression σ_o is the d.c. conductivity of the sample whose electron density is n, ω_o is the characteristic frequency of the electrons in the surface well and ω is the frequency the laser.

Ionizing radiation considered as the source of electrons in these excellent insulators could be substituted by field emission or photoemission from appropriate cathodes. The difficulties encountered with those possibilities are not particularly attractive specially since none of them can provide a sufficient number of electrons to allow a direct optical absorption to be carried out.

III. Conclusions

We have presented two proposals for measuring the effective mass of electrons injected into a liquid. In the first proposal we expect to find an oscillatory component of the photo emission as a function of the value of the applied magnetic field. This oscillatory behaviour is reminiscent of the de Haas van Alphen effect. In the second proposal we suggest the creation of an artificial two dimensional electron gas near the liquid-vapor interface and the study of the energy levels of the system.

The author gratefully acknowledges partial support from the United States Department of Energy contract DE ACO2 79 ER 10375, and enlightening explanations from Prof. W.E.Spicer regarding the models of photoemission from a semiconductor.

REFERENCES

1. R.Reininger, V.Asaf and I.T.Steinberger, Chem. Phys. Lett. 90: 287 (1982).
2. G.W.Gobeli and F.G.Allen Phys.Rev. 137:A245 (1965).

3. C.B.Duke TUNNELING IN SOLIDS, p.77, Academic Press,New York (1969).

4. L.S.Miller, S.Howe and W.E.Spear, Phys.Rev. 166:871 (1968).

5. L.Resca,R.Resta and S.Rodriguez, Sol. State. Comm. 26:849 (1978).

6. W.E.Spicer, Appl. Phys. 12:115 (1977); also C.N.Berglund and W.E.Spicer, Phys. Rev. 136:A1030 (1964).

7. G.W.Gobeli and F.G.Allen,Phys.Rev.127:141 (1962).

8. E.O.Kane,Phys.Rev.127:131 (1962).

9. E.O.Kane PROC.INT.CONF.ON THE PHYSICS OF SEMICONDUCTORS Kyoto (1966) Jnal. Phys. Soc. Japan 21S:37 (1966).

10. This is an adaptation of the expression in the eq. on p.77 ref.3.

11. J.R.Chelikowsky and M.L.Cohen, Phys. Rev. B14:556 (1976).

12. W.W.Tyler and H.H.Woodbury, Phys. Rev. 102:647 (1956).

13. Ithaco mod.164.

14. J.Poitrenaud and F.I.B. Williams, Phys. Rev. Lett. 29:1230 (1972).

15. L. Landau and E. Lifchitz, ELECTRODYNAMIQUE DES MILIEUX CONTINUS, p.60, ed. Mir, Moscow (1969).

16. P.Kneschaurek, A.Kamgar and J.F.Koch, Phys. Rev. B14:1610 (1976).

17. G.Ascarelli, J. Chem. Phys., 71:5030 (1979).

18. G.Ascarelli, J. Chem. Phys., 74:3082 (1981).

19. R.C.Munoz and G.Ascarelli, Phys. Rev. Lett., 51:215 (1983).

20. J.W.Hodby, J.A.Borders, F.C.Brown and S.Foner, Phys. Rev. Lett. 19:959 (1967); also J.W.Hodby in POLARONS IN IONIC CRYSTALS AND SEMICONDUCTORS p.389, J.Devreese ed. North Holland, Amsterdam (1972).

AUTHOR INDEX

INDEX